RENEWALS 458-4574
DATE DUE

WITHDRAWN
UTSA LIBRARIES

Cities, Transport and Communications

Also by Howard Dick

BALANCED DEVELOPMENT: East Java under the New Order (*co-editor with James J. Fox and Jamie Mackie*)

CORRUPTION IN ASIA: Rethinking the Governance Paradigm (*co-editor with Tim Lindsey*)

THE EMERGENCE OF A NATIONAL ECONOMY: An Economic History of Indonesia, 1800–2000 (*with Vincent J.H. Houben, Thomas Lindblad and Thee Kian Wie*)

THE INDONESIAN INTERISLAND SHIPPING INDUSTRY: An Analysis of Competition and Regulation

THE RISE AND FALL OF REVENUE FARMING: Business Elites and the Emergence of the Modern State in Southeast Asia (*co-editor with John Butcher*)

SURABAYA, CITY OF WORK: A Twentieth Century Socioeconomic History

Also by Peter J. Rimmer

ASEAN–AUSTRALIA TRANSPORT INTERCHANGE (*editor*)
HONG KONG'S FUTURE AS A REGIONAL TRANSPORT HUB

PACIFIC RIM DEVELOPMENT: Integration and Globalisation in East Asia (*editor*)

RIKISHA TO RAPID TRANSIT: Urban Public Transport Systems and Policy in Southeast Asia

THE UNDERSIDE OF MALAYSIAN HISTORY: Pullers, Prostitutes, Plantation Workers... (*co-editor with Lisa M. Allen*)

TRANSPORT IN THAILAND: The Railway Decision

Cities, Transport and Communications
The Integration of Southeast Asia since 1850

Howard Dick
Associate Professor
Department of Management
University of Melbourne
Australia

and

Peter J. Rimmer
Professor Emeritus and Visiting Fellow
Division of Pacific and Asian History
Research School of Pacific and Asian Studies
The Australian National University
Canberra
Australia

© Howard Dick and Peter J. Rimmer 2003

All rights reserved. No reproduction, copy or transmission of this publication may be made without written permission.

No paragraph of this publication may be reproduced, copied or transmitted save with written permission or in accordance with the provisions of the Copyright, Designs and Patents Act 1988, or under the terms of any licence permitting limited copying issued by the Copyright Licensing Agency, 90 Tottenham Court Road, London W1T 4LP.

Any person who does any unauthorised act in relation to this publication may be liable to criminal prosecution and civil claims for damages.

The authors have asserted their right to be identified as the authors of this work in accordance with the Copyright, Designs and Patents Act 1988.

First published 2003 by
PALGRAVE MACMILLAN
Houndmills, Basingstoke, Hampshire RG21 6XS and
175 Fifth Avenue, New York, N.Y. 10010
Companies and representatives throughout the world

PALGRAVE MACMILLAN is the global academic imprint of the Palgrave Macmillan division of St. Martin's Press, LLC and of Palgrave Macmillan Ltd. Macmillan® is a registered trademark in the United States, United Kingdom and other countries. Palgrave is a registered trademark in the European Union and other countries.

ISBN 0–333–55301–2

This book is printed on paper suitable for recycling and made from fully managed and sustained forest sources.

A catalogue record for this book is available from the British Library.

Library of Congress Cataloging in Publication Data
Dick, H.W. (Howard W.)
Cities, transport and communications : the integration of Southeast Asia since 1850/Howard Dick and Peter J. Rimmer.
    p. cm.
Includes bibliographical references and index.
ISBN 0–333–55301–2
    1. Metropolitan areas—Asia, Southeastern. 2. Cities and towns—Asia, Southeastern. 3. Transportation—Asia, Southeastern. 4. Telecommunication—Asia, Southeastern. 5. Asia, Southeastern—Economic conditions. I. Rimmer, Peter James. II. Title.

HT334.A785D53 2003
307.76′4′0959—dc21

2002044818

10 9 8 7 6 5 4 3 2 1
12 11 10 09 08 07 06 05 04 03

Printed and bound in Great Britain by
Antony Rowe Ltd, Chippenham and Eastbourne

To our families

Contents

List of Tables	viii
List of Figures	xi
Preface and Acknowledgements	xv
Glossary	xix
A Note on Place Names	xxii

Part I Patterns and Processes ... 1

1 Patterns: Networks and Urban Hierarchy ... 3
2 Processes: the Diffusion of Technology ... 37

Part II Hinterlands ... 73

3 The Archipelago ... 81
4 Islands: Java and Luzon ... 117
5 Rivers: Chao Phraya, Irrawaddy and Mekong ... 155
6 Peninsulas: Malaya and Annam ... 186

Part III Cities ... 217

7 World City: Singapore ... 229
8 Archipelagic Cities: Manila and Jakarta ... 257
9 River Cities: Bangkok, Rangoon and Saigon ... 289
10 First World City: Kuala Lumpur ... 319

Afterword	339
Bibliography	345
Index	376

List of Tables

1.1	Singapore: shipping movements by origin and destination, 1850–51	10
1.2	Southeast Asia: gross regional domestic product (GRDP) at official exchange rates and at purchasing power parity (PPP) by sub-region, 1995	26
1.3	Asian cities ranked in top 25 container ports, 1985, 1995 and 2000	29
1.4	Asian cities ranked in top 25 airports by air freight, 1985, 1995 and 2000	31
1.5	Asian cities ranked in top 25 airports by international passengers, 1985, 1995 and 2000	31
1.6	Asia-Pacific headquarters location of the world's top 25 container shipping, air transport and telecommunications firms, c.2000	32
2.1	Southeast Asia: telephone sets by country, 1930–1999	46
2.2	Southeast Asia: radio sets by country, 1955, 1970, 1997	46
2.3	Southeast Asia: television sets by country, 1960, 1975, 1999	49
2.4	Southeast Asia: growth in civil aviation passenger traffic, 1950–1999	60
2.5	Southeast Asia: growth in civil aviation freight traffic, 1950–1999	60
2.6	Southeast Asia: length of railway line in operation, 1870–2000	63
2.7	Southeast Asia: rail passenger traffic, 1910–1998	65
2.8	Southeast Asia: rail freight traffic, 1910–1998	65
2.9	Southeast Asia: number of registered passenger vehicles, 1930–1998	69
2.10	Southeast Asia: number of registered commercial vehicles, 1930–1998	69
II.1	Major export commodities associated with the five supra-national regions	77
3.1	Singapore and Java: exports and imports by destination, 1869	82
3.2	Netherlands Indies: inward shipping by main port, 1903–1938	89
3.3	Netherlands Indies: scale of interisland operations, NISN (1886) and KPM (1929)	92
3.4	Philippines: shipping fleet by number and net tonnage, 1930	103
3.5	Philippines: cargo (non-oil) and passengers shipped by region, 1994	112

3.6	Indonesia: ratio of interisland to foreign trade, 1914–1939 and 1972	114
3.7	Indonesia: distribution of manufacturing output and origin of inputs, 1987	115
4.1	Java: number of motor vehicles, 1900–1996	129
4.2	Java: rail passengers, 1911–1996	130
4.3	Java: rail freight, 1911–1996	130
4.4	Java: interurban transport task by mode, 1991	135
4.5	Manila railroad company passengers and freight, 1904–1991	152
5.1	Irrawaddy, Chao Phraya and Mekong rivers: vital statistics, c.2000	158
5.2	Burma, Indochina and Thailand: railway construction, 1869–1940	163
5.3	Burma, Indochina, Thailand: access per capita to rail, roads and telegraph lines, late 1930s	169
5.4	Thailand: expansion of the road network, 1950–2000	178
6.1	Federated Malay States: main commodities carried by rail, 1905–1906	194
6.2	Federated Malay States: road length, 1922–1927	203
6.3	Peninsular Malaysia: private vehicle registrations and length of roads by surface type, 1965–1990	208
III.1	ASEAN capital cities by population and economic size, 1995	222
7.1	Singapore: distribution of population by ethnic groups, 1824–1957	233
7.2	Singapore: division of public jinrikisha stands into clan districts, late 1910s	237
7.3	Singapore: vehicle types passing over selected bridges during traffic censuses, 1917, 1923 and 1930	238
7.4	Singapore: economic and social indicators, 1960–2000	241
7.5	Singapore: housing and development, residential properties by town, 1931–1999	245
7.6	Singapore: motor vehicle registrations and road length, 1965–2000	246
8.1	Manila and Metro Manila: population, 1903–2000	258
8.2	Jakarta: population, 1905–2000	258
8.3	City types, phases and trends of globalization and localization	258
8.4	Manila (NCR) and Luzon, Jakarta (DKI) and Java: passenger vehicle registrations by type, December 2000	272
8.5	Manila, Jakarta, Bangkok: modern building stock, mid-1999	286
9.1	Bangkok Metropolitan Administration (BMA) and Bangkok Metropolitan Region (BMR): area and population, 1913–2000	290

9.2	Bangkok: motor vehicle registrations by categories, 1950–2000	303
9.3	Division of responsibilities in Thailand's urban transport administration	304
9.4	Bangkok: number of trips by expressways, 1985–2000	310
9.5	Bangkok: proposed mass transit systems	312
10.1	Ipoh, Georgetown and Kuala Lumpur: population, 1896–1957	320
10.2	Greater Kuala Lumpur: Population by Conurbation, Klang Valley Corridor and agglomeration, 1947–2000	327

List of Figures

P.1	Definition of 'global and 'local' in relation to 'national'	xvi
1.1	Exports from Singapore to main destinations, 1850	11
1.2	Major trade routes, mid-nineteenth century	12
1.3	Rice movements, c.1870, c.1890, c.1912 and c.1929	14
1.4	Sugar movements, c.1870, c.1890, c.1912 and c.1929	15
1.5	Movements of salt, c.1929; dried and salted fish, c.1929; cattle; bullocks and buffalo, c.1929; and swine, c.1929	16
1.6	Coal movements, c.1870, c.1890, c.1912, c.1929	17
1.7	Liner shipping connections, June 1960	18
1.8	Asia-Pacific container ports, 1995	20
1.9	Asia-Pacific air routes, August 1995	22
1.10	Asia-Pacific telecommunications connections between country pairs, 1995	23
1.11	(a) Asia-Pacific gross domestic product, 1995; (b) Southeast Asia gross regional domestic product, 1995	25
1.12	City/hinterland schematic diagram, 1995	28
1.13	Estimated Asia-Pacific population, 1995	30
1.14	Port chains 1850, 1930, 2000	33
1.15	Interurban air connections showing the pivotal importance of Singapore and Hong Kong	34
1.16	The gateways to Southeast Asia	35
2.1	Main submarine cables and selected land links, c.1920	42
2.2	Mail steamers in Southeast Asia before opening of the Suez Canal, 1869 (a) world routes (b) Southeast Asian routes	52
2.3	Main air routes c.1938 showing date of connections (a) world routes; (b) Southeast Asian routes	56
II.1	Supra-national and sub-national regions	76
II.2	Monopoly and competitive transport states	78
3.1	Netherlands Indies: contract interisland mail routes, c.1864	85
3.2	Netherlands Indies: contract interisland mail routes c.1888	85
3.3	Netherlands Indies: contract interisland shipping routes, 1891	85
3.4	Straits Steamship Company: network, 1937	91
3.5	KPM: overseas lines, c.1939	91
3.6	Southeast Asia: KPM and Straits Steamship Co. networks, 1938	94
3.7	Indonesia: KPM interisland network by frequency of sailings, 1956	96
3.8	Indonesia: KPM interisland shipping routes, c.1940	99

3.9	Indonesia: Pelni interisland passenger shipping network, 1995–96	99
3.10	Port of Manila: inbound interisland shipping routes and frequencies, 1906	108
3.11	Port of Manila: inbound interisland shipping routes and frequencies, 16 May–15 June 1954	109
3.12	Port of Cebu: inbound interisland shipping routes and frequencies for 31-day period 16 May–15 June 1954	111
4.1	Java: Navigable rivers and post roads on relief map, c.1860	118
4.2	Java and Madura: (a) railway network to 1899 (b) railway and tramway network to 1925 by width of gauge	124
4.3	Java and Madura: (a) passenger traffic, 1929 (b) goods traffic, 1929	131
4.4	Java and Madura, 1939: (a) outward journeys; (b) inward journeys	133
4.5	Luzon: Navigable rivers on relief map, c.1900	137
4.6	Luzon: (a) railway construction, 1892–1914; (b) railway construction, 1915–1939	140
4.7	Luzon: desire-line chart: (a) passenger trips by road, 1992; (b) commodity trips by road, 1992	149
4.8	Luzon: travel time on the existing road network from Manila, 1992	150
5.1	Chao Phraya, Mekong and Irrawaddy river basins: transport patterns, c.1885	157
5.2	Lower Chao Phraya Basin: canals built since 1850	162
5.3	Chao Phraya, Mekong and Irrawaddy river basins: transport patterns, 1910	165
5.4	Chao Phraya, Mekong and Irrawaddy river basins: transport patterns, 1940	167
5.5	Chao Phraya, Mekong and Irrawaddy river basins: transport patterns, 1960	176
5.6	Thailand: transport costs, 1965	181
5.7	Chao Phraya, Mekong and Irrawaddy river basins: transport patterns, 2000	182
6.1	Malay Peninsula and Annam: transport patterns, c.1885	188
6.2	Malay Peninsula and Annam: transport patterns, 1910	193
6.3	Malay Peninsula and Annam: transport patterns, 1940	200
6.4	Malay Peninsula and central Vietnam: transport patterns, 1960	206
6.5	Malay Peninsula and central Vietnam: transport patterns, 2000	210
6.6	Malay Peninsula and central Vietnam: proposed transport corridors, 2010	212

6.7	Malay Peninsula and central Vietnam: projected urban developments and expressways, 2020	213
III.1	Southeast Asia: urban population, 1900 and 1930	220
III.2	Southeast Asia: urban population, 1960 and 1990	221
III.3	Southeast Asia: schematic diagram of Singapore's interurban connections, 1995	224
III.4	A model showing phases of convergence and divergence and the associated economic processes in the historical development of city types in Southeast Asia against the yardstick of city types in metropolitan countries	226
7.1	Singapore: urban growth, 1819–1969	230
7.2	Singapore: the 1822 Town Plan drawn to Raffles' specifications by Lt Phillip Jackson	231
7.3	Port of Singapore, 1939	234
7.4	Port of Singapore and expressway system, 2000	242
7.5	Singapore: new towns and the Mass Rapid Transit System, c.2000	247
7.6	Greater Singapore: spatial structure, c.2000	255
8.1	Manila: land use, c.1895	261
8.2	Manila: local hinterland and transport routes, 1900s	262
8.3	Manila: growth map, 1819–1971	268
8.4	Manila: transport network, 2000	273
8.5	Jakarta: land use, 1858	276
8.6	Jakarta: growth map, c.1600–mid-1930s	277
8.7	Jakarta: expressways, 1995	284
8.8	Jakarta: new towns, 1997	285
9.1	Bangkok, 1850	291
9.2	Bangkok: Canals and roads, c.1860s onwards	292
9.3	Bangkok: land use, c.1930	296
9.4	Bangkok: land use, c.1960	299
9.5	(a) Saigon: land use, c.1930s; (b) Cholon: land use, c.1930s	301
9.6	Rangoon: land use, c.1940	302
9.7	Bangkok: land use, c.1985	307
9.8	Bangkok: urban transport, 1990s (a) expressways (b) mass transit railway	311
9.9	Bangkok and vicinity: land use, c.1995	314
9.10	Central Thailand: Spatial Development Framework, c.2010	317
10.1	Kuala Lumpur city-region: development, 1910–2005	320
10.2	Kuala Lumpur: growth, 1860–1991	321
10.3	Kuala Lumpur, 1895	323
10.4	Kuala Lumpur: Structure Plan, 1982	331
10.5	Kuala Lumpur: railways, 2000	333
10.6	Kuala Lumpur City-region: mega-projects, late 1990s	335
A.1	Southeast Asia: contiguous urban space centred on Singapore	340

A.2 (a) Physical links between home, hotel, mall, office and airport; (b) time lapse diagram showing time spent between origin and destination including travel to airport terminal and flight time   341

A.3 Instantaneous telecommunications between city-cores has produced a 'pancake-like' urban structure   343

Preface and Acknowledgements

Since the 1980s, after a centuries-long eclipse, Southeast Asia has regained the status of one of the world's key economic regions. The Asian crisis of 1997–98 punctured expectations of relentless expansion but also showed that Southeast Asia was now big enough to shake the global economy. The region contains the world's fourth most populous and largest Muslim country, Indonesia, and three of the world's largest cities, Jakarta, Manila and Bangkok. At the very centre of the region, Singapore has become a world city and enjoys living standards comparable with Europe. Even Kuala Lumpur, for so long the smallest of the core Southeast Asian capitals, now boasts the Petronas Twin Towers as the world's two largest buildings – Bangkok's Baiyoke Tower is not far behind.

A macro view

This book seeks to challenge standard *national* perspectives on modern Southeast Asia. By convention, population and territory, land and sea, are grouped in nations, separated by black lines from their neighbours, which often are blanked out. The recent discourse of globalization involves recognition that the world – and Southeast Asia – is becoming more integrated. Faster transport and instantaneous telecommunications make people, goods and information much more mobile and national boundaries much more porous. Correspondingly, there is also more awareness of 'the local', denoting aspects of sub-national life that previously had been trivialized. Yet the black lines remain. 'Global' and 'local' are both still defined in relation to 'national'. We argue that it would be helpful to focus on the strategic role of capital cities in mediating and integrating most supra- and sub-national flows, notwithstanding some bypassing global–local interactions (Figure P.1). Of course, nations still matter, but the national paradigm should not become an intellectual straightjacket.

Stemming from this conceptual framework, our methodology – the simplest that we can devise – is to map Southeast Asia at three different scales: the international network of cities (Part I), sub-regional cities and hinterlands (Part II), and capital cities as systems in their own right (Part III). In terms of population, economic activity and flows of people, goods and information, these systems demonstrate both structure and dynamics. From both a historical and current perspective, the older infrastructure and organization of the *state* may be more important than the *nation* in which it has become embodied. Instead of just being assumed as given – or projected backwards – modern states and nation-states have emerged by historical

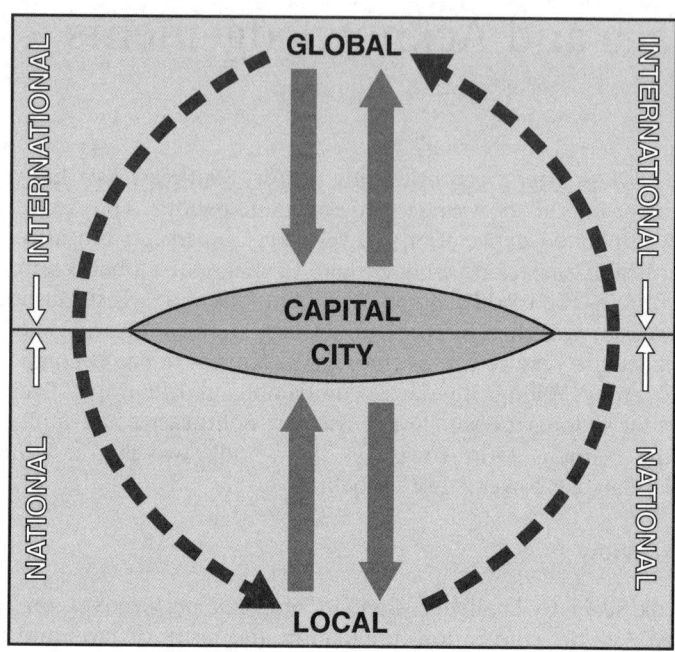

Figure P.1 Definition of 'global' and 'local' in relation to 'national' which affords a pivotal role to the capital city in integrating dominant supra- and sub-national flows and provides for direct global-local interactions that bypass the capital city

processes that were by no means inevitable. Our approach is firmly empirical. As a simple but powerful theme to identify spatial structure and change over time, we focus on transport and communications. Analytically we draw on the disciplines of history, economic history and economic geography. None of them are particularly fashionable but each has the great advantage of being a training in seeing what is there, unlike the reflexive disciplines of economics, political science and sociology which rationalize what theory suggests to be there and ignore whatever does not fit.

We have also chosen to elevate narrative over statistical analysis. The spatial evolution of Southeast Asia since the mid-nineteenth century is a story – or set of stories – that we wish to make more accessible. To summarize detailed and complex material, we have relied heavily on maps and figures to show how spatial patterns emerged. If other scholars are able to test our hypotheses more rigorously by data such as prices and wages, interest rates and capital flows, or trade intensities, we will be delighted. Much of the difficulty, as we have discovered, is to find sub-national time series that are comparable across the region.

We would have liked to explore in more depth the political forces and socioeconomic impacts of transport and communications on people and places. These vital aspects are woven into our stories but the detail is beyond the scope and length of this book. How people used the networks to build new and different lives, where the work force of rickshaw pullers, trishaw pedallers and jeepney drivers originated and how it was organized has been the subject of previous research by ourselves, our students and others (for example, Dick and Rimmer 1980; Dick 1981a,b; Rimmer 1982a,b, 1986, 1991; Roschlau 1986; Warren 1986). These and other micro studies highlighted the lack of any consistent macro-framework that gives a long-term perspective on temporal and spatial developments across Southeast Asia since the beginning of the modern era, which we take to be around 1850. This book attempts to meet that need. We hope it will encourage more and better work in both local and comparative (economic) history and geography.

Antecedents and acknowledgements

The study is the product of some twenty-five years collaboration and draws on our extensive fieldwork experience in Southeast Asia. Rimmer began in Thailand in the late-1960s as a geographer working on road–rail competition, Dick in Indonesia in 1973 as a transport economist working on interisland shipping. By 1980 we were both working on urban public transport and becoming more interested in the interaction between transport, planning and city development, giving rise to several joint articles and Rimmer's *Rikisha to Rapid Transit* (1986). In the late-1980s Dick joined a project studying regional development in East Java while Rimmer concentrated upon broader patterns of container shipping, airlines and telecommunications in the Asia-Pacific with a focus on Northeast Asia. We both became dissatisfied with the weak grasp on reality of the exploding literatures on globalization, economic liberalization and Asian development. Whatever their abstract validity, they were not grounded in the cities, towns and villages in which people led their daily lives. Yet the literatures on local development gave little insight into how the local connects to the global. Was the link really so difficult to make? It seemed time to go back to basics: the physical infrastructure of transport and communications, their modal networks and urban nodes.

Research for this book began in 1992 under the auspices of the Economic History of Southeast Asia (ECHOSEA) Project, set up under the direction of Tony Reid and Anne Booth in the Research School of Pacific and Asian Studies (RSPAS) at the Australian National University. Our study was one of a series commissioned by the Project and we have benefited greatly from interaction with the large group of participants.

Howard Dick's fieldwork and analysis was funded after 1992 by a Large Grant from the Australian Research Council. Peter Rimmer was supported by fieldwork funds of the Department of Human Geography, RSPAS, The

Australian National University (ANU). Logistical assistance was supplied by the secretarial and support staff of the Department of Human Geography and Division of Pacific and Asian History, both in RSPAS, The ANU, the Department of Management of The University of Melbourne and the Department of Economics of the University of Newcastle (NSW). Our thanks are due to Sittipong Dilokwanich (Mahidol University) for permission to use his maps of Bangkok.

Colleagues in the Department of Human Geography and the Division of Pacific and Asian History, RSPAS, The ANU gave specialist advice. Gavan Jones in the Research School of Social Sciences (ANU) assisted on demographic matters. Kennon Brezeale of the East–West Center, Honolulu supplied important references to Thai materials. Amarjit Kaur of The University of New England, provided additional information on transport in colonial Malaya.

Invaluable assistance has been offered by members of staff of the Australian National Library, especially those in the Asian Collections (Vacharin McFadden), The Australian National University Library, the Department of Transport and Regional Services Library; the National Library of the Philippines, National Library of Singapore, the Director of the National Library of Thailand (Chiraporn Jirapapha), the University of Singapore Library and the Algemeen Rijksarchief (Tweede Afdeeling) in The Hague, the KITLV Library in Leiden, the School of Oriental and African Studies (SOAS) and the Indian Office Library in London, the Newberry Library in Chicago, and the Library of Congress and United States National Archive in Washington, DC.

Grateful thanks are due for research assistance by Howard Dick to David Bulbeck, Louise Kinnaird, Malaya Papworth, Lorna Andreassen and Sonya Kelly; and by Peter Rimmer to Barbara Banks, Sandra Davenport, Elanna Lowes and Christine Tabart (Department of Human Geography, RSPAS, The ANU). We are both greatly indebted to our cartographers Ian Heyward and Kay Dancey (Cartographic Section, Research School of Pacific and Asian Studies, The ANU) for producing world-class maps.

Our Palgrave Macmillan editor, Nick Brock, has with care, patience and good humour piloted us safely through the reefs to the end of the long journey.

Finally, to our partners, Janet and Sue, and to our families we dedicate this book with unbounded love.

Melbourne and Canberra

Glossary

ADB	Asian Development Bank
ANU	The Australian National University
ASEAN	Association of Southeast Asian Nations
ASEAN10	Brunei, Cambodia, Indonesia, Laos, Malaysia, Myanmar (Burma), the Philippines, Singapore, Thailand and Vietnam.
bajaj	motorized becak
banca	dug-out canoe (Philippines)
becak	passenger trishaw (Jakarta)
bemo	(becak motor) three-wheeled passenger jitney (Jakarta)
BHQ	Business Headquarters
BI	British India Steam Navigation Company
BMA	Bangkok Metropolitan Administration
carromata	simple two-wheeled pony cart (Manila)
CBD	central business district
CITOS	Computer Integrated Terminal Operation System
DBS	Development Bank of Singapore
Desakota	Term used to describe the rural–urban transition in Asia (lit. Bahasa Indonesia *desa* = village and *kota* = town)
DKI	Daerah Ibu Kota (National Capital City Region) (Jakarta)
EBMR	Extended Bangkok Metropolitan Region
EBR	Extended Bangkok Region
EDB	Economic Development Board, Singapore
EDSA	Epifano de los Santos Avenue [Manila ring road]
EIC	East India Company
EIOSS	East Indies Ocean Steamship Company
ERP	Electronic Road Pricing
Estero	tidal creek or canal (Philippines)
Fl	Dutch or Netherlands Indies guilder
FMS	Federated Malay States
GMS	Greater Mekong Sub-region
GRDP	Gross Regional Domestic Product
ha	hectare
IATA	International Air Transport Association
JCJL	Java–China–Japan Line
jeepney	originally army jeeps rebuilt as passenger jitneys (Manila)
JTC	Jurong Town Corporation
KLCC	Kuala Lumpur City Center
KLIA	Kuala Lumpur International Airport
KLM	Koninklijke Luchtvaart Maatschappij (Royal Dutch Airlines)

KLSE	Kuala Lumpur Stock Exchange
KPM	Koninklijke Paketvaart Maatschappij (Royal Packet Company)
LRT	Light Rail Transit (Manila)
Marina	Maritime Industry Authority
MDC	Multimedia Development Corporation
Metrotren	Metropolitan Commuter Train Service (Manila)
MMC	Metro Manila Commission
MNE	Multinational Enterprise
MRT	Mass Rapid Transit
MSC	Multimedia Super Corridor
n.a.	not available (tables)
NCR	National Capital Region (Manila)
NDL	Norddeutscher Lloyd Bremen
NEP	National Economic Policy
NESDB	National Economic and Social Development, Thailand
NIEs	Newly Industrializing Economies
NISM	Nederlandsch-Indische Spoorweg Maatschappij
NISN	Netherlands Indies Steam Navigation Company (Nederlandsch-Indische Stoomvaart Maatschappij)
NSMO	Nederlandsch Stoomvaart Maatschappij 'Oceaan'
OHQ	Operational Headquarters
opelet(te)	small automobiles rebuilt as passenger jitneys (after Opel) (Jakarta)
P&O	Peninsular & Oriental Steam Navigation Company
PAL	Philippine Air Lines
Palapa	Indonesian telecommunications satellite
PATCO	Philippine Aerial Taxi Company
PNR	Philippine National Railways
PPP	Purchasing Power Parity
Prahu	A small Malay/Indonesian sailing vessel
PRRI	Pemerintah Revolusioner Republik Indonesia
PSA	Port of Singapore Authority
PUTRA	Projek Usahasama Transit Ringan Automatik
rickshaw	a light, two-wheeled vehicle pulled by a man with one or two passengers [Japanese *jin* = man, *riki* = power and *sha* = carriage]
RIL	Royal Interocean Lines
RL	(Koninklijke) Rotterdamsche Lloyd
RSPAS	Research School of Pacific and Asian Studies
Sdn Bhd	Sendirian Berhad: limited liability company (Malaysia)
SEA	Southeast Asia
SEATO	South East Asian Treaty Organisation
Sijori	Singapore–Johore–Riau
SingTel	Singapore Telecoms
SLORC	State Law and Order Restoration Council (Myanmar)

SMN	Stoomvaart Maatschappij 'Nederland'
SPDC	State Peace and Development Council (Myanmar)
SSS	Straits Steamship Company
STAR	Sistem Transit Aliran Ringan
teu	twenty-foot equivalent (container) units
tongkang	a Chinese-style barge for local trade or harbour lighterage
trishaw	a three-wheeled vehicle pedalled by one person and for one or two adult passengers. Known also as *becak* (Indonesia), *trisha* or *lancha* (Malaysia), tricycle (Philippines), *samlor* (Thailand) and *cyclo* (Vietnam)
UMNO	United Malays' National Organisation
UMS	Unfederated Malay States
URA	Urban Redevelopment Authority, Singapore
VOC	United East India Company (Dutch)

A Note on Place Names

Spellings in the book conform to the *Times Atlas* unless common international usage. On 18 June 1989 Burma officially changed its name to Myanmar and Rangoon became Yangon. The current names have been adopted when referring to developments since that date. However, throughout the book Cambodia is used rather than Kampuchea, Khong Falls rather than Khone Falls, Klang rather than Kelang, Malacca rather than Melaka, Sungei Kolok rather than Sungai Golok and Trengganu rather than Terengganu. Johore is preferred to Johor as the provincial name but the capital is referred to as Johor Baru. Alternative place names are shown in parentheses but over the period of the study Telok Anson's name was changed to Telok Intan only to revert back again.

Indonesian spelling is according to usage since 1972, thus Jakarta for Djakarta (colonial Batavia), Surabaya for Soerabaja and Aceh for Atjeh. The exceptions are persons and organizations adhering to the older Dutch style. Makassar was for some years known officially as Ujung Pandang.

Part I
Patterns and Processes

1
Patterns: Networks and Urban Hierarchy

Southeast Asia can be identified most readily as the ten member nations of the Association of South East Asian Nations (ASEAN). Indonesia, Malaysia, the Philippines, Singapore and Thailand formed the original grouping in 1967 as a basis for regional political cooperation. The tiny, oil-rich enclave of Brunei joined in 1980, socialist Vietnam in 1995, Laos and Myanmar (Burma) in 1997 and Cambodia in 1999. Papua New Guinea and East Timor enjoy observer status but not Australia or New Zealand. Although ASEAN is no more than a consultative forum and the level of economic cooperation is still modest, Southeast Asia has at last established a common political identity.

Before the 1940s none of these nations existed apart from Thailand (then Siam). At our somewhat arbitrary starting point of 1850, the world was even less familiar. After ignominious defeat in the Opium War (1840–42), China had been compelled to cede Hong Kong and to open Shanghai and several other 'treaty ports' to foreign trade. The Kingdom of Siam, which ruled Laos, parts of Cambodia and much of the Malay Peninsula, did not open its ports until the Bowring Treaty (1855), three years ahead of Japan. The separate Vietnamese kingdoms of Cochinchina, Annam and Tonkin were untroubled by French intervention. The United States had only just reached the Pacific by forcing Mexico to cede California, where the discovery of gold attracted the first wave of European migration. In the Philippines the Spanish maintained a desultory occupation, whose writ hardly extended beyond the main island of Luzon, leaving the Visayas and Mindanao to the mercy of pirate fleets. Among the sprawling islands that would eventually become Indonesia, the Dutch had established control over Java but ruled only nominally over most of the rest. The British had long been established in Bengal and from 1852 in Lower Burma. However, on the Malay Peninsula it had only the three small settlements of Penang (1786), Malacca (ceded by the Dutch in 1824) and Singapore (1819), plus the island of Labuan (1846) off northwest Kalimantan. The thin red line of British settlements extended beyond Singapore to the young Australasian colonies, some still in transition from their original role as remote gaols. It is therefore as illogical to make contemporary nations retrospective as to project the mid-nineteenth-century patchwork of kingdoms and colonies onto the twenty-first century.

National economies, the corollary of nation-states, are another artificial construct. The technology of national income accounting was developed no earlier than the 1940s to serve the policy needs of Keynesian macroeconomics and state intervention, whose rationale was disenchantment with the instability of a laissez-faire world economy and determination to avoid another global depression like the 1930s. This method of quantification was a very useful tool to manage fairly autonomous national economies, but it is less suited to understanding and managing the regional and global phenomena of a more integrated world. Nation and national economy may have become rather like a room of mirrors, reflecting back so much of their own conventions and habits of thought that it is impossible to make sense of what lies outside the room.

In physics or biology it is sound methodology to identify a system and its basic functions before attempting to study component sub-systems. Studying a sub-system in isolation is likely to be compromised by all kinds of implicit assumptions. Trying to explain the working of the system by aggregating knowledge of various components can also lead to curious results. Yet the nature and functioning of the global economy is approached in precisely this way. Every nation having its own flag and a seat in the United Nations produces its own independent national income accounts. Consolidation of these fragments, however, tells us little about the global economy as a structured and dynamic system. The generation and transfer of technology, the location of economic activity, consumption patterns, the global flows of goods, people and information set in motion by these processes, and the environmental impact can only be glimpsed.

There are enough signs of instability in the national economy paradigm. First, large multinational corporations are recognized to have as much economic weight and influence as smaller nation-states. Transnational organizations, groups and networks such as the overseas Chinese seem closer to the frontier of the emerging global economy and society. Secondly, Ohmae (1995) has argued from a marketing perspective that sub-national and transnational regions now deserve recognition in their own right. Thirdly, the discourse of the new information age suggests that information, and thus eventually power, money, investment, people and goods, will flow along different channels and concentrate in different ways. Fourthly, growing scientific awareness of the global environment demands understanding of global economic mechanisms as a matter of long-term human welfare.

This book therefore takes as its starting point not a given set of national economies but the working hypothesis that Southeast Asia is an open sub-system of the world economy. Its boundaries and internal structure are to be derived empirically. Most literature assumes that nation-states, like city-states, exist at a single point. For Singapore or Brunei this is indeed the case, but it is an absurd portrayal of Indonesia, the Philippines or Thailand. The problem, however, is to find data that are not conflated with national blocks of widely differing size. Fortunately time series data are available on the movement of goods, people and information, which may be understood as continuous flows of energy with precise spatial and temporal characteristics. For technological, economic and social reasons, these flows concentrate at the nodes of seaports, airports and 'teleports', which invariably coincide with cities.

The attempt to transcend the nationalization of space therefore leads not to formlessness but to a more specific unit of analysis. The next question is whether cities and hinterlands are appropriate units for studying globalization and economic integration. Decades ago, when Southeast Asia's economy was predominantly rural, this would not have been self-evident. Analysis of demographic and economic data, however, shows that cities are now unambiguously the main concentrations of population and economic activity and the source of most dynamism within their national economies. This is reflected in rapid rates of urbanization. In 1950 only 15 per cent of the people of Southeast Asia lived in cities; by the year 2000 it had increased to 37 per cent (UN 2001: 37).

Taking cities and their hinterlands rather than national economies as the fundamental units of analysis allows for multiple perspectives on Southeast Asia. First is the grand perspective of relations between cities that gives structure to Southeast Asia within the wider Asia-Pacific region and global economy (Part I). This chapter explores networks and hierarchies. Chapter 2 looks at change over time in terms of the impact of successive waves of new transport and communications technologies. Second is the sub-national perspective of the articulation of relations between cities and hinterlands as technologies have evolved from unmechanized to modern forms in step with the development of settlement and production (Part II). For sake of comparability, analysis is successively of the archipelago (Chapter 3), islands (Chapter 4), river basins (Chapter 5) and peninsulas (Chapter 6). The third perspective is that of cities themselves as they have evolved in embodied technology and scale as both physical and social environments and urban economic space (Chapters 7–10; Part III).

The methodology is hardly new. Although not specific to Southeast Asia, Berry (1964) wrote about 'a system of cities' and 'the city as a system'. Contemporaries explored the idea of the city as an organism. Yet these powerful ideas died off. What survived in geographical texts was the idea of urban hierarchy, a useful descriptive device but one having little explanatory power. Economists, who dominated the discourse of development after the 1960s, all but ignored cities and their hinterlands. A vigorous literature on trade, investment and economic integration acknowledged nothing below the national level but spatially amorphous industries. Only in the

1990s has there been some revival of interest in the economics of urban agglomeration (Fujita et al. 1999). As is so often the case, it is therefore necessary to glance back before moving on.

This chapter begins with a discussion of Southeast Asian identities as perceived within and outside the region. It then looks at the structure of the region from the viewpoint of Singapore as a central place. This suggests ironically that Southeast Asia was better articulated as a region in the mid-nineteenth than in the mid-twentieth century, but that the end of colonialism allowed the restoration of regional networks. Even so, in the 1990s, as in the 1850s, the commerce of Southeast Asia merges into that of neighbouring South Asia and especially Northeast Asia, so that no sharp differentiation is possible. To establish relativities, Southeast Asia is then scaled and disaggregated in terms of demography and economic activity. Finally, leading into Part III, there is an overview of the importance of Southeast Asia's cities and some exploration of urban space.

Regional identity

Regions are historically contingent. Identities, power centres, political boundaries and commercial networks are all subject to flux. Sometimes the common identity or experience of those who live within the region define it. On other occasions, outsiders define it as a matter of convenience.

Until recent decades the peoples of Southeast Asia perceived no common regional identity. The most embracing self-definition of the pre-colonial era was the Lands below the Winds (*Tanah dibawah Angin*), a term that was applied to the Malay Archipelago, extending as far north as the Gulf of Siam and southern Vietnam. This identity was in the first instance commercial because the direction of the monsoon winds determined the flows of ships, traders and cargoes (Reid 1988). Insofar as there was any transcending identity, it was religious: Islam in the case of the Malay Archipelago and Buddhism in the case of Mainland Southeast Asia. Language and culture otherwise defined identities, divisions exploited to the full by the colonial powers. The island of Java, for example, was divided into Javanese, Sundanese and Madurese, all classified as 'natives', alongside whom were 'other foreign Orientals' (mainly Chinese). Linguistically and culturally, 'Javanese' was much less than Java. In what became the modern Thai state, Thai-speaking people were the majority but by no means the entire population. The creation of national identities was the task of nationalist movements and had by no means been completed by independence. Some nationalisms cut across formerly well-recognized geographic identities. The Malay Archipelago, where Malay had long been a lingua franca and Islam a common religion, fragmented into Indonesia, Malaysia and the Philippines. The Golden Triangle on the borders of Thailand, Burma, Laos and China is a remnant of a much larger trading area where the boundaries of former kingdoms had been fluid (Walker 1999).

Western colonial powers promoted regional fragmentation. The Dutch and the British began their commerce with Southeast Asia at the beginning of the seventeenth century through the vehicles of the United East India Company (VOC) and the East India Company (EIC). The common references here to East India differentiated their sphere of activities from the better known West Indies. At this time there was no clear distinction between India and what would later become known as the (East) Indies. British India was understood broadly to encompass what is now Bangladesh (Bengal), Burma and, on the margin, the Straits Settlements of Penang, Malacca and Singapore. What lay beyond Singapore was generalized as the Far East. The Dutch referred to their possessions as Netherlands India (*Nederlandsch-Indië*), which the British translated with a subtle shift in emphasis as the Netherlands East Indies. The French referred to their possessions in proximity to China, as *l'Indochine*, which Americans occasionally referred to as the French East Indies. Approaching Asia across the Pacific from California and Hawaii, Americans thought more generally of East Asia. Until the Pacific War of 1941-45 there was no concept of a region of Southeast Asia that transcended these various colonial territories.[1]

Conventional wisdom has it that the term Southeast Asia came into use with the creation in 1943 of the Allied South East Asia Command. This confuses cause and reaction. The Japanese conquest and occupation of 1941-45 had already created the unity of Southeast Asia. Unlike the colonial powers, but in common with the Chinese, Japanese had an old regional category of Nanyo (Chinese Nanyang) or South Seas. However, Nanyo began in Taiwan and included Guam, Palau and New Guinea. By April 1942 the Japanese had completed a lightning conquest and carved up Southeast Asia into military and naval commands with a strategic centre in Singapore (Syonan).

The Southeast Asia Command thus came into being for the extraordinary political purpose of coordinating the Allied reconquest. Only this necessity gave rise for the first time to a common interest between the now dislodged colonial powers. From the outset, it was an uneasy collaboration (Aldrich 2000). British, French and Dutch expectations of a return to the colonial status quo were contested by a very different American vision. Like the Japanese, the Americans sought to hasten decolonization and, as in China, to maintain an open door with strong American influence. Self-government had been granted to the Philippines in 1935 and was followed by full independence in 1946. The Communist victory in China in 1949 reshuffled all the cards. America's Asian political agenda shifted to the Cold War goal of containing China and Southeast Asia became a soft frontier in the Cold War. After the French defeat in Vietnam in 1954, the South East Asian Treaty Organization (SEATO), including independent Thailand, was formed as a southern buffer in the containment of China. The United States took over from France the role of defending newly independent South Vietnam and brought in as allies Thailand, the Philippines, South Korea, Australia and New Zealand.

Within Southeast Asia, however, there was still no regional consensus. Indonesia was the key. Sukarno directed his political energies towards Pan-Asianism, first in building the relationship with India, then in 1955 hosting the Bandung summit of non-aligned nations, which China also attended, in direct challenge to the perceived neocolonial SEATO. The United States countered by sponsoring the abortive PRRI-Permesta rebellions of 1957–59 in Indonesia, while in 1963 Britain forged the neighbouring Federation of Malaysia (Kahin and Kahin 1995; Mackie 1974). With Ghana's Nkrumah, Sukarno attempted a now forgotten Asia–Africa initiative, then with domestic Communist Party support brokered a Jakarta–Phnom Penh–Beijing axis. The United States and Britain retaliated by at least tacitly supporting General Soeharto's takeover after the putsch of 30 September 1965.

Thus not until the formation of ASEAN in 1967 as a genuinely regional initiative can the idea of Southeast Asia be said to have taken root within the region. This was 23 years after the Southeast Asia Command and 26 years after the Japanese invasion. Formation of ASEAN in the middle of the Cold War marked for the first time a juncture of former imperial interests, now extending to Japan and Australia, and those of national governments. It was no natural or inevitable process but the outcome of a low-key but intense and protracted political struggle.

Structures of Southeast Asia

Southeast Asia is a vast region of land and sea. By the ASEAN10 definition, the land area of 4.5 million square kilometres is larger than the 15 economies in the European Union (3.2), equivalent to South Asia (4.3), and about half that of China (9.6) (UN 1995). However, the sea enclosed within these national boundaries constitutes an area as large again. The Malay term *Tanah Air* (Land and Sea) elegantly describes this vast archipelago of islands and peninsulas. Historically, it has been the sea rather than the land, much of it jungle, which has given the region its unique character. The Pacific, the Indian Ocean and the Arabian Sea, each a recognized sphere of trade and cultural interchange, all suffer the weakness of being hollow, without a centre: all significant activity lies on or behind the coastal rim. By contrast, as the world's largest archipelago Southeast Asia has been blessed by geography. Its large area of land provides an immense length of coastline that enabled scattered populations to enjoy excellent accessibility with pre-modern marine technology.

Moreover, the Southeast Asian archipelago could hardly lie in a more strategic location than as a barrage between China and India. Before modern transport and communications, trade between these two fabulously wealthy regions could avoid Southeast Asia only by overland caravan routes. Other than by porterage across the Thai–Malay Peninsula, the shortest and most sheltered route between China and India lay via the Straits of Malacca. Southeast Asia was therefore drawn into the mainstream of international commerce at a very early stage. In the seventh century the Kingdom of Sriwijaya on the site of modern Palembang became the first recognized power to participate in and grow rich from that trade, and its influence lasted for several hundred years (Wolters 1967). Its successor was Malacca, a city-state on the western Malay Peninsula, which thrived until its capture by the Portuguese in 1511 and played a more modest role as a staple port for Asian trade (Reid 1993). Rival local states were Brunei, facing the South China Sea on the north coast of Kalimantan, and Johore-Riau, in the vicinity of Singapore.

Of all the trading centres around the Straits of Malacca, Singapore proved to be the best located and most successful. Selection of the site was nevertheless the outcome of trial and error. From its base in Calcutta, the British East India Company had in 1786 occupied the island of Penang as an advance post on the far side of the Bay of Bengal. The aim was to safeguard the growing trade with China and to develop British trade in the Malay Archipelago. However, like the failed British settlement on Balambangan Island (1773–75) off northeast Kalimantan and the well-established Dutch port of Batavia (Jakarta), Penang was too far off-centre to become a central entrepot. British occupation of Java in the closing stages of the Napoleonic Wars (1811–16) alerted country traders to the trade potential of the archipelago, hitherto regulated from Batavia by the Dutch. After restoration of Dutch rule, the former British governor, Stamford Raffles, looked for alternative sites around the southern entrance to the Straits. In 1819 on his own initiative and by agreement with the local sultan he raised the British flag on the island of Singapore, presenting the home government with a fait accompli that in 1824 was reluctantly accepted by the Dutch (Chapter 7). The new trading settlement was brilliantly located for any vessels passing through the Straits of Malacca had to pass in sight of Singapore (Wong 1960: 195).

As the focal point of modern Southeast Asia, Singapore's networks of shipping and trade gave coherence to the region. In the mid-nineteenth century Singapore was a regional entrepot, not yet a world port. By tonnage the overwhelming proportion of all vessels (87 per cent) was engaged in intra-Asian trade: the Archipelago plus Siam and Cochinchina (southern Vietnam) accounted for 41 per cent, China 26 per cent and India 20 per cent (Table 1.1). Shipping beyond Asia to Europe, America and Australia was only 13 per cent.

Trade data is broadly consistent with this picture. The largest component of Singapore's trade was with the surrounding archipelago – 31 per cent with

Table 1.1 Singapore: shipping movements by origin and destination, 1850–51 (tons)

Origin	Western-rig (%)	Asian-rig (%)	Total	%
Intra-Asia	79	21	618,529	87.1
Archipelago	*71*	*29*	*239,685*	*33.8*
West Malaya	85	15	67,901	9.6
Sumatra/Riau	24	76	33,248	4.7
Java	94	6	42,484	6.0
Bali	67	33	19,139	2.7
Borneo	52	48	28,042	3.9
Celebes	—	100	3,692	0.5
East Malaya	88	12	20,405	2.9
Manila	100	—	19,125	2.7
Other Islands	—	100	5,649	0.8
Mainland SE Asia	*20*	*80*	*48,896*	*6.9*
Siam	47	53	20,988	3.0
Cochin-China	—	100	27,908	3.9
China	*88*	*12*	*186,224*	*26.2*
India	*100*	—	*143,724*	*20.2*
Bombay	100	—	72,826	10.2
Calcutta	100	—	62,304	8.8
Madras	100	—	8,594	1.2
Beyond Asia	100	—	91,422	12.9
UK/Europe	100	—	53,105	7.5
North America	100	—	27,821	3.9
Australia	100	—	10,496	1.5
Total	81	19	709,951	100.0

Source: Calculated from Wong (1960).

what is now Indonesia and Malaysia and another 7 per cent with what is now Burma, Thailand and southern Vietnam (Figure 1.1). However, the conveyor belt which drove the collection and distribution of products was the historic trade between India and China. Reflecting the strong British position in India, since the 1820s around 20 per cent of Singapore's trade had flowed across the Bay of Bengal, primarily to and from Calcutta. By 1850 the share of the Indian trade was entering a long-term decline but this was offset by steady growth of the China trade, which after 1840 found a new focal point in the British settlement of Hong Kong. Singapore thereby became more oriented towards the China trade and, with continuing immigration from South China, evolved into a Chinese town, closely linked into networks formerly based on Malacca, Penang and Batavia (Chapter 7). In 1850 all intra-Asian trade accounted for three-quarters of all Singapore's trade; Europe and in very small part North America and Australia accounted for most of the remainder.

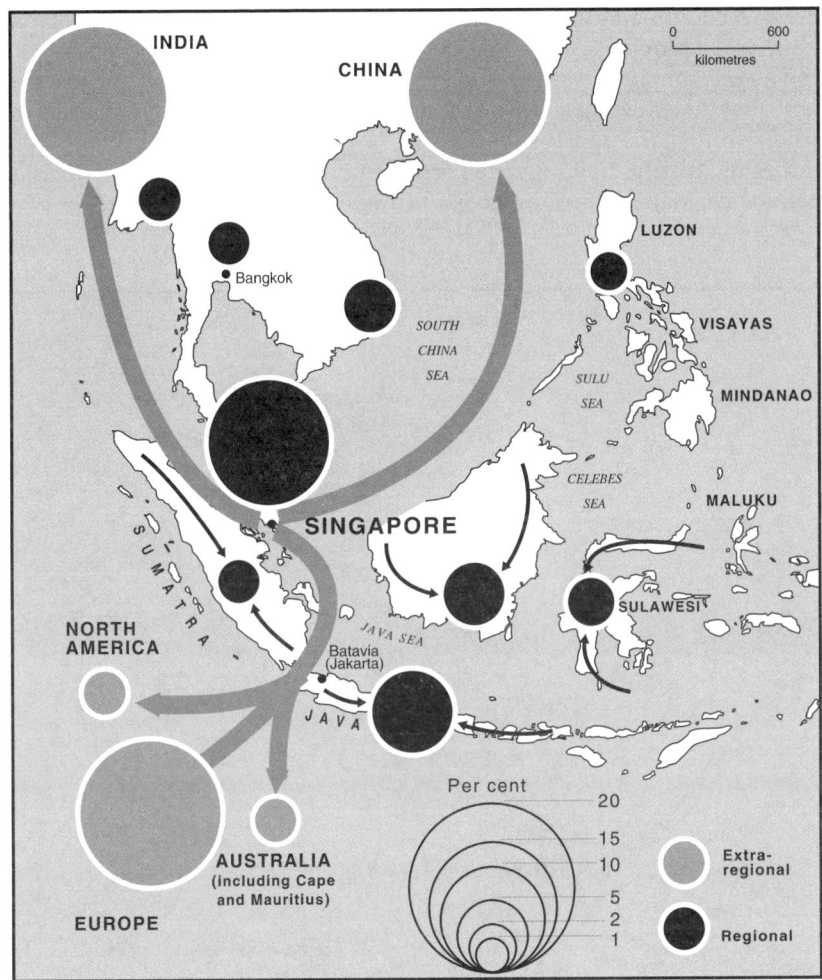

Figure 1.1 Exports from Singapore to main destinations, 1850 (*Source*: Wong 1960)

Singapore's pivotal role in intra-Asian trade stands in contrast with the situation of Java, the Dutch stronghold in Southeast Asia now turned into a vast export plantation. In 1850 three-quarters of Java's total trade (including with the other islands later part of Indonesia) was with Europe and four-fifths of its exports (Korthals Altes 1991). These exports consisted of coffee, sugar, tin and indigo, mainly to the Netherlands, with a much smaller return flow of imports, in which British manufactures featured prominently. China, Japan and India together accounted for only 3 per cent, the balance being made up almost entirely by local trade. Except for the small Japan trade still monopolized by the Dutch at Nagasaki, intra-Asian trade with India and China had shifted decisively to Singapore, which as a true entrepot drew together trade networks across the entire region. Java could be seen, without too much exaggeration, as an outlying island of the West Indies or Atlantic trading system rather than as an integral part of Asian commerce. Manila, then under Spanish rule, was still a backwater.

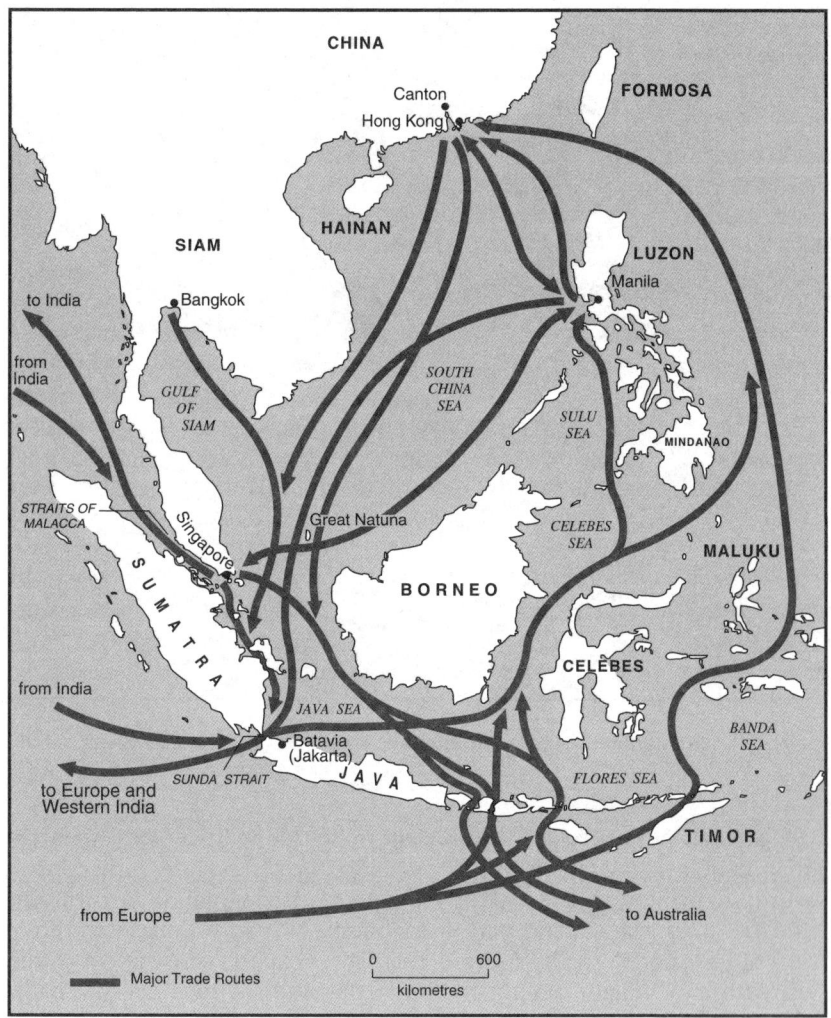

Figure 1.2 Major trade routes, mid-nineteenth century (*Source*: Wong 1960)

The pattern of shipping and trade began to shift in the mid-nineteenth century in response to technological change. The first regular steamship communications with Europe, the P&O mail via Ceylon and the then Isthmus of Suez, had been extended to the Straits Settlements and Hong Kong in 1845 but, except in the carriage of mails and first-class passengers, steamships were a curiosity. Trade still depended upon the seasonal monsoon winds. Being more versatile than local craft, square-rigged ships of European design had gained popularity among local Indian, Chinese and Arab owners, taking over the Indian trades and replacing junks in the China trade (Table 1.1). After Siam was opened to western trade in 1855, they would also quickly dominate that trade as well. Smaller local *prahu* nevertheless remained prominent in local trades within the Archipelago. Western-owned sailing ships in the long-distance China tea trade did not necessarily call at Singapore. Sunda Strait, close by Batavia and the straits east of Java were the main points of entry to the Indian Ocean en route to Europe or the east coast of North America (Figure 1.2).

Even Raffles could not have foreseen how Singapore's advantage would be enhanced by steamships and the opening of the Suez Canal, which after December 1869 made the Straits of Malacca the shortest passage between Europe, Asia and the Pacific. The free port of Singapore lay at the choke point of all that rerouted and booming traffic. Market forces now gave it a global role. With a pool of British, Chinese and Parsee entrepreneurs willing to invest capital, its shipping and trade flourished. Conversely, the free trade of Singapore and Hong Kong became more contested as the rest of Southeast Asia, except for Thailand, was brought under direct colonial control and forced into political and economic nexus with mother countries in Europe and North America. After 1890 the Dutch moved systematically to diminish the commercial influence of Singapore by redirecting the trade of the Outer Islands through Dutch-controlled ports (Chapter 3). In 1910 the Americans granted duty-free status to direct trade between the Philippines and the United States, which by 1930 absorbed 72 per cent of Philippine foreign trade (PIICC 1931). Lacking a strong centre and with marginalization of Chinese and Indian networks, Southeast Asia thus gradually lost coherence as a region.

As the western powers tightened their grip on Southeast Asia, Singapore's trade split more sharply into discrete circuits. Rapid growth of rubber exports to North America and Europe after 1900 boosted long-distance trade under European control. In this circuit Chinese participated only as local middlemen in accumulation and distribution. The Indian sphere especially waned, except in nearby Burma and in the specialist textile trade. Intra-Asian trade nevertheless survived, albeit almost invisible to European eyes. Strong links remained between South China and Southeast Asia: the mainstay was the two-way traffic in Chinese labour migration and the associated remittance business, supplemented by export of 'Straits produce' and rice

14 *Patterns and Processes*

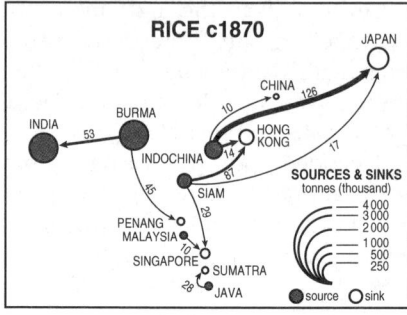
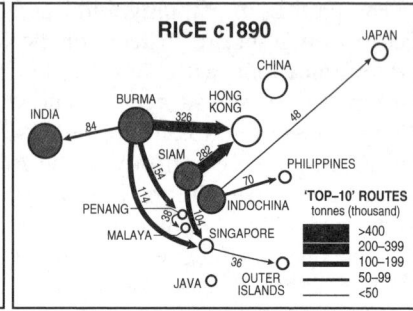
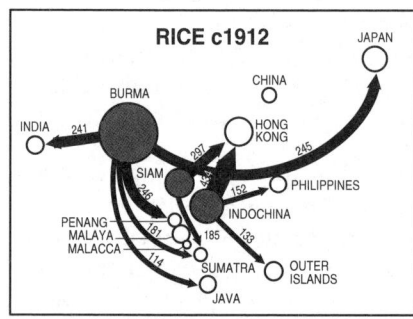
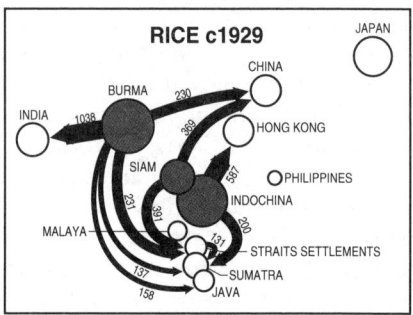

Figure 1.3 Rice movements, c.1870, c.1890, c.1912 and c.1929 (*Source*: annual trade reports from various colonies, Siam, China and Japan)

with return cargo of Chinese wares. Junks had long given way to steamers, mostly enjoying the security of European flags but still under control of Chinese brokers, agents and compradors. In 1930 five regular lines provided about ten sailings a month to Hong Kong, Swatow and Amoy (Xiamen); Chinese brokers were also prominent in trade with Bangkok and Saigon. Not until the 1940s, however, did principals in these networks begin to break out of the colonial straightjacket.

The intra-Asian provisioning trades became more important as export zones ceased to be self-sufficient in food. After the 1870s the delta regions of Lower Burma, the Central Thai plain and southern Indochina rapidly expanded rice output for intra-Asian and world markets through the entrepots of Singapore and Hong Kong (Figure 1.3). By the 1920s direct shipment to the Dutch archipelago had undercut Singapore's role (Huff 1994). On the return voyage ships carried Javanese sugar – Philippine sugar now went mainly to the United States (Figure 1.4).

Other important staple cargoes in the intra-Asian trade were salt, salted and dried fish, cattle (bullocks and buffaloes) and swine (Figure 1.5). In 1929 salt imports were focused on Burma. The main flows of dried and

Patterns: Networks and Urban Hierarchy 15

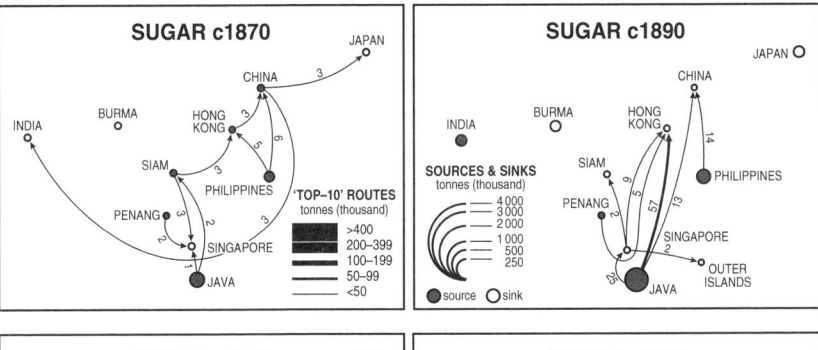

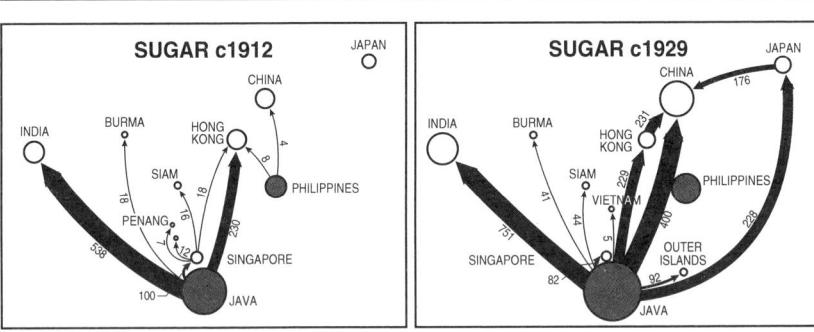

Figure 1.4 Sugar movements, *c.*1870, *c.*1890, *c.*1912 and *c.*1929 (*Source*: annual trade reports from various colonies, Siam, China and Japan)

salted fish were from the Outer Islands to Java and Indochina (Gulf of Thailand). Cattle, comprising bullocks and buffalo, were shipped as draught animals and for slaughter: despite frequent disruption by disease and quarantine the main flow was from the uplands of mainland Southeast Asia to the wet-rice regions. The main pig trade in Southeast Asia was from Indochina to Singapore – Hong Kong was supplied mainly from Southern China. Most of these trades apart from sugar were still handled through Chinese networks.

The coal trade was somewhat different. Proliferation of steamships, railways and factories gave rise to increasing demand for coal in Southeast Asia, which lacked good local supplies. At first coal was shipped at great expense from Britain but gradually India, Australia and Japan emerged as cheaper regional suppliers (Figure 1.6). In the interwar years industrialization turned Japan into a net importer, China and Indochina (Tonkin) emerged as important suppliers, and the Dutch archipelago became virtually self-sufficient.

The new colonial pattern was reflected in the development of shipping and, later, airline networks. In the case of shipping, the situation in

16 Patterns and Processes

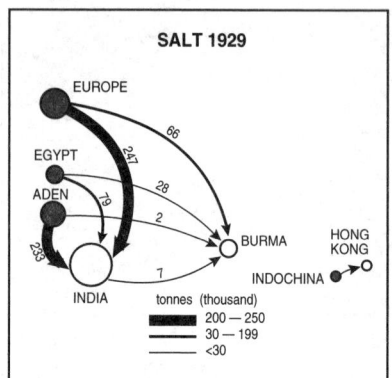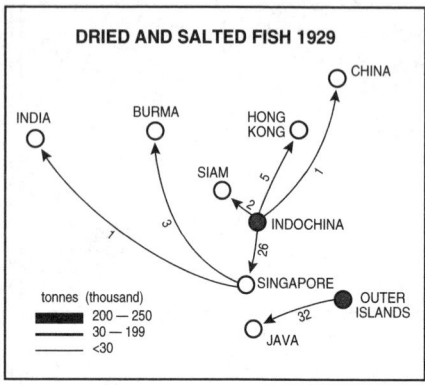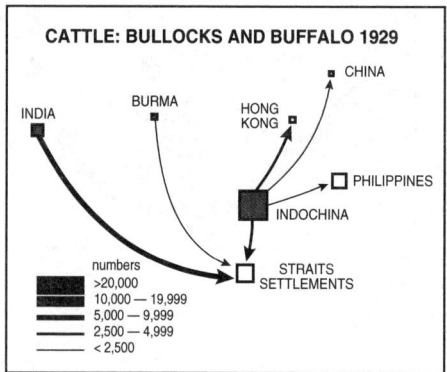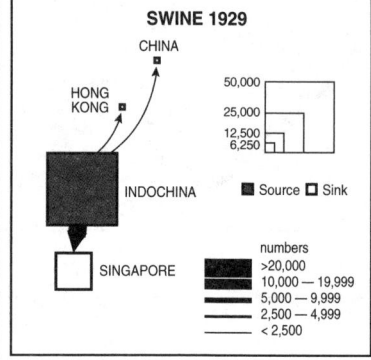

Figure 1.5 Movements of salt, c.1929; dried and salted fish, c.1929; cattle, bullocks and buffalo, c.1929; and swine, c.1929 (*Source*: annual trade reports from various colonies, Siam, China and Japan).

1960 – little changed since 1930 – may be compared with 1850 (Figure 1.7). The old intra-Asian route between Bombay/Calcutta and China still existed but, instead of opium, British and Japanese lines now carried large shipments of cotton to mills in Shanghai and Japan. Despite sailings as frequent as every other day, this was quite overshadowed by the new main trunk route from the East Coast of North America and Europe via the Suez Canal to the Straits, Hong Kong, China and Japan. Every day from Singapore there were at least two sailings in both directions, most frequently by British, German and Japanese lines. Dutch lines terminated in Indonesia, most French lines in Indochina. American lines called at both Singapore (for rubber) and Manila en route to and from the United States. From Manila via China and Japan there was also a fairly busy transpacific route to San Francisco and Los Angeles. Nevertheless, even the largest and fastest mail liners on Asian and transpacific routes were modest

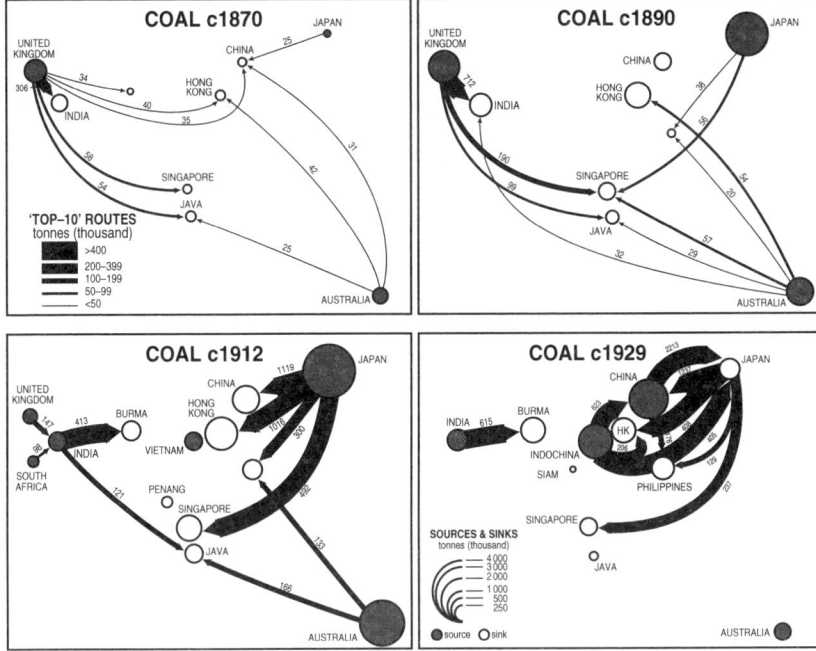

Figure 1.6 Coal movements, *c.*1870, *c.*1890, *c.*1912, *c.*1929 (*Source*: annual trade reports from various colonies, Siam, China and Japan)

compared with those on the Atlantic. The voyage from Europe to Asia was still a fairly long and leisurely one.

In the 1930s imperial connections were highlighted and reinforced by inauguration of long-distance air routes. The Dutch (1930), French (1931) and British (1933) opened overland air routes terminating in Jakarta, Saigon and Singapore respectively, followed by the American (Pan-Am) transpacific route island-hopping to Manila (1935). By modern standards these early flights were slow and expensive. Flying was possible only in daylight hours with frequent refuelling stops. Their main function was to speed the conveyance of mails and to promote national prestige. Airlines did not become commercially competitive with ocean liners until the 1960s (Chapter 2).

The basic colonial pattern of sea and air routes persisted into the 1970s. Decolonization of the Philippines (1946), Burma (1948), Indonesia (1945–49), Indochina (1945–54), Malaysia-Singapore (1946–63), the erosion of colonial economic control and the growth in trade with East Asia was reflected in the establishment of national carriers and the increased importance of Hong Kong and Japanese lines. The commercial framework of liner conferences and International Air Transport Association (IATA) agreements was nevertheless

18 *Patterns and Processes*

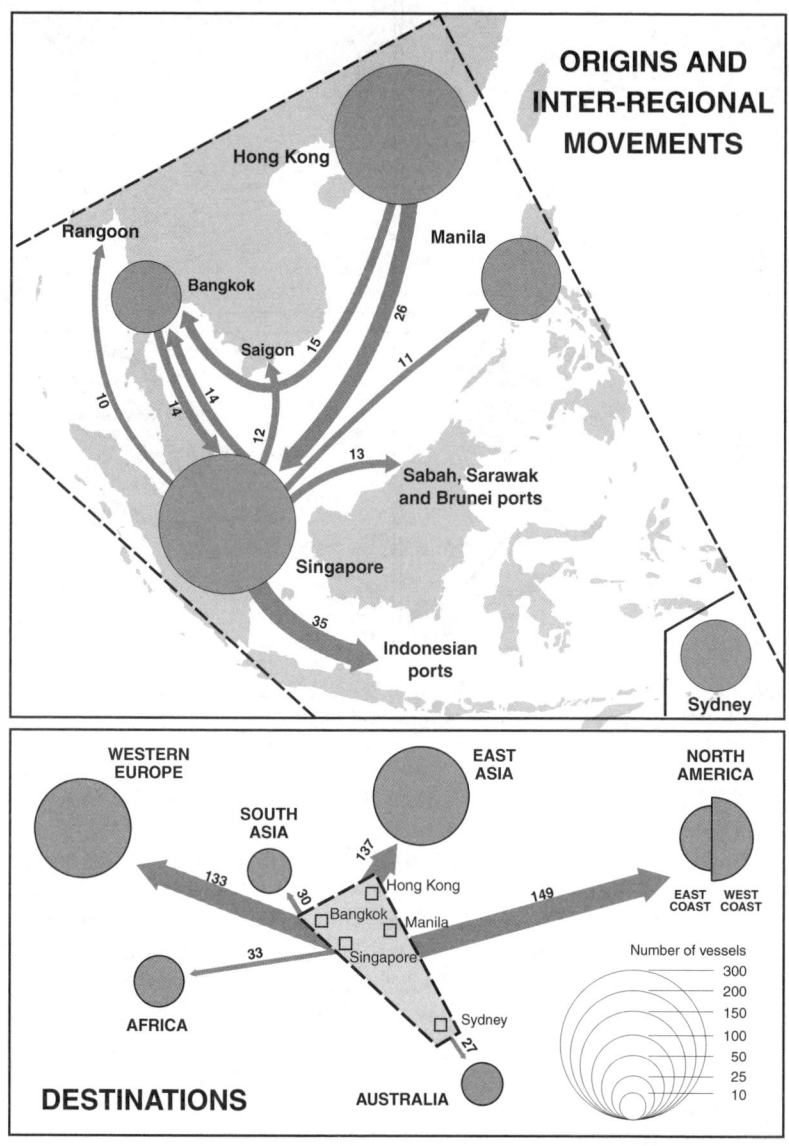

Figure 1.7 Liner shipping connections, June 1960 (a) movements within Southeast Asia (more than 10 sailings per month); (b) extra-regional movements from Bangkok, Hong Kong, Manila, Singapore and Sydney (*Source*: *Bangkok Post*; *Business Times*, Singapore; *Daily Commercial News*, Australia; *The Manila Times*; *South China Morning Post*, Hong Kong; and *The Straits Times*, Singapore). Information on sailings from Indonesian ports is incomplete.

flexible enough to accommodate these changes and share the traffic without displacing the main principals. Not until the late 1970s did it become apparent that a new era had dawned. Technology worked as a catalyst: 'Jumbo' jets brought about mass air travel, deepsea liner trades were transformed by containerization, and faster telecommunications hastened financial and commercial integration. Even more fundamentally, comparative advantage was shifting as industrialization spread from East to Southeast Asia and China. An invigorated Asia-Pacific economy overwhelmed traditional links with Europe. Except perhaps in the case of the Philippines and the United States, former colonies no longer maintain privileged trading relationships with their former colonial powers. Shipping companies and airlines from East and Southeast Asia now vie for global market share with European and American principals.

When container shipping was introduced to Southeast Asia in 1972 the main trunk route was still the long-distance steamer route from Europe via Suez and Singapore to Hong Kong and Japan. The transpacific route between East Asia and North America now overshadows this. By 2000 two-way Europe–Asia trade accounted for almost 5.8 million twenty-foot equivalent units (teu) compared with 10.9 million for the transpacific route and 3.6 million for the transatlantic route (NYK 2001; Rimmer 2002). East Asia has become the main generator of cargo with major ports in Japan (Osaka-Kobe, Tokyo-Yokohama), Korea (Pusan), China (Shanghai, Shenzhen), Taiwan (Kaohsiung) and Hong Kong. Singapore achieves equal importance with Hong Kong in Europe–Asia trade but plays a secondary role in transpacific trade (Rimmer 1998).

Containerization has led to a commercial reintegration of Southeast Asia focused on Singapore. To achieve the economies of scale of large ships and rapid turnaround, efficient container shipping requires hub-and-feeder networks to centralize cargo. In Southeast Asia Singapore was ideally located to fulfil this role and the Port of Singapore Authority acted promptly in the late 1960s to build the necessary infrastructure, as also did the port of Hong Kong. By the mid-1990s Singapore had become not just the leading port in Southeast Asia but the second largest general cargo port in the world (Figure 1.8). In 2000 Singapore (17 million teu) handled only slightly less than Hong Kong (17.8) and together with Pusan (7.5) and Kaohsiung (7.4) ranked ahead of Europe's busiest container port, Rotterdam (6.3) (CIY 2001). Much of the cargo shipped through these four Asian ports is transhipment. Singapore draws feeder cargo from the whole of Southeast Asia, dominating traffic with Indonesia and Malaysia and ranking ahead of Hong Kong and Kaohsiung in traffic with Bangkok and Saigon. In 2000 it was estimated that 40 per cent of Australian general cargo destined for Europe was also transhipped through Singapore. Most Philippine cargo, however, moves through the nearby ports of Hong Kong or Kaohsiung, which also capture a substantial share of transhipment cargo for North America.

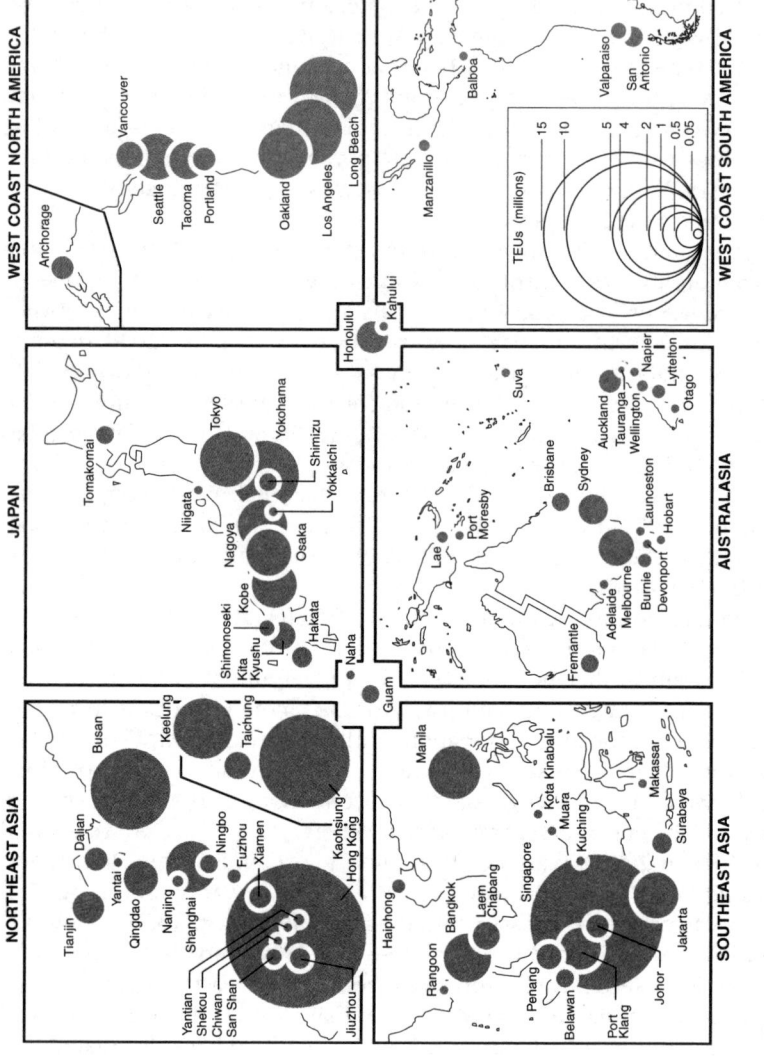

Figure 1.8 Asia-Pacific container ports, 1995 (*Source*: data from CIY 1997)

The international airline connections measured in number of flights around the Asia-Pacific Rim were much less focused on Southeast Asia (Figure 1.9). The strategic transport hub for the Asia-Pacific is indisputably Tokyo, which stands at the virtual centre, mediating communications between Asia and North America. The densest air traffic corridor (over 100 flights) is the arc between Taipei, Hong Kong, Bangkok, Singapore and Jakarta. Northeast Asia and Southeast Asia therefore have the highest intensity of interaction between regions around the Asia-Pacific. Southeast Asia itself, however, does not show up as an integrated region. The main corridor between Bangkok, West Malaysia, Singapore and Java is only weakly connected with the rest of the archipelago. Manila is oriented towards Hong Kong, Taipei and Tokyo rather than to Singapore, Jakarta or Kuala Lumpur. The rest of the Philippine and Indonesian archipelagoes has few international connections. Connections between Southeast Asia and Australia and India are also very weak. Singapore does not dominate air passenger traffic. Among the top fifty airports in 1998, Singapore (22 million passengers) ranked 35th in terminal passengers (international and domestic), well behind Tokyo Narita/Haneda (73 million) and Hong Kong (27 million). Within Southeast Asia, Singapore ranked behind Bangkok (24 million), which has maintained its standing as an efficient hub for regional and international flights (IATA 2000).

Southeast Asia attracts more minutes of telecommunications traffic (MiTT) than it generates on major routes. The United States is the key source of transpacific telecommunications flows but Japan, Hong Kong and Taiwan also play important subsidiary roles as net generators of traffic to Southeast Asia (Figure 1.10). Figures were unavailable for Singapore but the Philippines, Vietnam, Indonesia and Thailand all had net inflows. No Southeast Asian city was listed among the world's top-20 Internet hubs in 2000, Singapore was ranked thirty-third (Staple 1999, 2000).

Scaling Southeast Asia: demography and economy

Half of the world's population lives in an arc stretching from East Asia through Southeast Asia to South Asia. In national terms, China (1.3 billion in 2000) and India (1 billion) predominate. Southeast Asia's population of 500 million is nevertheless of the same order of magnitude as that of the European Union and almost twice that of the United States. Despite regions of very high population density such as Java and Luzon, Southeast Asia's average population density of 112.5 per square km is not especially high and would be even lower if area was calculated as both land and sea. Land comprises only about half Southeast Asia's total area, the balance being enclosed seas.

The most salient feature of the economic history of Southeast Asia over the past two centuries has been the enormous increase in population. By fairly crude reckoning, the population around 1800 would have been

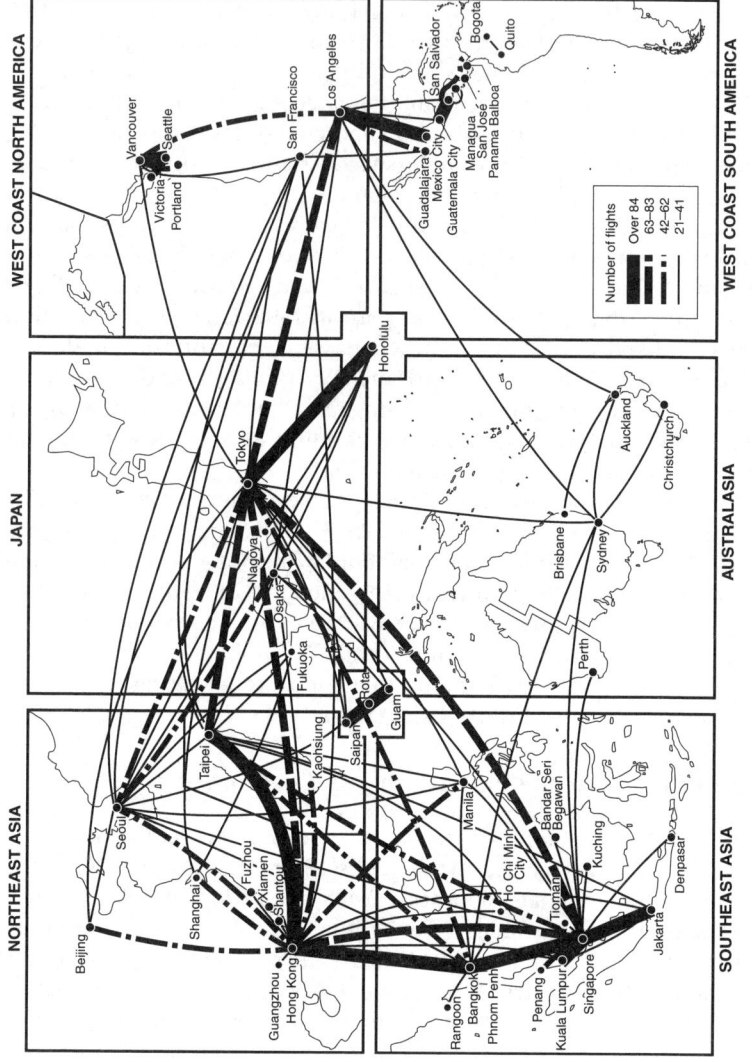

Figure 1.9 Asia-Pacific air routes, August 1995 (*Source*: data from ABC 1995)

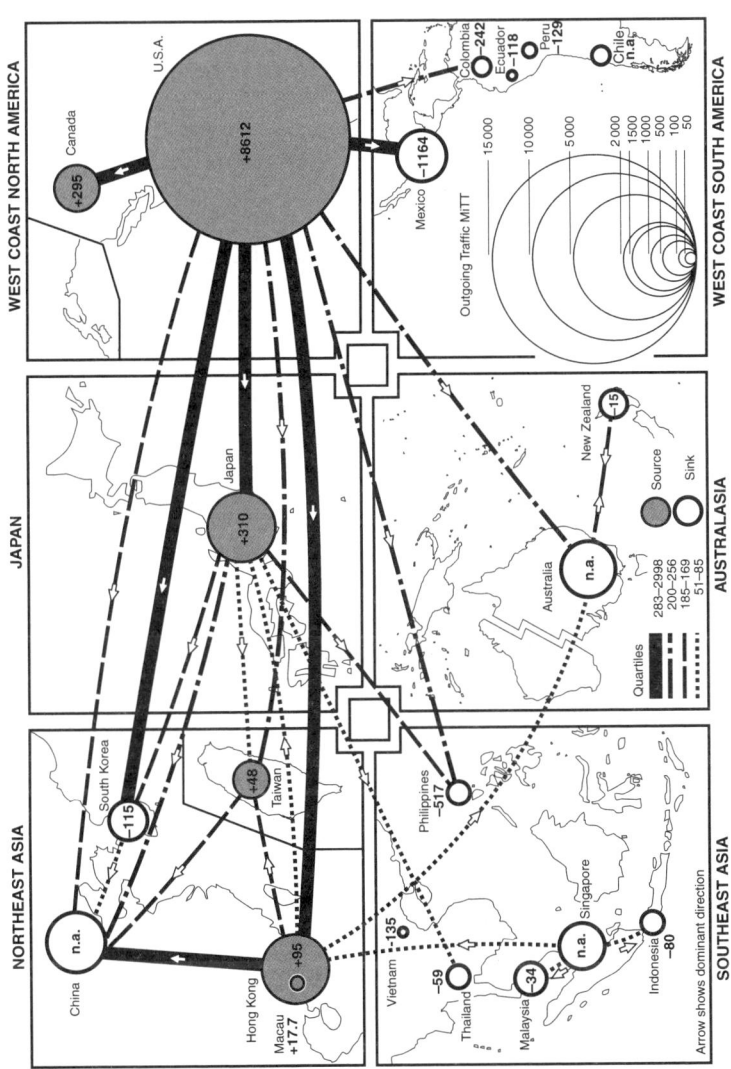

Figure 1.10 Asia-Pacific telecommunications connections between country pairs, 1995 (*Source*: data from Staple 1998)

somewhere between 30 and 40 million (Reid 1988: 14). In other words, at that time the entire region had been inhabited by a population barely twice that of contemporary Malaysia! Since much of that population was concentrated in a few areas, notably Java, the Red River delta of northern Vietnam and Upper Burma, most of SEA remained jungle with a fringe of coastal and riverine settlements. As late as 1900 SEA's population was still not much over 80 million (Elson 1997: 76). Most of the increase has occurred during the present century.

Southeast Asia's level of urbanization has varied markedly over recent centuries. Reid (1988: 67–8, 303; 1992: 473–4) argues that before 1700 Southeast Asia was actually highly urbanized by international standards, but that this urban-centred world of Southeast Asia succumbed to European pressure. When rapid population growth began in the mid-nineteenth century, it was associated with the opening of land for agricultural production, mainly for the world market. Java (sugar, coffee, tobacco), Luzon (tobacco, hemp, sugar), Malaya (rubber), Sumatra (rubber, coffee, tobacco) and more recently Mindanao (copra, hemp, fruit) were the leading producers of plantation and peasant crops for First World markets, while Lower Burma, central Thailand and southern Vietnam produced rice for regional subsistence (Elson 1997). However, despite settlement of extensive hinterlands, urban population growth was very modest. Scaled in terms of population, Southeast Asia's cities were still very small. In 1940 the largest city was Bangkok with a population of 685,000 followed by Manila-Quezon City (672,000), Jakarta (c.600,000), Singapore (585,000), Rangoon (501,000) and Saigon (excluding Cholon) (256,000) (Mitchell 1998). No Southeast Asian city compared in size with Tokyo-Yokohama (7.6 million), Osaka-Kobe (4.2 million), Calcutta (3.5 million) or Shanghai (3.3 million).

Urban population growth accelerated during the Japanese occupation (1942–45) and continued to grow rapidly through the turmoil of the 1940s and subsequent decolonization until an urbanization level of almost 15 per cent was reached in 1950 (UN 2001). This was the heyday of the Third World city, portrayed as bloated and dysfunctional by comparison with the shady and well-ordered late-colonial city. At root was the breakdown of colonial control because migrants could now move without constraint from villages into the larger cities. The incentives were better opportunities of employment and income, underpinned by the liberality of newly independent governments in spending on capital cities. After the 1960s rural–urban migration became associated with industrialization, which was concentrated in and around the main cities of the newly industrialized countries. By 2030 almost 56 per cent of Southeast Asia's population is predicted to be urbanized (UN 2001: 43).

As the economic weight shifts from the transatlantic to the transpacific economy, East and South Asia are acquiring economic importance more commensurate with their 50 per cent share of world population. Until the

Patterns: Networks and Urban Hierarchy 25

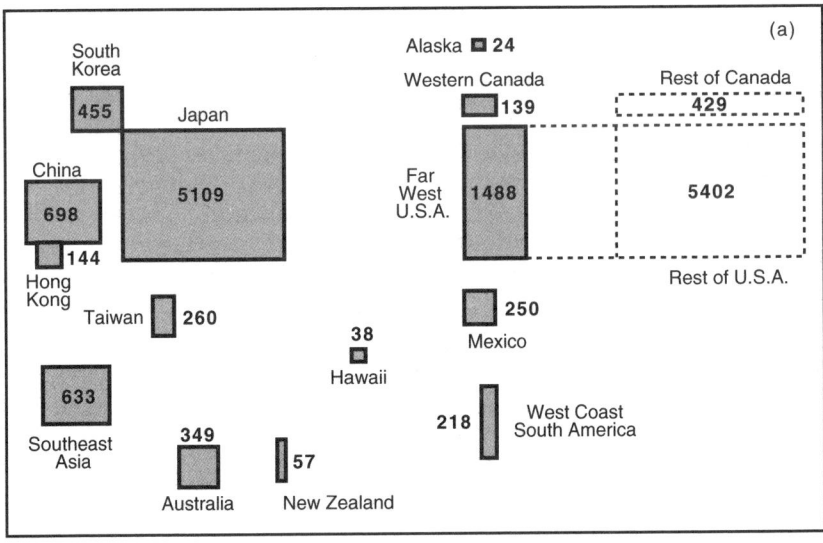

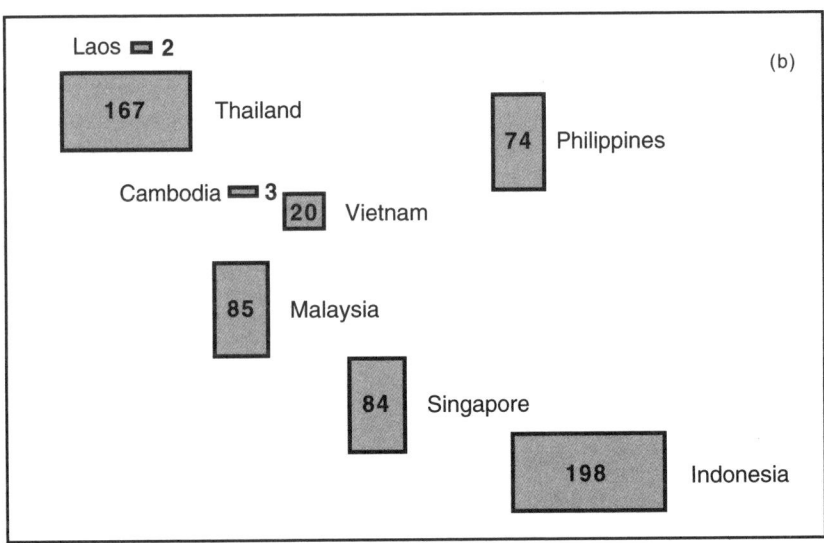

Figure 1.11 (a) Asia-Pacific gross domestic product, 1995; (b) Southeast Asia gross regional domestic product, 1995 (*Source*: based on data from World Bank 1997)

1980s China and India were desperately poor countries. However, China has since been able to sustain spectacularly high rates of economic growth and since its recent liberalization India also shows signs of following this example. In Southeast Asia rapid economic growth began in Singapore during the late 1960s followed in the 1970s by Malaysia, Thailand and Indonesia and, more recently, the Philippines and Vietnam. These high rates of growth have transformed the economic map of the Asia Pacific (Figure 1.11a/b). While the economic axis continued to lie between Japan and the West Coast of North America China and Southeast Asia are both becoming substantial economic blocs in their own right. Were GDP figures adjusted to purchasing power parities (PPP), China and Southeast Asia would both be larger relative to Japan. One way of appreciating the significance of these changes is to compare the economic weight of Southeast Asia with that of Australia. In 1967 the GDP of ASEAN was about 25 per cent smaller than that of Australia. By 1998 in conventional figures it was 40 per cent larger and in PPP terms almost four times larger.

A more interesting perspective on Southeast Asia than national blocks is a disaggregation in terms of the economic weight of component sub-regions. Table 1.2 sets out the Gross Regional Domestic Product (GRDP) figures for 1995, just before the Asian crisis. Excluding Burma, northern Thailand and Indochina for which sub-regional data are not available, the rest of Southeast Asia divides into three economic segments. A main corridor stretching from the Central Plain of Thailand through Peninsula Malaysia, Sumatra and Singapore to Java and Bali constitutes

Table 1.2 Southeast Asia:[1] gross regional domestic product (GRDP) at official exchange rates and at purchasing power parity (PPP) by sub-region, 1995

Sub-region	$ billion	%	$ b. (PPP)	%
Central Thailand	110	19	290	19
S. Thai peninsula	15	3	39	3
West Malaysia	72	13	152	10
Singapore	84	15	84	5
Sumatra	44	8	162	11
Java/Bali	121	21	448	29
CORE S.E. ASIA	446	79	1175	77
Kalimantan (exc. Brunei)	31	5	95	6
Eastern Indonesia	16	3	59	4
Visayas/Mindanao	27	5	72	5
PERIPHERAL S.E. ASIA	74	13	226	15
Luzon (exc. Bicol)	46	8	119	8
SOUTHEAST ASIA[1]	566	100	1,520	100

Note: 1. Excludes Indochina, north and northeast Thailand and Myanmar.
Source: National statistical yearbooks; World Bank 1997

three-quarters of the whole economy and may be regarded as the core. Second is a vast island periphery of Kalimantan, Eastern Indonesia, the Visayas and Mindanao constituting 15 per cent of the whole and third is the island of Luzon, a modest 8 per cent (about the same size as nearby Hong Kong). These proportions vary little between the standard GRDP figures valued at official exchange rates and those adjusted to purchasing power parity (PPP) but there are interesting variations in composition. The PPP figures reduce the relative weight of Singapore and Hong Kong while increasing that of Indonesia, Malaysia and, to a lesser extent, the Philippines and Thailand.

The PPP figures can be translated into a schematic map in which each main sub-region is shown as a box whose area is scaled in proportion to Singapore (Figure 1.12). This map further highlights differences between core and periphery. The core forms a corridor on the western side of the map and lies as one extended side of a Singapore–Bangkok–Hong Kong triangle. The other two sides, one through the South China Sea and the other a short distance through the economic backwater of Indochina, are transport routes rather than economic corridors. Singapore is the focal point of the southern end of the corridor. Much play has been made of the Growth Triangle between Singapore, Johor (Malaysia) and Riau (Indonesia), but this is little more than a city of Greater Singapore. More substantive is the great cluster around Singapore of the Malaysian Peninsula, Sumatra, Java and Kalimantan (including East Malaysia). This cluster makes up about 60 per cent (excluding Indochina) of the economy of Southeast Asia and is knitted together by central links with Singapore as well as by links around the circumference.

The periphery consists of eastern Indonesia and the southern Philippines. Although each has a weight roughly equivalent to that of Singapore, economic activity is spread over a vast maritime area and is highly fragmented. Even the largest islands of Sulawesi and Mindanao, though each notionally worth around 40 per cent of Singapore, consist of several sub-regions which have only very poor communications with each other. To some extent this is true also of Kalimantan, here combining the East Malaysian states of Sarawak and Sabah with the four Indonesian provinces, but somewhat arbitrarily this island is counted as a block in its own right. Unlike eastern Indonesia and the southern Philippines, whose transport and communications are national rather than international, Kalimantan connects fairly well with Java, Singapore and West Malaysia. Though still part of the periphery, it is the most accessible part of it and combines rich natural resources with sparse population.

The diagram also highlights the outlying position of the Philippines. Its main island of Luzon lies on the very edge of Southeast Asia with only weak transport and communications with the rest of Southeast Asia (see below). Domestic links with the Visayas and Mindanao do not extend with any

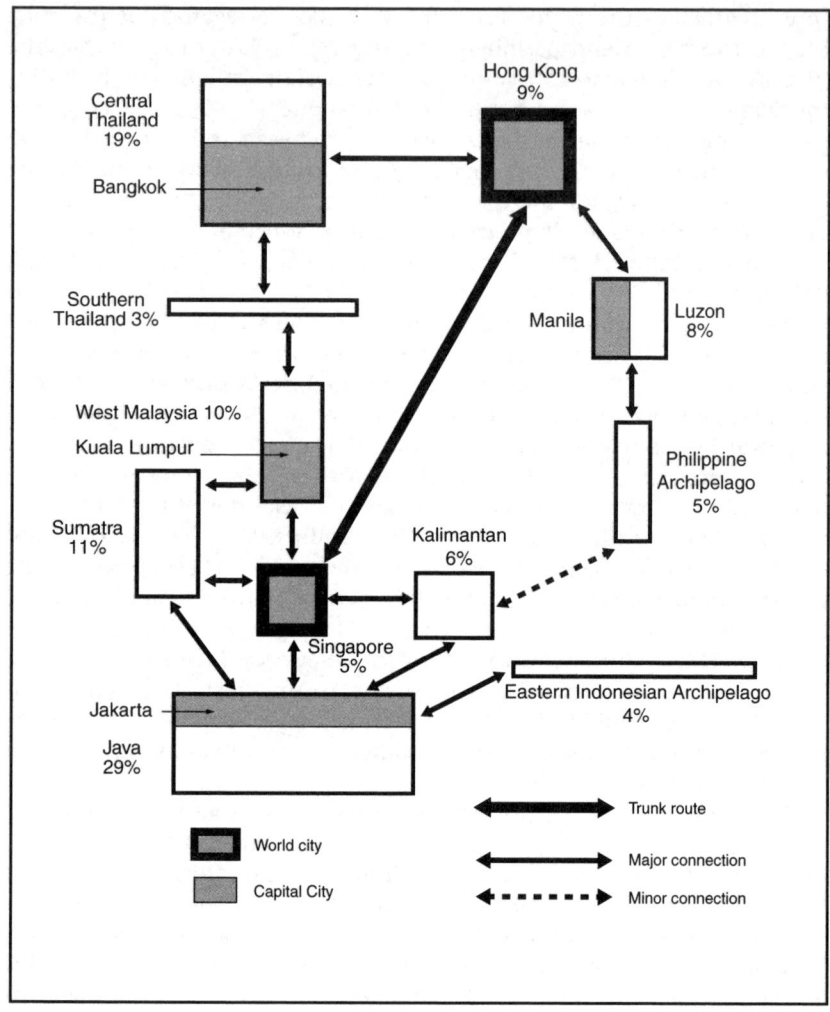

Figure 1.12 City/hinterland schematic diagram, 1995 (*Source*: Tables 1.2, III.1)

consequence into eastern Indonesia or East Malaysia, even allowing for smuggling across the border to Kalimantan. The orientation of Luzon and Manila are towards Hong Kong, Taiwan and Japan (Chapter 3).

Not shown on the diagram is the mainland periphery of Burma's Shan states, northern/northeast Thailand and northwestern Indochina, also the backdoor to China's Yunnan province. This sub-region around the so-called 'Golden Triangle' enjoys a turbulent prosperity based on the opium poppy and the drug trade. Since the end of the Cold War, closer cooperation across borders and better roads are creating a viable Mekong sub-region (Chapter 5).

World cities

Measured by population, Southeast Asia will have twelve urban agglomerations with populations over two million by 2015 (Figure 1.13). Of these Bangkok, Jakarta and Manila will rank among the world's top 25 urban agglomerations, which will also include six in cities in Northeast Asia (China, Korea and Japan) and Los Angeles (UN 1995, 2001).

Singapore and Hong Kong do not have such large populations but are recognized as 'world cities' because of their vital service functions, not least in banking and finance. There is no single criterion of a world city. Such status arises from a multiplicity of high-level functions being concentrated in one place. For example, Singapore, along with Hong Kong and Tokyo, qualify as global financial centres. Another important test involves strength of linkages with the rest of the world across various forms of transport (Rimmer 1999, 2002). Tables 1.3–1.5 rank Asia-Pacific cities in terms of their throughput of containers, air freight and air passengers. In 2000 Singapore and Bangkok, like Hong Kong/Yantian, Tokyo/Yokohama, Seoul/Pusan and Taipei/Keelung, featured on all three counts. Kuala Lumpur/Port Klang, like

Table 1.3 Asian cities ranked in top 25 container ports, 1985, 1995 and 2000

Asia-Pacific	1985		1995		2000	
	teu	rank	teu	rank	teu	rank
Hong Kong	2,289	3	12,550	1	18,100	1
Singapore	1,699	6	10,800	2	17,040	2
Pusan (Seoul)	1,115	12	4,503	5	7,540	3
Kaohsiung	1,901	4	5,232	3	7,426	4
Shanghai	NR	—	1,527	19	5,613	6
Port Klang	NR	—	NR	—	3,207	11
Tokyo	1,004	14	2,177	12	2,899	14
Manila	505	23	1,688	16	2,868	15
Jakarta	NR	—	1,465	23	2,476	19
Yokohama	1,327	9	2,757	8	2,317	21
Kobe	1,857	5	1,457	24	2,266	22
Bangkok	NR	—	1,433	25	2,195	23
Yantian	NR	—	NR	—	2,148	24
Keelung (Taipei)	1,158	11	2,170	13	NR	NR
Nagoya	NR	—	1,477	22	NR	NR

Note: NR = Not ranked. Southeast Asian ports in bold.
Source: CIY (1987–2002).

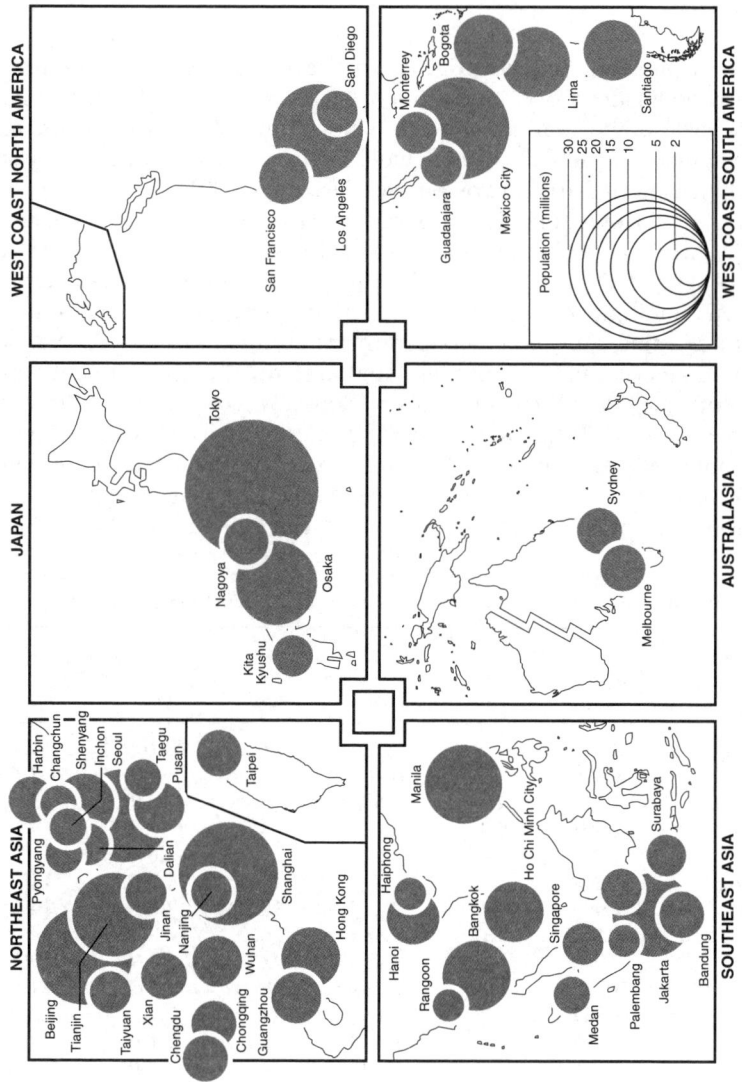

Figure 1.13 Estimated Asia-Pacific population, 1995 (*Source:* Based on data from UN 1995)

Table 1.4 Asian cities ranked in top 25 airports by air freight, 1985, 1995 and 2000 ('000 tonnes)

Asia-Pacific	1985		1995		2000	
	tonnes	rank	tonnes	rank	tonnes	rank
Hong Kong	420	7	1,458	2	2,241	1
Tokyo	726	2	1,606	1	1,876	2
Singapore	299	10	1,104	6	1,683	3
Seoul	308	9	1,014	8	1,592	4
Taipei	223	12	734	12	1,196	11
Osaka	NR	—	417	17	851	14
Bangkok	171	16	619	14	828	15
Kuala Lumpur	NR	—	NR	—	479	18
Manila	NR	—	275	23	NR	—

Note: NR = Not ranked. Southeast Asian airports in bold.
Source: ICAO (1986–2001).

Table 1.5 Asian cities ranked in top 25 airports by international passengers, 1985, 1995 and 2000

Asia-Pacific	1985 Passengers		1995 Passengers		2000 Passengers	
	million	rank	million	rank	million	rank
Hong Kong	9.9	7	27.4	3	32.1	5
Singapore	8.7	9	21.7	6	27.0	7
Tokyo	9.2	8	21.5	7	24.0	8
Bangkok	5.4	21	15.1	10	20.9	11
Seoul	NR	—	13.4	13	17.9	13
Taipei	4.7	24	12.6	15	16.7	14

Note: NR = Not ranked. Southeast Asian cities in bold.
Source: ICAO (1986–2001).

Osaka/Kobe, ranked in container shipping and air freight but not as an air passenger hub. The remaining centres, Jakarta and Manila, like Kaohsiung, Shanghai and Nagoya, ranked only as container ports. There are no time series data for telecommunications hubs.

Another way of assessing world-city status is to count the headquarters location of business multinationals (Rimmer 1999, 2002). Table 1.6 sets out the headquarters location of the world's top 25 transport and communications organizations around 2000. Tokyo is the only city with multiple representation across container shipping, air passenger and telecommunications. Hong Kong and Singapore both have one company in

Table 1.6 Asia-Pacific headquarters location of the world's top 25 container shipping, air transport and telecommunications firms, c.2000

	Container shipping	Airlines	Tele-communications
Tokyo	NYK Line (8) Mitsui OSK Lines (12) K Line (17)	Japan Airlines (7)	KDD (26)
Singapore	APL (5)	Singapore Airlines (6)	Singapore Telecom (15)
Hong Kong	OOCL (13)	Cathay Pacific (10)	C&W Hong Kong (12)
Seoul	Hanjin Shipping Co.(6) Hyundai Merchant Marine (14)	Korean Airlines (19)	—
Taipei	Evergeen/Uniglory Marine Corporation (2) Yangming Marine Transport Co (16)	—	—
Beijing/Tianjin	Cosco (7)	—	China Telecom(10)
Shanghai	China Shipping Container Lines (15)	—	
Sydney	—	Qantas (23)	Telstra (19)
Kuala Lumpur	—	Malaysia Airlines System (21)	
Bangkok		Thai Airways (13)	

Note: Liner shipping ranking based on size of fleet capacity; airline ranking based on scheduled international passengers carried; telecommunications: figures are based on outgoing traffic for 1999.
Source: Fossey (2000), IATA (2001) and Staple (2001).

the top 25 for each mode. Seoul and Beijing have headquarters in two modes. Taipei and Shanghai have only shipping headquarters and Sydney, Kuala Lumpur and Bangkok only airline headquarters.

Conclusion

Southeast Asia's evolving role and structure can be represented in regional perspective by an axis of intra-Asian trade which we refer to colloquially as Asia's 'Main Street' (Figure 1.14). In the mid-nineteenth century Main Street was Calcutta–Penang–Singapore–Hong Kong–Canton. The P&O mail steamers shuttled between Bombay and Hong Kong, exchanging mails in Ceylon with the main Suez–Calcutta line. Several other lines of fast steamers plied the opium run from Bombay or Calcutta to the Straits and China. Calcutta was the centre of British power and the hub of commerce for all Asia. Japan did not yet figure.

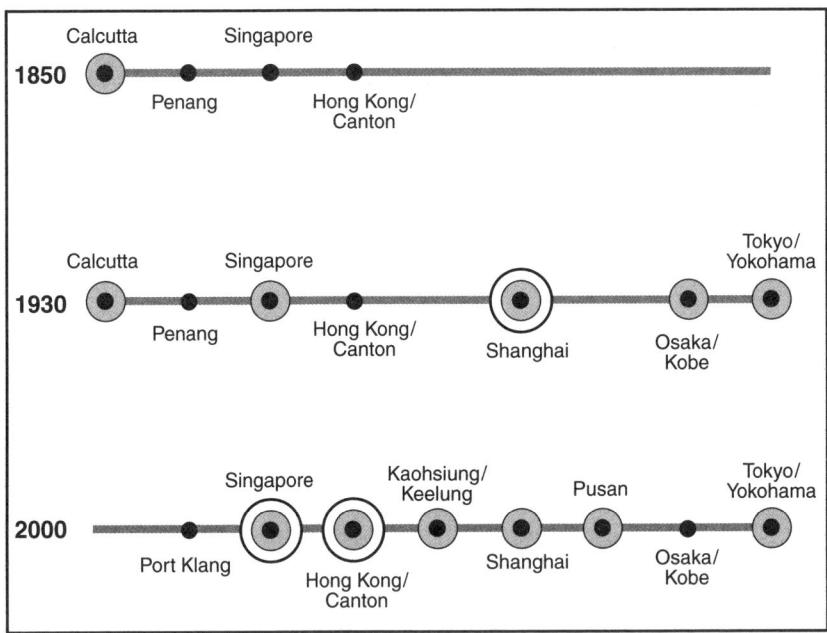

Figure 1.14 Port chains, 1850, 1930, 2000. The size of circle approximates the ranking of ports at each particular point in time

By 1930 the locus of Main Street had moved substantially eastwards. Calcutta had lost its pre-eminence and within India was challenged by the textile capital of Bombay. The main flow of shipping and trade was now from the East Coast of the United States and Europe via the Suez Canal to the Straits and East Asia. If any Asian city could then claim international status it was Shanghai, whose population of over three million was growing quickly under the effects of industrialization and the pressure of refugees from strife-torn China. Shanghai was almost two cities in one, an open, cosmopolitan treaty port lightly governed under foreign laws beside a densely settled Chinese city (Murphey 1953). Shanghai's economic and cultural influence reached well into Southeast Asia. By contrast Singapore and Hong Kong, though way-ports for most shipping between Europe and China/Japan, were still modest and compact trading cities with populations of around 560,000 and 360,000 respectively (Mitchell 1998). Tokyo-Yokohama (5.6 million) and Osaka-Kobe (3.2 million) had much larger populations and led the rest of Asia in industrialization, but their regional influence was confined to the expanding Japanese empire and parts of North China. This era drew to a sudden close with the Japanese invasion of

China and the seizure of Shanghai in 1937. Chinese and foreign capitalists, sometimes with whole factories, began to relocate to Hong Kong, which in the 1950s became the 'boom city' of Asia. Along with the rest of China, Shanghai remained closed to private trade and investment until the 1980s.

'Main Street 2000' is Singapore–Hong Kong–Tokyo, with Shanghai bidding for re-inclusion. As the Indian economy languished so also did Calcutta, growing in population but losing its international status. Meanwhile Japan's rapid and sustained postwar expansion pushed Tokyo-Yokohama to world ranking in population, economic importance and international connections. Japan thus continued to tilt Asia's economic weight to the east, as did China's economic resurgence in the 1990s. Within Southeast Asia, Penang has been relegated to little more than a local port under competition from Port Klang.

As reflected in inter-urban air connections, the prime Singapore–Hong Kong link is supported by the emergence of a new regional arc between Bangkok, Kuala Lumpur, Singapore and Jakarta with an extension to Surabaya (Figure 1.15). There are also minor links integrating Hanoi and Saigon. However, Manila still stands apart with its primary link to

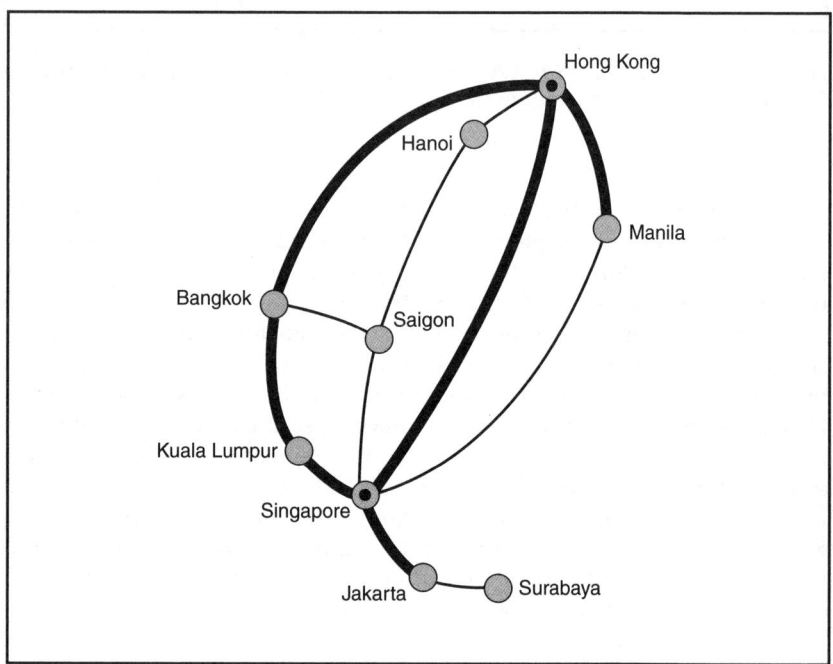

Figure 1.15 Interurban air connections showing the pivotal importance of Singapore and Hong Kong

Hong Kong and lesser links to Taiwan and Japan. Economically the Philippines belongs more closely to a Greater South China region but to omit from any definition of Southeast Asia one of the founding members of ASEAN would be unduly controversial.

Transport flow data therefore point to difficulties with the conventional view of Southeast Asia as a coherent region. An alternative approach is to view Southeast Asia as an open region, overlapping other adjacent regions. Thus Tokyo, Hong Kong and even Taipei, though not usually recognized as part of Southeast Asia, may be regarded as regional *gateways*. A similar approach may be taken in the west to Calcutta and Colombo, and in the south, perhaps, also to Sydney and Perth. Schematically, the situation may be portrayed as an inverted triangle in Figure 1.16. Such a perspective accords with actual networks of telecommunications, air and sea traffic and locates Southeast Asia within the wider Asia-Pacific region.

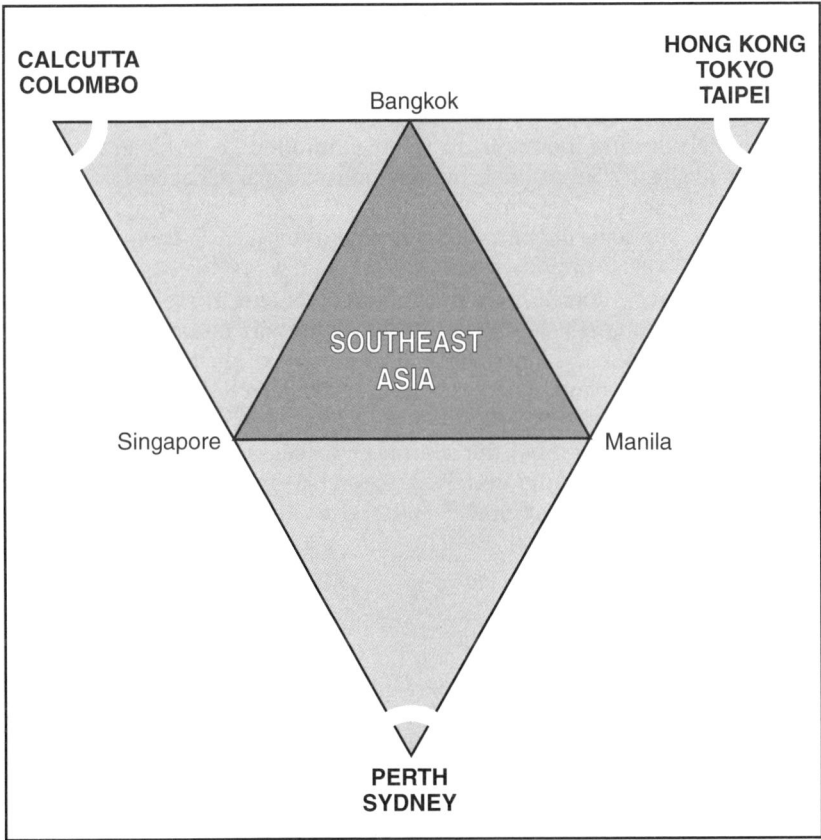

Figure 1.16 The gateways to Southeast Asia

As world cities Singapore and Hong Kong anchor Southeast Asia in the mainstream of global capitalism. In finance, logistics, telecommunications and scientific research and as regional centres for American and Japanese multinationals, they are both First World cities. It is nonsense to suggest that their high incomes per capita should exclude them from 'real' Southeast Asia. By analogy with physics, Singapore and Hong Kong function like dynamos transferring energy from the centres of the world economy to adjacent Southeast Asian cities, which in turn function as smaller dynamos for their own hinterlands. Energy also flows back the other way. It would therefore be more accurate to say that Singapore exchanges energy with the main cities of Southeast Asia, in most cases much more than they exchange energy with each other. 'Energy' is used here as the broadest possible term to encompass flows of goods, people and information, all of which interact through the common medium of money in market institutions. Singapore's ability to function as a dynamo is enhanced by its historic role as a free port and now also as an open market for financial transactions and direct foreign investment. Singapore was not damaged by the financial crisis which hit Thailand, Malaysia, Indonesia and the Philippines in 1997–98 and, despite some weakening of its exchange rate, its institutions remained sound. Though Singapore's role is challenged by neighbouring Malaysia and Indonesia, its spillover into Johore and Riau makes the economy of Greater Singapore genuinely transnational (Chapter 7).

Asia's 'Main Street', including Shanghai, will be the highway of the twenty-first century. The trade liberalization agreements of the Asia-Pacific Economic Community (APEC) provide a multilateral framework for expanding trade and investment, as increasingly does the ASEAN Plus Three (China, Japan, South Korea) grouping (Tay et al. 2001). China has proposed a free-trade agreement with ASEAN. Since 2001 Singapore has signed free-trade agreements with New Zealand, Japan, Australia and the United States and is negotiating with ASEAN, China, South Korea and Mexico. The framework is multilateral and bilateral but the commerce will as ever flow through urban nodes and networks. Some addresses may change, but these main urban economies will continue to grow even larger and more dominant.

Notes

1. R. Emerson, L.A. Mills and V. Thompson published *Government and Nationalism in Southeast Asia* (1942) under the auspices of the Institute of Pacific Relations in New York, followed by K.M. Panikkar, *The Future of South-East Asia* (1943) published in London. However Karl Pelzer's magnum opus referred matter of factly to the 'Asiatic Tropics' (Pelzer 1945).

2
Processes: the Diffusion of Technology

'Globalization' has become part of modern discourse but still admits no precise definition. Like 'Progress' or 'Development', it is one of those catch-all expressions that means both everything and nothing. Broadly it denotes increasing mutual interdependence in an apparently shrinking world. In fact, as historians and geographers both recognize, technological change has been shrinking the world for the past two centuries (Allen and Hamnett 1995).

The technologies of transport and communications establish human relationships between time and space and thereby constrain the possibilities for all social activity. A new technology that increases the speed or reduces the cost of transport and communications opens up new possibilities. In relation to this process, Janelle (1969) proposed the term 'time–space convergence', Harvey (1989) the more economic notion of 'time–space compression'. Imagining our modern world without aircraft or telecommunications is to envisage its collapse. If Southeast Asia in 1850 were imagined as a flat surface, improvements in transport and communications could be modelled as a succession of forces that have pushed, twisted and stretched it into sharp relief. Jungles have become fertile hinterlands and then industrial zones. Remote highlands have been connected by railways and highways to the economic mainstream. Towns have burgeoned into great cities. Conversely, rivers have silted up and once-thriving ports and hinterlands have fallen into decay. Equivalent geological processes have taken billions of years.

New technologies often define the age. An invention such as the telegraph, railway, aeroplane or television that vastly extends the possibilities for human activity, generates tremendous excitement and for a time focuses the creative energies, sometimes on a global scale. Ambitious projects are conceived, capital is mobilized, new organizations are formed, activities relocated, and new industries established. There is observed much of the dynamism that societies show in times of war. Recently this phenomenon has been manifest in the diffusion of telecommunications, the Internet, and computer graphics, new global languages and activities that are shaping a new wave of 'globalization'. The personal computer – essentially an advanced typewriter – became through connection to telephone cables a means of exchanging information almost instantaneously around the entire world. From the pioneer users, academics and researchers, access quickly spread to the community at large. A virtual world has emerged that is transforming identity and human interaction, marketing and public relations, eroding the tyranny of autocratic governments, and creating new opportunities for crime.

The 'information revolution' is a contemporary analogy for the 'industrial revolution' in nineteenth-century transport and communications. Much more is involved than society applying a more efficient means of transmitting information. Society is reinventing itself. So it was for the steamship, the railway, the tram, the motor car and the aeroplane, the telegraph, the telephone, the wireless, the television and now the computer. Each of these inventions had a dramatic and sustained impact in its own right. In waves and clusters, they brought about remarkable transformations, even though the technologies themselves quite soon become familiar parts of the urban and rural landscape.

Southeast Asia's status as a colonized and undeveloped part of the world economy gave rise to the lazy assumption that before recent industrialization the region was trapped in a state of backwardness. This is completely ahistorical. The people of Southeast Asia have always have been literally on the move and eager to innovate. Restlessness did not begin with the steamship, railway and automobile. In modern transport and communications, the first commercial prototypes appeared in Europe or North America, but it was seldom long before they appeared in Asia. The more capital-intensive modes were usually applied by western firms or governments to facilitate aggressive imperial expansion. In the course of the nineteenth and twentieth centuries the world did not merely shrink: it was shrunk to a European mould. Nevertheless, the pattern of diffusion was uneven. Some technologies were applied more readily than others and some places were tied in much sooner than others into the emerging world economy. There was also a good deal of spontaneous innovation by local capital and the small-scale sector. The outcome was not only a dramatic spatial restructuring but also more marked spatial differentiation that is still reflected in the contemporary urban hierarchy.

This chapter seeks not to recount the evolution of the new transport and communications technologies, but to show the complicated pattern of diffusion by mode and geographic area.[1] Two different patterns emerged. First, in the case of mail steamers, telegraphs and airlines, there was almost organic network development. Once the network had reached Asia, there was a strong incentive for each colony or kingdom to affiliate, then to articulate a domestic network. In these cases, technology diffused rapidly.

Secondly, there were self-standing networks, of which the best examples were railways and roads. Despite a few cross-border links, these networks primarily served local hinterlands. The full cost of establishing and maintaining the network therefore had to be borne locally, which was an economic disincentive. The rate and level of investment thus varied greatly across the region. Some colonies were almost bypassed by the railway age. Hence, although all main cities of Southeast Asia were eventually connected to international transport and communications networks, the rate and form of dispersion into their hinterlands varied tremendously, with profound consequences for economic and administrative development. After a very brief overview of the technological revolution, this chapter will first examine the international networks of telecommunications, steamships and airlines, then the autonomous networks of railways, roads and urban public transport. The aim is to provide a chronological framework for the following geographically specific chapters.

The technological revolution

The new transport and communications technologies of the past two hundred years flowed from a series of scientific breakthroughs that were gradually refined to the point of commercial application. These may be grouped into the categories of steam power, electrical power, and the internal combustion engine. Steam power dominated the transport revolution of the nineteenth century, being applied successfully after 1807 to ships and then by the 1820s to railways and tramways. While the Industrial Revolution is commonly recognized as the application of steam power to manufacturing, more far-reaching in its economic and social impact was the application of steam power to the movement of goods, people and information (mails). Industrial steam power was concentrated in a fairly narrow range of industries, such as textiles and metalworking, and in a few specific locations, notably the United Kingdom. The nineteenth-century Transport Revolution, however, was a global phenomenon. In Asia, Africa and Latin America the technology of production was only marginally affected but the spatial economy was rapidly transformed.

Electricity had its impact primarily in the field of communications. In the mid-nineteenth century, electric batteries permitted the development of long-distance telegraphs followed by telephones (1880s) and wireless (1900s). Invention of the electric motor (1871) allowed direct application to traction in the form of electric tramways. Commercial application followed improvements in the early to mid-1880s of transformers, which transmitted current at high voltage from large generating units in standing power stations. By the 1890s electricity was recognized as the cleanest and most efficient power source for urban tramways and underground rail systems and by the 1920s as alternating current was being used in Germany for intercity rail lines. Electric traction improved the efficiency of transport modes formerly reliant upon steam power but did not lead to any new forms of transport.

The internal combustion engine has probably been the most influential technology of the twentieth century. Applied experimentally in the 1890s to small boats and horseless carriages, in the 1900s it gave rise to the age of the commercial automobile and in the 1920s, after some years of experimentation, to the age of civil aviation. In the form of the diesel, it began to take over marine propulsion in the 1920s and railway locomotives in the 1950s. Despite the shift from coal to oil as the main source of fuel, in the late-twentieth century the steam boiler and the once-revolutionary steam turbine have been all but relegated to the generation of electricity.

These inventions did not fall from the sky in disembodied form. Notwithstanding complex antecedents, extending back to Arabia, India and China, they all came to fruition in Europe and North America in conjunction with what used to be known as the Industrial Revolution or what Scott (1998) prefers to call 'high modernism'. Had they emerged in China, India or Southeast Asia, the history of the modern world would have been very different. As well recognized, this historic fact gave a particular trajectory to the diffusion of these technologies, namely from the core nations of Europe and North America to the periphery, including Asia, that was colonized and exploited by means of them. As encapsulated in Headrick's (1981, 1988) excellent titles, they were indeed 'The Tools of Empire' and 'The Tentacles of Progress.' Adas (1989) refers to 'the machine as civilizer'. The narrative of their diffusion is the narrative of western imperialism between the mid-nineteenth and mid-twentieth centuries, after which they became instruments of Cold War rivalry.

Telecommunications networks

Until the mid-nineteenth century information was not readily conveyed in disembodied form. The message still had to travel with the messenger. Over long distances this could be done fastest by standing relays of horses. At the turn of the nineteenth century Napoleon merely copied what the Romans had done almost two thousand years before, and was copied in turn by his representative in Java, Governor-General Daendels, in laying out the Great Post Road (Chapter 4). The most sophisticated form of manual transmission was semaphore (literally 'sign language'), which used flags on movable arms to denote letters of the alphabet and could thereby code actual text. Semaphore stations on high ground were much used in the nineteenth century to signal the arrival of ships. The Spanish in Luzon used a series of relay stations to convey such information from the lookout station at Bolinao to Manila.

Modern telecommunications developed with remarkable speed after 1844, when Samuel Morse used an electric telegraph wire to transmit a coded message of dots and dashes – later known as Morse code – between Washington, DC and Baltimore. Landlines were easy to lay. The challenge with long distance telegraphy was to maintain an electric cable under water. Gutta percha was found to provide effective sheathing but it was still a matter of costly trial and error as to how the cables were to be laid. In 1851, on the second attempt, a submarine cable was laid across the Channel between Dover and Calais and then, two years later, between England and Ireland. Attempts in the late-1850s to lay a transatlantic cable were a costly failure, but in 1866 the task was finally achieved by using Brunel's monstrous steamship *Great Eastern*, the only ship afloat large and powerful enough for the task.

For Britain a telegraph link to India was almost as important as that across the Atlantic. The first cable was joined to India via the Red Sea as early as 1859 but it was too light, poorly sheathed, badly laid and failed to transmit a single word (Headrick 1988: 100). Efforts were then directed to connecting a landline from Constantinople through the Middle East to Karachi, which enabled the first messages to be exchanged between Britain and India in January 1865, just before the transatlantic link. Relays across the Ottoman Empire nevertheless took several days and were often garbled by delivery. Something more reliable was needed. A British parliamentary select committee led to two new ventures. Siemens' Indo-European Telegraph Company was commissioned to lay a landline from Berlin across the Ukraine to Odessa and thence via Teheran and the Persian Gulf to India. This line was completed in January 1870, within weeks of the opening of the Suez Canal, and reduced transmission times to a matter of hours (Headrick 1988: 101). Six months later the British Indian Submarine Telegraph Company completed a second link, entirely under British control, via Gibraltar, Malta, Egypt and the Red Sea to Bombay.

Telegraph links to Southeast Asia followed almost immediately upon those to India and benefited from the preceding period of experimentation. At Bombay the submarine telegraph connected across the subcontinent via the existing domestic line. From Madras another submarine cable was laid across the Bay of Bengal to Penang and Singapore and by the end of 1871 had been extended via Saigon to Hong Kong, Shanghai, Nagasaki and Vladivostok, where it completed a circuit back to Europe via Siberia (Figure 2.1). All of this was brought into commercial service on 1 January 1872. A southern extension from Singapore to Java was in 1872 connected at Darwin to a transcontinental line through the desert from southern Australia – the wooden poles were soon eaten out by termites and had to be replaced by metal. Thus, within only five to six years of the transatlantic link between Europe and the United States, the main cities of Asia and Australia were also brought into telegraphic communication with Europe. Only the Philippines lagged behind, with the dedicated submarine cable from Hong Kong to Luzon not being connected until May 1880. Thereafter

Figure 2.1(a)

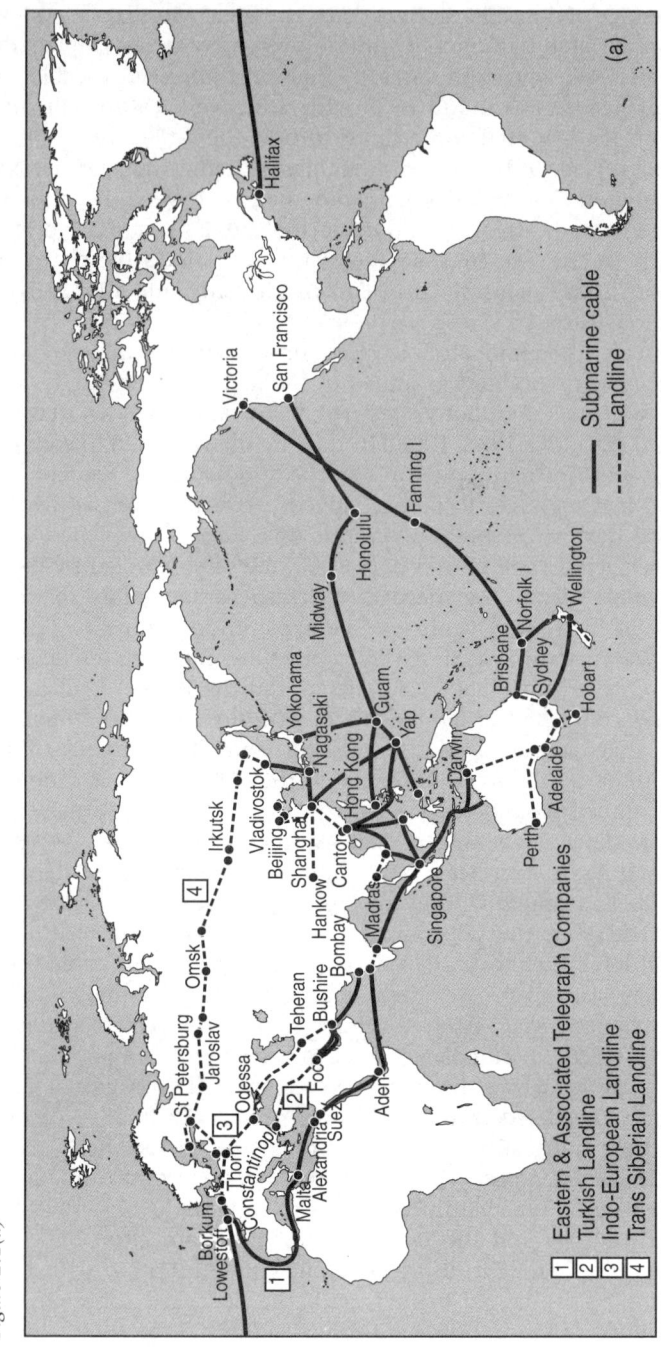

Figure 2.1(b)

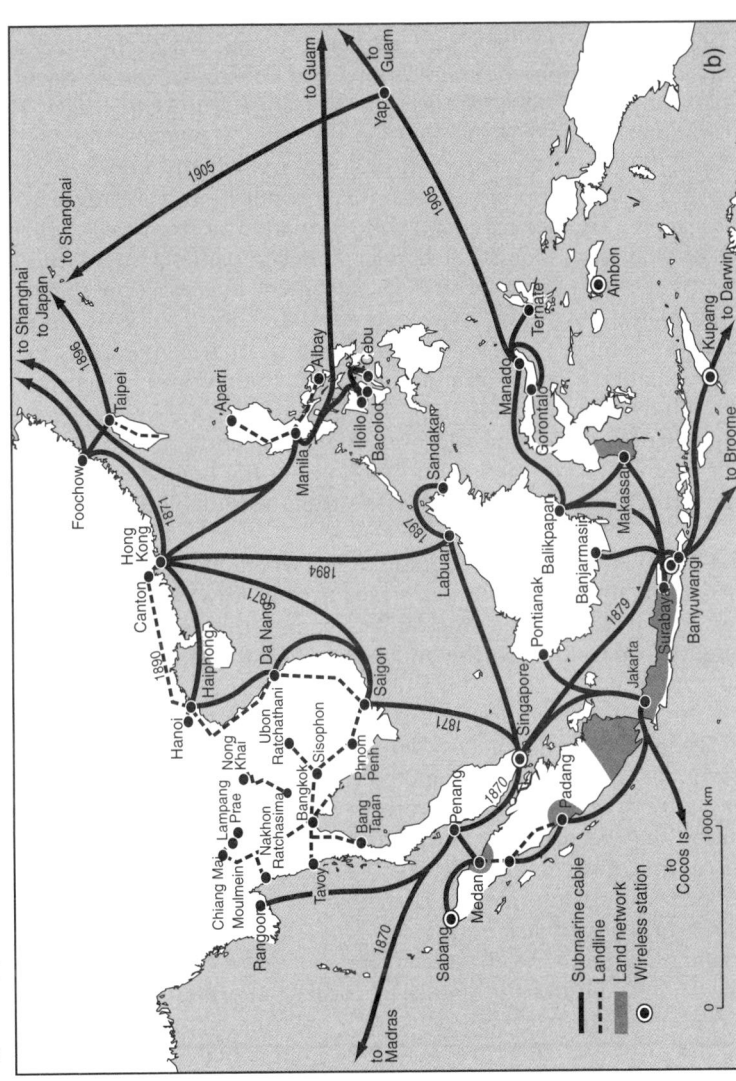

Figure 2.1 Main submarine cables and selected land links, c.1920 showing date of connections (a) world routes; and (b) Southeast Asian routes (*Source*: Data derived from Ahvenainen 1986)

it was a matter of completing the transpacific cables and shortening and improving the various links.

One reason for the rapid spread of telegraphic networks was that self-contained local systems had already been laid out in various parts of Asia. As early as 1853, not quite ten years after Morse's successful demonstration, a landline had been laid between Calcutta and Agra, and the availability to the British of telegraph communications was a factor in the defeat of the Indian Mutiny (1857–58). In Australia the capital cities also began to be linked up after the mid-1850s. In 1856 the Dutch opened a line between Jakarta and the administrative centre of Bogor, more than 15 years before the railway, and in 1859 also made a shortlived connection (via Bangka Island) with Singapore. Another reason was the formation of rival, well-capitalized and technologically sophisticated telegraph companies that in the late-nineteenth century fought for strategic advantage in controlling the world's new information highways (Ahvenainen 1981). By the 1900s the main network was complete and these early multinational corporations formed cartel agreements to set tariffs and share traffic to their mutual benefit.

Telegraphic communication did not supplant the sea-borne mails; rather, it complemented them. The telegraph could efficiently transmit brief messages of urgent news or market information but the limited capacity of the lines and the high cost per word made them unsuitable for exchange of correspondence. Messages had to be kept short, often by abbreviation, then coded for conveyance to the telegraph office, where the operator tapped out in Morse code a message that was picked up and forwarded on by repeater stations until it reached the office at destination, there to be typed out and delivered in coded form to the addressee. Even if transmission took over 24 hours and required detailed matters to be followed up by despatches and mail, it was still an amazing improvement on the situation before 1870. Transmission of information had become virtually instantaneous, with the time saved measurable in weeks and even months. Subsequent savings have been a matter of hours. The achievement of the recent telecommunications 'revolution' has been not so much the faster speed of transmission as the greater volume of what may efficiently be forwarded. By means of fax, email and modems, most information exchange can now be shifted onto high-speed networks at very low cost.

The highest-speed network tends to define the age. It sets the limits of the possible and establishes the points of contact, the pressure points of the world economy. The telegraph network turned the map of Asia inside out, pulling the main port cities into almost 24-hour contact with the centres of the world economy in Europe and America. To be a Briton in Singapore, Rangoon or Hong Kong, a Dutchman in Java, a Frenchman in Saigon or an American in Manila was now to feel in contact and affinity with that homeland, rather than being an exile on the edge of some alien, mestizo society. Commercial advertisements often carried the telegraphic address and identified the commercial codebook, such as Bentleys, used for abbreviating and encoding messages. Access to the telegraph network, like web sites today, identified those businesses that were international.

Within Asia's cities, telephones soon became a means of routine communication. Almost immediately after Alexander Graham Bell's patent (1876), the telephone was demonstrated in Melbourne (1877) and Singapore (1878). Lines were operating in Melbourne (1879) and Brisbane (1880) even before Paris (1881). There soon followed Shanghai (1881), Hong Kong (1882), Jakarta (1882), Tokyo (1883), Surabaya and Semarang (1884), Taiping (1886), Bangkok (1889) and the Philippines (1890). By 1892 telephones were in general use throughout Java and by 1896 intercity links were in place between Jakarta, Semarang and Surabaya. Interlocal telephones became more efficient than telegraphs. Nevertheless, the number of sets in use was fairly small. In 1940 the Netherlands Indies still had only 51,000 connections, including extensions, and on the basis of 1950 figures there were unlikely to have been more than 100,000 telephones across the whole of Southeast Asia (Table 2.1). Most of these would have been used by Europeans and Chinese businessmen. Popular diffusion has been a fairly recent phenomenon. By 1970 there were a million sets and by 1990 almost 6 million.

Another dramatic innovation was wireless. Guglielmo Marconi's experiments at sea in 1899 led to rapid development of ship-to-ship and ship-to-shore radio (Baker 1970). Hitherto ships at sea had to all intents and purposes been in outer space, and not uncommonly they went missing without trace. Telegraphs were of no benefit because there was no way to link them to the system. By the time of the First World War, marine radio had become a standard item of equipment aboard passenger liners. In 1912 the *Titanic* was already fitted with radio, as were other ships in the vicinity, but the distress signals were not received because operators on the other ships were off duty.

The other immediate benefit of radio was radio-telegraphy, which through bypassing the cable monopolies had the potential drastically to reduce the cost of transmitting telegraphs. Having no vested interests in international cable networks, the Dutch were the first to act. In 1909 work began on four powerful shore stations, one at Sabang off the northern tip of Sumatra, designed to serve shipping in the Indian Ocean, and the others at Situbondo (East Java), Ambon (Maluku) and Kupang (Timor) intended as an economical way of servicing the vast eastern archipelago, which was almost without telegraph links (Figure 2.1). In 1911 the French government, who had been long frustrated by their dependence upon British telegraph, approved ambitious plans for radio links with their Southeast Asian colonies. The following year the British embarked on a similar project. The First World War broke out before the powerful radio transmitters could be installed but in 1914 the British followed the Dutch lead by opening local radio traffic between Sandakan and Kota Kinabalu in British North Borneo (Sabah),

Table 2.1 Southeast Asia: telephone sets by country, 1930–1999 ('000)

	Indonesia	Malaysia	Singapore	Philippines	Thailand	Vietnam	Burma	Total
1930	46		9	n.a.	3	n.a.	n.a.	(58)
1950	43	24		20	(6)	12[1]	(5)	(110)
1970	201	167	147	310	153	35[1]	26	1,039
1990	1,066	1,586	1,054	610	1,054	99	70	5,539
1999	6,080	4,431	(1,936)	2,892	5,200	2,800	249	(23,588)
Mobile	2,220	2,990	(2,440)	2,850	2,400	300	11	(13,211)

Note: 1. South Vietnam only. Thailand (1951) figure. Total in 1999 excludes Brunei 79,000 (not including mobile 66,000), Cambodia 28,000 (89,000) and Laos with 34,000 (9,000). Singapore (2000) figure.
Source: Based on Mitchell (1998), EBI (2002).

Table 2.2 Southeast Asia: radio sets by country, 1955, 1970, 1997 ('000)

	Indonesia	Malaysia	Singapore	Philippines	Thailand	Vietnam	Burma	Total
1950	500	157	89	217	108	n.a.	2	(1,073)
1970	2,550	430	237	1,500	2,775	1,300[1]	15	8,807
1997	31,500	9,100	2,550	11,500	13,960	8,200	4,200	81,010

Note: 1. South Vietnam only. Total in 1997 excludes Brunei 93,000, Cambodia 1.34 million and Laos 730,000 sets.
Source: Based on Mitchell (1998), EBI (2002).

replacing the submarine cable. In the same year an emergency naval radio station was commissioned at Singapore, followed in 1915 by a commercial wireless station (Makepeace et al. 1921: 154). Two years later a radio service was opened between Singapore and Sarawak, which had been without a telegraph link. In 1920 the Governor-General of Indochina took his own initiative to contract for the long-foreshadowed land station at Saigon, which was fully commissioned in 1924 and quickly captured from the telegraph companies half of the traffic to and from France (Headrick 1988: 135). A local network distributed messages within Indochina.

Shortwave radio brought about a series of rapid improvements in telecommunications after the mid-1920s. By reducing the size and power of land stations, it gave radio a competitive edge over the cable telegraph and lowered the costs of international traffic. The Netherlands became a leader in radio and electronics through the Philips Corporation and rivalled the French in developing more sophisticated communications with Southeast Asia. Philips Broadcasting Company introduced a short-wave radio-telegraph link between the Netherlands and the Indies in 1927, almost simultaneously with that between Australia and Britain and between France and Indochina in 1926–27 (Headrick 1981: 213; 1988: 136). International radio-telephony was pioneered by the French in 1928 and followed almost immediately by the Dutch. By 1940 it accounted for almost three-quarters of domestic telegrams in the Netherlands Indies and a third of international telegrams (CBS 1947). The third innovation was public shortwave broadcasting. In May 1931 Radio-Coloniale opened a service to Indochina, followed towards the end of 1932 by the BBC (Headrick 1988: 136).

By the eve of the Second World War, Southeast Asia was thus joined to Europe and North America by an almost seamless web of telecommunications. Although the movement of people and goods was still fairly slow, information now flowed almost instantaneously by telegraph, radio-telegraph, radio-telephone and shortwave radio. For Europeans, who could afford the cost, the consequence was a drastic shrinkage in the psychic distance between the colonies and mother countries in the northern hemisphere. Ironically this did not strengthen colonial rule. Albeit at second hand, information quickly filtered down to educated local elites, especially via daily newspapers. In the more globalized and restless world of the 1920s and 1930s, colonial rule might appear as a right to Europeans but increasingly it was resented as an imposition and an anachronism by those whose talents were trivialized and whose upward mobility was blocked.

After the Second World War radios and later television gradually became items of mass consumption. One million radio sets were licensed across Southeast Asia by 1955. Boosted by introduction of much cheaper transistor radios, official figures soared to almost ten million in 1970 and 70 million by 1992 (Table 2.2). The number of unlicensed radios is unknown but Indonesian intercensal figures for 1995 recorded 91 per cent of Jakarta households owning at least one radio or radio-cassette player. Commercial television resumed in Britain and the United States in 1946, reached the Philippines in 1953, Thailand in 1955, Australia in 1956, Indonesia in 1962 and Singapore and Malaysia at the end of 1963 (http://www.tvhistory). By 1992 official figures recorded 28 million sets across Southeast Asia (Table 2.3). Indonesia accounted for 11.5 million of these: by 1995 in Jakarta 83 per cent of households owned a set, though in rural areas the percentage was low (BPS, Supas 1995a,b); in Manila 75 per cent of households owned a set in 1990 (Philippines, Census 1992). Governments promoted television as a means of mass access to state-controlled media, to facilitate transmission across the archipelago. Only fourteen years after the United States had launched its first telecommunications (Telstar) satellite, in 1976 Indonesia became the first developing nation to have a satellite placed in orbit. The Palapa satellite greatly improved telecommunications and television throughout the vast archipelago and very soon access agreements were signed with the Philippines, Malaysia and Thailand (Hudson 1990: 185–6). Public broadcasting monopolies, along with censorship of the press and the Internet, became one of the battlegrounds in the struggle for democracy.

Steamship networks

Steam engines had been designed as giant pumps to raise water from mines. In 1781–82 James Watt patented a technique to apply the power of a vertically acting piston to rotate a horizontal shaft. These Boulton & Watt engines were the first capable of driving machinery as a substitute for waterwheels or windmills and are regarded as ushering in the Industrial Revolution. In 1802 a prototype stern-wheel paddle steamer ran trials at Glasgow and five years later the first commercial steamship was placed on the Hudson River in the state of New York. In Britain commercial steam shipping began in 1812 with a passenger ferry on the Clyde, the centre of Scotland's business and engineering. In 1821 the British Post Office replaced sailing packets by steamers across the Irish Sea and the Channel (Robinson 1964: 123). By the late 1820s steamers were an accepted means of coastal communication, allowing a regular timetable for mails and passengers.

The first steamers appeared in Asia quite soon after their introduction to commercial service in Britain. In July 1823 the *Diana* was launched at Calcutta and fitted with a steam engine brought out by sailing ship. Her intended purpose had been to tow sailing ships up the Hooghly River to Calcutta but on outbreak of the first Anglo-Burmese War in 1824 she showed her worth as a despatch vessel and naval tug. In December 1825 the steamer *Enterprize* reached Calcutta via the Cape and took up similar local duties. In September 1828 the *Hooghly* made a trial voyage on the Ganges, taking just 23 days as far as Allahabad compared with the usual two or three months by unpowered vessels sailed, rowed or towed against the current.

Table 2.3 Southeast Asia: television sets by country, 1960, 1975, 1999 ('000)

	Indonesia	Malaysia	Singapore	Philippines	Thailand	Vietnam	Burma	Total[1]
1960	nil	nil	nil	38	60	nil	nil	98
1975	300	452	280	711	725	500	0	2,968
1999	30,000	3,800	1,200	8,200	17,600	14,500	323	75,623

Note: 1. Total in 1999 excludes Brunei 205,000, Cambodia 98,000 and Laos 51,000 sets.
Source: Based on Mitchell (1998), EBI (2002).

Regular steamer service began in 1834 with the first of four specially designed iron-hulled shallow-draft steamers fitted to tow a passenger barge (Bernstein 1987). In the other British colony of New South Wales, the first two steamers appeared in 1831 and by the early 1840s there were regular lines between Sydney, Melbourne and Tasmania (Bach 1976).

In Southeast Asia the pioneer steamer was the *Van der Capellen*, commissioned in Surabaya in November 1825 by a British trading house for a packet service along the north coast of Java (ENI 1921: 111). She was replaced in 1840 by the larger *Koningin der Nederlanden* but apart from occasional steamers calling at Singapore en route to or from China there was no other local commercial steam navigation. In this early period the main significance of steamships was their use as naval auxiliaries in suppressing piracy. The increase in shipping and trade after the foundation of Singapore in 1819 had encouraged piracy on the sea-lanes between India, the Straits Settlements and South China (Anderson 1997). From Sulu pirate fleets sallied forth each year to a regular schedule, setting up settlements around the coasts of Malaya and Riau where they could refit and sell their goods and slaves (Warren 1981). Because their prahus were fast, formidably armed, and sailed by men expert in those waters they were usually able to evade, outsail and even threaten the slow, lumbering European warships. Steamers, however, could approach at speed regardless of wind and if necessary tow armed ships into position. In 1837 a naval steamer was placed on permanent station in the Straits (Wright 1970: 8). In 1843 Admiral Keppel swept the seas from Singapore to Borneo, destroying known pirate lairs and attacking pirate craft. The Dutch conducted similar operations in their own waters. In the Spanish waters steam gunboats did not come into service until 1860, after which piracy was quickly subdued in Luzon and the Visayas (Warren 1981).

In 1845 arrival of the first P&O steamer provided Southeast Asia with an international mail link, the product of some twenty years of endeavour to speed communication between Britain and India. In the 1820s urgent despatches from London to the Mediterranean had sped by coach to Dover, steamer across the Channel to Calais, coach via Paris to Marseilles and thence by fast sailing packet to Egypt. In 1829 the East India Company (EIC) despatched a steamer from Bombay to Suez to link into this courier service by an overland connection to Alexandria (Gibbs 1963). The Admiralty introduced steam packets between Malta and Alexandria in 1835, and the Bombay connection became more regular. Reliability was improved after 1839 when Britain occupied Aden, at the mouth of the Red Sea, to replace Mocha as a coaling station. Dissatisfied with the high cost of naval packets, in 1840 the British government called tenders from private firms for the carriage of mails. Cunard won the tender for the Transatlantic mail and the restyled Peninsular & *Oriental* Steam Navigation Company (P&O), hitherto serving the Iberian Peninsula, won the tender for a mail line to Calcutta, leaving the EIC to continue serving Bombay. The India route was by steamer to Alexandria, overland to Suez and by connecting steamer from Suez to Calcutta: the first sailing from Suez was early in 1843. Success of the Indian mail soon led to a demand to extend the line to British settlements further east. In 1845 the small P&O steamer *Lady Mary Wood* commenced a monthly service from Galle (Ceylon) – the transfer point with the Suez–Calcutta steamer – to Penang, Singapore, Hong Kong and, after 1850, Shanghai. From 1842 a steamer was also provided from Singapore to the East Coast of Australia, but this lapsed when ships were requisitioned for the Crimean War.

By 1852 the P&O mail had therefore achieved the historic feat of bringing India, Southeast Asia, China and Australia into a single network of scheduled communication with Europe (Gibbs 1963). First-class passengers and some valuable cargo such as bullion were carried, but primarily these ships were carrying information. Preceding the telegraph by two to three decades, these mail steamers were the Internet of the mid-nineteenth century. Although the network directly served only British settlements, the Netherlands Indies government from the outset provided a steamer to connect with the P&O steamer at Singapore, giving a 48–50-day transit time for mails from London to Batavia (ENI 1921: 115). This led in 1850 to a contract for an interisland steam packet service (Figure 2.2; Chapter 3). In early 1857 P&O placed its own steamer in service to Manila, first from Singapore, then from 1858 the mails over the shorter distance from Hong Kong. In 1858, three years after the Bowring Treaty, the Siamese steamer *Chow Phya* opened a mail connection at Singapore. Following the occupation of Saigon in 1859, French steamers began to connect at Singapore in 1861 as a temporary arrangement before a subsidized French mail line was inaugurated in October 1862. Messageries Imperiales (later Messageries Maritimes, MM), which already served the Mediterranean, opened a parallel mail line from Suez to Singapore, Saigon, Hong Kong and Shanghai, soon extended to Yokohama. In 1867 the American-flag Pacific Mail reached Yokohama from San Francisco, which two years later on completion of the transcontinental railway was joined to New York and the Atlantic mail. Thus, even before opening of the Suez Canal and completion of the telegraph to Singapore, Java and Australia, virtually the whole of maritime Asia had already been brought within a network of mail lines that spanned the globe. Jules Verne's 'Around the world in eighty days' was now achievable (Figure 2.2).

The opening of the Suez Canal in December 1869 dramatically altered the economics of steamships between Europe and Asia. Hitherto the heavily subsidized mail companies, P&O and MM, had maintained staging posts between Port Said and Suez for the overland carriage of mails and first class passengers, but such a journey and the expense was hardly feasible for cargo or indeed most passengers, who had no other choice but to persevere with the three-to-four-month voyage around the stormy Cape of Good Hope. Steamers were out of the question because of their high coal consumption

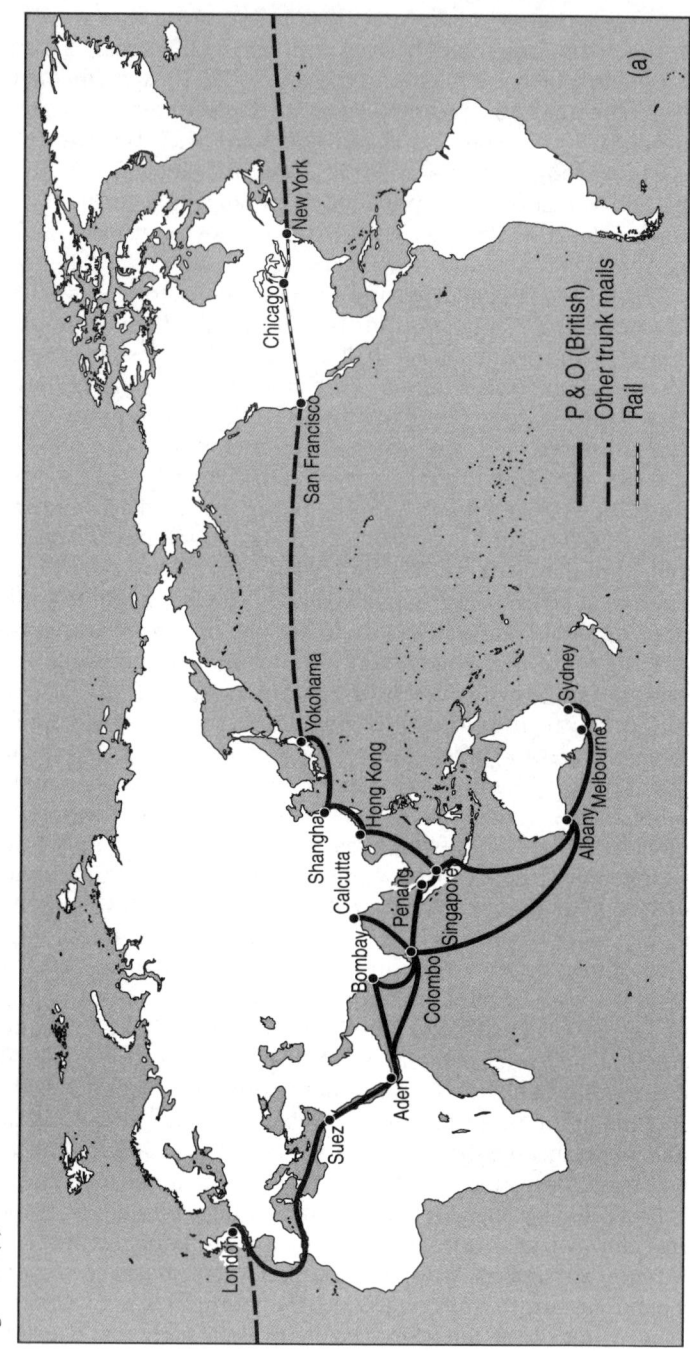

Figure 2.2(a)

Figure 2.2(b)

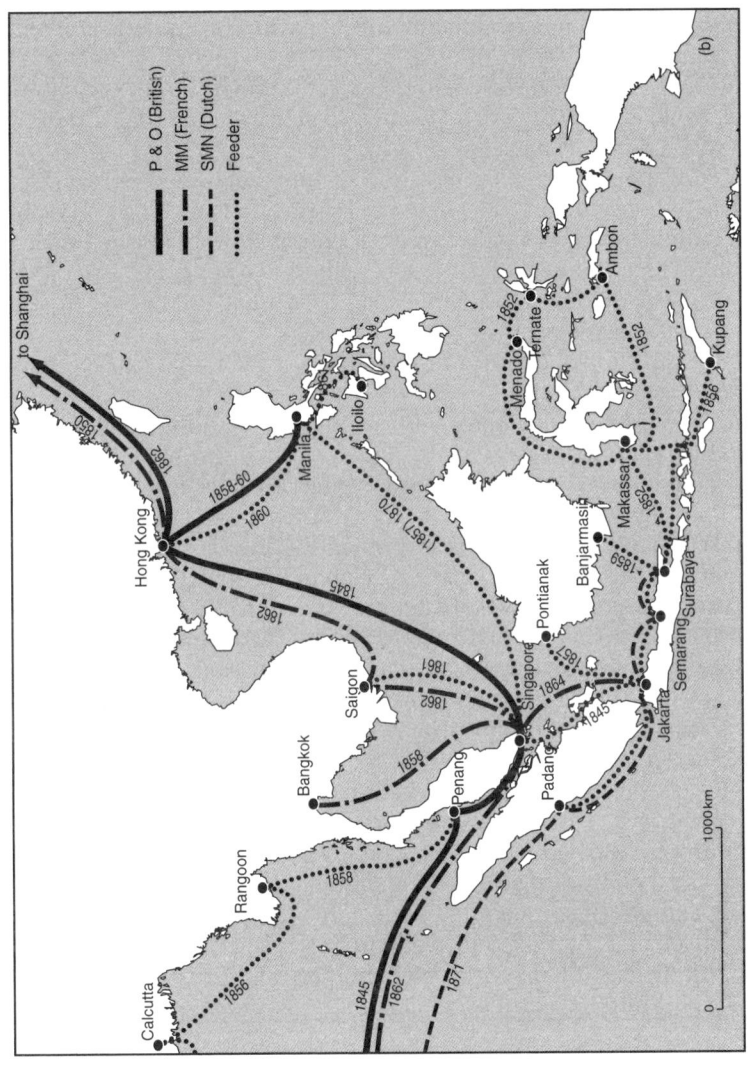

Figure 2.2 Mail steamers in Southeast Asia before opening of the Suez Canal, 1869 (a) world routes (b) Southeast Asian routes (Source: various)

and the cost of coaling along such a long route. This challenge was taken up by marine engineer Alfred Holt, who designed a compound steam engine that, by recycling the steam from the main piston into a second lower-pressure cylinder, achieved great economies in coal consumption. In 1866 his *Agamemnon*, the first of three sisters ordered by Holt's Ocean Steamship Company, sailed non-stop around the Cape to Singapore and China to load what was for that time a massive and highly profitable cargo (Hyde 1957). After 1869 she could sail direct through the Suez Canal. So, of course, could almost any other steamer, coaling as necessary at Suez, Aden, Bombay, Colombo, Penang and Singapore.

The consequence was a proliferation of steamship lines between Europe and Asia. Whereas before 1870 only the British and French had been able to afford national networks, other colonial powers now established their own subsidized direct connection between the mother country and Asian colony: the Dutch-flag Stoomvaart Maatschappij 'Nederland' (SMN) began sailings in 1871 and the Spanish-flag Olano, Larrinaga & Co. in 1873. The combination of the shorter distance to Asia through the Canal, cheaper coaling and improvements in the efficiency of marine engines also allowed unsubsidized British lines to compete with sailing vessels in the carriage of cargo and ordinary classes of passengers. Holt's Ocean Steamship Company soon had competitors. Henderson opened a direct steamer line to Rangoon in 1871; the Orient Line to Sydney in 1877. National rivalries ensured that the subsidized mail lines were also augmented by the German Norddeutscher Lloyd after 1886 and the Japanese Nippon Yusen Kaisha in 1896. The Italians were represented by the Rubbatino line, while in 1897 the Danish East Asiatic Company opened a direct line between Europe and Bangkok.

Competition, improvements in propulsion, and larger ships all helped to drive down ocean freight rates. This trend was particularly marked on the Atlantic but Asian trades also benefited, especially after the opening of the Suez Canal in 1869 (North 1958; Fletcher 1958). Apart from the subsidized mail lines, steamships had hitherto been rather a novelty, especially along the coasts of Malaya, Thailand, Indochina and the Philippines. In the 1870s a stream of new and secondhand steamers sailed east through the Canal for local service, rapidly displacing sailing ships, junks and prahus from the main carrying and passenger trades. As a free port, Singapore became the entrepot for much of this shipping with local Chinese merchants being prominent among the owners and managers. Not until the twentieth century did western capital gain control over the local shipping of Singapore (Chapter 3).

Airline networks

Despite Orville and Wilbur Wright's pioneering flight in 1903, commercial aviation awaited the technological improvements of the First World War. Discharged air force pilots took the initiative in pioneering long-distance flights. In 1919 there occurred both the first transatlantic flight and Ross Smith's epic flight from England to Australia via Singapore. The latter's 28-day journey was hardly faster than the P&O mail but it awakened people to the possibilities of air travel. During the 1920s a dense multi-airline network took shape in Europe while American airlines gradually spanned the continent and began to extend into the Caribbean and Latin America (Davies 1964). Flights to Asia, however, faced long treacherous stages across deserts and seas without emergency airstrips. Except for experimental flights and a precocious airmail service in up-country Thailand, aviation development in Asia was a phenomenon of the 1930s.

Commercial aviation in Southeast Asia was pioneered by the Dutch. In mid-1927 the Koninklijke Luchtvaart Maatschappij (KLM) completed a trial return flight between Amsterdam and Jakarta. After a series of flights in late 1928 and 1929, a fortnightly mail service was inaugurated in September 1930 and a year later upgraded to a weekly through passenger service (Davies 1964: 173–4). Because flying was restricted to daylight hours, the journey took ten days but was still less than half that of the fastest mail liner. By 1937 KLM had reduced flight time to a reliable six days with 11-passenger DC3s flying twice a week. In January 1931 Air Orient, soon to become Air France, followed the Dutch lead by extending its Marseilles–Syria–Baghdad route to an airmail link with Saigon (Figure 2.3).

Britain's long-distance commercial aviation lagged well behind. In 1926 Imperial Airways had opened an air link from Cairo to Basra on the Persian Gulf, which in March 1929 finally reached India at Karachi. Not until December 1934, almost five years after the Dutch, did Imperial Airways reach Singapore, there connecting with a Qantas flight via Java and Timor to Australia. The entire journey from London to Sydney took 12.5 days, compared with four weeks by sea. From 1938 luxury flying boats with cabins and deck chairs placed the service on a sound basis. When KLM made an Australian connection in mid-1938 in eight days from Europe, Imperial reduced its flight time to 8.5 days without change of aircraft. Meanwhile giant flying boats had allowed Pan-American Airways in November 1935 to meet the even greater challenge of a regular transpacific flight between San Francisco and Manila, soon extended through to Hong Kong. In 1939 the German airline Lufthansa belatedly followed upon the Dutch, French and British with a short-lived service to Bangkok.

On the eve of the Second World War international aviation was nevertheless still in its infancy. Few passengers could afford the high cost of air travel and airmail, though faster than seamail, was still much slower than the telegraph. The impact was perhaps more psychological than economic. These new air routes were above all flights of imagination. In a decade when the human spirit was beset by material crisis, they not only symbolized Man's conquest of distance but also by their speed and regularity gave Europeans and Americans in Asia a confidence of being part of metropolitan

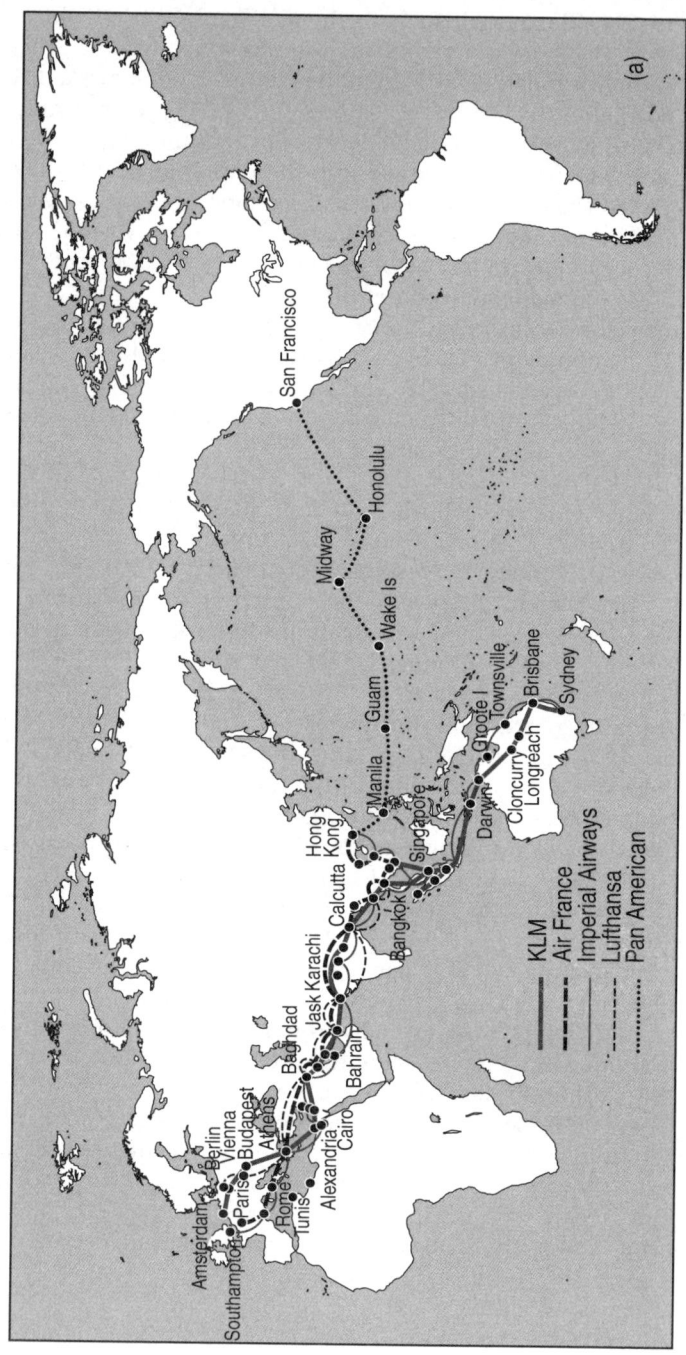

Figure 2.3(a)

Figure 2.3(b)

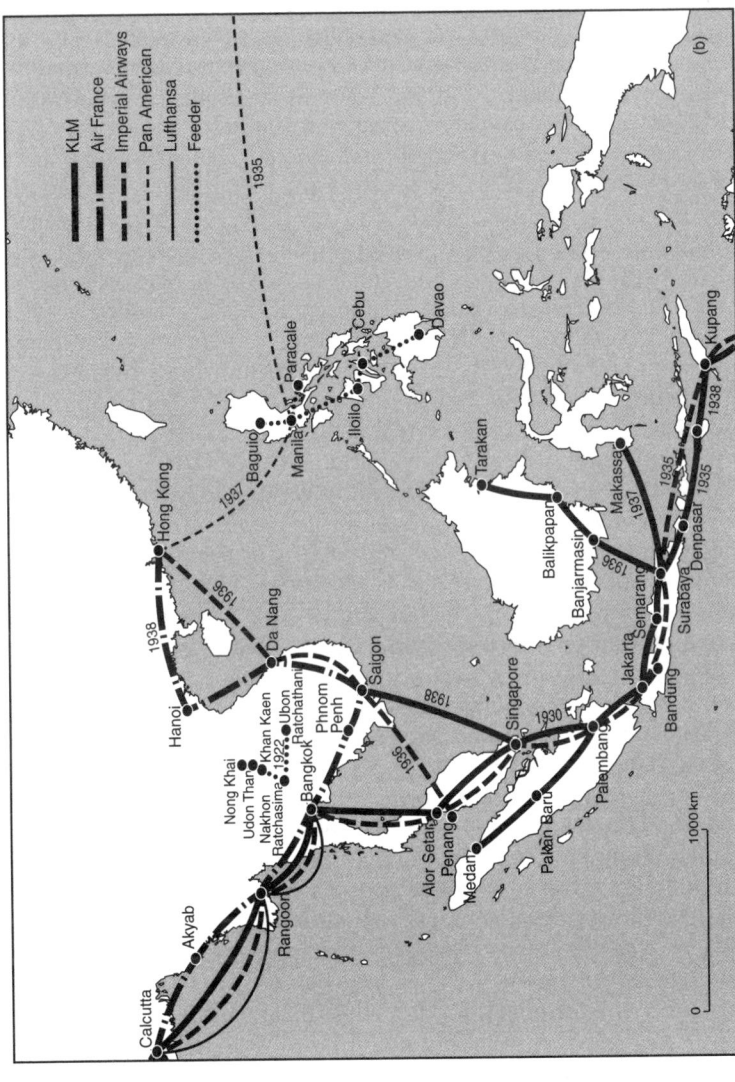

Figure 2.3 Main air routes c.1938 showing date of connections (a) world routes; (b) Southeast Asian routes
(Source: Davies 1964; Rimmer with Davenport 1996)

society. Long-distance airlines were the last great triumph of western imperialism. Ironically air power – Japanese air power – would soon sweep those empires away.

Domestic aviation had a more immediate and direct impact. Apart from an airmail and passenger service in the Philippines which lasted for a few months in 1921, and leaving aside Australia, the first regular service began in 1922 in Thailand from the railhead of Korat (Nakhon Ratchasima) to the northeast towns of Ubon, Udon and Nong Khai (Figure 2.3). At the end of 1928 KNILM, a KLM subsidiary, commenced domestic air services in Java and within two years had extended the network to Palembang, Medan and Singapore. In December 1931, well before the inauguration of transpacific aviation, the Philippine Aerial Taxi Company (PATCO) began a local service from Manila to the hill station of Baguio, while two years later the Iloilo–Negros Air Express Co. opened an airmail service from Manila to Ioilo and Negros, soon extended to Cebu and Davao. In Malaya domestic aviation began in 1937 with regular flights by Wearne Brothers between Singapore, Kuala Lumpur and Penang. By the eve of the Pacific War, most of the main cities of Southeast Asia had therefore been brought into a provisional network. The number of passengers was small, but these pioneer domestic airlines significantly reduced the isolation of previously remote districts such as the northeast of Thailand and the Philippine island of Mindanao. Air transport combined with interlocal telephone links and wireless to knit together the main urban centres of each country or colony, reinforcing what was in many parts of Southeast Asia a growing sense of national identity.

After 1945 with bigger and better aircraft and a swarm of demobilized pilots, domestic air services were soon restored. Airlines quickly secured most local first-class traffic for officials and businessmen, relegating passenger ships to the role of mass public transport for deck or dormitory passengers. In colonial Malaya the coastal shipping firm Straits Steamship Company established Malayan Airways (1947) instead of investing in a new passenger fleet; a year later the associated Swire shipping interests bought out Hong Kong's Cathay Pacific Airways (1946). The newly independent nations of the Philippines (Philippine Airlines, PAL – formerly PATCO) and Indonesia (Garuda Indonesian Airways) followed India in setting up national carriers with shareholdings and/or technical assistance from European and American airlines. The corridor Calcutta–Bangkok/Singapore–Hong Kong–Seoul–Japan was very well served.

In the 1960s the introduction of Boeing 707 jets made aircraft more economical than passenger ships as long-distance carriers. Air India and PAL had opened long-distance routes as early as 1948. In the 1970s the national carriers Air Thai, Malaysian Airlines (MAS), PAL and Garuda all challenged the established airlines by undercutting fares set by the IATA cartel. The decisive shift took place after the advent in 1970 of the 'Jumbo' jet, that could carry as many passengers as a medium-sized liner and with very much smaller crews. Well-capitalized Singapore Airlines (SIA) embraced the new aircraft from the outset; by the 1980s most Asian carriers were flying Jumbos on long-distance routes. They were therefore well positioned to benefit from the growth in purchasing power within Southeast Asia, which now became an important origin and destination in its own right for both business and tourist traffic. The explosive growth in traffic is shown in Tables 2.4–2.5, the pattern of flights in Figure 1.9.

Autonomous networks: railways

Mechanical railways evolved from wagons drawn by horses along wooden rails to haul material from mines. British collieries began to experiment with steam engines in the 1810s but even heavy cast-iron rails proved too brittle. In 1820 a technique was found to roll much stronger wrought iron rails in 20-foot lengths. The corollary was a light mobile boiler and engine that developed sufficient traction. These two inventions were brought together by the Newcastle engineer George Stephenson, who superintended the opening in 1825 of the first proper steam railway to carry coal from the West Durham coalfield near Darlington to the adjacent port of Stockton. With 'Stephenson's Rocket' he then won the competition to design an engine for the first commercial passenger railway, the Liverpool & Manchester, which opened five years later. This alternative to coach travel on bumpy and dusty roads found immediate popularity. Development in France and the United States was not much behind. By the end of the 1830s railways were being laid throughout Europe and as far as Russia (St Petersburg) (Bagwell 1988: 80).

Railways were introduced quite early to Asia and, like steamships, along the pathways of the British Empire. The Great Indian Peninsula Railway Company was formed in 1844 with George Stephenson among the directors and his son Robert as consulting engineer (Sahni 1953). Much time was taken to secure incorporation and raise the huge capital but construction began in 1851 and the first short line was opened near Bombay two years later, followed by lines at Calcutta (1854) and Madras (1856). The great ambition of a transcontinental link was realized in 1870 and a year later Madras was joined to the trunk network (Headrick 1988: 65–7). In Australia the first short urban lines in Melbourne, Sydney and Adelaide were also completed in the mid-1850s (Blainey 1966).

Southeast Asia's first railway opened on Java in 1867. In the mid-nineteenth century Java was the only hinterland in Southeast Asia, yet it was intensively developed for export production. Not only could it benefit from railways; it also generated enough revenue to finance them. This calculus did not apply to the British port city settlements of Singapore and Penang or to the Spanish colony of the Philippines. The first railway concession with interest-rate guarantee was approved by the Dutch parliament in 1863 for

Table 2.4 Southeast Asia: growth in civil aviation passenger traffic, 1950–1999 (million passenger/km)

	Indonesia	Malaysia	Singapore	Philippines	Thailand	Vietnam	Burma	Total[1]
1950	148	n.a.	n.a.	187	19	n.a.	47	(401)
1960	259	39	286	63	38	54	739	1,478
1975	2556	1633	5104	2842	4268	120	163	16,686
1995	24,028	23,431	48,400	14,374	27,053	2303	147	139,736
1999	12,389	33,708	64,528	10,292	38,345	3831	438	163,531

Note: 1. South Vietnam only. Total in 1999 excludes Brunei 2,803 million, Cambodia 42 million and Laos 48 million passenger/km.
Source: Based on Mitchell (1998), EBI (2002).

Table 2.5 Southeast Asia: growth in civil aviation freight traffic, 1950–1999 (million metric ton/km)

	Indonesia	Malaysia	Singapore	Philippines	Thailand	Vietnam	Burma	Total
1950	5	n.a.	n.a.	7	0.5	n.a.	n.a.	(12.5)
1960	4	1	0	4	1	0.5	1	11.5
1975	43	27	147	100	102	1	1	421
1999	341	1426	5482	241	1670	98	3	9261

Note: 1. million ton/km Total in 1999 excludes Brunei 75 million, Cambodia 0.4 million and Laos 5 million passenger/km.
Source: Based on Mitchell (1998), EBI (2002).

a short railway between the two inland royal capitals of Yogyakarta and Surakarta (Solo), later extended to the port of Semarang, and another in 1864 for a railway between Jakarta and the summer capital of Bogor. The first section of track was opened in 1867, some 28 years after the first railway in the Netherlands and the two lines were completed in 1873 (Bagwell 1988; Chapter 4).

Java's railway development lagged more than a decade behind India and Australia but coincided with that in Japan. Only two years after the Meiji Restoration (1868) Japan's first line was opened between Tokyo and the port of Yokohama, followed in 1874 by the line from Osaka to Kobe. The two main cities of Tokyo and Osaka were joined in 1889, two years later the line reached the northern tip of Honshu, and by 1901 extended to Shimonoseki at the southern tip. In Japan railways were a strategic imperative, whereas in Java the emphasis was commercial, to connect the productive interior of the island to the main northcoast ports of Jakarta, Semarang and Surabaya. Jakarta and Surabaya were notionally joined by rail in 1894 but a one-day through connection was delayed until 1929, the year when the two cities were linked by domestic airline flights. By 1900 Java nevertheless had a substantial rail network, especially in the centre and east of the island, which was supplemented by a rapidly expanding narrow-gauge tramway network (Chapter 4).

The main Philippine island of Luzon was much later to enter the railway era and never gained a comprehensive network. Spain was itself a late starter in Europe, not building its first railway until 1848 (Bagwell 1988), by which time construction had almost begun in India and Australia. In 1875 royal authority was given to draw up a railway plan for the Philippines but another 12 years elapsed before a concession was let to a British firm. The Manila Railway Company's 195-kilometre line from Manila north through the central plain to the port of Dagupan was completed in November 1892. No other track was built before the American conquest in 1898. The American government began with an almost ideological faith in the potential of railways but the achievements were very modest (Chapter 4). The main southern line from Manila, whose construction began in 1909 was not completed until 1938, by which time the railway was being overwhelmed by bus and truck competition. In the central Visayas in 1909 the Philippine Railway Company opened two short lines across the plain of Panay and along the east coast of Cebu but high construction costs and poor patronage caused the rest of the concession to be abandoned. The planned Visayan network of rail and ferry links never eventuated and railways were never seriously contemplated for Mindanao. Roads became the basic transport infrastructure of the Philippines.

Burma, Thailand, Malaya and Vietnam had less need of railway systems because their long rivers and/or coastlines gave ready access to cheap water transport. Perhaps because of British rule, in 1877 Burma nevertheless became the second country in Southeast Asia after Java to begin railway construction. The northern line more or less duplicated the transport artery of the Irrawaddy, reaching the former upriver capital of Mandalay in 1889 and Myitkyina, some 700 kilometres from Rangoon and just beyond the head of navigation, in 1898. Ten years later there were branch lines from Mandalay to Lashio near the Chinese border, to the lower Irrawaddy port of Bassein and to the coast port of Martaban/Moulmein. Rangoon was therefore well connected to the rest of the country both by water and by rail (Chapter 5).

Without the impetus or revenues of a colonial government, the kingdom of Thailand did not begin its railway programme until the 1890s. The first main line reached Korat on the isolated northeast plateau in 1900 but subsequent progress was slow. The long southern line down the peninsula reached the Malayan border and connected with the British system in 1918 while the northern line did not reach its terminus at Chiang Mai until 1921. The main north–south trunk route was nevertheless completed well before the truncated American system on Luzon. An eastern extension reached the Cambodian border in 1926; the line from Korat was pushed on to Ubon on the Mekong in 1930 and to Udon via Khon Kaen in 1941. This was at least a system, however skeletal (Chapter 5).

The Indochina rail system began in 1885 with the opening of a short line from Saigon to Mytho on the Mekong. However, it was almost another thirty years before further work was carried out in the south. Reflecting the French obsession to make Indochina the southern gateway to China, all effort was directed to opening rail links from Hanoi to southwestern China. A narrow-gauge railway reached the border in 1894. In 1898 a 200 million franc program authorized construction of the Red River railway from Hanoi across high mountain ranges to the capital of China's Yunnan province. A remarkable feat of engineering, the line opened in 1910 but traffic never fulfilled expectations. Meanwhile the long foreshadowed Transindochinois trunk line languished. Not until 1936 were the northern, central and southern sections finally joined together, enabling passengers to travel between Hanoi and Saigon in a rather uncomfortable 40 hours. A voyage by coastal passenger steamer was still much to be preferred. Besides, there seems to have been little demand for travel between the two French capitals, which belonged to virtually separate colonies. Similarly, no rail connection was made between Saigon and Phnom Penh, which was served by fast river steamer (Chapter 5).

Malaya was one British colony where railways were introduced quite late but British control was not established until the 1870s. The first lines from the mid-1880s were short spurs connecting the main towns of the peninsula to their swampy coastal ports. In the 1900s these towns were gradually joined together in a north–south trunk line. Penang (Prai) was connected to Kuala Lumpur in 1903 and in 1909 the line reached Johore, opposite Singapore. A northern extension met the Thai system at the border at

Table 2.6 Southeast Asia: length of railway line in operation, 1870–2000 (km)

	Indonesia	Malaya	Philippines	Thailand	Indochina		Burma	Total[1]
1870	99	nil	nil	nil	nil		nil	99
1900	3,574	457	196	125	182		1,093	5,627
1930	7,395	1,953	1,306	2,862	2,395		3,107	19,018
1960	6,640	2,387	1,020	2,100	1,364		2,991	16,502
1990	6,708	1,777	921	3,728	*Vietnam*	*Cambodia*	2,892	(16,026)
2000	6,458	2,227	492	4,071	2,652	602	3,955	20,457

Note: Indochina figures in 1960 refer to South Vietnam only. The 1990 total excludes Vietnam and Cambodia. The 2000 total excludes 19 kilometres of rail in Brunei and 117 kilometres in Singapore.
Source: Mitchell (1998), JIG (2002).

Padang Besar in 1918: trains could then run though to Bangkok on some days of the week but the commercial significance was marginal. A line through Pahang made a second connection to the Thai system in the northeast state of Kelantan in 1931 without much affecting the isolation of the East Coast (Chapter 6).

The impact of railways on Southeast Asia is not easily assessed. Their construction was for the most part an early-twentieth-century phenomenon. In 1900 there were still only around 5000 kilometres of track across the whole vast region, and this modest total was dominated by Java (Table 2.6). By the late 1930s, after completion of the Transindochinois, track length had increased to 20,000 kilometres, a modest total compared with 32,500 kilometres in the United Kingdom alone and 25,000 in Japan (Mitchell 1998). Thereafter construction all but ceased. The opening of a few new lines was offset by closure of others, so that track length was almost the same in the 1990s. The number of passengers and the tonnage of freight was also much the same in 1990 as in 1930 (Tables 2.7–2.8). Growth in traffic was taken up by other modes.

Southeast Asia's railway systems did more to integrate port cities and their hinterlands than to make international connections, as in Europe, or transnational connections, as in the United States and Canada. International links were made, most notably between the Malayan, Thai and Cambodian systems and between northern Vietnam and southern China, but none of these became important traffic arteries. Except in the case of the Hanoi–Yunnan railway, which was built to tap foreign trade, the international connections were incidental. Only the Thai system, skeletal though it was, provided some national integration like the Indian or Japanese systems. That was hardly a coincidence. As the one independent nation in prewar Southeast Asia, Thailand felt under pressure to assert its territorial integrity. The Indonesian and Philippine systems were confined to specific islands, while the Peninsular Malaysian system barely touched the East Coast. The simple explanation is that colonial governments sought to economize on infrastructure. Given the availability of low-cost water transport, there was little expectation of return to railway entrepreneurs without government subsidy. The dashed hopes of railway entrepreneurs in the Philippines proved the point. Most private lines were in time taken over by the state. Thus instead of a railway revolution, as in Europe or North America, there were localized impacts (Chapters 4–6).

This economic assessment is nevertheless only part of the story. Regardless of the length, breadth or density of track or the amount of traffic, railways became part of the fabric of modern life. In Southeast Asia, as elsewhere, railway clocks and timetables gave time a new dimension to life and a new concept of certainty. Fare structures calibrated a class system in units of currency. Railways pioneered modern work organisation. By 1929 the state

Table 2.7 Southeast Asia: rail passenger traffic, 1910–1998 (million)

	Indonesia	Malaya	Philippines	Thailand	Indochina		Burma	Total[1]
1910	72	9	2	3	10		n.a.	96
1930	133	12	(13)	5	11		29	203
1960	158	7.7	9.5	40	2.6		40	258
1990	56	8.5	1	85	Vietnam	Cambodia	49	(200)
1998	167	(5.9)	0.5	62	5.7	0.4	n.a.	(242)

Note: Indochina figures in 1960 are for South Vietnam only; Philippines figures in parentheses are for 1929; 1990s figures exclude Metrotren. Figures shown in parentheses for Malaysia are taken from 1996.
Source: Based on Mitchell (1998), BPS (2001), JIG (2002).

Table 2.8 Southeast Asia: rail freight traffic, 1910–1998 (million tonnes)

	Indonesia	Malaya	Philippines	Thailand	Indochina		Burma	Total[1]
1910	7.6	0.7	0.1	0.3	0.6		n.a.	9.3
1930	15.8	2.2	1.6	1.2	1.0		5.0	27
1960	6.6	3.6	1.4	3.7	0.4		3.1	19
1990	12.5	4.6	0.0	7.9	Vietnam	Cambodia	1.7	(27)
1998	5.6	(5.4)	0.0	11	4.8	0.3	n.a.	(27)

Note: Indochina figures in 1960 are for south Vietnam only. Total in 1999 excludes Cambodia 294,000 tonnes. Malaysian figure in parentheses is for 1996.
Source: Based on Mitchell (1998), BPS (2001), JIG (2002).

railways of colonial Indonesia employed 45,000 people (SSNI 1933: 17). Drivers, crew, and station staff were the elite among the indigenous workforce, formed trade unions and pioneered strike action. Train travel and the opportunities for unstructured encounters became part of the plots of a new vernacular literature for an emerging urban middle class (G. Lockhart, pers. comm.). A person did not even have to travel by train to be influenced by the railway ethos. From city and town to local village, the railway station was a centre of activity and a physical link to the wider world. In guerila struggles for independence, rail track and rolling stock became prime targets for sabotage. In such ways, railways insinuated themselves into the *mentalité* of Southeast Asian nationalism.

Autonomous networks: roads

Roads are a less obvious form of technological diffusion than railways. Despite heavy reliance on waterways, parts of Southeast Asia such as Java certainly had roads before the colonial era, although nowhere paved as in China – not even Daendels' post road met that standard. There were also networks of pathways by which people and goods could circulate. After the mid-nineteenth century, the quickening of trade and the spread of railway networks generated more road traffic to and from rivers and railway stations. Vehicles were typically oxcarts for goods and carriages or pony carts for passengers. In Java, the demand for feeder transport called forth investment in narrow-gauge tramways. Sugar estates also laid out their own rails to bring cane to the mills, and timber concessionaires throughout Southeast Asia often laid temporary tramlines to haul logs from forest to stream. Mines in Europe had long used horse-drawn tramways in a similar way. While light rails could be relocated with shifts in cultivation or extraction, tramways were much less flexible than motorized vehicles and often had to be worked in conjunction with teams of oxen or, in the case of the teak industry, elephants.

The first automobiles arrived in Java in 1894, Singapore 1896, Bangkok 1897, North Sumatra 1902 and Rangoon 1905. Throughout the 1900s, however, they remained a curiosity. In the 1910s early mass-produced automobiles, notably the black T-model Ford, became status symbols for wealthy Europeans and Chinese but for want of reliability and passable roads were confined to the main cities. After the First World War larger and more efficient buses and trucks came onto the market. Initially railway owners looked with favour upon these vehicles as low-cost feeders and a means of serving low-density routes that would not justify rail lines. During the 1920s, however, buses and trucks began to undermine rail monopolies by carrying passengers and goods at cheaper rates. Governments began to examine ways of regulating road transport to protect the profitability of railways (Chapters 4–6).

The impact of the new motor age was most uneven across Southeast Asia. In the Philippines, where the rail network was weakly articulated and

American officials were imbued with the spirit of the automobile age, buses and trucks provided an ideal means of improving rural–urban communications, not just in Luzon but also in frontier territories such as Mindanao. It was some years before the techniques were mastered of constructing and maintaining roads under stress of torrential rains and typhoons but by 1930 the settled parts of Luzon had a first-class road network, extending into and through the highlands (Chapter 4). A road network projected for Mindanao but little had been achieved before the Pacific War.

Roads also substituted for railways in Indochina, which on the eve of the First World War had consisted of fragments unconnected by land transport. In 1912 the colonial government launched a programme to build some 9,000 kilometres of roads, of which the most important was the 2,600 kilometre 'Mandarin Road' from Langson on the Chinese frontier via Hanoi, Hue and Saigon to Phnom Penh and Battambang (in the mid-1920s extended to the railhead on the Thai border) (Desfeuilles 1927). The programme was continued after the First World War with emphasis upon roads into the mountainous and hitherto isolated hill tribe areas. As early as 1923 a British consular official praised Indochina's 'excellent road system' (Gorton 1923). Although the Red River and Mekong deltas continued to rely upon water transport, roads linked up hinterlands and gave the highlands ready access to the coast (Robequain 1944). Ironically, Indochina's road network is now the worst in Southeast Asia, having suffered badly from political turmoil since 1945 (Chapter 6).

In the Outer Islands of Indonesia, roads were constructed to fill in the gaps between coastal, river and, in Sumatra, rail systems. A unified Sumatran railway system was projected (Reitsma 1925), but roads turned out to be much cheaper to construct for the small volume of traffic. As early as 1907 the government had introduced a busline (*autodienst*) from Palembang across the dividing range to the small west-coast port of Bengkulu, and this was gradually extended into a network. Pathways were upgraded to link the highlands of Sumatra to the west coast and to the heads of navigation of the rivers that flowed down to Malacca Strait. It also became possible to travel by road from Padang to Medan. Not until the 1980s, however, was through bus and truck traffic competitive from Java to Padang and Medan, using vehicular ferries across Sunda Strait. In the 1990s the three main islands of Sumatra, Kalimantan and Sulawesi all acquired a skeletal paved highway system (Rimmer et al. 1994: 180–1).

In Peninsular Malaya, a dense network of roads was gradually completed on the west coast, where road–rail competition became severe in the interwar years. Some developmental roads were built to open up remote areas such as Pahang but the isolated east coast continued to rely upon water transport (Kaur 1985a; Chapter 6). In Thailand highway transport remained unimportant until the 1950s. Roads had begun to be laid out in the 1920s from up-country railheads to the Chinese border and the Mekong River but no roads led out of Bangkok until the mid-1930s. Completion of the Friendship Highway from Saraburi to Nakhon Ratchasima in 1958 may be taken as the start of Thailand's motor age, which was greatly stimulated by the American presence during the Vietnam War. In Burma, the upgrading of cart roads into a modern highway system began in the 1920s but even in the 1950s this road system, such as it was, served mainly to feed traffic to the rivers and railways (Nyoe and Khin 1957). This was also true of the famous Burma Road from the railhead at Lashio to Kunming, which on opening in 1939 as a wartime backdoor to China became Burma's first international land link. Although in 1942 Burma had a better road system than Thailand, wartime damage, subsequent insurgency and the economic stagnation of recent decades have reversed this situation. Today Burma's road system is not much better than that of Vietnam.

Despite the uneven spread and quality of road infrastructure, the trend in vehicle numbers has been one of steady increase. On the eve of the Second World War there were just over 100,000 passenger cars and around 50,000 registered commercial vehicles throughout the whole of Southeast Asia: by 1995 there were about eight million cars and over six million commercial vehicles (Tables 2.9–2.10). Here is the emphatic explanation for the stagnation of the railways. Southeast Asia's postwar economic development has been facilitated by road transport. Location and mobility have shifted decisively from waterways and railways to roads. Southeast Asian societies are now wheel-based. Investment in road infrastructure, however, has been unable to keep up with the demand, leading to increasing congestion in urban areas. Presently low rates of vehicle ownership suggest that it is likely to worsen.

Autonomous networks: urban public transport

In street tramways, the main Southeast Asian cities were near the frontier of international technology. The prototype was the horsetram, an omnibus pulled by horses along fixed rails and operating to a fixed schedule. Introduced to New York and Boston in 1852 and to London (on an experimental basis) and Sydney in 1861, horsetrams began operating in Jakarta as early as 1869, followed by Manila (1881), Tokyo (1882) and Bangkok (1889). In a tropical climate it was hard work for the animals. A steamtram, namely a boiler and light engine on wheels able to pull one or two carriages, promised to be more comfortable and reliable and to justify the higher capital cost. Jakarta again led the way in 1882, followed by Rangoon (1884), Singapore (1885), Sydney (1886), Surabaya (1889) and Penang (1893).

Meanwhile, more modern and capital-intensive technologies were being introduced in Europe and North America. In 1873 hilly San Francisco introduced a cable tram, which was tried in Sydney and became a substantial network in Melbourne. The wonder of the age, however, was electric traction. A laden tram gliding smoothly through crowded city streets was

Table 2.9 Southeast Asia: number of registered passenger vehicles, 1930–1998 ('000)

	Indonesia	Malaysia	Singapore	Philippines	Thailand	Vietnam	Burma	Total
1930	58		16	23	4	n.a.	n.a.	(101)
1950	31	27	18	45	10	6[1]	8	145
1975	383	472	149	383	266	n.a.	37	(1690)
1995	2107	2589	364	627	1913	n.a.	27	(7627)
1998	2773	3853	375	749	(1661)	n.a.	n.a.	(9411)

Note: 1. South Vietnam only. Total in 1995 excludes Brunei 51,000, Cambodia 42,000 and Laos 17,000 vehicles. Total in 1998 excludes Brunei 58,000. No information is available for Cambodia and Laos. Thailand figure in parentheses is for 1999.
Source: Based on Mitchell (1998), IRF (2001).

Table 2.10 Southeast Asia: number of registered commercial vehicles, 1930–1998 ('000)

	Indonesia	Malaysia	Singapore	Philippines	Thailand	Vietnam	Burma	Total
1930	20		3	14	4	n.a.	n.a.	(41)
1950	29	16	8	55	13	4[1]	12	137
1975	231	158	46	273	246	n.a.	40	(994)
1995	2025	466	137	1219	2736	n.a.	63	(6646)
1998	2221	691	141	1509	(2885)	n.a.	n.a.	(7447)

Note: 1. South Vietnam only. Total in 1995 excludes Brunei 16,000, Cambodia 14,000 and Laos 17,000 vehicles. Total in 1998 excludes Brunei 14,000 vehicles. No information is available for Cambodia and Laos. Figure in parentheses for Thailand is for 1999.
Source: Based on Mitchell (1998), IRF (2001).

enormously impressive. No more piles of horse manure and urine, no more belching smoke, just ugly overhead cables. A demonstration electric tram ran in Melbourne during the Great Exhibition of 1888 and over the next decade was copied around Australia, as also in the mid-1890s by Madras and in 1895 by Kyoto (Japan). In Southeast Asia, Bangkok was precocious with an electric tramway opening in 1893, Jakarta followed in April 1899, three months before the first line in the Netherlands (Duparc 1972: 11). There followed Calcutta (1902), Osaka (1903), Hong Kong and Mandalay (1904), Singapore, Penang and Tokyo (1905), Rangoon and Manila (1906) and Saigon (1912). The main cities of Southeast Asia were therefore contemporary with those of Europe and North America. The new silent trams encouraged the trend towards suburban living for wealthy Europeans and more importantly, in terms of passenger numbers, could handle the daily flux of workers to and from the business district and port.

Road–rail competition also became apparent in urban public transport (Rimmer 1986a). In the late-nineteenth-century tramways faced competition from coolie-drawn rickshaws or, in Java and the Philippines, pony carts. Automobiles at first replaced the carriages of wealthy Europeans and Chinese but in the interwar years as jitneys and taxis began to compete directly with tramways on main trunk lines. Competition became vigorous in the 1930s, when the depression led to a strong demand for low-cost transport. Tramways also lost patronage to bicycles, whose falling real price made them a popular form of private transport. Wartime damage was a further blow, from which the Manila tramway never recovered. Jakarta closed most of its tramlines in 1960, Bangkok in 1965. Asian cities were therefore not far behind American and Australian cities in turning from trams to buses. At the time this economized on infrastructure, but in the long run left cities unable to cope with the explosion in population and motorized traffic. Bangkok became the nightmare of the automobile age. Belatedly mass transit rail systems are now being constructed at immense cost (Chapters 8–10).

Conclusion

The rapid diffusion of transport and communications technologies to Southeast Asia in the late nineteenth and twentieth centuries hardly supports a crude model of backwardness. On the contrary, it is striking how often the most modern advances were quickly applied. This was part of the dynamics of imperialism. The leading colonial powers, Britain, France, the Netherlands and the United States, were the centres of industrial revolution technology and of ongoing research and development. They also had a very strong incentive to use their new technological powers to shorten the economic distance with their Asian colonies and to facilitate their exploitation, or what would nowadays be termed development. Great power rivalries intensified these pressures. Spain, together with Portugal the most technologically backward of the imperial powers and the least vigorous in transferring modern technology, was also in 1898 the first to lose its colonies – both the Philippines and Cuba.

Yet if technology transfer was often rapid, it was also uneven across Southeast Asia and patchy in its diffusion within colonies and kingdoms. How can we explain the rapid diffusion of electric tramways and the slow diffusion of railways? The key is that cities were the point of contact between Southeast Asia and the world. Telegraphs, steamship mails and passengers, and later aircraft all touched there. Information flowed from one side of the world to the other within 24 hours, and what was in the public domain was quickly published as news in the daily press. Westerners were disproportionately represented in the large cities, and from the late-nineteenth century their numbers grew rapidly. Having access to capital, they naturally sought to create in these cities some version of the modern world, asserting their modernity and supremacy over the rest of colonial society. Cities were naturally the first to receive electricity, initially for tramways and streetlighting, later for electric fans, lifts and telephone exchanges, printing presses and cinema projectors.

What demands explanation is therefore the slow diffusion of technology to hinterlands. One obstacle was low productivity and lack of mass purchasing power, which restricted the size of the market and its growth potential. Another was the high initial capital cost of railways. Because it was difficult to mobilize local capital on a large scale and foreign capital was focused on export production, network infrastructure often became the responsibility of a parsimonious state. However, the competitive resilience of the small-scale sector must also be taken into account. 'Traditional' boats and carts adapted innovations in technology and organization. In rural Luzon in the 1990s, most of the technologies observed by the Americans in the 1900s were still to be found, including wheel-less sleds dragged by buffalo, horse carts and dug-out canoes (*banca*). Yet few of these technologies are any longer strictly 'traditional'. Bicycles, motor vehicles and marine engines were taken up with enthusiasm and modified in ingenious ways. Dug-out canoes are powered by outboard engines; on Luzon outrigger carts with ball-bearing wheels are poled along little-used railway lines. There was never any reluctance to innovate. The outcome has been a complex technological matrix that defies any crude dualistic model of a modern, large-scale and 'traditional' small-scale sector.

Nevertheless, there is a fundamental dualism between the high-speed transport of aircraft and tollroads – as yet Southeast Asia has no high-speed rail corridors – and slow-speed sea, rail and road transport. This modal split is also a socioeconomic divide. 'Time is money.' Those who buy high-speed transport are those who can afford to do so. In communications the divide is not quite so sharp, because IDD or STD telephone calls or Internet access, for example, can be bought in very small units, but the pattern is much the same. As elaborated in Part III, high-speed transport and communications are creating a single global market of urban, middle-class elites. Poverty remains local.

Note

1. Unless specifically acknowledged, basic technical information is drawn from *Encyclopaedia Britannica* (1993) and dates from Chapters 3–10.

Part II
Hinterlands

Geography gives logic to the otherwise fragmented national economic histories of Southeast Asia. Over the very long term since 1850 the region's economic development may best be studied in terms of geographical units that are both larger and smaller than now familiar national economies. In this way the shape and structure of these national economies can be seen gradually to emerge.

Figure II.1 identifies four regions that form the basis of the subsequent chapters on hinterlands and together encompass most of Southeast Asia. *The Archipelago* or macro-region of maritime Southeast Asia encompasses the islands and ports within an arc stretching from Luzon Strait north of the Philippines through Indonesia to the Kra Isthmus in southern Thailand (Chapter 3). Within this arc the major *islands* of Java and Luzon are considered separately as geographically similar cores of the emerging national economies of Indonesia and the Philippines (Chapter 4). *River Basins* encompass the rice plains of the Chao Phraya, Irrawaddy and Mekong, which become the cores of the emerging national economies of Thailand and Burma and of the regional economy of southern Vietnam – the Red River, whose development history is more akin to southern China, is not part of this comparison (Chapter 5). *The Peninsula* stretches from the Kra Isthmus through Malaya to Singapore and is contrasted with the peninsula-like coastal strip of Annam (Chapter 6). Although this core of the emerging national economies of Malaysia and Singapore is treated as a special case, its historical experience resembles that of the islands of Java and Luzon rather than of the mainland river basins.

Three phases of economic development recur through all four chapters. Before 1850 there was a network of trading cities whose hinterlands were sparsely populated except along navigable waterways and accessible coastlines. From the mid-nineteenth century frontiers were opened up by clearing the jungle for land-extensive, export-oriented agriculture. The core regions specialized on a narrow range of commodities (Table II.1). Sugar, coffee, tobacco and rubber were the main export commodities in Java; sugar, abaca and copra in Luzon. Rice and teak were prominent in the River Basins and tin, rubber and oil palm in the Peninsula. These commodities gave coherence to the economic development of the individual regions. Eventually accelerated population growth, closure of the land frontier and increasing density of settlement resulted in labour surplus and widespread poverty in much of these areas. By the early twentieth century there was already out-migration from the northern valleys and northeast regions of Siam, central Java and Madura and the Ilocos region of northwest Luzon. This phase of agricultural-led development continued with much diminished returns into the early postwar years of national independence.

Since the 1970s, a third phase of industrialization has shifted the economic centre of gravity back towards main cities, which have been transformed into extended metropolitan regions with hybrid rural–urban (*desakota*)

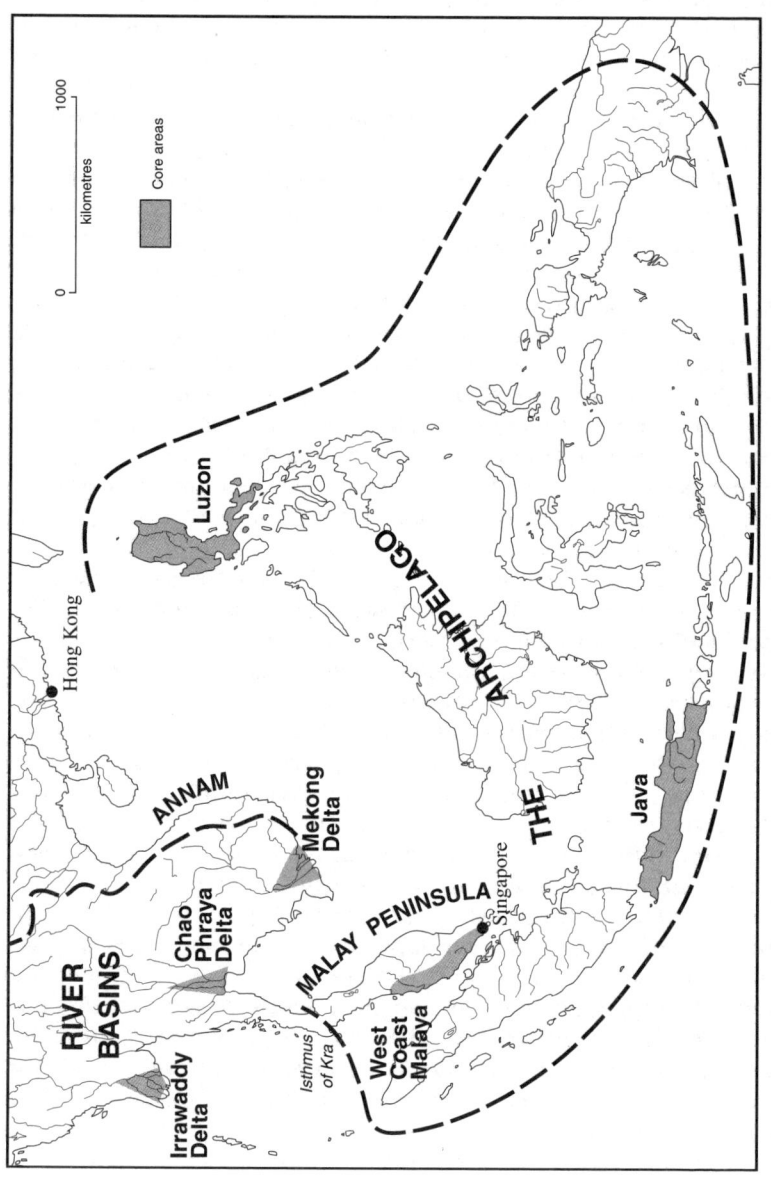

Figure II.1 Supra-national and sub-national regions

Table II.1 Major export commodities associated with the five supra-national regions

Supra-region	Export commodities
Archipelago	Various
Islands—Java	Sugar, tobacco, rubber
—Luzon	Sugar, abaca, copra
River Basins	Rice, teak
Peninsula	Tin, rubber, oil palm

characteristics (see Part III). Connections by road between these urban regions and hinterlands have been much improved but the hinterland share of national production has declined with that of the agricultural sector. Hinterlands are now primarily sources of surplus labour to urban regions and are sustained by remittances. Part II therefore focuses on the articulation of hinterlands in their heyday between the mid-nineteenth and mid-twentieth centuries, leaving the new urban regions to be examined in Part III.

The extent and rate of hinterland development between the 1850s and 1950s was determined by accessibility to international markets. Hitherto isolated tropical areas with abundant uncultivated land and an underemployed labour force enjoyed a 'vent for surplus' (Myint 1971: 120–39). Production could at first extend no further than porterage distance from natural waterways. Modern transport technologies extended the frontier by reducing the steepness of the transport cost gradients. Individual chapters show the dramatic impact of the shifts in line-haul modes from water (coastal shipping, rivers and canals) to rail, to highways (later expressways) and to air.

A recurrent theme is monopoly and competition. When all flows are handled by a single line-haul mode (inland waterway, railway or coastal shipping) and each landing, station or port has a captive hinterland, there is complete monopoly (Figure II.2). The area of monopoly is extended by the development of feeder roads and consolidated where a denser feeder network allows consolidation of transport facilities at larger landings, stations or ports. Conversely, incipient competition occurs where trunk roads from other regions give spot access and localized competition occurs between inland waterway, rail, coastal shipping and road transport. Sporadic competition happens where trunk road transport links the inland waterway, railway or coast at several points and ties feeder roads into the long-distance highway network. General competition occurs where trunk roads parallel line-haul modes with traffic losses according to the nature and availability of traffic. By the 1960s many areas outside the core sub-national regions were experiencing general competition but there were still pockets where competition from road transport was incipient or sporadic, allowing the inland waterway, rail transport or coastal shipping to retain a substantial share of the traffic.

78 Hinterlands

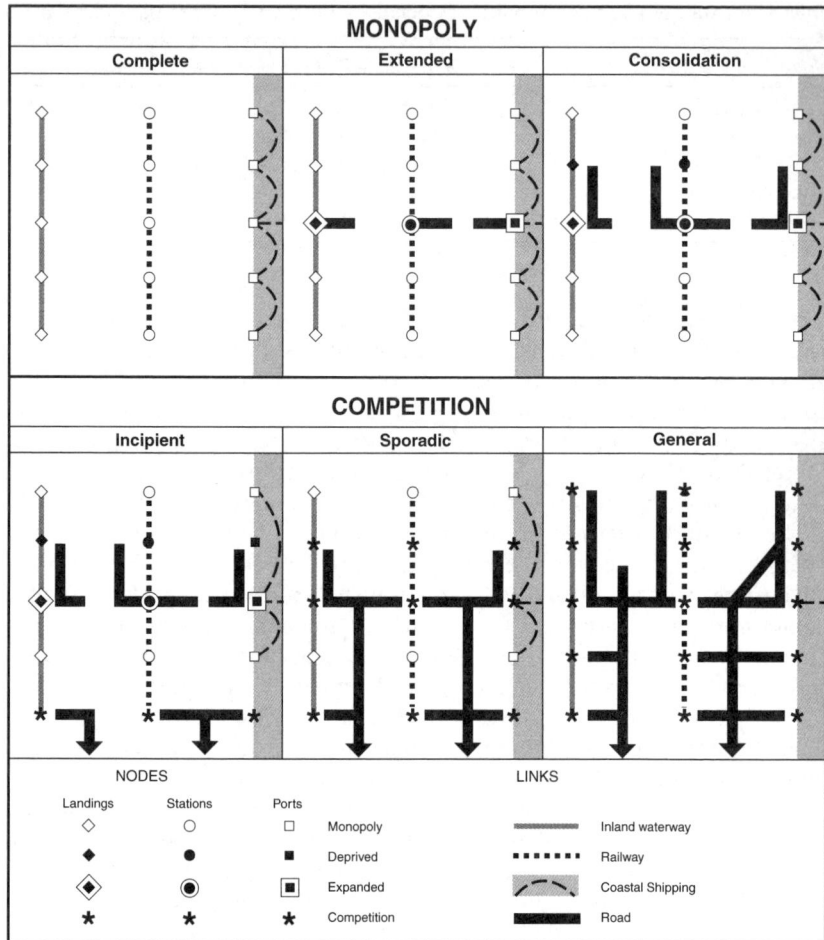

Figure II.2 Monopoly and competitive transport states (*Source*: Based on Rimmer 1971a: 105)

Transport organization, regulation and efficiency are also important. By the late-colonial period, railways in Java, Luzon and Malaya were already experiencing strong price competition from road transport and, as in Europe and North America, turning to government for regulatory protection. After independence the efficiency of state-owned railways and the efficacy of such protection both declined markedly. Underinvestment in road infrastructure did not prevent a large shift in traffic despite worsening travelling conditions. Whatever the historical trajectory, railways have long been eclipsed throughout Southeast Asia.

Each hinterland developed according to its own time path and some were incorporated into incipient nation-states much sooner than others. Thus the core regions of Java, Luzon, the Central Plain of Thailand and the West Coast of Malaya came to play a dominant role in national politics, commerce and society for reasons that had much to do with their head start in modern economic development. Other sub-regions which were prominent in the mid-nineteenth century such as Aceh or South Sulawesi in Indonesia, Cebu or Iloilo in the central Philippines, Chiang Mai in northern Thailand or Hue in central Vietnam have struggled against relegation to the periphery.

3
The Archipelago

Island Southeast Asia is the world's largest archipelago. To a well-found ship and an experienced crew, the seas used to be commons, more easily traversed than the jungles behind the coasts and main rivers. However, colonialism and nationalism have fragmented the land and seas of what used to be called the Malay Archipelago into the separate nations of Indonesia, Malaysia, the Philippines and the micro-states of Singapore, Brunei and East Timor. Over the past two centuries there has been an epic struggle between territorial states that have sought by mercantilist policies to capture and control people and wealth and the integrating commercial forces of free trade based upon Singapore. From 1819 to 1949 the Netherlands was the main antagonist, in the 1960s and 1970s Indonesia, in the 1980s and 1990s Malaysia. In what became the Philippines, first Spanish and then American rulers carried on a more muted conflict with the British free-trade colony of Hong Kong. Although trade from the southwestern and Muslim part of Sulu and Mindanao has continued to flow to Singapore, the rest of the Philippine segment of the archipelago is still oriented more towards East Asia than to Southeast Asia.

This chapter has three sections. The first section shows how the interaction of colonial/national policy and rival commercial shipping interests shaped regional trade networks against a background of technological change, especially the switch from sail to steamships and, since the 1970s, containerization. The leitmotif is the suppression and revival of Singapore as a maritime centre. The second section focuses on the Philippines, which in the colonial era had a very different maritime regime from Indonesia and is best studied comparatively. The final section assesses the impact of evolving networks of shipping and trade upon economic structure, in particular the extent to which Indonesia and the Philippines have emerged as integrated national economies with consequent cores and peripheries.

Nationalizing the archipelago

Around 1800 the Malay Archipelago was still a vigorous, seamless economy not yet closed off by political boundaries, nor much bothered by ethnic identities (Reid 1988, 1993). Malay was the lingua franca; Islam was on most coasts a common religion. Trade had been much disrupted by Dutch and Spanish conquest and monopolies but beyond Java and Luzon these powers controlled too little territory and too few ports to be able to enforce cabotage. Recent research suggests that the 'native' trade of the Archipelago was more buoyant than has been recognized (Knaap 1996). The growing British influence for free trade also stimulated Indian, Arab and Chinese networks, especially during the British occupation of the Netherlands Indies (1811–16).

The Singapore network

From its foundation in 1819 Singapore was an immediate success (Wong 1960). Its regime of free trade, rule of law and security of property was modern, but none the less consistent with that of pre-colonial entrepots such as Malacca. British policy was to draw Asian traders to Singapore and thereby allow British merchants access to Asian networks and Asian markets. Trade with Europe and North America accounted for the minor part of total trade. As late as 1869, on the eve of the opening of the Suez Canal, three-quarters of Singapore's exports and two-thirds of its imports were within Asia, mostly within Southeast Asia and a third within the Archipelago itself (Table 3.1). After 1850, as the Indian trade declined relative to that of Hong Kong and China, Singapore became predominantly a Chinese business centre under benevolent British rule (Chapter 7).

In the mid-nineteenth century the staples of trade were very different from those of the twentieth century. Exports were an array of products from

Table 3.1 Singapore and Java: exports and imports by destination, 1869

Destination	Exports (%)		Imports (%)	
	Singapore	Java	Singapore	Java
Archipelago	33.5	12	31	36
Other Southeast Asia	15.5	1	11	1
Total Southeast Asia	**49**	**13**	**42**	**37**
China	13	1	12	2
India	12	—	12	2
Total Asia	**74**	**14**	**66**	**41**
Europe/North America	24	84	32	59
Other	2	2	2	—
Total	100	100	100	100

Source: Calculated from Wong (1960).

the jungle (resins, waxes, sago, gutta percha, rattans, sappanwood) and from the sea (dried fish, shark fin, pearl shell, beche-de-mer). Except for sago and gutta percha (insulation for submarine cables), most of this so-called Straits Produce was shipped to China. From small mines and plantations there was also tin, pepper, spices and the dyestuff gambier. Some production was localized: tin came mainly from the short rivers of the West Coast of Malaya, pepper from Sumatra, sappanwood and sandalwood from the eastern Archipelago. Otherwise specialization was limited. From Aceh at the northern tip of Sumatra to Sulawesi and the Sulu in the east, small prahus, western-rigged sailing ships and later small steamers brought a similar range of local products and returned with a similar assortment of textiles, metalware, ceramics and coin from Britain, China and India. Time was not of the essence for many vessels, especially those from the eastern Archipelago, made only one voyage a year with the monsoon winds.

Singapore was also the focus of a vital Southeast Asian circuit. It was the distribution centre for counter flows of rice and Javanese sugar, as also salt and dried fish (Thailand) (Figures 1.3–1.6). Livestock were mainly to victual Singapore itself (Figure 1.6). All these trades were controlled by Chinese networks based in Singapore (Huff 1994). The densely settled wet-rice regions were also concentrations of purchasing power that drew in imports, especially of textiles and opium. Thus Singapore's main local trading partner after the West Malay Peninsula was Java, including the clandestine opium trade of North Bali (Lindblad 1996).

After 1900 the rubber boom reoriented the trade of Singapore towards Europe and North America, whose automobile industry was the source of this new demand. Rubber and its reciprocal import trade tightened the nexus between Singapore/Penang and the Inner Archipelago (Eastern Sumatra, the Malay Peninsula, and West and South Kalimantan), while as a result of Dutch policy the Outer Archipelago (Western Sumatra, East Kalimantan and Eastern Indonesia) was virtually detached from Singapore's sphere. Direct shipment between Java and mainland Southeast Asia, as also between Java and China, cut Singapore out of the rice/sugar trade. Yet despite this contraction of Singapore's intra-Asian trade, the networks remained capable of revitalization once Dutch pressure was removed after the Second World War.

Dutch mercantilism

During the nineteenth century the Dutch began to transform the Netherlands Indies into a territorial state with Java as its core (Dick et al. 2002). The former United East India Company (VOC) had maintained a much wider Asian empire, with settlements extending from India to Japan. Loss of the spice island of Ceylon in 1795, British intrusion at Penang (1786), Singapore (1819), Sarawak (1841) and Labuan (1846), and the Anglo-Dutch treaty of 1824, which swapped Dutch Malacca on the Malay Peninsula for the British settlement of Bencoolen in Sumatra, pushed the Dutch back into a defensive perimeter corresponding to the boundaries of modern Indonesia. Even this was not secure. During the Java War (1825–30), the Dutch were almost driven out, and triumphed only at the cost of bankrupting their own country. The response was to turn Java into a source of revenue. After 1830 the Cultivation System mobilized the land and labour of Java to produce crops such as coffee, sugar, tobacco and indigo for export to the Netherlands (Elson 1994). At the same time, textile mills were set up in the Netherlands to export to Java under tariff protection from British manufactures. Neither trade was intended to pass via Singapore or through British hands. The Netherlands Trading Company (NHM) was given the monopoly of shipment, under the Dutch flag. By 1869 some 84 per cent of Java's exports were shipped to Europe and North America, most directly to the Netherlands (Table 3.1).

Steamers provided a means for closer links between Java and the rest of the Dutch archipelago. After 1825 there was a steam packet along the north coast of Java and from mid-1845 a naval steamer collected the P&O mails in Singapore for carriage to Batavia, but Dutch settlements outside Java were still dependent upon sailing vessels. Steamships could not operate at a profit without a large subsidy. After lengthy study of routes and costs, in 1850 the government awarded a mail contract with fixed rates for government passengers and cargo for a subsidy of Fl 160,000 per annum to a syndicate headed by a retired naval officer Willem Cores de Vries, who had been in charge of the study (Campo 1992). Operations began early in 1852 with the first of four ships. Mails, cabin passengers, soldiers and valuable freight such as opium could now be shipped monthly to the main Dutch settlements in the archipelago. In the absence of any rival bid, in 1859 the syndicate was able to renew at a much higher subsidy of Fl 500,000 per annum for a network now extending to Kalimantan and Timor (Figure 3.1).

In 1863 a British syndicate associated with the British India Steam Navigation Company won the mail contract for a subsidy of just under Fl 300,000. Incorporated as the Netherlands Indies Steam Navigation Company (NISN), the new firm commenced operations on 1 January 1866. Under the new ten-year contract, the north coast of Java was served weekly, Singapore, Padang and Palembang fortnightly, and other Outer Islands ports monthly (Waal 1879). The NISN secured renewal of the contract for 15 years from 1 January 1876 by agreeing almost to halve the subsidy per mile but the outbreak of the Aceh War allowed the company to compensate by carriage of extra government passengers, troops and freight. Even before the war, most passengers had been on government account: by 1878 official passengers were 70 per cent. Dividends increased steadily from 10 per cent in 1873 to 20 per cent five years later but the government could not have run steamers more cheaply (Waal 1880). Moreover, the NISN network and fleet expanded well beyond that of the former syndicate, especially around Sumatra (Figure 3.2). Non-contract lines extended further afield, including experimental lines to China and Japan and to Australia.

The Archipelago 85

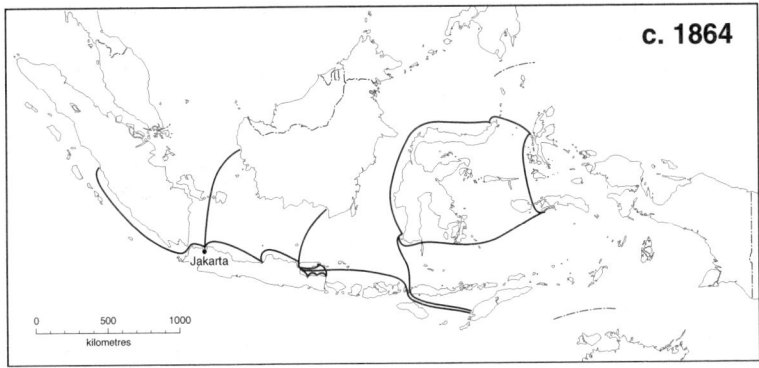

Figure 3.1 Netherlands Indies: contract interisland mail routes, c.1864 (A.W. Cores de Vries Syndicate)

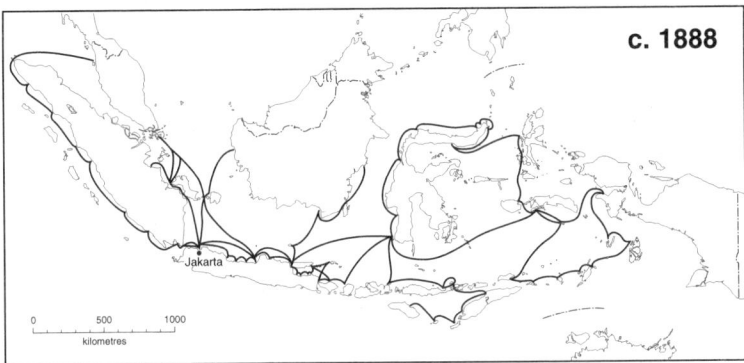

Figure 3.2 Netherlands Indies: contract interisland mail routes c.1888 (Netherlands Indies Steam Navigation Co.)

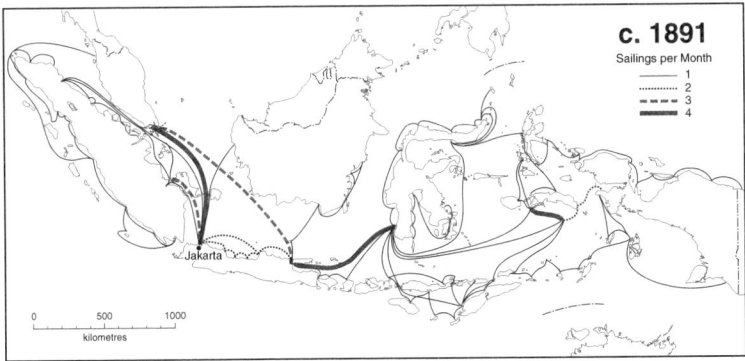

Figure 3.3 Netherlands Indies: contract interisland shipping routes, 1891 (Koninklijke Paketvaart Mij) (*Source*: Boer and Westermann 1941)

The opening of the Suez Canal made possible direct Dutch-flag steam navigation between the Netherlands and Java, bypassing Singapore. Formed under royal patronage in Amsterdam in May 1870, the Stoomvaart Maatschappij 'Nederland' (SMN) was unlucky to lose two of its original ships but by 1875 was at last able to provide a contracted monthly mail service and from 1879 fortnightly (Brugmans 1950). After mid-1893 weekly sailings were provided in a joint service between SMN and Rotterdamsche Lloyd (RL), alternating between Amsterdam and Rotterdam. Passengers wishing fast passage took the train overland and embarked at Naples (later Marseilles). Steamers called at Padang (West Sumatra) en route to Java ports, there connecting with NISN steamers for the Outer Islands. This direct connection eliminated dependence upon feeder connections in Singapore with the P&O or the French Compagnie des Messageries Maritimes and enhanced Dutch awareness of ownership of their colony.

In the carriage of cargo, however, steamers reinforced the entrepot role of Singapore (Wong 1978). By the mid-1860s small steamboats were running between Singapore and Penang via the tin ports of the West Coast. The Suez Canal allowed small steamships to be sailed East for sale to local buyers, while British commission houses in Singapore took orders from local firms for new ships to be built in British yards. Steamships soon captured the more profitable cargo from western-rigged sailing ships, Chinese tongkang, and Malay prahus. Not the least attraction of steamships was their virtual immunity from piracy. Despite the NISN's efficient interisland network, most exports from islands other than Java were carried by small local steamers into Singapore and Penang, trades in which non-contract NISN ships also participated. Most of these Chinese-owned ships flew the British flag, had at least a nominal British master – not infrequently an alcoholic – but were managed by the Chinese supercargo (*chinchew*).

In the 1880s European lines also began to introduce small feeder ships to the Archipelago to collect and distribute cargo for their deepsea ships calling at Penang and Singapore en route to and from China and Japan. The most aggressive was the British firm of Alfred Holt & Company, whose Ocean Steamship Company (Blue Funnel Line) had in 1866 pioneered the carrying trade by steamship between Europe and China (Hyde 1957; Falkus 1990). Holts sought especially to capture the valuable tobacco trade of North Sumatra (Deli) and the rice trade of Bangkok. After commencing a subsidized German mail service between Europe and East Asia in 1886, Norddeutscher Lloyd (NDL) soon did likewise. Both firms relied heavily upon local agents in Singapore. Holts' agents were Mansfield & Company, whose principal was Dutchman Th. C. Bogaardt; NDL turned to the well-established firm of Behn Meyer & Company. By 1890 their steamers were venturing as far afield as Sulawesi, the Maluku and Sulu.

The prominence of British-flag shipping in the archipelago was a great frustration to Dutch shipping interests. Convenience and national loyalty allowed them to monopolize the mails and first- and second-class passengers but, except for Java's plantation exports, they were marginal in the carrying trade, which flowed like water downhill to Singapore and Penang (Dick 1996). Declaration of free ports in the Outer Islands, beginning with Tanjung Pinang (Riau) off Singapore in 1829 and followed by Pontianak and Sambas (1833), Lampung (1839), Makassar (1847), Menado (1849) and Ambon, Banda and Ternate (1854), had had little effect, trade there still being less free than in Singapore and without the physical infrastructure and business networks (Knaap 1989; Wong 1960). The situation was especially frustrating for the two main Dutch lines, fighting tough competition from Holts and NDL for the trade of what they saw as their own archipelago. Accusing the 'foreign' NISN of using its non-contract steamers to siphon off trade to Singapore, the three leading principals combined a jingoistic public relations campaign with assiduous lobbying to prevent the NISN from renewing the mail contract from 1 January 1891 (Campo 1992). On the protectionist grounds that they would build a new fleet of ships in Dutch yards, they were awarded the contract in July 1888, notwithstanding a higher bid (Boer and Westermann 1941).

The Koninklijke Paketvaart Maatschappij (Royal Packet Company or KPM), formed in September 1888 by the same interests that controlled the SMN and RL, was forged by big business and government as the economic instrument of Dutch imperialism (Campo 1992).[1] Its original fleet of 13 ships was somewhat smaller than that of the NISN but completely new and able to serve a more extensive network (Figure 3.3). The capital was large enough to build a fleet, buy out the rival NISN, and sustain prolonged warfare with the British and Germans. The KPM's main competitive weapon was the 'through bill of lading'. By arrangement with the parent lines, exporters in the Outer Islands would be cited a through rate to the final destination that was no more expensive than transhipment via Singapore: the KPM as first carrier and the mail lines as second carrier then split the revenue. The first battleground was the valuable tobacco trade of Deli, hitherto shared between Holts (about two-thirds) and NDL (about one-third). When the KPM offered throughshipment on a new Deli/Batavia line, Holts retaliated by opening an Amsterdam/Java line, forming the Ocean Steamship subsidiary Nederlandsch Stoomvaart Maatschappij 'Oceaan' (NSMO) to operate deepsea ships under the Dutch flag, and setting up a joint venture with their Singapore agents to operate a fleet of local feeder vessels (the East Indies Ocean Steamship Company, EIOSS), some under the Dutch flag (Boer 1998). In February 1892 Holts came to a pool agreement with the Dutch lines for working the Amsterdam–Java line but fierce rivalry continued in the Sumatran trades until March 1893, when Holts agreed to withdraw their feeder from the West Coast of Sumatra in return for the Dutch lines restricting their operations in the Straits of Malacca. Henceforth the Dutch mail lines jointly served Java ports via Padang and Sunda Strait, while Holts sailed via Penang and Singapore.

Part of the urgency for a settlement was increasing pressure from the well-subsidized NDL. German commercial networks were well established in Southeast Asia and in Singapore NDL had excellent agents in Behn Meyer & Company. NDL targeted the Deli tobacco trade, the Bangkok rice trade, and a new tobacco trade in Sandakan (Sabah), a convenient way-port en route to German New Guinea. Holts therefore faced vigorous competition on all fronts in Southeast Asia and after Bogaardt's retirement seem to have had difficulty in managing their large local fleet. In 1899 the maritime world was stunned to learn that the entire EIOSS fleet had been sold to the NDL in return for a favourable transhipment agreement (Falkus 1990). Holts kept rights to two feeders on the Deli tobacco run. This brought the Dutch lines into the front line of conflict with the NDL and prompted the formation in 1900 of the Batavia Freight Conference to regulate the Java trade of the Trio Lines (SMN, RL and NSMO) (Boer 1997: 33).

The settlement with Holts allowed the KPM to challenge British-flag Chinese steamers based in Singapore. Here the KPM gained leverage from the colonial government's 'pacification' of the Outer Islands. When Chinese principals or their local agents were denied preferential contracts with the local ruler, they lost control of the cargo base and became vulnerable to competitive pressure. The through bill of lading was ineffective in the fight for local traffic, which involved offloading in Singapore for sorting, processing and repacking, but the KPM network could cross-subsidize a rate war on one route from profits on others. Conquest of Lombok in 1894, followed by Bali in 1906, eliminated Chinese commercial strongholds in eastern Indonesia. By 1914 no Chinese ships sailed further east than Surabaya. When the NDL fleet was interned at the outbreak of the First World War, Eastern Indonesia became to all intents a Dutch lake.

Consolidation of Dutch control over shipping and trade was underpinned by massive public investment in deepsea ports. Apart from its remarkably strategic location, Singapore was also a well-protected harbour equipped with deepsea wharves and godowns, coaling facilities, and dockyards. In Java, by contrast, Surabaya was the only natural harbour. At Jakarta and Semarang, ships had to anchor well offshore, from where passengers were ferried and cargo lightered up a shallow river to the Customs house, a slow, wet and rather dangerous operation in the west monsoon. When the large mail steamers of the SMN came into service the delays became intolerable. In 1877 work began on a deepsea steamer port at Tanjung Priok, nine kilometres east of old Batavia, which opened to traffic in 1886 (ENI 1921: 69–70). In the Outer Islands the pioneer scheme was at Sabang, a small island off the northern tip of Sumatra, where in the 1890s a deepsea harbour was constructed with berthing, coal staithes and drydock (ENI 1918: 71).

The intention was to make Sabang the entrepot for the foreign trade of Aceh, where the Dutch were still fighting a bitter war, and to restrict British-flag shipping out of Penang, which was seen as 'trading with the enemy' (Reid 1969: 268–70; Campo 1992: 157–62). In 1903 Sabang was declared a free port but without commercial networks never became more than a coaling station and transfer point for mails.

Despite the disappointment of Sabang, the colonial government held to the conviction that modern deepsea ports were the key to trade diversion. In 1909 as the conquest of the Outer Islands was being completed, the Netherlands engineers Kraus and de Jongh were commissioned to investigate the design of transhipment ports suitable to berth the largest ships (Kraus and Jongh 1910). Tanjung Priok was extended with new harbour basins. At Surabaya, hitherto a roadstead port, a basin with deepwater railway wharves was opened in 1917 and completed in 1925 (Dick 2002). In the eastern part of the archipelago, new deepsea wharves with huge godowns were opened in 1918 at Makassar, which since the 1900s had become the stapling point for the copra trade (Boer and Westermann 1941: 233–7). In the west, a new deepsea port was opened in 1920 at Belawan (just below Medan) to capture the East Coast of Sumatra's valuable plantation exports, especially tobacco, hitherto transhipped through Penang and Singapore; the Dutch mail steamers were also rerouted from Padang to Belawan. Shipping statistics show the rapid expansion of Belawan and Makassar (Table 3.2). Modest facilities were built at other ports around the archipelago, including private investment in bulk-handling facilities in oil and coal ports, but these four ports, each designed to facilitate transhipment, remained the pivots of Dutch trade. There the KPM interisland network intersected with the deepsea lines, supported by a sophisticated financial infrastructure of banks, trading houses, agencies and insurance companies.

Withdrawal of the local NDL fleet in August 1914 at the outbreak of the First World War tested the agreement between Holts and the Dutch lines as Holts moved to secure its transhipment trade. In 1913 Holts had placed two new feeders on the Singapore–Deli line. Now it ordered three large ships in Hong Kong for the Singapore–Sandakan line. Still wary of reviving a directly

Table 3.2 Netherlands Indies: inward shipping by main port, 1903–1938 ('000 m^3)

Port	1903	1929	1938
Jakarta	4.1	20.4	17
Surabaya	3.8	19.5	14
Belawan	0.7	10.1	10
Makassar	1.2	8.0	7

Source: Knaap (1989: Table 7).

managed entity like the former EIOSS, Holts injected a 50 per cent increase in equity and three new ships to a modest Anglo-Chinese firm, the Straits Steamship Company (SSS), which Bogaardt had organized in Singapore in 1890 to merge various interests in the coastal steamship trade between Singapore and Penang (Tregonning 1967). Although the venture had started slowly, in the 1900s an energetic new manager seized the opportunities of the rubber boom. Between 1903 and 1911 the company invested in a 'white fleet' of fast, shallow-draft ships designed to compete with the railways in the West Coast trade (see also Chapter 6). It also expanded to the East Coast. In 1912, shortly after the British had taken control of the former Thai provinces of Kelantan and Trengganu, SSS negotiated with the controlling Danish East Asiatic Company and the Thai Royal Family to take up an interest in the Siam Steam Navigation Company, which provided regular services between Singapore, the East Coast of Malaya, the South Coast of Thailand and Bangkok (Tregonning 1967: 40–1). After withdrawal of the NDL, SSS on behalf of Holts also placed its own steamers in the Singapore–Bangkok trade. In 1920 by acquisition of the large but motley fleet of the Chinese-owned Eastern Shipping Company of Penang, SSS extended to North Sumatra, Aceh, western Thai tin ports, and southern Burma. In the mid-1920s, it challenged the KPM with a line to Pontianak (West Kalimantan), leading to agreement over spheres of operation. The network as it existed in the 1930s is shown in Figure 3.4.

Holts' expansion in Southeast Asia brought British lines into conflict with each other, leading in May 1925 to the Victoria Point Agreement which permanently defined their rights in Asia (Tregonning 1967: 59–61). The key was the division between Holts/SSS and the British India Steam Navigation Company (BI), which in 1913 had merged with P&O. Since 1856, BI had controlled the coastal trades of India and the Bay of Bengal and from there expanded eastwards into the indentured labour traffic between India and Malaya and liner trades via Singapore to China/Japan and Australia. As SSS had moved into Burma, BI had moved into the Singapore–Bangkok trade, while P&O already ran two feeders between Singapore and North Sumatra. Victoria Point at the southernmost tip of Burma was acknowledged as the dividing line between the India/Burma sphere of BI and the Southeast Asian sphere of Holts/SSS but each company kept existing rights in the other's sphere. Other parties to the agreement were the Holts' affiliated China Navigation Company, whose lines extended from the China Coast to the Straits, Bangkok, Saigon, Philippines and Australia, and its main rival, Jardine Matheson's Indo-China Steam Navigation Company.

The KPM responded to Holts and Chinese rivals with a strategy of 'if you can't beat them, join them'. Because Chinese traders were unwilling to tranship on through bills of lading, the KPM would carry their rubber to Singapore. From 1907 small, shallow-draught, low-cost coasters had been commissioned for Sumatran river ports. The tactic was now to reduce

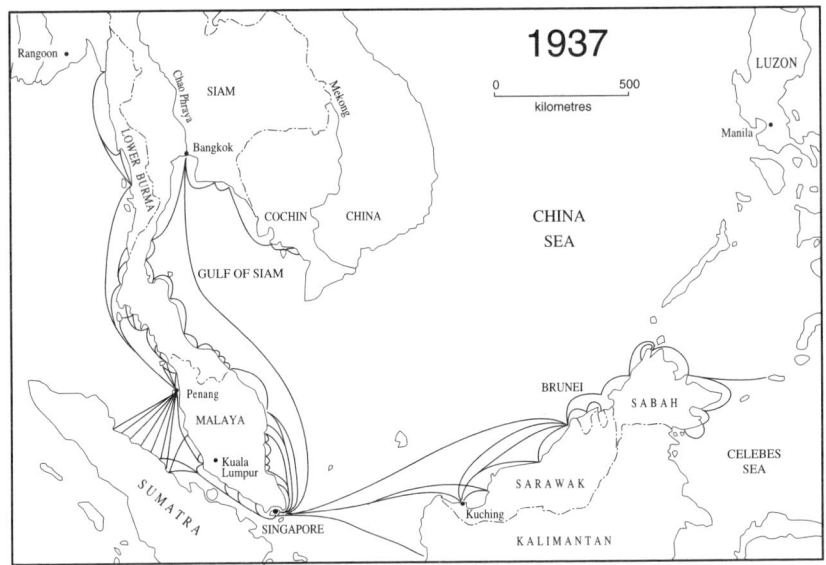

Figure 3.4 Straits Steamship Company: network, 1937 (*Source*: SSS 1938)

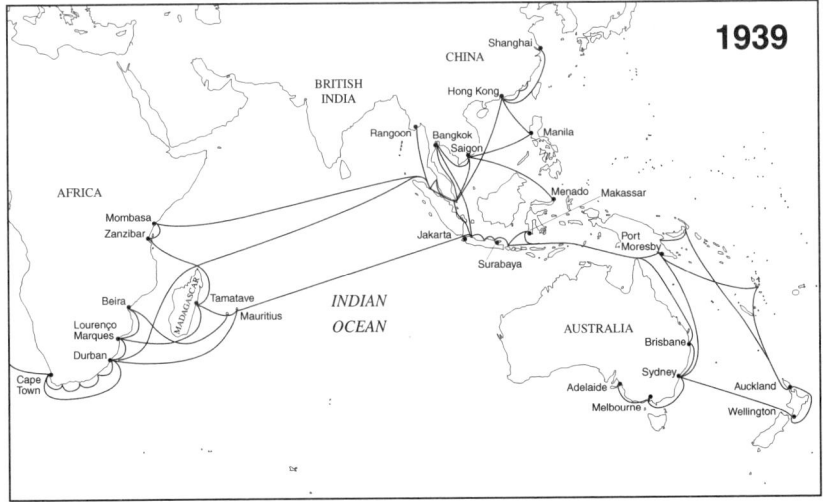

Figure 3.5 KPM: overseas lines, *c*.1939 (*Source*: Boer and Westermann 1941)

Chinese competition to a single firm in each trade and enforce a pool agreement at agreed tariffs. To all intents, Singapore became the KPM's fifth base port. From impressive adjacent offices, KPM and Holts managers looked out across the shipping in the roads, where as many small KPM ships could be seen as in Tanjung Priok or Surabaya. Singapore thus remained a transhipment port for Indonesia but only for the zone of the rubber trade extending from Belawan/Medan as far as Pontianak and Banjarmasin in West and South Kalimantan. The trade of Aceh continued to flow mainly to Penang, though on a much smaller scale.

At its peak in 1929 the KPM ran a vast network of Indonesian services, among which Singapore may be included as an extraterritorial port. This schedule was printed, like a train timetable, a year in advance and adhered to strictly. The scope of the KPM's operations in 1929, the second last year of the Great Archipelago contract, may be compared with that of the NISN at its peak in 1886 (Table 3.3). Capacity and number of passengers had both increased by an order of magnitude. Besides express, luxurious, white-hulled passenger liners on the trunk routes, 'tween-deck workhorses for deck passengers, livestock and cargo, and shallow-draft river boats, the fleet included freighters that carried bulk commodities like coal, salt and cased oil. Directly or through affiliates the KPM also had interests in stevedoring, a coal mine, coal bunkering, towage, shipyards, workshops and hotels (Boer and Westermann 1941).

The KPM's reach extended well beyond the Archipelago (Figure 3.5). Singapore had long been the distribution point for rice and sugar. In 1910 the KPM opened a direct Java–Siam Line, followed in 1915 by Deli–Rangoon, in 1928 Saigon–Java and in 1929 Saigon–Moluccas (Boer et al. 1994: 66). To supply Chinese indentured labour, in 1915 a Deli–Straits–South China Line was added at the request of Sumatran plantation interests. After 1908 there had been a monthly line between Java and Australia. In 1931 as a trade development initiative, a monthly line was opened from Indonesia to South Africa, later extended both to China and to South America. And this was only half the network. In 1902 the KPM and its principals had formed the Java–China–Japan Line (JCJL) as a separate company to open a subsidized direct line between Java, the Outer Islands and East Asia, hitherto served by transhipment through Singapore. KPM and JCJL in turn assisted their principals

Table 3.3 Netherlands Indies: scale of interisland operations, NISN (1886) and KPM (1929)

Company	Year	Ships	Tonnage	'000 miles	'000 pass.
NISN	1886	32	28,000	772	103
KPM	1929	137	268,000	4,495	1203

Source: Knaap (1989: 69, 71).

in the formation of the Java–Bengal Line (1906) and the Java–Pacific Line (1915) to the United States. By 1939 the KPM itself had become the largest Dutch shipping firm; with JCJL and its two principals it owned almost a million tons of shipping (Mulder 1991: 23). This huge shipping empire, financed and controlled from the Netherlands but increasingly managed from the Indies, had become a feat of technology and organization in which capital transcended space, tying the Netherlands Indies and the rest of Southeast Asia into the trade of six continents.

The Dutch achievement was all but matched by Holts and its affiliates. Holts' main lines reached from Europe to Southeast and East Asia and across the Pacific to North America, as well as to South Africa and Australia. Within Southeast Asia, Straits Steamships, subsidiaries and affiliates served Burma, Sumatra, Malaya, Kalimantan, Thailand and, after taking over the Chinese Ho Hong Steamship Company in 1932, also Hong Kong, South China and Rangoon. Holts' own feeders linked Singapore to North Sumatra and Western Australia while the affiliated China Navigation Company served Southeast Asia, Hong Kong, China, Japan and eastern Australia.

These maritime networks structured the commercial space of Southeast Asia and set the parameters by which its producers and consumers were incorporated into the world economy. All the lines and combines, British, Dutch, French, German, Danish and Italian, were tied together in carefully negotiated cartel (conference) agreements that set the level and structure of rates and often allocated the cargo route by route. While the mother countries of Britain, Netherlands and France still lacked an overall vision of Southeast Asia, the managers of these great lines, knowing in fine detail the origin–destination matrix of cargo and passengers, had in a practical way already constituted Southeast Asia and transcended it (Figure 3.6).

One consequence of this extraordinary structure was the almost complete suppression of local Chinese shipowning. British, German and Dutch shipowners had realised in the 1880s that in order to control foreign trade they had to control the local coasting trade. By the interwar years this became such an obsession to the KPM and SSS that a Chinese going about his business with one small ship was still seen as a competitive threat to the whole vast shipping empire. During the 1920s, when rubber prices were fairly high, Chinese shipowners could hold their own against fierce predatory tactics. Because of their weak capital base, however, the 1930s depression was a disaster. The last significant Chinese shipping firm in Dutch waters, the Heap Eng Moh Steamship Company, owned by the late 'sugar king' Oei Tiong Ham of Semarang but operating out of Singapore under the British flag, was secretly taken over by the KPM in 1928, but left under Chinese management, as also in 1936 was the Soon Bee Steamship Company (ÁRA.KPM: 908). SSS adopted the same strategy in 1932 when it took over the Ho Hong Steamship Company after the failure of the parent Ho Hong bank, thereby gaining entrée to the China–Straits and local

1938

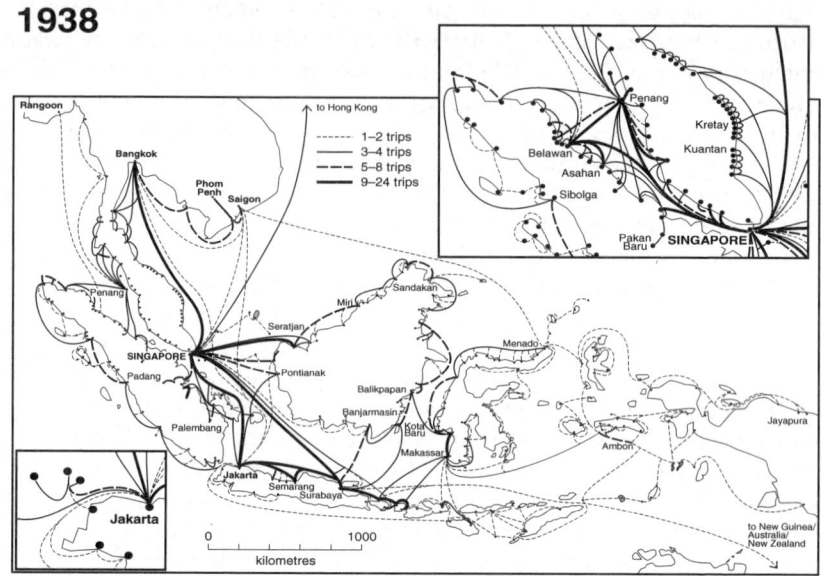

Figure 3.6 Southeast Asia: KPM and Straits Steamship Co. networks by frequency of sailings per month outward, 1938 (*Source*: Boer and Westermann 1941)

Sumatran trades (Tregonning 1967: 65–8). Chinese trade networks remained intact, but their control of shipping and the leverage that went with it was all but eliminated. By the late 1930s the local shipping of Singapore was completely under European control.

That left the KPM to worry about the pinpricks of competition from the 'mosquito fleet' of little wooden motor vessels and traditional sailing prahus, both of which enjoyed a revival during the 1930s depression (Dick 1987a,b). Under the revised 1936 shipping regulations, the KPM was able to introduce its own mosquito fleet by employing local officers on powered coasters smaller than 500m^3 (175 gross tons). Since 1922 SSS had taken advantage of a similar category under British jurisdiction to run 75-tonners (about 200 gross tons) with local manning in small trades around the Straits of Malacca (Tregonning 1967: 79–85).

Postwar reorientations (1945–1970s)

The Japanese occupation and its aftermath of decolonization broke the stranglehold of the two great shipping empires, although the Dutch combine was first to feel the impact. Loss of about half the KPM fleet was quickly made good but the economic and commercial environment could not be restored. As Indonesians fought for independence, the KPM and JCJL agreed on a new postwar strategy. On 1 July 1947 JCJL merged with the 'foreign' lines of the KPM to form a new entity, Koninklijke Java-China Paketvaart Lijnen trading as Royal Interocean Lines (RIL). Jakarta was ruled out as headquarters and, after some debate, Hong Kong was chosen in preference to Singapore (Boer 1994: 70). RIL was therefore not an Indonesian or even a Southeast Asian venture. The rump KPM confined its operations to Indonesian waters along with Singapore and the Malay Peninsula but was no longer strategically committed to Indonesia. After the transfer of sovereignty in December 1949, the KPM peremptorily rejected the new government's proposal for a joint venture on the model of Garuda/KLM airlines (Dick 1987a). Under a policy of disengagement (*omschakeling*), the fleet was to be kept employed in Indonesia as long and as profitably as possible, but investment was to be outside the country by building ships for management by RIL. Despite erosion of the former monopoly, deterioration of the Indonesian economy, and tetchy relations with government, the end came sooner than expected. On 3 December 1957 communist-led trade unions seized the KPM's offices, ships and installations in Indonesia. Except for residual lines from Singapore to Dutch New Guinea and East Timor, it was the end of the company's interisland operations. On 1 January 1967 the remnant KPM was merged into RIL, which in 1977 became part of the global Dutch entity Nedlloyd Lines (see below).[2]

Holts and SSS were under less competitive pressure to restructure their operations and, being firmly locked into international conference arrangements, had less scope to do so. The immediate change was the abandonment of passenger services on the Malayan coast, where SSS and Holts instead took the initiative of opening Malayan Airways in 1947 under a ten-year franchise (Tregonning 1967: 242–5; also Chapter 6). Freight transhipment through Singapore, including much illegal trade from Indonesia, remained buoyant until Indonesia's suspension of the trade during the Confrontation campaign (1963–66) against Malaysia. SSS then abandoned feeder services to Sumatra and Kalimantan, leaving Holts to carry on direct Dutch-flag liner services between Europe and Indonesia alongside Hapag-Lloyd (formerly NDL) and some new Indonesian firms. SSS and its Malaysian subsidiary (Perkapalan Kris Sdn Bhd) continued cargo-passenger services from the Peninsula and Singapore to East Malaysia (Sarawak and Sabah) until the business was sold to the Singaporean government's Keppel group in 1983.

As European shipping interests contracted, there was a dramatic revival in local Chinese shipping. Seeking to earn American dollars from the sale of rubber, in 1946 the British government in Singapore had provided local Chinese with wartime landing craft that could be converted to mercantile use for trading across to Sumatra and Kalimantan in defiance of the Dutch blockade against the Republik Indonesia. After the transfer of sovereignty in December 1949, these shipowners prospered from the Korean War rubber boom (Twang 1998). Buying secondhand ships, they became tough competitors in Sumatran and Kalimantan ports, where the gains from smuggling, often in cooperation with local military commanders, made shipment via Singapore more attractive than via Java ports. After 1958 some owners were able to buy laid-up KPM ships and put them back in service to Indonesia. Within a few years, Singaporean shipping was as dominant in Indonesian trades as before the KPM assault in the 1890s. Many principals had Indonesian-Chinese backgrounds and worked through relatives in Indonesia.

Within Indonesia, suspension of KPM sailings in December 1957 marked the end of the scheduled interisland shipping network (Figure 3.7). The largest surviving operator, the new state shipping corporation Pelni, was still too ill-equipped and inexperienced to take over the role of the KPM. Private companies proliferated but none had more than a few ships. Responsibility for coordinating liner shipping thus devolved upon a new Ministry of Shipping, later reconstituted as the Directorate-General of Sea Communications, which took over the stylish KPM head office in central Jakarta (Dick 1987a). Ships were allocated to routes in an extensive liner network but there was no enforcement: dispensations were just a supplement to the salary of civil servants. With much old tonnage hastily chartered from shipowners in Singapore and Hong Kong, capacity increased but utilization fell sharply. The worsening foreign exchange crisis inhibited the supply of spare parts and raised the cost of repairs.

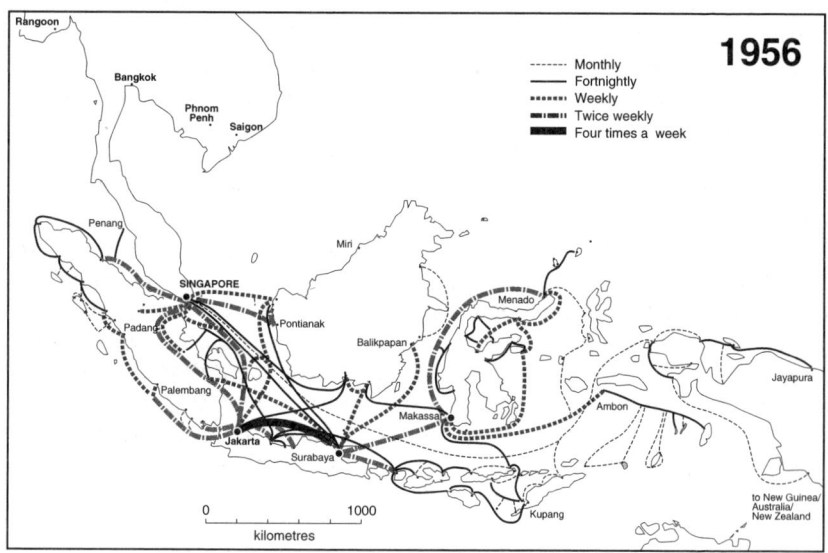

Figure 3.7 Indonesia: KPM interisland network, 1956 by frequency of vessels per month inward (*Source*: KPM 1955)

After 1966, the New Order government faced the daunting task of restoring the vital national infrastructure of interisland shipping. Under the umbrella of the Intergovernmental Group on Indonesia (IGGI), Dutch aid played an important role. The Netherlands Dredging & Engineering Company (NEDECO) was given the task of rehabilitating ports. Pelni was given an injection of funds and a team of foreign advisers with KPM backgrounds. Private shipping companies were allowed to utilize foreign exchange to import vessels or repair existing ones. The policy regime reverted to that of 1958, but now under a Directorate-General of Sea Communications. Assisted by a team of Dutch advisers under a former head of the KPM, in 1969 the Directorate-General introduced a revised liner network known as the Regular Liner System. The aim was by bureaucratic coordination to simulate an efficient KPM-like system from Pelni and approximately fifty small operators. Progress was made but in the long run the barriers to entry of new firms, capacity controls, route licensing and official tariffs entrenched inefficiency and perpetuated corruption (Dick 1987a).

The container revolution

Since its introduction to Southeast Asia in 1972, container shipping has had as dramatic an impact as steamships and the opening of the Suez Canal. The practice of pre-loading cargo in twenty-foot units (teu) for quick mechanical handling found favour among European and American shipowners in the 1960s as a means of speeding port turnaround and increasing profitability. Adoption on a large scale, however, required massive investment in new ships and equipment beyond the capacity of any one company. In the mid-1960s British lines combined into two large consortia to containerize the Europe–Australia trade, paving the way for the even larger investment in the Europe–Asia trade, for which old rivals Holts and P&O combined in the Overseas Containers Ltd (OCL) consortium (Jennings 1980: 60–3). To achieve the necessary economies of scale, it was decided that Singapore would be the only port of call in Southeast Asia, followed by Hong Kong and Japanese ports. Other ports were served by small feeder ships.

The overwhelming success of containerization led to more widespread changes than had been anticipated in the late 1960s. First, private shipowners gradually consolidated into national entities. The P&O group, like the Dutch Nedlloyd group, absorbed all its affiliates, including BI, and in 1986 bought out its main partner, Holts, which disappeared from the shipping industry (Falkus 1990). In 1998 the maritime world was amazed to learn that P&O had in turn merged with Nedlloyd to form P&O-Nedlloyd, a transnational entity like Royal Dutch Shell. The British and Dutch rivals, which had contested the Southeast Asian trades in the 1890s and divided it between them in the 1920s and 1930s, thereby merged into one great multinational firm. Old differences were buried because of competitive pressure from new entrants. In Southeast Asia there emerged the state enterprises Neptune

Orient Lines of Singapore and Malaysian International Shipping Corporation, while Hong Kong, Taiwan, South Korea and China also gave rise to aggressive, low-cost shipping lines willing to invest heavily in container shipping.

Secondly, containerization helped to reintegrate the archipelago around Singapore and, to a lesser extent, Hong Kong. The Port of Singapore Authority did not rest on its laurels but continued to invest heavily in expansion and increased productivity. By 2000 throughput was 17 million teu (CIY 2001). This involved huge economies of both scale and scope. Before containerization, cargo from elsewhere in Southeast Asia was unloaded, removed to a trader's godown, and then be repacked with other consignments for shipment to the port of destination. Now the 'box' does not leave the wharf, being discharged from a feeder vessel, stacked, and reloaded onto a mother ship within 48 hours. At the intersection of shipping routes to every part of the world, Singapore offered transhipment any day to almost any main port. Mother ships as large as 7000 teu now provide throughshipment via Singapore more cheaply than direct shipment from ports as far away as Sydney and Melbourne. Singapore's maritime hinterland therefore now extends well beyond the archipelago.

Thirdly, containerization eventually undermined Indonesia's protectionist and highly regulated maritime regime. Until 1985 the Directorate-General of Sea Communications had restricted transhipment of containers between Java and Singapore to one state-owned enterprise. Ignoring the technological tide, the Directorate-General formalised a 'gateway policy', whereby Indonesia's liner exports could be shipped only through the ports of Belawan, Jakarta, Surabaya and Makassar (Dick 1987a). This was even more restrictive than Dutch policy between the wars. It protected Indonesian-flag operators of conventional ships, but at the expense of Indonesian exporters and importers, who had to pay higher freight for inferior service. Conventional ships would spend several weeks loading cargo piece-by-piece, port-by-port in Indonesia and then incur high costs for break-bulk discharge in European or North American ports.

The non-oil export drive which followed the collapse of oil prices in the early 1980s overturned this outmoded policy. Presidential Decree No. 4 of 1985 also allowed freedom of shipping to and from Indonesian ports. For the first time since the 1890s, Indonesian trade was free to flow in least-cost channels, which for the most part meant via Singapore. Feeder vessels were immediately placed in service between Singapore and Indonesian ports. Access by Indonesian shippers to very low deepsea freight rates and frequent and reliable sailings via Singapore was a significant element in the subsequent export boom. Three years after shippers had been allowed to tranship via Singapore, bypassing high-cost and unreliable interisland shipping, the decision was taken to reduce the regulatory and cost burden on interisland shipping by abolishing route allocation and tariff controls, freeing shipping firms to operate according to commercial considerations. To provide a network of fast interisland links, Pelni was assisted by government funding to acquire a fleet of modern express passenger liners that now serve a network not so different from that of the former KPM (Figures 3.8, 3.9). Domestic non-bulk cargo shipping, however, has languished. Under a maritime regime which is now less highly regulated but still internationally uncompetitive, leading Indonesian shipping firms prefer to be based in Singapore.

As the Indonesian government relaxed its restrictions on shipping and trade, an increasingly nationalistic Malaysian government raised theirs. At stake was firstly the transhipment of Malaysia's foreign trade through Singapore and, secondly, the role of Singapore in the 'domestic' trade between

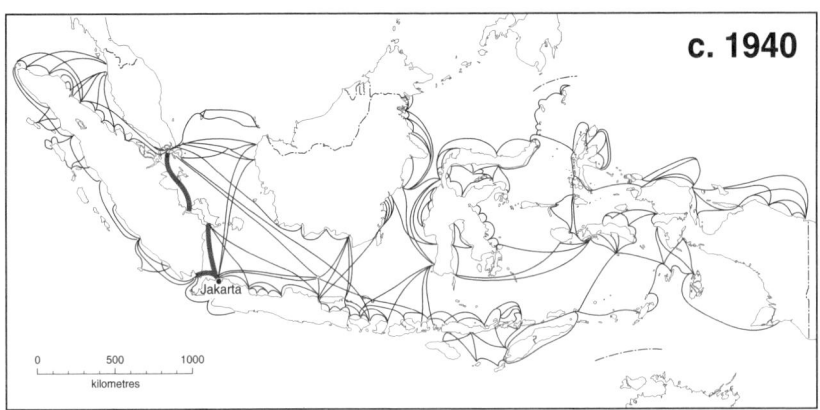

Figure 3.8 Indonesia: KPM interisland shipping routes, c.1940 (*Source*: Boer and Westermann 1941)

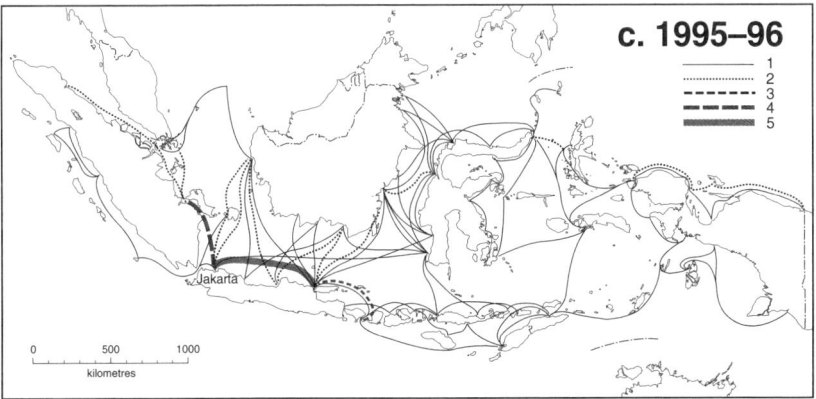

Figure 3.9 Indonesia: Pelni interisland passenger shipping network, 1995–96 (*Source*: Pelni)

West and East Malaysia. Port Klang (formerly Port Swettenham) had long been a deepsea port of call for rubber, tin and palm oil, but after containerization in 1971 most of that trade was drawn off in transhipment via Singapore at cheaper through rates. For nationalist reasons, in the 1990s the Malaysian government aggressively promoted Port Klang as a rival international container port, while the Thai government developed Laem Chabang on the Eastern Seaboard. Fierce competition between rival container consortia has lead to more frequent direct calls at these secondary ports, but Singapore still dominates. In 2000 the container throughput of Hong Kong (18 million teu) and Singapore (17 million) dwarfed that of Jakarta (3.3 million), Port Klang (3.2 million) and Laem Chabang (2.2 million) (BT 14 March 2001).

The current trend towards global logistics reinforces Singapore's competitive advantage, despite its high cost structure. What now matters is not just physical port facilities but the interface with warehousing, insurance, finance, air transport and telecommunications, all integrated by high-speed information flows. Mercantilist port policies are overwhelmed by sheer force of self-interest. Even the rapid development of the new Malaysian port of Tanjung Pelepas, which in 2001 handled over two million teu, confirms this (BT 4 January 2002). Located opposite Singapore on Johore Strait, it is a competing terminal but part of a common multinational hub, like Singaporean-owned shipyards which have relocated onto adjacent islands in Indonesia's province of Riau. Singapore is too small for its future to be other than multinational (see Chapter 7).

The Philippine Archipelago

The Spanish period (to 1898)

Philippine ports were late to be opened to international trade. Manila was not opened until 1834, well after the establishment of Singapore. The main regional ports of Iloilo and Zamboanga (1855) and Cebu (1860) were opened in the same decade as Siam and Japan. New export staples then began to emerge, of which the most notable was Manila hemp or abaca, grown mainly in the Bicol district of the southern Luzon peninsula (Owen 1984). In the Visayas, sugar production developed on the island of Negros across the strait from Iloilo. Development was nevertheless seriously held back by systematic raiding by Muslim pirate fleets from southern Mindanao and Sulu. Coastal settlements lived under continual threat of depradation. Not until the introduction of shallow-draft steam gunboats in 1861 did life and property become secure (Warren 1981).

Until the 1860s communications between the islands was still by sailing ship. During the northeast monsoon (November–March) the journey from Manila to Iloilo in the middle of the archipelago took between four and six days by square-rigged ship, and ten to 15 days in the other direction

(MacMicking 1967: 240). A month could elapse without news from Manila. Ports along the exposed north coast of Luzon closed during the stormy months of September to November, a problem not overcome by the first under-powered steamers (MacMicking 1967: 240; Man 1984: 96–110).

Private commercial steam shipping developed much more slowly than in the Netherlands Indies. From 1854 the P&O mails were collected in Hong Kong by a naval auxiliary but the Spanish government's offer of a modest annual subsidy to carry them beyond Manila was not taken up (Bowring 1859: 303). In 1861 a small wooden steamer briefly offered an irregular service between Manila and Iloilo but interisland steam shipping is better dated from 1868, when a British entrepreneur, Mr Neil McLeod, introduced an iron-hulled steamer that apparently paid for herself in the first year (Philcom 1900, II: 40). In 1872 steamers began to carry abaca from Bicol and by 1881 had captured half the trade (Owen 1984: 97). Principals were mainly British and American trading houses, but ships were registered in the name of associates, relatives or employees in order to fly the Spanish flag and meet cabotage law. These firms carried their own cargoes at their own convenience and only any surplus of capacity was offered to the market. Passengers and private mails had to wait for the first available steamer and put up with whatever conditions they found on board (Worcester 1898: 315–17).

To establish a reliable common carrier, in 1890 the McLeod interests formed Compania Maritima as a specialist shipping company and contracted under subsidy for several mail lines. This led to fierce competition with the shipping of other trading houses (Philcom 1900, II: 296). In 1894 commercial wisdom prevailed and the capital of Compania Maritima was increased to $2.5 million to incorporate its main Manila-based rivals and win a revised mail contract, which took effect in 1897 (Philcom 1900, II: 290). By 1901 the company had a fleet of 19 steamers of 13,400 tons, about the same tonnage as the original KPM fleet. The Spanish corporation Tabacalera and various small firms remained independent. Unfortunately this new liner system had scarcely got into its stride before the Filipino revolution broke with full force, soon to be followed by the American conquest.

The American period (1898–1946)

When the Americans burst upon the sleepy Philippines, shipping was the one field of transport that was modern and fairly well organized. Ironically their military occupation was highly disruptive. The Americans brought their own military transports, the Army Quartermaster chartered in as needed for government cargoes, and the Insular Purchasing Agent managed local distribution. In October 1901 under Act 266 a coastguard service was set up with responsibility not only to maintain lighthouses and prevent smuggling but also to carry official mails, passengers and cargo and supplementary private traffic to provincial outports. For these purposes a fleet of 14 wooden-hulled steam cutters was built in Shanghai. The private sector was all but ignored. Common carrier mail contracts and regular schedules lapsed, productivity declined, and the demand for tonnage rose sharply. Shipping was nevertheless highly profitable until November 1902, when an act was introduced to prevent price-fixing and price discrimination (USNA, BIA 674: 863/49). It also gave dispensation to foreign-flag ships to run in interisland trade. By 1903 there was overcapacity, rates fell, and ships were laid up.

In 1903 the government appointed a Committee on Coastwise Laws to inquire into the working of the Customs laws and regulations upon interisland shipping. A degree of paranoia was still manifest: 'by far the greater majority of these vessels are officered and manned by persons other than citizens of the US, and [...] the [Customs] supervision now being had over these vessels is necessary for the best interests of the Insular Government' (USNA, BIA 674: 863/49: 6–7). Nevertheless, the knowledge which the Committee gained of the private sector paved the way for a more commercially oriented approach. Cabotage was applied more strictly after July 1904 by ending dispensations for foreign-flag ships and withdrawing chartered US Army ships. In March 1905 Act 1310 was passed 'to encourage and aid the Philippine coastwise trade and to secure the carriage of mails, government freight and passengers by commercial vessels under contract, to effect uniform reasonable rates for the Government and the Public, and to increase the safety standards and service of contracting vessels'. A Superintendent of Interisland Transport took over the functions of the Purchasing Agent and called tenders for 21 common carrier routes for mails, government passengers and freight, and common carrier service (USNA, BIA 674: 12957/2, 27). Eleven routes were let in August 1906, leaving the Coastguard to serve routes where there was no interest from private common carriers. By trial and error the Americans thus found their way back to a system rather like that which the Spanish government had established in 1894. Unlike the Netherlands Indies, however, there was no single dominant firm.

While order was restored, in the long run this regime became a source of great inefficiency. A contracted carrier enjoyed a franchised monopoly of its route, since new entrants required a 'certificate of public convenience' before they could enter the trade. Monitoring was placed in the hands of a tribunal, including the Collector of Customs and a shipowner representative, to deal with any complaints or requests for variations of contract. In October 1907 these powers passed to a Board of Rate Regulation, which in turn became the Board of Public Utility Commissioners. While the Boards devoted much time and wisdom to the minutiae of tariff setting, the interisland shipping industry stagnated. Franchises allowed companies to occupy valuable routes with old and antiquated ships without threat of competition (USNA, BIA 674: 2893/40).

Nor was this the worst of the government's meddling. By Act 3084 of 1923 the Philippine legislature decreed that 'foreign nationals' – meaning Spanish and Chinese – would no longer be allowed to replace ships in service as of 1918 or to add to their number. American firms showed no interest in investing in Philippine shipping except as ancillary to some industries such as forestry. Protected by their friends in the legislature, Filipino shipowners had no incentive to do so. A more stifling constraint could hardly be imagined. The life of many ancient ships was thereby prolonged and shippers suffered badly.

The one encouraging trend was the diffusion of the marine motor, which greatly improved the efficiency of small ships of less than about 400 tons. Small steamers were costly to operate because so much space was taken up by boilers and bunkers, besides which most coal had to be imported. Oil engines were ideal for small wooden vessels that could be built cheaply in the Philippines from local timber. By 1914 such engines were being advertised for sale in Manila, as also were small outboard motors (MT 26/7/14). Motor vessels began to be registered in large numbers during and after the First World War and, especially in the Visayas and around Mindanao, brought marked improvement in the regularity and cost of transport, also reducing the need to subsidize larger steamships to serve small settlements. Nevertheless, sail shipping retained a niche, though a diminishing one. In 1930 unpowered craft accounted for more than half all registered tonnage and still dominated bay and river shipping, including lighterage (Table 3.4).

The poor condition of the shipping fleet was a scandal. When the cutter *Negros* sank on 26 May 1927 with loss of 100 lives, media and public attention at last focused on the lack of safety, high rates and poor service of the interisland fleet. The resultant committee of inquiry drew scathing submissions (PIGG 1927). The Davao Chamber of Commerce described interisland ships as, with few exceptions, small, dirty and without suitable

Table 3.4 Philippines: shipping fleet by number and net tonnage, 1930

	Coastal		Bay/River		Total	
	no.	tons	no.	tons	no.	tons
Steamers	93	34,600	1	40	94	34,640
Motorized	130	14,150	18	430	148	14,580
Launches	199	3,020	369	3,120	568	6,140
Powered	*422*	*51,770*	*388*	*3,590*	*810*	*55,360*
Sail	170	20,550	918	38,700	1,088	59,250
Misc.[1]	1,111	23,950	583	5,000	1,694	28,950
Unpowered	*1,281*	*44,500*	*1,501*	*43,700*	*2,782*	*88,200*
Total	1,703	96,260	1,889	47,300	3,592	143,560
%		67		33		100

Note: 1. Lorchas, lighters, barges, cascoes, etc.
Source: PIICC (1931).

passenger accommodation, so that passengers had to ride with cattle, pigs and chickens and often atop a deck load of gasoline and case oil; they were generally overloaded and often refused local cargo. Hardly the KPM! Comprehensively damning was a report of the Spanish Chamber of Commerce of the Philippines. High costs were attributed to ships claimed on average to be forty years old, partly attributable to the replacement ban on 'foreign' shipowners. Repairs were costly because of a 50 per cent duty on repairs outside the Philippines. Another problem was overmanning, partly because of the regulations and partly because ship's labour had often to be used to handle cargo in outports. Finally, cargo was delayed by acute port congestion, especially in the Pasig River in Manila.

The committee reported other alarming features that were obviously common knowledge and may still be observed today. First, not only were companies in the habit of selling more tickets than allowed under the ship's certificate, but crew then issued their own tickets to passengers who were never listed. An example was cited of one ship that carried 600–700 persons with lifeboats and rafts for only 200. On most ships the lifesaving equipment was never tested and usually unserviceable. Besides overcrowding there were poor toilet facilities, insanitary conditions, and indifferent food. Overcrowding and overloading of deck cargo exacerbated the instability of many vessels caused by the addition of upper passenger decks and removal of ballast.

After weighing the evidence, the Committee wisely decided that the underlying cause was excessive regulation. Two bills were recommended. The first, to remove the prohibition on replacement of foreign-owned ships, was narrowly defeated in the legislature (PIGG 1927: 21). The second, to set aside regulation by the Public Service Commission, except in the setting of maximum fares and rates, took effect from December 1927. Henceforth each shipowner could decide what ships he wished to employ, on which routes, and, subject to the maximum fares and freights, how much to charge. Other recommendations were more rigorous seaworthiness and sanitary inspection, removal of the impost on overseas repairs, reintroduction of subsidies, and full payment for carriage of mails (PIGG 1927: 104). This partial deregulation led to gradual rehabilitation as owners began to order new tonnage. There were also new entrants. Express passenger liners, new and secondhand, were added to the main lines, so that by the late 1930s it was at last possible to travel interisland at speed and in comfort. By 1937 there were also two or three flights a week to Cebu and Iloilo and once a week on to Davao, though in 1939 these small planes carried only 22,000 commercial passengers within or beyond Luzon (PIHC 1939: 113; 1943: 88).

Just as the Spanish mail network had functioned only briefly before the American conquest, so the fine fleet of the 1930s came to a premature end. Almost the entire fleet was destroyed between 1941 and 1945 (Wernstedt

1957: 17). Some ships were sunk by Japanese action; many were scuttled at Manila in December 1941; those salvaged were later sunk by American forces. As islands reverted to autarchy, spasmodic interisland shipping devolved upon sailing craft (Hartendorp 1958).

Independence

Physical capacity was restored almost immediately by American largesse in the form of an emergency fleet of warbuilt ex-military supply ships (Wernstedt 1957: 19–22; Hartendorp 1958: 198–9). Over time and with great ingenuity these small freighters were lengthened and reconstructed with spartan passenger decks. They remained the backbone of the interisland fleet until secondhand Japanese ferries became available in the 1960s. On the operational side, however, most of the problems of in the 1920s now reappeared. After independence in 1946 the Philippine government reverted to the form of regulation practised before 1927. Passenger fares and freight rates were set by the quasi-judicial Public Utilities Commission/Board of Transportation, which also reinstituted route and capacity licensing. Along with import controls on ships, the predictable effect was to deter innovation, allow inefficiency, and ignore the worsening seaworthiness of an aging fleet.

In June 1974 during the Marcos years, regulation of the maritime sector was devolved to a new body, the Maritime Industry Authority (Marina), which exercised a similar panoply of powers to Indonesia's Directorate-General of Sea Communications and was to some extent modelled upon it. The new authority took over responsibility for both interisland and deepsea shipping, the shipbuilding and repair industry and manpower training (Marina 1994). Marina's immediate task was to overcome the crisis of the aging interisland fleet. A US$20 million rehabilitation loan from the World Bank, parallel with that provided to Indonesia, was supported by exemptions from import duties and taxes on imported ships. In 1985 the capacity-licensing powers of the Board of Transportation were devolved to Marina, and two years later registration, licensing, safety and manning regulation that since 1901 had been carried out by the Philippine Coastguard.

Deregulation

Despite retonnaging, seaworthiness and safety continued to be observed mainly in the breach. Secondhand ships were rebuilt to carry the maximum number of passengers, most without cabins, in rows of double bunks both below and above deck. Overbooking was common and tickets were sold freely aboard ship, leading to gross overloading. In December 1987 the *Dona Paz* with over 3000 passengers collided at night with a coastal tanker: the sea itself caught fire and only 26 were saved (MN 1988: 114, 243). This was the world's worst peacetime maritime disaster, far worse than the *Titanic*.

As in 1927, heavy loss of life led to a review of interisland shipping. In 1989 the Presidential Task Force on Interisland Shipping recommended deregulation of route and capacity licensing. This became policy in 1992, leading to freedom of entry and greater commercial autonomy in setting fares and freight. Competition led to the disappearance of some long-established operators and the merger of others. Survivors invested vigorously in new tonnage, including much larger, faster and more sophisticated ships. The secondhand Japanese ferries that had replaced warbuilt tonnage in the 1970s but since been much knocked about were in turn replaced by larger and faster vehicular ferries, now obsolete in Japan, that could carry cars and lorries beneath the passenger decks. Vehicular ferries were supplemented by container carriers, which by the 1990s had become standard in the trade between Manila and Mindanao.[3] Most interisland general cargo, even bananas, cattle and pigs, is now carried in some form of container, as also is international transhipment cargo to and from Philippine outports.

Another trend has been the differentiation of general and bulk cargo. Until the 1960s virtually all cargo was handled in assorted pieces, whether in wooden crates (groceries, textiles), bags (rice, sugar), drums (oil products, coconut oil), bunches (bananas) or loose stow (abaca, coconuts or copra chips, rattan). Ship's derricks were sometimes used but most cargo was manhandled up and down the gangplank on men's shoulders, as still in outports on smaller vessels. With industrialization, however, materials such as cement, fertilizer, grain, oil products and chemicals were handled more economically by the shipload, often at private ports with dedicated mechanical terminals. Secondhand coastal bulk carriers and tankers became available at secondhand prices from Japan, where they were as common as trucks. Including international trade, bulk cargoes now account for over half the country's entire volume of shipments, the most important component being crude oil and oil products to and from the refineries at Bataan and Batangas. The same trend was apparent in Indonesia.

Despite the enhanced vitality of the corporate sector, the Philippine interisland fleet is still amorphous. In 1994 of the 4,650 vessels in commercial service (excluding the huge fishing fleet), 87 per cent were of less than 500 gross tons and 61 per cent of less than 100 tons; the average size of 292 tons was that of a respectable harbour ferry; 59 per cent of the fleet – mostly the smaller vessels – were wooden (Marina 1996). These figures are consistent with the bustle that may be observed in any small port around the Philippines, at least in the sheltered waters to the south of Manila. Wherever there is water, there are boats, if only dug-out canoes (*banca*) with an outboard motor. Even more so than Indonesia, the Philippines is a society that lives not just in proximity to but also in association with the sea.

National economic integration

The Philippines and Indonesia were not natural or pre-existing national units. Having been consolidated as territories by colonial conquest, they achieved independence in the 1940s as nation-states with democratic institutions, a central bureaucracy and a common currency, macroeconomic policy and trade regime. Integrated economies with intra-national specialization and trade took much longer to emerge. Wernstedt and Spencer (1967: 131–2) observed of the American period that 'little was done to integrate the island world into one regional system'; Paauw (1963) made a similar observation about colonial Indonesia. Integration has been a fairly recent process and the resultant economic structures of the Philippines and Indonesia look quite different.

Philippines

The twin poles of the Philippine Archipelago have long been Manila and Cebu. In 1565 the Spanish made their first stronghold at Cebu, where Magellan had met his death, and had their prime interest been the commerce of the archipelago, this central location might well have remained their base. The lure of the China trade and Manila's grand natural harbour dictated otherwise. After 1571 this brought about a bifurcation between foreign trade based on Manila and domestic trade focused on Cebu. Except in the suppression of piracy, the Spanish showed remarkably little interest in the rest of the archipelago. Despite the southern outpost of Zamboanga, most of Mindanao remained jungle until the American period, as did the western island of Palawan. The southwestern islands of Muslim Sulu, notorious as the nest of pirates, continued to resist foreign encroachment and to trade autonomously with Borneo and Singapore (Warren 1981). As late as 1906 Manila's domestic shipping movements were still primarily around the coast of Luzon (Figure 3.10). Economic integration with the rest of the archipelago was minimal.

Manila's dominant position as the main port for foreign trade increased under American rule. By 1900 the United States already absorbed 11 per cent of Philippine foreign trade but most passengers and cargo were still handled by transhipment through Hong Kong. To counter British influence, the Payne–Aldrich Act of 1909 offered duty-free entry to goods carried directly to or from the United States. Sugar exports, by tonnage the main bulk cargo, benefited greatly. As in the Netherlands Indies, the redirection of trade was facilitated by massive port works, especially in Manila. Like most Southeast Asian ports, Manila had been a roadstead. Coastal vessels could berth in the Pasig River but deepsea ships had to anchor offshore and handle cargo by lighters. During the 1900s the colonial government gave high priority to constructing piers to berth the largest deepsea ships. Between 1901 and 1932 Manila absorbed almost two-thirds of all port investment, with most of the

108 Hinterlands

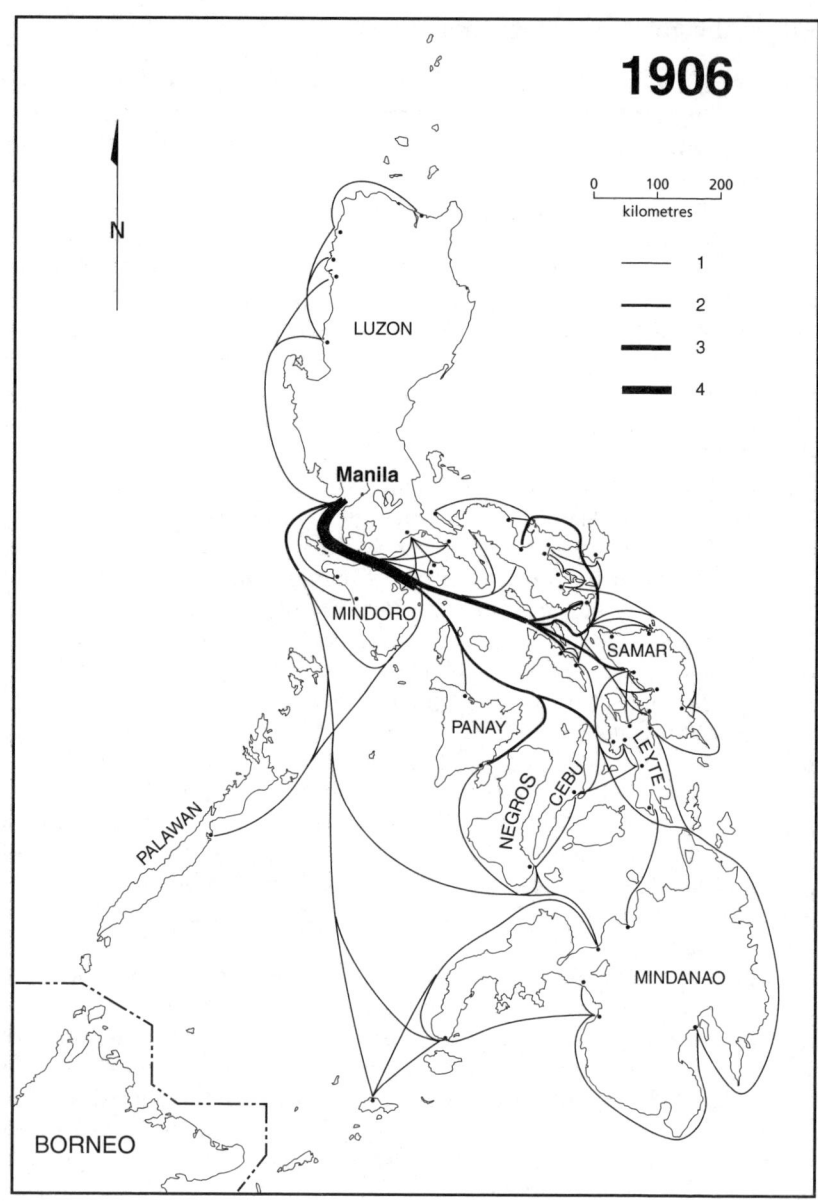

Figure 3.10 Port of Manila: inbound interisland shipping routes and frequencies, 1906 (*Source*: Based on Wernstedt 1957: 15)

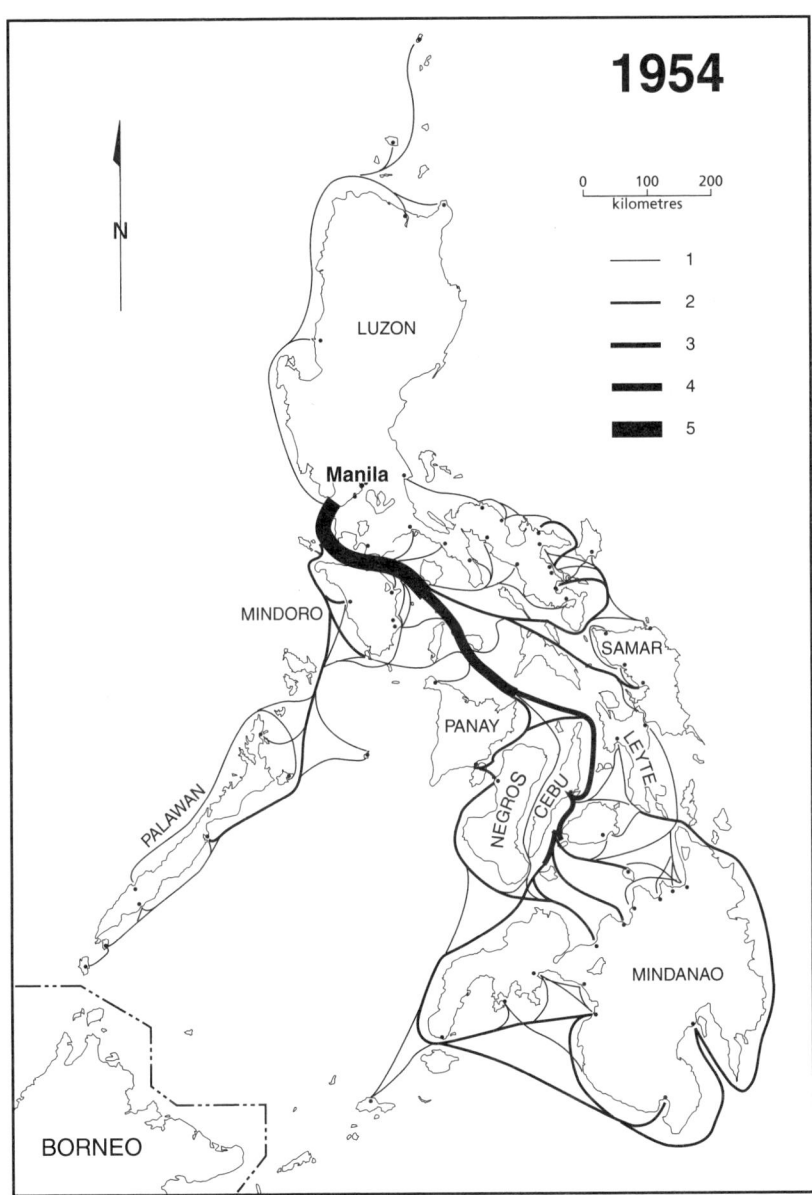

Figure 3.11 Port of Manila: inbound interisland shipping routes and frequencies, 16 May–15 June 1954 (*Source*: Based on Wernstedt 1957: 54)

balance spent on Cebu and Iloilo (Cavender 1933: 8). These improvements quickly attracted direct calls at Manila by American, British, German and Japanese shipping lines at the expense of transhipment between Manila and Hong Kong. By 1930 Manila was a port of call for 45 deepsea lines and accounted for around 80 per cent of customs receipts (MHB 1930; PIICC 1931). Local merchants still dreamed of Manila becoming an entrepot to rival Hong Kong, waxing lyrical over the 'Pearl of the Orient'. In fact Manila's port costs were two to three times more expensive and it had no prospect of becoming a free port. It did, however, very effectively channel trade towards the United States, whose proportion reached 72 per cent (PIICC 1931).

Domestic trade focused on Cebu and Iloilo. In 1930 these busy hubs shared 49 per cent of interisland ship movements by tonnage compared with 26 per cent for Manila; by number of movements the figures were 70 and 16 per cent respectively (PIBCI c.1932). Iloilo developed by linkages with the sugar industry of the adjacent island of Negros (McCoy 1982). Cebu fed off the development of the frontier economy of Mindanao, which American forces conquered in the 1900s and gradually brought under civilian administration (Tan 1967). Mindanao's principal industry was timber but high wartime commodity prices after 1914 and better transport stimulated immigration from the Visayas and the opening up of land for copra, hemp, corn and cattle, while around Davao Japanese settlers began to develop a substantial presence (PIGG 1918: 91–3; Beckett 1982; Edgerton 1982).

In the mid-1950s the American geographer Wernstedt carried out the first detailed empirical study of Philippine interisland shipping and trade, which remains an invaluable baseline data source. Figure 3.11 gives an overview of the interisland trade pattern. The main axis lay between Manila and Cebu. Manila had the largest hinterland, consisting of its own island of Luzon, Mindoro and the remote and sparsely populated island of Palawan that reached almost to the border with then British North Borneo (Wernstedt 1957: Map 46). Being so mountainous, Cebu looked outwards to a maritime hinterland, beyond the neighbouring island of Bohol to Samar and Leyte but also mediating traffic to ports as far as eastern Negros and northern Mindanao (Figure 3.12). Most shipping bound for Zamboanga and Davao passed through the port of Cebu, decades of emigration from Cebu and Bohol had generated a lively intercourse of passengers, and much of the corn to feed the barren island of Cebu was now drawn from there. Iloilo had once ranked ahead of Cebu in number of movements but its traffic declined in the 1930s when the Negros sugar mills switched from lighterage to direct shipment (McCoy 1982). Iloilo's hinterland had contracted to its own island of Panay and western Negros.

Since the 1950s, the main axis of the Philippine economy has extended from Manila–Cebu to Manila–Cebu–Mindanao. The frontier economy of Mindanao developed as abundant cheap labour from the Visayas and capital

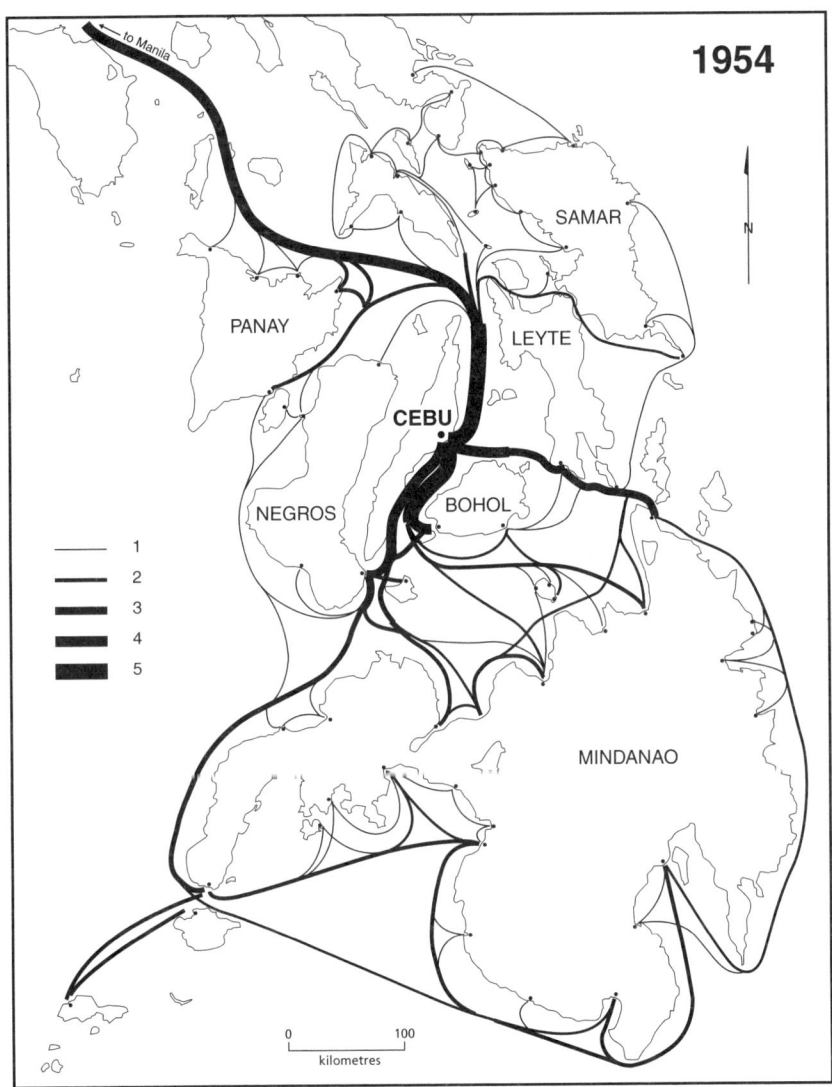

Figure 3.12 Port of Cebu: inbound interisland shipping routes and frequencies for 31-day period 16 May–15 June 1954 (*Source*: Based on Wernstedt 1957: 40)

from Manila and overseas flowed into resource-based exports. Mindanao's share of national population increased from about 8 per cent in 1900 to 22 per cent in the 1990s (PSYB 1995). Progress would have been even faster but for the insecurity caused since 1972 by the Muslim rebellion in the hinterland around Zamboanga and Cotabato.

In the mid-1980s Cebu's economy also began to boom after several decades of stagnation. The opening of the Mactan Export Processing Zone in 1986 and the associated international airport and container port stimulated labour-intensive industrialization. By the mid-1990s Cebu had basic links into international networks through direct feeder shipping links to Hong Kong, Kaohsiung and Singapore and direct daily flights to Singapore, Hong Kong and Japan and several times per week to Kuala Lumpur (Cebu 1996). Domestic connections included 16 daily flights to Manila, two to Davao and one each to General Santos (Mindanao), Iloilo and Bacolod (Negros). Manufactured exports, tourism and associated business thus enabled Cebu to re-establish itself as a second city, albeit still in the shadow of Manila. Yet Cebu barely registers on a world scale: in 1999 it handled only 0.4 million international airline passengers, 0.4 million teu and just 60,000 air freight tonnes (CIY 2001; www.cebu-airport.de.tt).

Reflecting Cebu's central location, the Visayas may be regarded from a national perspective as a set of inner rather than outer islands. In the case of passenger movements by seaport, in 1994 Luzon ports (mainly Manila) accounted for less than one-fifth of the total, the balance being shared between the Visayas and adjacent Mindanao (Table 3.5). Of interisland cargo by tonnage and also domestic container movements, the Visayas and Mindanao together shipped just over half. Only in international container movements did Luzon (Manila) dominate, with 88 per cent of the total. Thus Manila continues to be the main international gateway but the Visayas – and especially Cebu – play a central role in the domestic circulation of people and commodities.

Roads also play a role in economic integration. A 1930 set of road maps showed each island as a distinct entity: not one had a well-articulated road system (Bachrach 1930). Luzon had the densest network but the southern Bicol peninsula was still unconnected by road or rail (Chapter 5). Mindanao's apparent network was not much more than a blueprint. The first national highway programme followed self-government in 1935 and was the brainchild of President Quezon, who sought to open new land for agricultural settlement and thereby better distribute population across the archipelago.

Table 3.5 Philippines: cargo (non-oil) and passengers shipped by region, 1994 (percentage)

Region	Passengers (domestic)	Cargo (domestic)	Containers (domestic)	Containers (international)
Luzon	19	47	45	88
Visayas	44	31	27	6
Mindanao	37	22	28	6
Total	100	100	100	100

Source: Calculated from Philippines Ports Authority (1995).

For lack of funds, however, little was achieved before the Japanese invasion (PBPW 1950: 28–30, 37). In the 1970s under the Marcos administration a national highway system came closer to realization with construction of the Pan-Philippine (Maharlika) Highway. Funded under Japanese aid, the two-lane highway stretched 2,000 miles from Aparri in North Luzon to Davao, Cotabato and Zamboanga in southern Mindanao, with ferry connections between Luzon and Samar, a bridge between Samar and Leyte, and another ferry from Leyte to the northern tip of Mindanao (PA 1973, 1975). Yet the symbolism of national unity was more apparent than transport utility. Even now there is still no demand for people or goods to travel by road from one end of the Philippines to the other. Demand is very uneven along the network. Busy sections are badly damaged by heavy vehicles, especially after typhoons and heavy rain. In Mindanao there is threat of guerilla action. The National Highway is not yet a safe and well-maintained artery.

Nevertheless, there are signs that the Philippine Archipelago is at last knitting together as a single economy. Vehicular ferries have become the norm for both short ferry crossings and trunk routes. For those able to pay the premium fare, jetfoils and high-speed catamarans have reduced journey times from hours to minutes. Terminals are being relocated to shorten these road–sea–road links. As highways are upgraded, the distinct circuits of land and sea transport will increasingly merge. Ultimately the Philippines' distinct archipelagic economic geography may all but disappear, as has been achieved in Denmark and Japan.

Development of the Visayas and Mindanao is therefore leading to a more balanced and integrated domestic economy. This can be seen from figures on regional domestic product. By the mid-1990s the economy of the Philippines had three roughly equal parts: the capital city region of Manila, its immediate hinterland of Luzon (excepting the southern peninsula of Bicol), and the rest of the archipelago (PSYB 1995). The main island of Luzon (30 per cent) and the 'other islands' (37 per cent) – dividing almost equally into the Visayas and Mindanao – were therefore fairly equal tributaries to Metro Manila.

Indonesia[4]

The Netherlands Indies was a patchwork, not a unitary state. In the 1890s and 1900s through a mix of negotiation and conquest the Outer Islands were brought into a framework of Dutch rule. This included territorial administration, a legal code, a common currency and scheduled steamship connections with Java. Nevertheless, administration was mostly by indirect rule and, as already seen, commercial orientation was only gradually turned away from Singapore. This was superficial integration. Most districts remained export–import economies, having little relationship with adjacent ports or with Java. Interisland trade with Java was only a very small proportion of total trade (Table 3.6).

Table 3.6 Indonesia: ratio of interisland to foreign trade, 1914–1939 (f million) and 1972 (Rp. billion) (%)

Year	Interisland trade[1]	Foreign trade	Ratio (%)
1914	62	1,114	5.5
1921	247	2,440	10
1929	310	2,593	12
1939	211	1,291	17
1972	372	1,422	26

Note: 1. Between Java and Outer Islands only.
Source: Korthals Altes (1991); BPS, Statistical Pocketbook; Rosendale (1981).

The potential for specialization and trade between labour-abundant Java and the resource-rich Outer Islands could not be realized as long as the export boom allowed manufactures to be imported more cheaply than they could be produced on Java. This situation changed temporarily during the First World War, when normal trade was interrupted, and permanently after 1933 when, in response to the world depression and the pressure of cheap Japanese imports, the colonial government imposed tariffs and quotas on selected imports, including textiles (Segers 1987). Release of idle sugar fields also turned Java from a rice-deficit to a rice-surplus economy. There was now ample scope for trade between Java and the Outer Islands (Table 3.6). Unfortunately the trend towards economic integration was interrupted in 1942 by the Japanese occupation, which forced districts to revert to self-sufficiency, and then by the turmoil of revolution (1945–49).

Further progress towards economic integration awaited the resumption of Java's industrialization in the 1970s. Domestic trade was then stimulated both through input–output linkages and growth in local purchasing power. Value data on interisland trade were not published beyond 1972 (Table 3.6) but some insight can be gained from input–output estimates by origin and destination for the manufacturing sector in 1987 (Table 3.7). The distribution of final manufacturing output from Java was heavily biased towards interisland trade, although inputs were still weighted towards imports; the Outer Islands now traded extensively with Java in both general and bulk cargo. Thus no longer did interisland trade consist primarily of transhipment of exports and imports. Here at last was the substance of a national economy.

Nevertheless, the Indonesian archipelago still divides uneasily into a favoured western part of Sumatra, Java and Kalimantan and the vast, fragmented periphery of Eastern Indonesia. The former clusters around a Singapore–Jakarta axis and, together with Malaysia, constitutes the economic core of Southeast Asia (Chapter 1). Eastern Indonesia, about half the nation's area of land and sea, is officially recognized as a lagging region. Its original basis of prosperity – cloves and spices – was destroyed by the Dutch. After the 1890s it enjoyed an economic revival based on world demand for copra but

Table 3.7 Indonesia: distribution of manufacturing output and origin of inputs, 1987 (%)

Java	Intra-island	Interisland	Foreign	Total
Output destination	79.6	14.7	5.7	100
Input origin	64.1	12.6	23.3	100
Sumatra				
Output destination	64.5	21.2	14.3	100
Input origin	63.9	20.0	16.1	100
Kalimantan				
Output destination	24.5	36.3	39.2	100
Input origin	73.8	11.2	15.0	100

Source: Bappenas (1991).

in the 1960s this staple succumbed to low prices, political turmoil and economic dislocation. Despite localized resource exports, most of Eastern Indonesia has since languished. Unlike the Visayas and Cebu, these islands have no centre. The only large city, Makassar (about one million in population), serves as the main gateway but has no daily international flights and only incidental direct international container shipping connections. Eastern Indonesia also faces ethnic and separatist strife in Sulawesi, Maluku and Irian Jaya. Recent political decentralization will benefit resource-rich provinces but is likely to impede the process of integration. The future of many islands will depend upon out-migration and remittances.

Conclusion

Since the mid-nineteenth century great waves of imperialism, decolonization, nation-building and recent industrialization have had far-reaching effects on the Malay Archipelago, now encompassing six nations. Yet it remains a vast maritime and trading network, still centred around Singapore and Hong Kong. The revolution of containerization swept away many of the mercantilist restrictions that had been imposed by colonial and national governments. In shipping, trade and business networks, the archipelago is once more an integrated region that extends well beyond the limits of what is conventionally regarded as Southeast Asia.

Ironies abound. The maritime supremacy of the former colonial powers has shrunk into a single Anglo-Dutch combine fighting for its survival against vigorous Asian competitors. More tonnage is registered in Singapore than in Britain, once the great maritime power. The Philippines, whose flag used rarely to be seen in international ports, has become a major flag of registry and around 200,000 Filipino officers and crew work overseas (MT 10 April 2000). Indonesia, which in the colonial period was part of a global shipping network, struggles to maintain a national fleet, mainly because of the burden of incompetent state administration.

National economies have emerged in Indonesia and the Philippines. In both cases, it has been a slow, evolutionary and incomplete process. Java and Luzon are still the only well-integrated islands (Chapter 4). The other islands were never articulated by railway networks and only recently have acquired skeletal two-lane highway systems, much of which carries only light traffic. In Sumatra and Kalimantan, the great rivers flowing from the interior to the sea still define more clearly than highways the main economic hinterlands. Most trade moves from hinterlands to coastal ports and from there is carried on by sea. However, the industrialization of Java and Luzon and resource-based development in the outer islands have stimulated rapid growth in interisland trade according to internal comparative advantage: labour-intensive manufactures are shipped outwards in exchange for resource-based inputs. The integration of domestic production and consumption and the consequent rise in mass purchasing power will drive economic growth if macroeconomic conditions permit.

Overall, the Philippines enjoys the more favourable economic geography. Western Indonesia benefits from proximity to Singapore and Malaysia but Eastern Indonesia floats in an economic vacuum. By contrast, remote Mindanao is more akin in size and resources to Sumatra than to the East Indonesian archipelago; via the second industrializing growth pole of Cebu much of the island is well integrated with the economic core of Manila-Luzon. The Philippine periphery is the poor, typhoon-prone east-central islands of Samar and Leyte, whose hemp and copra no longer enjoy a ready world market. Even in these cases, however, proximity to and good transport links with Luzon, Cebu and northern Mindanao facilitate emigration and remittances. Southwestern Mindanao and Sulu are the extreme periphery but here, as in Indonesia's province of Aceh, the problem is not so much lack of resources as their inequitable distribution, giving rise to bitter insurgencies with a religious edge that revive memories of protracted colonial wars. They remind us that historically the formation of both the Philippines and Indonesia was a contested and violent process. Globalization is loosening central government control before the gains of recent economic development have been fairly distributed.

Notes

1. Campo (1992) and my own work on the KPM draws on the company's extensive archive now deposited in the ARA, especially the *Jaarverslag van de Directie in Indië* (annual).
2. In 1908 the interlocked SMN, RL, KPM and JCPL had formed the holding company Nederlandsche Scheepvaart Unie (NSU), which in 1969 became the managing entity and in 1977 as Nedlloyd Lijnen a single concern (Boer et al. 1994).
3. Here the Philippines was well ahead of Indonesia: Aboitiz Lines introduced its first specialized container ship in 1978.
4. The argument of this section is elaborated in Dick et al. (2002).

4
Islands: Java and Luzon

In the pre-modern era when water-based transportation was the cheapest and most efficient mode, islands were advantaged over more extensive, land-locked territories. Access to the hinterland from many points around an encircling coastline was multiplied by connection with navigable rivers, bays or lakes. In Southeast Asia, Java and Luzon are the two most populous islands and the core of emerging national economies. They share obvious physical similarities in their long coastlines, few navigable rivers, and distinct lowland/highland zones; rich, volcanic soils sustain extensive agricultural hinterlands, oriented towards wet rice crops. Rapid population growth eventually gave rise to abundant cheap labour, which has become the basis for recent industrialization. Each island is home to the national capital, which has become the focus of economic activity. Systematic comparison is therefore tempting, and it is curious that this is not to be found elsewhere in the literature.

Java and Luzon had themselves to be consolidated as administrative and economic cores before wider national economies could be feasible. Transport and communications were crucial, with the striking difference that Java's infrastructure was based on railways while that of Luzon centred around roads. Although the difference has become less marked in recent years as Java's traffic has shifted from rail to road, these different paths of investment had profound consequences. The chapter will look first at Java, then at Luzon, and in conclusion will draw out the comparison.

Java

At the beginning of the nineteenth century, when its population was probably around five million, Java was not yet an economic unit but a set of loosely connected economies and societies under a ramshackle colonial state (Houben 2002: 56–63). Its economic and cultural heartland was the self-governing principalities (*Vorstenlanden*), rump of the powerful sixteenth–seventeenth-century kingdom of Mataram, around the court cities of Surakarta (Solo) and Yogyakarta (Figure 4.1). The prosperous northern

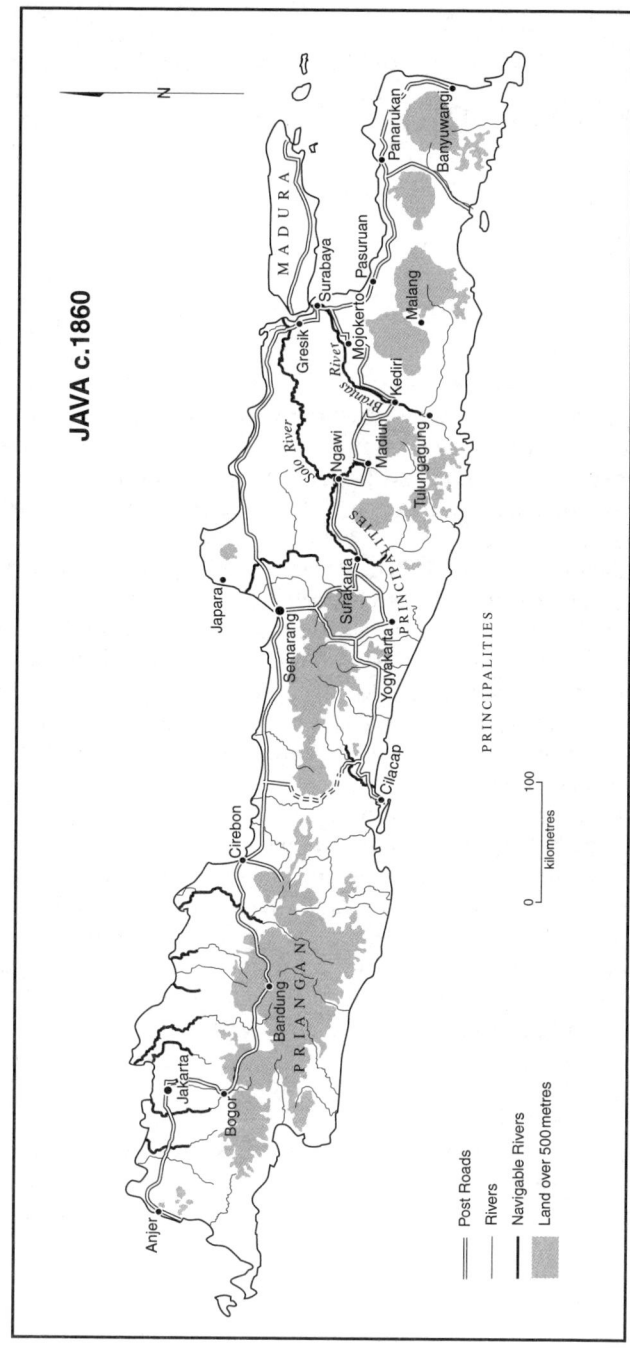

Figure 4.1 Java: Navigable rivers and post roads on relief map c.1860 (*Source*: Versteeg 1860 and other sources)

coastal plain had already come under colonial rule along with the dry offshore island of Madura. Except for Surabaya and nearby Gresik there were no sheltered harbours, but tidal creeks provided safe anchorage for sailing craft and became nodes for the shipment of products and development of towns. Apart from the sleepy colonial capital of Jakarta (then Batavia) and the nearby summer capital of Buitzenzorg (Bogor), the western part of the island was of little economic importance except for some state coffee cultivation in the Priangan highlands. Like the mountainous eastern part of the island, much was still jungle with some slash-and-burn agriculture (Boomgaard 1989).

Nevertheless, this was no society of remote, village-bound subsistence farmers. There was money and markets. Notwithstanding very real dangers, people were used to travelling in search of new land, for work, or on pilgrimage. The main transport constraint was the time and cost of movement by land. Broad roads between the principalities and Semarang allowed high officials to travel by coach (Schrieke 1957: 105–20; OMW 1907a). Ordinary people walked. Seasonal labourers (*bujang*) walked from central Java to the Priangan and even to Jakarta. High-value products such as coffee were carried by human porters.

Proto-modern networks

Transport and communications relied heavily on water. The two great rivers and axes of settlement were the Solo and the Brantas (Figure 4.1). Navigable from Gresik as far upstream as the court city of Surakarta (Solo) and by tributary to Madiun, the Solo River (Bengawan Solo) was the main artery for the movement of goods between the interior of central Java and the coast. Prahus carrying up to 100 *koyang* (138 tonnes) could hurry downstream from Solo to Gresik in six to eight days, but the return trip required poling or haulage and could take as long as three or four months (Kussendrager 1841: 231). Productivity of the large river fleet must therefore have been quite low. Some tributaries were navigable for very small craft; others allowed goods to be floated downstream on disposable bamboo rafts. Teak and firewood were sent in rafts to Gresik for distribution to sugar mills in the Brantas delta (Umbgrove 1862).

The Brantas was navigable to just below Tulungagung, only a short distance from the south coast of Java, and in the wet season by a tributary further east to Trenggalek (Kussendrager 1841: 264). From Tulungagung the Brantas flowed northwards through the busy market town of Kediri and then through a widening plain traced a wide easterly arc to the fertile Sidoarjo delta. In the 1830s, when the Cultivation System began to stimulate the cultivation of export crops, many sugar mills were established in the delta, from where the industry gradually moved upstream as far as Kediri. Bullock carts (*pedati*) carried sugar to the nearest river stage, from where it shipped cheaply to warehouses in Surabaya; prahus returned with firewood, lime, pots and woven mats and baskets for sugar manufacture and packing (Umbgrove 1862).

Modern land transport and communications emerged simultaneously with the modern colonial state. The model for both was Napoleonic France: prefectures, a salaried civil bureaucracy, codified law, a network of straight roads, and a system of posts to all parts of the Empire. The Dutch King, Ludwig Napoleon, appointed Governor-General Hermann Daendels (1808–11) with a mandate for reform. Daendels began by ordering construction of a post road from Anjer on the western tip of Java to Banyuwangi in the east via the summer capital of Bogor, about the distance between Paris and Marseilles (Figure 4.1). Wherever possible the road was to be 15 metres wide, straight and with a good camber and gutters to carry away water and permit all-weather use (ENI 1921: 743). Apart from a new and difficult section through the Priangan highlands, most work involved widening and improving existing roads and was carried out by corvée labour (OMW 1907a: 4). The main improvement was not the road itself, however, but the organization of stages: horses were changed every five to six miles, which allowed a regular gallop of about ten miles per hour (Kinloch 1852/1987: 39, 51). This change occurred expeditiously beneath a roofed structure, initially of bamboo and thatch, which extended the full width of the road. A foreman was responsible for the horses being properly tended and ready when required. So that passengers could eat and sleep in comfort after a day's hard travelling, 12 inns were provided between Bogor and Surabaya, interspersed with smaller rest houses (*pasanggrahan*) (ENI 1919: 466).

As a 'tool of empire', the essential purpose of the post road was to facilitate regular, fast and reliable communications between officials of the state apparatus (Haan 1910: 484–91). From 1810 the state itself provided a weekly – after 1828 a twice-weekly – postwagon drawn by a team of four horses for the conveyance of mails and official passengers. Private contractors provided connections from Semarang to the principalities and from Surabaya on to Panarukan, from where a postrider completed the final stage along the rough and narrow road to Banyuwangi. Later the government also provided postriders for connection to and between officials in district and sub-district centres. By 1830 travel time over the 800 kilometres between Jakarta and Surabaya had been reduced from at least 14 days in the dry season and 21 days in the wet season to 9–10 days in either season for the post, 6–7 days for an individual despatch, and 4–5 days for an express despatch (ENI 1921: 744). Private carriages might also use the post roads by the sanction of the Governor-General in Council but there was little such traffic beyond the main coastal towns (Kinloch 1852: 108).

Under the stimulus of the Cultivation System, both the road network and traffic grew rapidly after 1830. Corvée labour was mobilized to produce export crops, resulting in much increased movement of goods by land. In the early years of the System, much labour was requisitioned just to carry crops from the hills or fields to the government warehouses (Elson 1994). From there more porters, packhorses and carts were requisitioned to carry the product to the nearest waterway or port. Corvée also provided district officers with the means to improve and maintain roads and paths, even to the extent of requiring neat fences and hedges along the side of the road, a custom that persists in rural areas. Gradually there emerged a specialized trade of carters, using simple buffalo or ox carts (Fernando 1996). Under pressure of demand, rates in central Java rose quite sharply (OMW 1907a: 15). Coffee was valuable enough to bear high costs of land transport but there continued to be complaints about the transit time. In 1840 officials in central Java complained that ships were waiting three to five months at Semarang for coffee to be delivered from inland warehouses. Despite the abundance of corvée labour, transport was obviously still a constraint.

By the eve of the railway revolution, the road network had become fairly sophisticated (Figure 4.1). In 1854 the government had decided on laying an inland post road from the Priangan southeast to the principalities, then via Ngawi and Kediri down the Brantas valley to Surabaya (OMW 1907a: 7–8). In more populous districts, a network of wagon roads connected from the post roads to district centres, the lowest level of the administrative hierarchy. Official guides specified distances and travel times along the stages of all these roads, with official rest houses in the middle of long sections. In hilly or mountainous districts there were networks of trails and paths (*binnenweg*) suitable only for pedestrians or riders and used mainly by local people. Over time many of these were upgraded to wagon roads, which became synonymous with the term *binnenweg* as distinct from local or village (*desa*) paths.

Maps are unreliable as to the quality of roads. Foreign travellers such as Kussendrager (1841), Kinloch (1852/1987) and d'Almeida (1864) paid tribute to the post roads. Kussendrager (1841: 232) also praised the broad highway from Semarang to Solo, the 'very fine' carriageway from there to Yogyakarta, and various smaller roads leading off into the countryside. Of the more easterly residency of Madiun, he observed the network of roads linking the various regencies and districts, and of Kediri that the jungle paths which formerly connected villages had given way to good roads (Kussendrager 1841: 263–5). The contrast with Luzon, where travellers mostly emphasized the difficulty of roads and their impassibility in the wet season, was marked. Nevertheless, these roads were still lightly trafficked. Of the interior of east Java, d'Almeida (1864: 289) reported that 'we seldom came across a carriage of any description – except in the large towns – and only buffalo carts by daylight'.

The high-speed network of post roads and highways was prohibited for the movement of goods. At the outset of the Cultivation System, the typical Javanese buffalo cart still used solid wooden wheels, which soon lost shape and did great damage to roads (OMW 1907a: 15–16n). These were relegated to parallel cartways. In 1827 Governor-General du Bus de Gisignies noted the contrast between the fine post road and the Javanese wading through the mud with his buffaloes and cart (OMW 1907a: 10). Solid-wheeled carts were gradually phased out in favour of more substantial – and gaily decorated – ox carts with large, spoked wagon-wheels. Vries (1931: 150) dated the first such ox carts in Pasuruan to the early 1820s, by which time packhorses were also in use to carry goods to and from the highlands. It thereby became possible in 1857 to declare the post road open to all vehicles and animals and to order that the separating dykes be broken (OMW 1907a: 6). Nevertheless, as late as 1871 the resident of Japara reported that both European and Javanese officials were reluctant to remove the dykes for fear that heavy traffic would spoil the roads (OMW 1907a: 7n). Reports from around 1880 over the condition of the highway between Jakarta and Tangerang suggest that their concerns were not unfounded: to avoid the mudholes, riders and pedestrians were said to prefer to go along the dyke or the even narrower bunds between the ricefields (OMW 1907a: 25). In central and east Java enforcing the sugar mills' obligation to repair damage from their heavy carts was a perennial problem.

Road quality dictated a hierarchy of traffic. In districts with only paths, porters were still the usual means. A typical load seems to have been not quite 40 kilograms (60 kattis), the cost depending upon the availability of labour. Heavier loads could be carried by packhorse, sometimes in caravans (OMW 1907a: 15). In flat districts with roads, wheeled vehicles were used. Passenger carriages with springs were drawn by horses; for freight larger unsprung carts were drawn by horse, oxen and sometimes buffalo or men, especially in mountainous terrain. Cattle were driven on the hoof: in 1874 Javanese were reported to take 28 days to drive cattle from Semarang to Jakarta (TNI 1874).

The most spectacular pre-railway growth was in mails (OMW 1907a: 29–30). Netherlands postal regulations were introduced to Java in 1863 with provision for uniform 10 cents postage, postal remittances, and post boxes and daily clearance and delivery in the cities. A daily post was provided between Batavia, Semarang (for the principalities) and Surabaya, other towns being served three times a week. District and sub-district towns off the main highways were served until 1882 by runners under corvée labour (*heerendiesten*), after which a proper district post was implemented. Patronage increased from around 200,000 letters in 1846 to three million by 1876, excluding 0.25 million postcards (a recent innovation) and another 1.8 million items of newspapers and printed matter. This growth in traffic occurred despite the opening in 1856/57 of the telegraph lines between Batavia, Bogor, Semarang and Surabaya and their linking in December 1870 to the international network.

Railways

Because of the network of natural waterways and feeder roads, Java had less urgent need of railways than elsewhere in Southeast Asia. Nevertheless, in running Java more or less as a state enterprise, the Dutch were acutely aware of costs of production and international competitiveness. The rationale of the Cultivation System had been to reduce the costs of export production in Java to a level competitive with the West Indies. Railways offered scope for substantial savings in transport cost by breaking the nexus between settlement, export production and river and coastal shipping. The first initiatives in the 1860s were short private railways from coastal ports into the interior, specifically from Jakarta to Bogor and from Semarang to Solo and Yogyakarta, both of which were completed in 1873 – the corresponding state-owned line in east Java, from Surabaya and Pasuruan to Malang, opened six years later (Figure 4.2a). As the trade of the principalities was redirected from the Solo River to nearby Semarang, each main port was able to tap the trade of its immediate hinterland, segmenting the island into three sub-regions.

In the 1880s work began to break the east–west transport constraint by linking the three separate lines into a basic network (Figure 4.2a). In 1884 the Jakarta–Bogor line reached Bandung, while a westward line from Surabaya up the Lower Brantas valley cut across to Madiun and Solo to connect with the central Java system of the private Nederlandsch-Indische Spoorweg Maatschappij (NISM). In 1894 the corresponding link was completed from the now fully state-owned West Java system to the NISM line at Yogyakarta. Five years later when a third rail placed inside the wider NISM gauge allowed a through connection between Jakarta and Surabaya. In the late 1890s the state rail network was extended and filled out in a second wave of construction, especially into the southern plantation districts. By 1903, when a branch line reached Banyuwangi, three years after the connection from Jakarta to Anjer, the railway network had surpassed that of the old post roads. The broad-gauge rail network was then virtually complete.

The other important railway development of the late 1890s was the laying out of narrow-gauge rural tramways as feeders to the main system (Figure 4.2b). The densest network was in the lower Brantas valley, where the Oost Java, Modjokerto, Kediri and Madioen steamtram companies provided connections to most sugar mills, allowing cultivation to extend in a broader and more dense band along the Brantas, without regard to proximity to the river or the main rail line. The Pasuruan and Probolinggo steamtram companies performed a similar role on the north coast and the Malang steamtram company in the highlands. Upland estates also benefited as railheads were moved closer to the point of production, reducing the distance for expensive cartage.

The most decisive impact of railways was the shift of goods movement from river to rail transport. The reasons for this change were both ecological

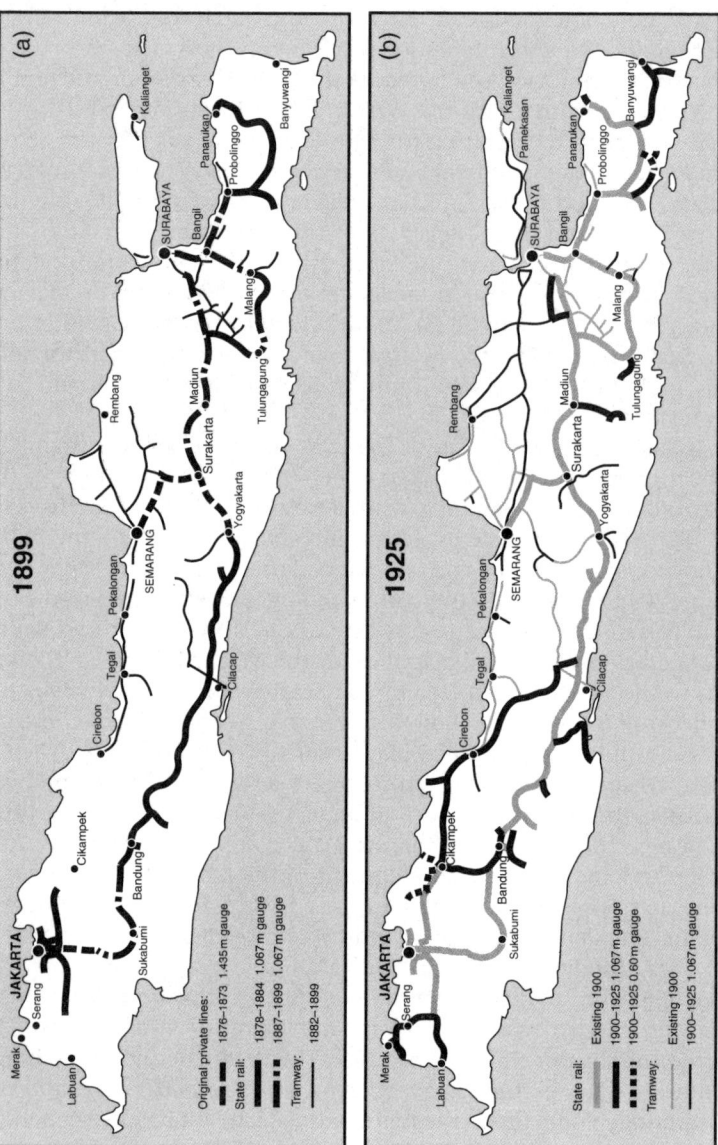

Figure 4.2 Java and Madura: (a) railway network to 1899 (*Source:* Reitsma 1925) (b) railway and tramway network to 1925 by width of gauge (*Source:* Reitsma 1925)

and commercial. Clearance of lowlands and hillsides, drainage of swamps, and damage to riverbanks caused erosion and siltation, accelerated wet season run-off, and diminished dry season flows. By the 1870s the transport arteries of the Solo and Brantas rivers had become hazardous to navigate in the wet season and sometimes unnavigable in the dry. The 1874 sugar crop of the Madiun district was still awaiting shipment at the end of the following year because severe drought had closed the Solo River. Irrigation works steadily reduced dry season flows (OMW 1906: 16–17). Without railways, the hinterland of sugar production would have contracted back to the coast and lower Brantas Valley.

Despite running more or less parallel to the main rivers, railways were also a faster and more reliable mode of transport, even though shipment by river prahu was nominally much cheaper. Low freight rates by river prahu were offset by the risks of pilferage and water damage (OMW 1906: 13). Export commodities such as sugar, for which quality was an important consideration, therefore shifted to rail. Sugar could be consolidated into wagon loads, and the volume of traffic justified rail spurs to the mills. By the early 1890s the Bengawan Solo had ceased to be a transport artery for the district of Surakarta and was of little importance to Madiun (KV 1892: App. C). By the mid-1900s it was reported that only five mills close to Surabaya still shipped their sugar harvest by river prahu (KKNS 1905: 46). River prahus were relegated to carrying village produce or very low-value items such as wood, firewood, charcoal, and coconuts (OMW 1906). A few traditionally built prahus still ply on the lower reaches of the Brantas and Solo rivers, larger ones for sand and gravel and smaller ones as cross-river ferries.

From road to rail the shift was much less dramatic. Rail was competitive for full wagon loads of export crops but local crops, handicrafts and import goods continued to be carried in small consignments by traditional forms of transport. Besides high tariffs, rail haul incurred feeder and transhipment costs at both ends of the journey, which had to be reckoned in the door-to-door calculus along with the formalities of documentation and the greater risk of damage in handling. Moreover, rail competition forced a reduction in the costs of competing modes. The resident of Japara reported that in competition with the narrow-gauge tramway small local carts had reduced charges to only one-third of their previous level. Elsewhere charges probably fell by rather less but still roughly matched those for rail *before* calculation of feeder and transhipment costs. Local carters seem to have charged what the traffic would bear, subject to the opportunity cost of labour.

Railways stimulated popular mobility less than might have been expected. Most Javanese trips were with goods to market. When the fare structure encouraged it, many Javanese quickly switched from foot to travel by train, but over very short distances. In 1873 the private NISM introduced a very successful 4th class at a fare of just one cent per kilometre between adjacent stations, half that of 3rd class (OMW 1907b: 1). However, the cost of carrying goods in excess of the modest free allowance was still prohibitive (KV 1892: C5). Discounted return fares found little favour because traders needed to sell their goods before they could afford the return ticket. Villagers who sought to travel longer distances in search of work were even less able to afford the fare. A few cents was the upper limit of purchasing power. Since markets were located in accordance with the journey time of women carrying goods on their back, railways therefore did not obviously increase the propensity of movement. Java was by no means a village-bound society but it was a society of many very short movements, which connected with other short movements, each a separate transaction, like links of a chain.

The policy of the state railways was actually to discourage popular mobility. In 1886 the Inspector-General of Railways succinctly explained the official ambivalence:

> Although lowering fares would undoubtedly lead to more traffic in the long-run, it must not be thought desirable to introduce a fourth class at a lower fare because the Native, who as a cultivator is occupied locally, should not be transformed into a wanderer...the purpose of the passenger tariff must not be to mobilise the whole Indonesian population but to give the better-off part of the population, the traders and tradespeople, opportunity to make use of the faster means of public transport to travel over greater distances. (OMW 1907b: 2)

Not until 1900, after years of opposition, did the government finally allow the State Railways to follow the NISM's early lead and introduce a general one-cent per kilometre fare for short journeys (OMW 1907b: 5–6). By then freight had come to dominate railway earnings (SSNI 1933: App. 2).

The residency reports of 1891 gave a consistent view that railways stimulated trade but not village agriculture (KV 1892). In other words, by widening the market the railways created a 'vent for surplus', raising the proportion of household time directed towards the market economy but without much change in the technology, scale or intensity of production. Thus somewhat more rice and dry-season crops were sold on the market, along with fruits, coconuts, coconut oil, palm sugar, chickens, eggs and some handicrafts but these were adjustments rather than any transformation. Consistent with the trend since the Cultivation System, there was also continuing growth in off-farm employment and greater ease of circular migration.

Railways also integrated marginal districts more closely with the rest of the island. During the railway boom of the late 1890s, several narrow-gauge tramway companies were formed to connect the main rail system to drier, hilly or remote areas that were outside the mainstream of economic life. These areas produced cash crops such as corn and cassava, tobacco, fruit and vegetables, as well as cattle, chickens and fish. Besides government teak plantations, there was also timber and firewood. Because the economic base was limited, the drylands also provided a seasonal workforce for the sugar mills, coffee estates, and the wharves. Serving thin traffic obliged tramways to look to local trade and set schedules to fit in with the timing of markets. In such ways the benefits of railways were more widely distributed. Companies serving poorer districts nevertheless struggled to earn enough to maintain dividends.

The benefit of railways to the plantation economy was unambiguous. At a critical time in the 1880s and 1890s, when the sugar industry found itself squeezed by a collapse in international prices caused by rapid growth in European beet output, railways relaxed the constraint of unreliable shipment by river and expensive carriage by land (Elson 1984: 137–8). Without this modern and efficient infrastructure, the Javanese sugar export industry called into being by the Cultivation System would probably have been a 50–60-year wonder. Instead, the huge investment in railways and tramways allowed output to be concentrated in larger and more modern factories. Fundamentally, it was a matter of scale. The productive potential of the new industrial revolution technology could be released only with a transport system of equivalent scale to link the factory with world markets. Growth in output could not have been accommodated by traditional carts and river prahus at a price competitive on world markets.

Railways had an even more dramatic impact in opening up hitherto sparsely populated upland areas to plantations, what Burger (1939) described so perceptively as 'the unlocking of Java's interior for world commerce'. The Agrarian Law of 1870 had allowed uncultivated land to be taken up by foreign planters on 75-year heritable lease (*erfpacht*), but economic viability depended on access to the export market. The first area to be opened up was the highlands immediately south of Surabaya with the completion of the railway to Malang in 1879. Five years later the Kediri line reached Blitar in the southern hills. The devastating coffee blight of the 1880s failed to cripple the industry because rail gave access to the virgin slopes of the highlands and southern hills. By 1891 Blitar was a boom town in a hinterland dominated by coffee estates (KV 1892: App. C). The peak of the boom, the last great expansion of East Java's export frontier, occurred towards the end of the decade. The Blitar line was joined to the southern extension of the Malang line in 1897, the same year in which the newly formed Malang Stoomtram Mij opened its first section of line (Figure 4.2a). These lines provided access to a large upland region, which grew rapidly in population (Elson 1984: 166) and soon became Java's main locus of coffee cultivation – by the early 1920s it held a similar importance in regard to rubber (DLNH 1926).

State sponsorship and ownership of railways was therefore no less critical to the development of plantation capitalism in the late-nineteenth century than state direction had been to the mobilization of land and labour in mid-century. By the 1900s the State Railways, like the KPM in interisland trade (Chapter 3), had become a giant, quasi-monopolistic logistics agency that, in concert with private railway and tramway companies, arbitrated the production and trade of most of the island of Java. By trial and error and much tough political bargaining, the level and structure of passenger fares and freight rates was gradually adjusted to achieve a desired mix of traffic: other modes had to make the most of the opportunities that were left. Because of the profitability of wagon-load freight, great effort was devoted to securing the sugar traffic, if possible by exclusive contract. Although colonial ministers in The Hague stoutly resisted tariff reductions that might reduce state revenue or to oblige the state to make payments against the 5 per cent dividend guaranteed to the private NISM, discounts were offered when sugar prices were low and after 1905 contract freight rates were linked to world prices (OMW 1907b: 50–60). Freight earnings thus increased much faster than passenger earnings: between 1878 and 1882 they were equivalent but by 1894 freight earnings were double passenger earnings and this relativity persisted until the 1910s (SSNI 1933: App. 2).

Competition between rail and road led to an evolving balance with 'traditional' modes of transport. The initiative lay with the railways and tramways, which could choose their business and set rates to dominate longer distance freight and passenger movement. Rates were adjusted also for village products and small quantities of accompanied goods carried free of charge but road traffic and walking remained competitive over short distances, especially in serving the regular needs of the village-based economy. However, it was not a zero sum game. First, economic expansion combined with falling real costs of all modes of transport increased personal mobility and the circulation of commodities, so that the demand for transport continued to increase. Secondly, rail and road modes were complementary as well as competitive. Growth in rail traffic generated demand for road vehicles to and from the stations, besides which there was more lively traffic between villages and local markets. On balance the number of road vehicles grew steadily. This applied particularly to pony-traps suited to carrying passengers with accompanying goods. There were also innovations in function. For example, in 1904 the resident of Mojokerto in the Lower Brantas reported that pony-traps were being used like the omnibus, not for single hire but picking up and putting down passengers along the road at very low fares – all this a generation before motor buses (OMW 1906: 11).

If some unmechanized modes retained vitality, others, like river prahus, were forced 'down market'. In the case of ox carts, the resident of Mojokerto identified the simple equation that expansion of the plantation area increased the number of carts, whereas the laying of rails for cane transport diminished it (OMW: 1906: 6–7). Since the 1830s when such carts had first been requisitioned, the mainstay had been cane transport from fields to the factory. As the rail network intensified, however, and mills laid mini-gauge rails even into the fields, this demand collapsed except among older and smaller mills, none of which survived the 1910s. In the lowland plain carts were relegated to the transport of low-value local goods which were either heavy, such as bricks and tiles, or of very large volume, such as wooden or bamboo furniture, timber or firewood. In the highlands ox carts were better able to hold their own until the era of motorized transport. For reasons of terrain, packhorses and even porters also continued to be used.

The automobile age

The first automobiles, such as that introduced to Java by Sultan Pakubuwana of Surakarta in 1894 (Knaap 1989: 86–7), were simply 'horseless carriages' for convenience of movement around town. Travel further afield was faster, more comfortable and more reliable by first-class rail or, along the coast, by KPM steamer. As late as 1910 there were still only 1000 vehicles throughout Java and Madura and before the First World War they were incidental to traffic, even in the main coastal cities (Table 4.1; ENI 1921: 746). By 1913 and 1914, however, imports were running at around one thousand vehicles per year. The Department of Public Works produced a general highway plan for Java in 1913 and the following year began upgrading the north coast road from Jakarta to Cirebon and the old southern post road between Bandung, Cilacap and Yogyakarta (ENI 1921: 746). In 1924 an Automobile Club was established. By 1929 the number of vehicles had increased to 45,000 and imports were running at more than 10,000 units per annum. Demand was so great that in 1927 General Motors made the decision to locate its new assembly plant not in Singapore but in Jakarta.

The First World War led to rapid improvement in the size and reliability of buses and trucks, which began to be imported to the Indies on an experimental basis. In 1923/24 rail and tramway companies began to report competition from shared taxis and buses. The tramway association attributed this upsurge to several factors (ARA, OJS 24: VNIST, Nota 23/12/26). First, rail tariffs were very high because the 25 per cent increase in 1920 at the peak of the postwar boom had not been reviewed in light of subsequent deflation. Second, imported light buses had become very cheap: with few overheads, they could carry around 20 passengers for as little as one cent per kilometre and still make a modest profit. Third, buses offered a more frequent service and stopped on demand anywhere along the road, reducing the need for passengers to walk or pay for transport to and from the railway station.

Table 4.1 Java: number of motor vehicles, 1900–1996 (exc. motor cycles)

Year	1900	1910	1929	1941	1966	1996
Number	15	1,000	45,000	58,000	140,000	2 million

Source: *Indische Verslag*, Indonesian Statistical Yearbook.

Rail and tramway operators sought government protection against this new competition, claiming a public service role and pointing to the burden of high overhead costs. Unlike in the Philippines, there was no regulatory body and no franchise barriers to entry. In July 1925, two years after a similar initiative in the Netherlands, the Governor-General responded by setting up a top-level Motor Traffic Committee to investigate the extent of bus and truck competition and suggest policy measures (ARA, OJS 24 Motorverkeer, GG 9/7/25). Yet the wheels ground very slowly. Despite strong pressure from the State Railways, other government departments saw the benefit to the population of cheap fares. Not until 1933 did there emerge a draft Road Traffic Act and this did not come into force until 1 January 1937. Meanwhile, rail and tramway companies had been forced to rethink their commercial strategies. One strategy was to reduce fares and improve frequencies. Another was to start up subsidiary bus services as extensions of and feeders to the rail network. Nevertheless, the economics were against rail. For Java as a whole, the railway and tramway companies never surpassed the peak passenger carriage of 166 million in 1920 (Table 4.2). In 1939 no more passengers were carried than in 1911, although the average distance of each trip was greater.

In the late 1920s railways were optimistic that they could restructure their business towards freight. Mass passenger transport over short distances at very low fares was not a profitable business and could if necessary be ceded to road transport. Whereas the rapid expansion in output of the plantation sector, especially sugar, and also of the oil industry held out good prospects for freight. In 1929 railway tonnage of 16.1 million tonnes comfortably surpassed the 1920 peak of 12.1 million tonnes (Table 4.3). During the depression, however, rail freight collapsed to just 5.4 million tonnes and by 1939 was still no more than had been carried in 1911/12. Since the number of trucks appears scarcely to have changed during the 1930s, it may be presumed that trucks carried a larger share of a smaller total traffic.

Table 4.2 Java: rail passengers, 1911–1996 (million)

Year	1911	1920	1929	1939	1952	1960	1970s	1996
Passengers	75	166	130	76	97	143	20	149

Source: *Indische Verslag*, Indonesian Statistical Yearbook.

Table 4.3 Java: rail freight, 1911–1996 (million tonnes)

Year	1911	1920	1929	1939	1960s	1996
Tonnes	8	12	16	8	2	7

Source: *Indische Verslag*, Indonesian Statistical Yearbook.

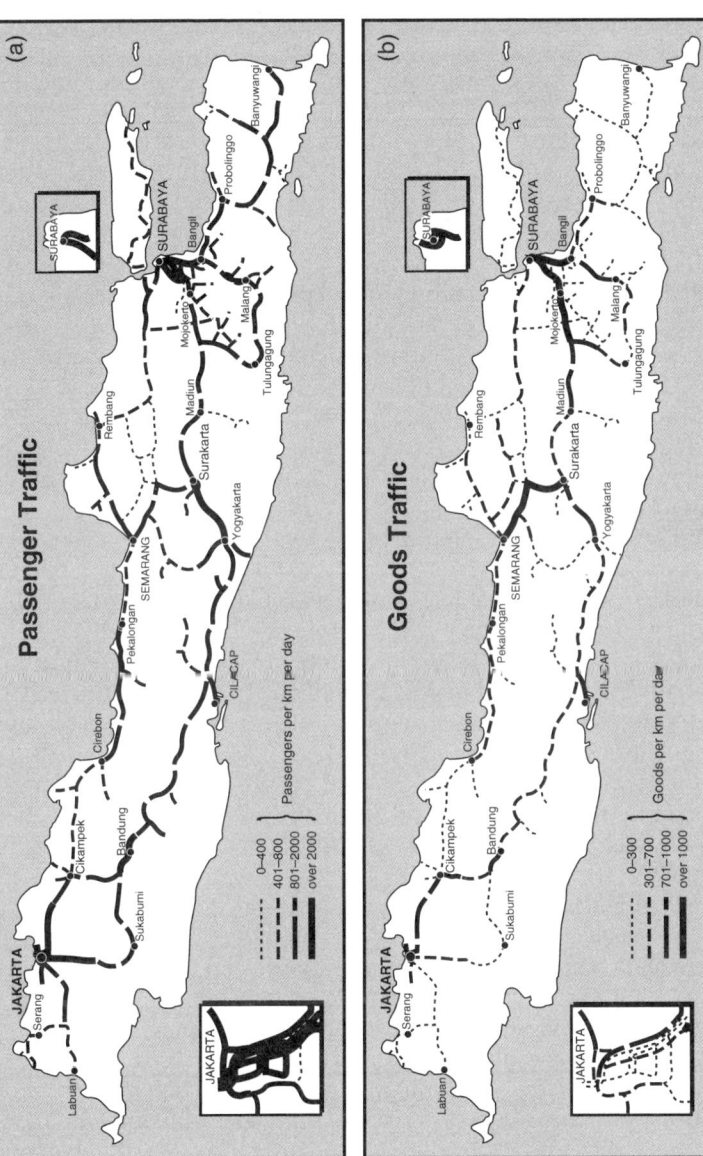

Figure 4.3 Java and Madura: (a) passenger traffic 1929 (b) goods traffic, 1929 (*Source*: SSNI 1933)

On the eve of the Second World War, Java nevertheless remained preeminently a rail-based society. On 1 January 1941 the whole of Java and Madura could boast only 58,000 motorized vehicles, including 10,000 motor cycles (Table 4.1). The European residents might drive around town and village people might ride buses to and from the local market but the transport *system* was still the railways. The annual state railway timetable, including connections with private lines, was, like the annual KPM timetable for interisland shipping, the time–distance matrix of society. Figure 4.4 shows the main network links by frequency of service in 1939. The pattern is consistent with somewhat earlier data for passenger and goods movement (Figure 4.3). Traffic was concentrated in short-distance flows between hinterlands and adjacent cities and ports. Faster and more frequent rail services stimulated the growth of the prosperous, Europeanized, highland city of Bandung and encouraged the decentralization of government departments, including the state railways. By contrast, intercity movement between West, Central and East Java remained quite modest. Not until November 1929 was a daily express introduced between Jakarta and Surabaya, prompted by opening of an air link for passengers and mails (KRN 1930: 13). The busiest air route, however, was between Jakarta and Bandung, with 2 or 3 flights per day and almost 8,000 passengers in 1939 (IV 1941: 427).

The 1950s were the Indian summer of the Java railways. Recorded passenger movement increased from 66 million in 1952 to a peak of 143 million in 1960 (Table 4.2). Assuming that fare evasion was at least 15 per cent of traffic, this would have equated to the record carriage of 1920, albeit from a much larger population. By the mid-1970s, in the face of vigorous bus competition, passenger traffic had collapsed to only 20 million. To withstand intensifying road-based competition, railways needed efficient internal organization. Instead the Railway Department (Jawatan Kereta Api), which by 1958 had absorbed the former State Railways, residual private railways and tramways and the urban networks of Jakarta and Surabaya, became a vast employment relief organization. Schedules, customs and rituals were faithfully maintained but profitability ceased to matter because the government automatically made up any financial deficit. The government did not, however, commit itself to the massive investment needed to modernize the system. The network gradually contracted from its 1931 peak of 5,500 kilometres. Aid programmes allowed rolling stock to be replaced on main lines but elsewhere vintage technology survived as though in a working museum.

In reality the railways could no longer serve more than a fraction of Java's booming passenger traffic. Rehabilitation of Jakarta's suburban network helped patronage to increase from 60 million in 1991 to 149 million in 1996 (Table 4.2), impressive enough by prewar standards but at 1990s levels of population and mobility just a niche operation. In the case of freight the decline of rail transport was even more dramatic. At its best in 1960/61 the railways carried only half the depressed 1939 level of eight million tonnes and by the

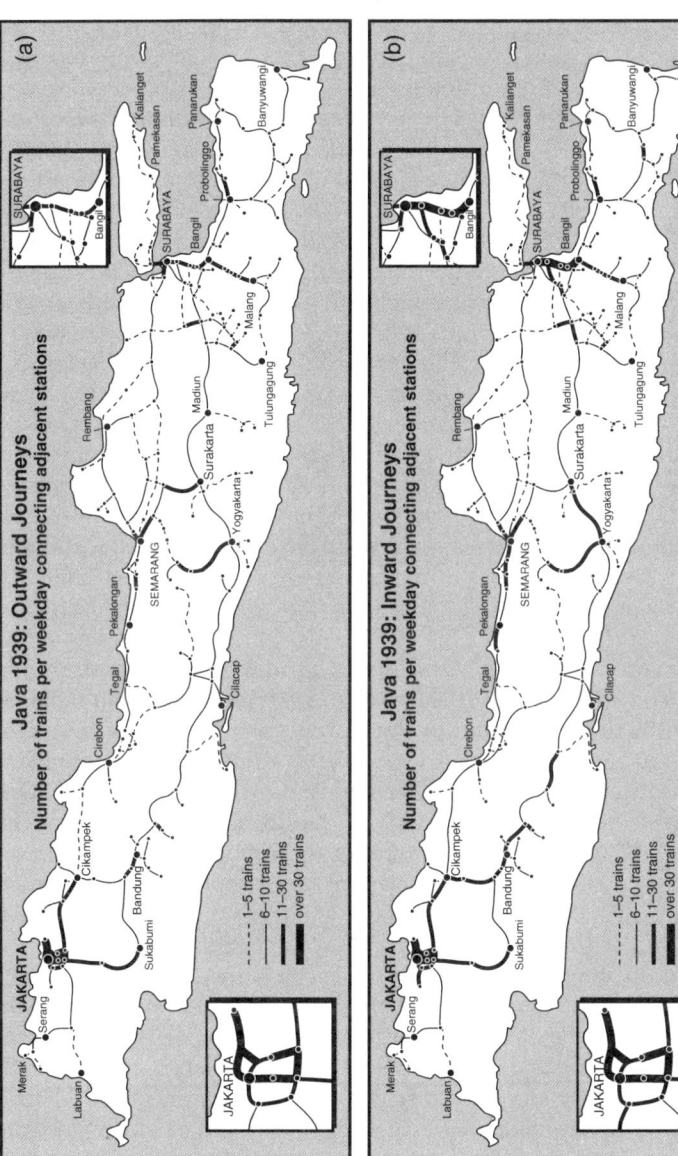

Figure 4.4 Java and Madura, 1939: (a) outward journeys; (b) inward journeys (*Source:* Calculated from Officieele Reisgids 1939)

late 1960s this had fallen away to only two million tonnes (Table 4.3). Because of massive pilferage the railways were no longer trusted with less than car-load shipments and had to rely upon carriage of state-controlled bulk commodities, mainly petroleum products and fertilizer. By 1996 economic growth and rehabilitation had allowed rail freight to recover to 6.7 million tonnes but was still no better than 1930s depression levels.

Over short distances the transport revolution of the 1940s and 1950s was actually the bicycle. In 1905 there was mention of the increasing popularity of bicycles but they did not become an item of mass consumption until the cheap Japanese models of the 1930s (KKNS 1905). Between 1931 and 1940 about 240,000 bicycles were imported to the Netherlands Indies, compared with motor vehicle registrations for the entire colony of 88,000 (CBS 1947). The bicycle became to Indonesians what the car was to Europeans. In the cities it allowed people to commute to work; in the country with some simple adaptations it became a very efficient means of carrying small quantities of goods from village to market. During the Japanese occupation (1942–45), when motorized public transport disappeared, bicycle technology gave rise to the *becak*, a three-wheeled pedicab. After 1945 they became ubiquitous at the expense of the pony cart (Chapter 8). Only under the New Order did rising real incomes see the bicycle give way to Japanese-made motor cycles. The number of motor cycles registered in Java exploded from 216,000 in 1966 to 8.4 million in 2000 – two-thirds of the total number of registered vehicles – and new motor cycles were being assembled in Indonesia at the rate of a million each year (ISYB 2001).

Medium-distance transport between village and market town shifted further to buses in the 1950s. Heavy American-built engines and chassis were imported as the base for locally made wooden bodies with a large roof rack for baskets and bundles of goods, a very efficient local adaptation. Villagers would walk or ride a bicycle or pony cart to the main road, then catch the market bus, going early morning, returning in the middle of the day. After the mid-1970s these buses began to lose business to light commercial vehicles, small pick-ups equivalent to the Philippine jeepney, which ran more frequently than buses and could detour off the main roads. They were also versatile enough to carry a ton of rice or fertilizer.

Railways faced tough competition even in long-distance transport. Intercity bus operators competed in quality of service, offering soft seating, air-conditioning and taped music. Another innovation of the 1970s was overnight express intercity buses (*bis malam*), with tickets sold through travel agents. Every city and town received central government subsidies to construct bus terminals, which became much busier interchanges than the railway stations. Over medium distances buses in turn faced competition from light commercial vehicles known as Colts, which provided faster – but rarely safer – point-to-point carriage to passengers with little accompanying baggage. By the 1950s government officials and businessmen could afford to

Table 4.4 Java: interurban transport task by mode, 1991

	Road[1]	Rail	Sea	Total
Passengers (b. pass-km)	133.1	9.2	0.2	142.5
%	93.4	6.5	—	100
Freight (b. tonne-km)	27.6	1.1	42.1	70.8
%	39	1.6	59.5	100

Note: 1. National and provincial highways only.
Source: World Bank, *Strategic Urban Roads Infrastructure*, 1996, Table 3.1.

fly intercity by Garuda – by 1997 there were 40 shuttle flights a day each way between Jakarta and Surabaya.

Thus by the 1980s there had been a revolution in personal mobility in Java (Dick and Forbes 1992). Facilitated by central government funding of local roads and bridges, villagers could now travel almost on demand to the nearest market town and, if they wished, connect through to Surabaya or Bali, Jakarta or even Sumatra. As two-lane roads became more and more crowded, the rate of travel was not fast, neither was it particularly safe, but road transport provided a seamless web of connections through the island. This was the era of popular mobility that railways had promised and government had feared in the 1880s. Eventually the population of Java did become itinerant, as attested by rapid urbanization and the prevalence of circular migration (Hugo 1996). Ironically, by that time the railways, for all their massive infrastructure and organisation, had become marginal to the transport task (Table 4.4). Passengers and freight continued to move over much the same network links but now by road instead of by rail.

The latest phase in the development of Java's transport infrastructure is construction of a toll-road system suitable for rapid long-distance and heavy-vehicle movements. As in the late-colonial period, most first-class roads or highways are still no more than sealed two-lane strips with a verge on either side and a row of trees with whitewashed trunks as a night marker. Overtaking is hazardous and speed reduces to that of the slowest vehicle. Traffic growth of 5–10 per cent per annum has led to congestion, especially in North Java (World Bank 1996: 3). Moreover, heavy vehicles loaded with containers do much damage to unstrengthened pavements. Conceived by Sukarno in the early 1960s, the first toll road was opened in 1969 between the outskirts of Jakarta and Jagorawi near Bogor. In the 1980s there followed links to Cengkareng airport, west towards the Sumatra ferry terminal on Sunda Strait and east via Bekasi and Krawang to Cikempek, all connecting into the Jakarta ring road. Elsewhere were just short urban bypasses in Surabaya and Semarang. Thus by the mid-1990s the Jakarta agglomeration not only had half of Java's registered automobiles but also most of the 370 kilometres of toll roads (Chapter 8). Plans were well advanced to build another 510 kilometres of toll roads by 2006 to link together the main cities of Java (World Bank 1996), but these projects were deferred or abandoned following the Asian crisis and the downfall of President Soeharto, whose children were heavily involved as principals.

Luzon

In the mid-nineteenth century, the several cultivated and populous parts of the island were so poorly linked by land along its 1,100-kilometre length that Luzon was to all intents an archipelago. 'Roads' were dry-season tracks and trails negotiable only by porters and pack animals. Communications therefore relied heavily upon coastal shipping and rivers, few of which were navigable for any distance (Bureau of Navigation 1912). Manila itself enjoyed close access to two rich hinterlands (Figure 4.5). Its short Pasig River provided a waterway southeast to Laguna de Bay, a shallow freshwater lake whose villages supplied the city with rice, vegetables, fruit, and fish. North of Manila Bay the Pampanga delta gave access to the lower section of the rice-growing Central Luzon Plain, where formal settlement had begun under various religious orders in the late-sixteenth century. The northernmost section of the plain drained by the Agno River into Lingayen Gulf and the South China Sea. From there, beyond the very narrow but populous coastal plain of La Union and Ilocos, lay the rich hinterland of the Cagayan Valley in the remote northeast. Navigable in its lower reaches by coastal shipping and for some 250 kilometres by canoe (*banca*), the Cagayan River was the island's longest traffic artery and after the 1780s the key to export of the famous cigar-leaf tobacco (Jesus 1980). Luzon's other main productive region, the Bicol Peninsula, known for the export of abaca or Manila hemp, stretched southeast like a long, spiky tail.

The tempo of economic activity quickened in the mid-nineteenth century as the Philippines was gradually opened to foreign trade, first Manila in 1834, then Sual on Lingayen Gulf in 1855. Rice and sugar from the Central Plain and abaca from Bicol found a ready international market. Extension of tobacco cultivation in the Cagayan Valley also enlarged the domestic market for rice. The virgin lands of interior of the Central Plain attracted migration from overpopulated La Union and Ilocos and by 1887 the population had reached one million (McLennan 1980). Districts which in the mid-nineteenth century were virtually uninhabited had reached densities of over 50 per square kilometre, forming a band of settlement from the shores of Manila Bay to Lingayen Gulf. Rice, muscovado sugar and cattle-raising were the main activities.

This development took place without much direct benefit from modern transport. Admittedly in the 1870s steam shipping brought the ports at either end of the Central Plain into regular connection with Manila. Steam launches provided daily sailings up river to Bacolor, just below San

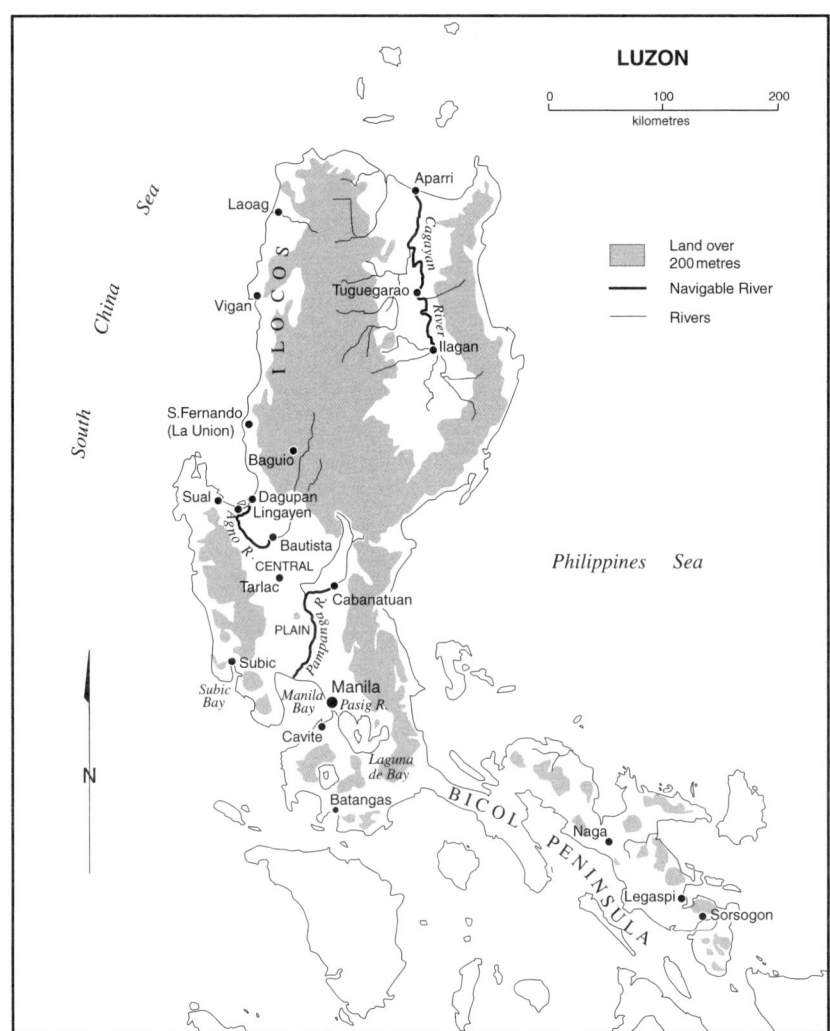

Figure 4.5 Luzon: Navigable rivers on relief map, c.1900 (*Source*: *Atlas of the Philippine Islands* 1900)

Fernando, as also to points around Manila Bay and up the Pasig River to Laguna de Bay (Larkin 1972: 100; EC). Larger coastal steamers sailed every few days to Dagupan and on to Ilocos ports and the Cagayan River. Within the Central Plain, however, shipment still relied upon small wooden craft powered by sail, pole and oars. On the Pampanga River, barges (*cascoes*) of 50 tons capacity could ascend as high as Cabanatuan in Nueva Ecija province between July and October, then float down on the floodwaters laden with unhusked rice (*palay*) (McLennan 1980: 245). In the dry season small boats of seven tons had difficulty ascending so far, although 30-ton vessels could still reach Arayat, about half way, in the dry season.

American prejudice was that the Spanish built very few decent roads, but Nature and War had wrought great damage before the inventory was made. Roadways and many stone arch bridges survived. Modern roadworks can be dated to around 1830, when Governor-General Pascual Enrile caused post roads to be constructed, much as Daendels in Java some twenty years earlier (PBPW 1950: 14). In 1868 a royal decree transferred control of public highways from the provinces to a Bureau of Public Works under the Governor-General and standards were set out for road widths and surfaces. The quality of construction and maintenance, however, depended very much on local initiative. Apart from earth and gravel, the only hard surfaces were either adobe blocks, which wore quickly, or cobblestones, which were expensive to lay but also very rough. Except at times of official inspection, there seems to have been no central coordination to ensure that roads were kept up to standard. Traffic consisted mainly of pedestrians, riders, packhorses and occasional notables carried in palanquin (known by the Americans as 'hammocks'). The improvement of roads allowed greater use of buffalo (*carabao*) carts to haul products to the closest waterway, though the narrow wooden wheels were probably the main cause of damage to road surfaces (Owen 1984: 93–4).

The Atlas published in 1900 shows a fairly dense road network just south of Manila in Cavite, Batangas and Laguna, and in the central and northern parts of the Central Plain, the southern part being served by waterways of the delta (Atlas 1900). There was also the Camino Real (Royal Highway) that completed a great loop up the coast of Ilocos and back to the head of the Cagayan Valley. Beyond Laguna, however, there were only isolated sections of road, most notably in Bicol. Supplementing these roads were local paths and trails. The passable road network for carriages or carts must, by contemporary accounts, have been a good deal less than shown on the map, especially in the rainy season.

To the end of the Spanish period the rhythm of the local economy was therefore still set by the seasons. In the wet and windy monsoon season from July to October steamers had great difficulty in gaining entry to bar-bound rivers or in anchoring off exposed lee shores. Land transport all but ceased as the roads became impassable, though goods could be floated down the rising rivers. Bowring (1859: 338) vividly described the dramatic shift between seasons:

> On arrival of the N.E. monsoon commercial enterprise begins and many shipments take place; the roads are passable, the warehouses filled with goods; this lasts till the end of June or July. Then come on the heavy

rains: the vessels for the coasting trade are laid up for the season; the rivers overflow; most of the temporary bridges are carried away by the floods; everybody is occupied by what the Spaniards call their 'interior life'; they settle the accounts of the past year and prepare for that which is to come...

Transport problems were an irritant but no pressing constraint because the main crops of rice and sugar were seasonal, as was the movement of labour.

The railway

Railway development in Luzon lagged over twenty years behind Java and seemed at first almost too late to have any significant impact on the development of the island. Construction began in 1887 under a Spanish government guarantee of an 8 per cent return on capital and the 196-kilometre Manila–Dagupan line opened for traffic in November 1892. It was well located, bringing Manila into direct communication by land with a longitudinal cross-section of the Central Plain (Figure 4.6a). Several decades earlier the middle sections would have traversed very thinly populated country but by the 1890s, after several decades of migration from coastal Ilocos, it followed an existing axis of settlement. The railway's immediate achievement was to redirect the Manila traffic of the northern part of the plain from the port of Dagupan and coastal shipping. Each day regardless of the weather two trains each way completed the journey in just under eight hours, avoiding a sometimes rough passage and the uncertainties of the bar at Dagupan, which quickly declined as a port. The hitherto land-locked province of Tarlac noticeably benefited (Corpuz 1989: 83–5). Elsewhere the railway seems to have had no obvious impact beyond a strip about 10 kilometres on either side of the track, where in 1899 the manager estimated a doubling of production (Philcom 1900, II: 314). Between 1893, the first full year of operations, and 1896–97, the last two normal years, traffic earnings increased about 50 per cent (Philcom 1908, I: 336). Feeder access was still too poor to generate a wider diffusion of benefits. The economic context was also unfavourable: by the 1890s the Central Plain was entering a period of depression, exacerbated by natural disasters such as cattle disease (rinderpest). Between 1898 and 1900 the railway was sabotaged by local resistance to American occupation and required extensive repairs by the United States Army (BIA 13931A/994).

In the 1900s the railway began to lead economic development in the Central Plain, but carrying passengers (about 90 per cent third class) rather than freight. Over the period 1910–14, passenger traffic averaged around 60 per cent of total revenue compared with only about one-third for freight (BIA 13931/156). This new pattern had two main aspects. First, the construction of branch lines and spurs increased the traffic catchment. Secondly, rehabilitation of feeder roads widened the zone served by each

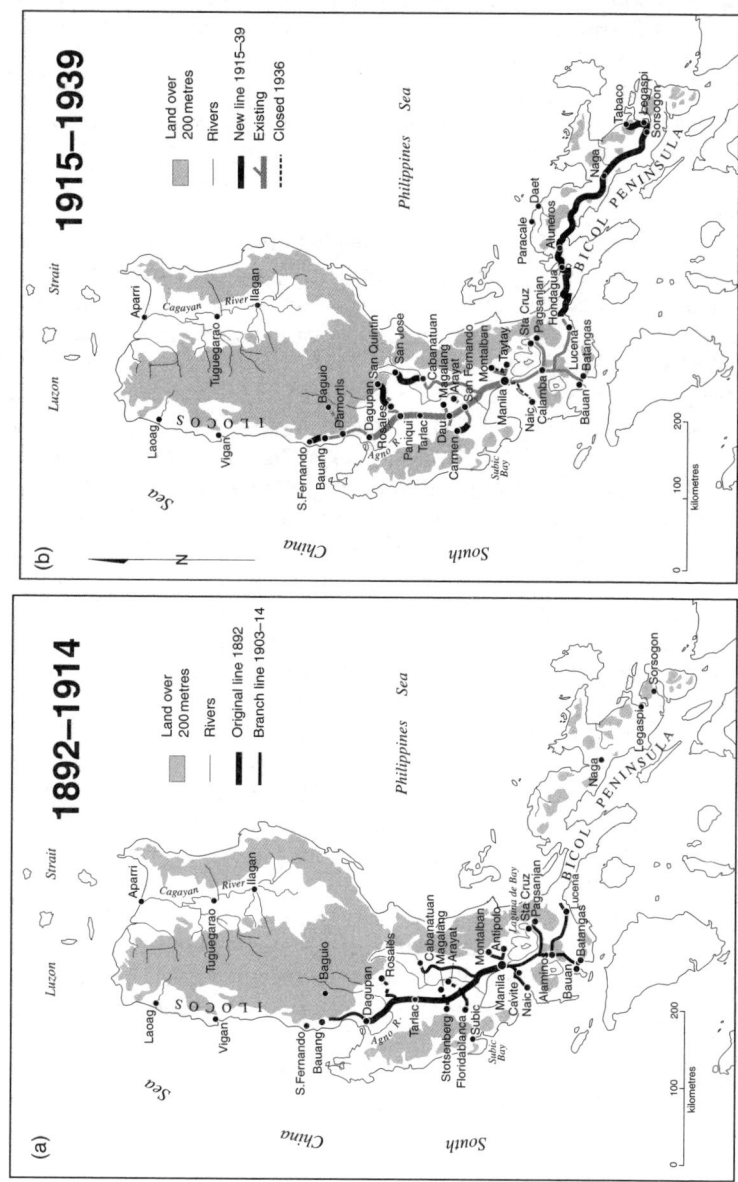

Figure 4.6 Luzon: (a) railway construction, 1892–1914 (*Source*: MRC, 1914); (b) railway construction, 1915–1939 (*Source*: MRC 1939)

line. This more finely articulated infrastructure encouraged immigration and more intensive land use, which in turn called forth capital. Steam-driven rice mills located along the railway. In the 1920s large sugar factories (centrals) were established in a 'dependent relationship' with the railway (Corpuz 1989: 260). Production and trade also diversified into poultry, vegetables and fruit for the buoyant Manila market (Philcom 1908: 378). Some administrative centres were relocated from river-based to rail-based towns, which grew faster in population (Corpuz 1989: 77–83).

Expansion of the railway network was also timely in removing what would soon have become a great impediment. Settlement of the Central Plain was accompanied by deforestation, not only clearance for cultivation but also cutting of timber as firewood, especially to feed the sugar mills. As in Java, this accelerated wet season run-off and led to the silting up of once reliable waterways (Larkin 1972: 101). In 1892 Cabanatuan, a centre of the rice trade, could still be reached by *cascoes* in the wet season and *bancas* in the dry; ten years later this was no longer feasible (McLennan 1980: 187–9, 198, 326n52). Fortunately the railway reached Cabanatuan in 1905. Shorter lines, such as from San Fernando northeast to Arayat and southwest to Floridablanca, tapped sugar-growing districts (Figure 4.6b). Railways thus ensured that the hinterland expanded instead of contracting.

Nevertheless, railway revenue did not grow quickly enough to service the large debt incurred to finance the construction costs of new lines. To accelerate work on the southern and Baguio lines, in 1910 the Insular Government had taken advantage of a 1905 provision to guarantee 4 per cent bonds and authorize special loans up to P7 million (Harrison 1916: 4–5). During 1914 the financial viability of the MRC came into question (BIA 13931A/288A). Referring to 'the long series of scandals and frauds' in construction as 'an object lesson in how private ownership of railroads may not be ideal', Governor-General Harrison recommended taking the railway into government ownership (BIA 13931A/289). The disruption following outbreak of the First World War forced the company to negotiations and in January 1916 provisional agreement was announced for the Insular Government to buy out the Manila Railroad Company for P 8 million – in effect, the value of its exposure (Harrison 1916: 6–7). In January 1917 the British-owned but American-registered Manila Railroad Company came formally into public ownership as Philippines' 'greatest public utility' (Harrison 1916: 1).

Roads

If the civilizing mission of the Spanish was to bring Christianity to the Philippines, that of the Americans was to build infrastructure. Used to the enterprise, technology and sheer energy of their own restless society, they were appalled at how little the Spanish had improved their colony and with unbounded confidence set about making up for lost time. 'We are giving to these people railroads, ports, and irrigation systems, in the belief that by stimulating these centers of activity where the goods are handled most generally we can best encourage commerce...' (BIA 2399/5: GG, 20/9/10). The Spanish had forfeited their moral claim by failing to improve the country. The Americans would tolerate the old religion and govern in cooperation with the local elite but they would insist upon firm administration of justice – and they would build. The District Engineer became a man of great authority.

By the mid-1900s when the American government began systematically to rehabilitate and extend the road system, a decade of rain and floods, war and neglect had destroyed most of the Spanish heritage. Preliminary work of mapping and temporary repair of ferries and bridges was carried out between 1898 and 1900 by the Engineer Corps of the American Army. In September 1900 the first act passed by the Philippine Commission allocated $2 million for repair of roads and bridges under military supervision. Nevertheless, in May 1903 a detailed report on the condition of roads in Luzon showed that the Insular Government faced an enormous task (BIA 2146/17).

Roads meant something different from what is now understood. Outside Manila there was as yet no motorized traffic. The hierarchy of traffic was pedestrian, pony, pack train, bull cart and horse-drawn wagon. This corresponded to a hierarchy of roads. Networks of paths served people on foot, including porters (*cargadores*); trails were negotiable on horseback and by trains of pack animals roped together. The ubiquitous two-wheeled bull cart drawn by an ox needed wider roads with a hardened surface. However, in the wet season the narrow, steel-rimmed wooden tyres rutted and eventually cut through the pavement, trapping water and allowing the ox-hooves to churn it into a quagmire. In consequence narrow-tyred carts were banned in 1903 (Philcom 1903: 152–3). Wagons drawn by teams of horses were used to supply the American army and as the heaviest type of vehicle had the greatest need for good roads. This meant not only hard pavements but also bridges and ferries to avoid rough and uncertain fords in which wagons could easily capsize. In this decade first-class roads were known by the Americans as wagon roads, which turned out to be suited also to early motorized cars, buses and trucks. Nevertheless, high economic returns could be gained from much cheaper investments in upgrading paths to trails or trails to cart-roads.

In 1905 a Bureau of Public Works (BPW) was established with a district engineer in charge of works in each province (PBPW 1950: 37–8). Its first head, W. Cameron Forbes, realized that sustained progress required not only massive investment in road infrastructure, which the Spanish had already attempted, but also a system of road management to maintain the improvements. This was underpinned by two other aspects of his vision. First, provinces and municipalities, which bore the burden of management, were given access to road funds by earmarking the head tax (*cedula*). In 1907 provinces were authorized to double the rate (the 'double *cedula'*) and use the increment for public works; incentives were a partial refund of provincial taxes plus an immediate special grant for roadworks (PBPW 1950: 25). Secondly, the mechanism for ongoing maintenance became the *caminero*, a workman paid by local government to keep under repair about one kilometre of road (Philcom 1908: 216, 224). To enhance their status, *camineros* were issued with badges and insignia and given prizes for the best performance (Philcom 1909: 143). Forbes himself became known as 'El Caminero' (Halsema 1991: 26). In 1909 the Secretary of Commerce & Police pointed out that in Pangasinan and Ilocos road improvements had allowed cartloads and their daily haul to double or triple (Philcom 1909: 142). Nevertheless, because head taxes were proportional to population, the more densely populated provinces in the Central Plain and Bicol had the best infrastructure (BIA 2146/38). It was some years before sections of improved roads were connected into a general road system or an all-weather trunk road completed the full length of the island.

One example of the rate of progress is the province of Pampanga in the lower delta of the Central Plain. In 1910 a citizen of the main town of Lubao complained about the 'deplorable condition of most roads' (Larkin 1972: 205–6). Pathways were passable only in the dry season and many bridges were damaged. In 1912 the province had only 33 miles of first-class roads (Larkin 1972: 247). The first automobile appeared a year later (Larkin 1972: 288). An extensive 24-hour traffic count in March 1915 suggested that the traffic mix was beginning to change: pedestrians still accounted for three-quarters of all movements, horses and buffalo (*carabao*) another few per cent, but of vehicles horse-drawn passenger carts (*carromata*) were 60 per cent, bull carts 31 per cent and bicycles, motor cycles and automobiles just 8 per cent (QBBPW April 1915). Yet by 1916 San Fernando could be reached more quickly from Manila by car (two hours) than by train (2.5 hours) (Halsema 1916). Feeder roads were under construction to the railway and between towns. By 1921 all but one town was joined to the main highway by all-weather roads.

The number of vehicles increased much faster than the length of roads. America being the home of the automobile, it was to be expected that American residents in the Philippines would share the national enthusiasm for the new technology, and that the innovation would be emulated by wealthy local people. The first auto seems to have been imported to Manila in 1903 (Corpuz 1989: 105n) but at first these expensive new toys were confined to Manila and environs. Vehicle registration became mandatory in March 1910 and at the beginning of 1911 the chief of police advised that there were 440 automobiles in Manila and perhaps another 60 in the provinces (BIA 18343/22). In the same year the Manila Automobile Club was formed under the patronage of the Governor-General and with leading Filipinos Osmena, Roxas and Singson as honorary vice-presidents (FER, 5/12: 440). Annual car imports increased from around 400 in 1911–12 to a peak of 4,000 in 1926, so that by 1929 there were in Luzon 16,640 cars plus 7,410 trucks (PIICC 1931). The Shell oil company supplied petrol from around 170 dealers in most significant towns outside Manila (Bachrach 1930: 14). Booklets of sectional road maps showing road conditions throughout the archipelago were published and regularly updated.

By the mid-1920s the volume of motorized and heavy vehicle traffic had led to a crisis in road management, with almost two-thirds of the road budget being absorbed in routine maintenance (PIGG 1926: 225). The level of outlay justified the heavy initial cost of tarred surfacing, which had been tried on an experimental basis, but the problem was to find the necessary funds. One new source was a tax on imported motor fuels, which came into force in 1927 with 25 per cent of receipts earmarked for roadworks. 'User-pays' contributions thereby rose from 10 per cent in 1926 (registration and licence fees only) to 34 per cent in 1928 (PIGG 1926: 224–5, 1928: 216–17). From 1928 the other new source was the issue of provincial bonds.

Opening up the highlands

The Spanish government made little attempt to impose formal administration over the remote highlands of northwestern Luzon. Within a year of the conquest, the new American administration was investigating the establishment of a hill station some 5,000 feet above sea level at Baguio (Philcom 1900, II: 323–37). As in India and Java, hill stations offered Europeans an escape from the rigours of the tropics, to which they were thought to be physically unsuited (Philcom 1903, I: 58). The problem was one of access. Beset by formidable engineering problems, the 45-kilometre Benguet Road took four years to complete, hugely over budget, and still needed continual maintenance for wet season landslides, exacerbated by typhoons (Philcom 1909: 64; Corpuz 1990: 197–8). Nevertheless, in 1909 automobiles replaced mule transport, reducing the journey time up the mountain to two hours and making it feasible feasible to designate Baguio as the summer capital for the hot months of February to June (Philcom 1909: 64–5, 145). Annual relocation of entire government bureaux was decided to be too extravagant, but the 'City of Pines' quickly became established as the preferred resort for the Manila elite. From late 1928 the seasonal Baguio Express to Damortis allowed the 266-kilometre trip by rail and bus to be completed in a comfortable 8.5 hours from Manila, compared with 2–2.5 days by rail and horseback at the turn of the century (Bachrach 1930: 29; Philcom 1900, II: 324, 337).

The success of Baguio as a colonial idyll stimulated the development of the rest of the highlands. As peace was established and the highlands brought into regular communication with the lowlands, production, trade, and tourism followed. Once producing little besides cattle, the area around Baguio and La Trinidad was given government assistance for planting cool-climate fruit, vegetables and flowers. Mining also increased. The patrol officers, known as *apo*, who established formal administration over the highlands of Mountain Province gave a high priority to upgrading the mountain dwellers' pathways into horse trails and simple roads, drawing on local labour as tax-in-kind (Jenista 1987: 135–46). This was seen as fundamental to suppressing tribal conflicts and imposing rough law-and-order. As the Secretary of the Interior stated, 'it is hardly possible to exaggerate the pacifying and civilizing influence which trail construction has exercised' (Philcom 1909: 128). Trails also economized on scarce labour. Establishment of American authority over the highlands increased the demand for human porterage: in 1907/08 the lieutenant-governor of Bontoc sub-province reckoned the official tax-in-kind requirement for *cargadores* to be about 500 per week (Philcom 1908, II: 337). Use of pack-horses relieved the tax burden and freed labour for construction works (Philcom 1909: 128). By 1931 the *New York Times* (26 April 1931) could introduce to readers the beauty of the rice terraces and the accessibility by automobile of a region where only a generation earlier Spanish officials had feared to go (NYT 1931).

Road–rail competition

Cars, trucks and buses were at first just feeders to the railway, but by 1920 the Manila Railroad Company (MRC) was losing traffic to the new mode (MRC 1920: 7). Because road transport was not taxed for the construction and maintenance costs of roads, the MRC was at a marked competitive disadvantage. Transport licences prohibited road operators from offering direct competition on parallel routes (BIA 26640/43), but poaching was difficult to police, especially where stations had poor all-weather road access. At the same time, the MRC was restrained from aggressive pricing by the Public Utility (later Service) Commission. These commercial pressures gradually forced MRC to an accommodation with road operators that allowed for traffic to shift between modes. First, the railway specialized in those forms of carriage where it had a competitive advantage, specifically long hauls, full wagonloads and express delivery. Secondly, the MRC came to formal and informal working agreements with the main bus companies to handle local feeder traffic and on-carriage. The outcome by the late 1930s was an integrated and fairly efficient transport network throughout Luzon. Except for completion of the southern line in January 1938, the MRC did not seek to expand its network and even closed some short lines.

Specialization meant a sharper focus on the two main commodities of the central plain, sugar and rice, whose share increased from half of non-express freight in 1916 to almost two-thirds in 1937–38 (MRC 1938). However, whereas their proportions were roughly equal in 1916, by 1937–38 that of sugar had increased, whereas that of rice had markedly declined. Sugar remained a staple because the MRC provided dedicated transport infrastructure. When the Payne–Aldrich Act of 1909 opened the American market to duty-free Philippine sugar, the new demand was not for the traditional heavy muscovado but for fine-grained centrifugal sugar. The first large modern centrals of Calamba (1913) near Los Baños and Pampanga (1919) at Floridablanca in the Central Plain were built with rail access and enjoyed special bulk rates (YBPI 1920: 156). In 1938 the railway carried 534,000 revenue tons of cane to the mills and 232,000 tons of sugar to the port of Manila (MRC 1938). Rice mills were also located beside the railway but the shorter distances from field to mill and the smaller size of shipments from the mill increasingly favoured trucks. By the mid-1930s about one-third of rice delivered to Manila was by truck, the other two-thirds by rail (McLennan 1980: 250). Other significant commodities carried by rail in 1938 were manufactures (181,000 tons), timber and forest products (151,000), copra (95,000), and minerals (37,000) (MRC 1938). Abaca from Bicol continued to be shipped to Manila by sea.

By the late 1930s the state-owned Manila Railroad Company (MRC) had become an inter-modal transport network that encompassed virtually the whole of Luzon. In 1928 a pioneer transhipment arrangement was made with the Northern Luzon Transportation Company to serve Ilocos from the rail terminus at San Fernando (La Union); under a revised rail schedule it became possible to travel in one day between Manila and the northernmost town of Laoag (MRC 1928). Two years later the Railway took over from the Department of Public Works the Benguet Auto Line providing the road link from the stations of Damortis and Bauang to the highland resort of Baguio. After opening of the branch railway through to San Jose (Nueva Ecija) in February 1939, through service to towns in the Cagayan Valley was provided by ticketing with the Rural Bus Company. Within the Central Plain, local bus companies gave connections to towns in Pangasinan and isolated Bataan. In the environs of Manila, formation in 1932 of the subsidiary Luzon Bus Lines allowed the Railway to provide feeder bus services, paving the way for the closure four years later of several short, uneconomic local lines (BIA 13931A/994: 13–14). In the south, extensions through to Sorsogon were provided after 1932 by arrangement with the Alatco Bus Company; in 1938 similar arrangements were made from Daet to the east coast mining towns such as Paracale, supplementing the Railway's shipping line to East Coast ports and islands (MRC 1938). In January 1938 the almost 500 kilometres of the two southern lines were finally joined from Manila to Legaspi, permitting running of a twice-daily Bicol Express in 24 hours without the interruption of a ferry voyage. By then, however, first-class passengers could fly to Naga or Legaspi – as also to Baguio – with the Philippine Aerial Taxi Company (PATCO).

In September 1938 the MRC, now an instrument of self-government under Manuel Quezon, diversified beyond Luzon into the interisland trunk route from Manila to Cebu/Iloilo and Zamboanga by taking over operation of the express steamer *Mayon*. From Zamboanga a small connecting steamer sailed on to Cotabato, where the subsidiary Mindanao Motor Line conveyed passengers and baggage to Cagayan de Oro, Iligan and Davao (MRC 1938; BIA 13931D/19). MRC also owned and managed the luxury Manila Hotel and, after 1937, handled stevedoring at the deepsea wharves of the Manila Port Terminal. This integrated national transport organization was swept away in early 1942.

Postwar

The Japanese occupation of 1942–45 caused immense damage to the land transport infrastructure of the Philippines and Luzon in particular. The most vulnerable part of the system was bridges, so essential to communications in the wet season. Many bridges were destroyed at the beginning of 1942 during the brief resistance to the Japanese invasion. Subsequent guerilla actions caused further damage and most of what remained intact or had been repaired by the Japanese was destroyed during 1945 in the hard-fought reoccupation. Much prewar equipment had been destroyed, records were lost, personnel scattered and funds dissipated. Road surfaces had been little damaged by military action but suffered badly from lack of maintenance. The American reoccupation actually made the situation worse: the great influx of vehicles, including many heavy trucks, probably caused more damage to the poorly surfaced and badly maintained pavements than the war itself (PBPW 1950: 143).

Because the economy was now more heavily dependent upon land transport, on the eve of independence in 1946 the crisis was perhaps greater than at the beginning of the original American occupation. The crucial difference from 1900 was that after 1945 the Americans were able to restore normal communications within a very short time by a massive effort in military and civilian aid (Hartendorp 1958: 379). There was no lack of vehicles, since military surplus vehicles, especially trucks, were abundant and cheap. By 1949 truck registrations were almost double the level of 1941, whereas automobile numbers had barely recovered (PBPW 1950: 128–9). The bottleneck was lack of bridges and all-weather roads. Temporary bridges had to suffice until replaced by permanent structures of concrete and steel, a programme that took more than a decade. Nevertheless, in 1946 vehicle miles were perhaps one-third above the 1941 total and by 1948 had almost doubled (PBPW 1950: Figure 14).

Increasing revenues from petrol tax and vehicle registration provided funds for further roadworks. By 1950 an almost continuous paved road ran from Manila through the Central Plain to Dagupan and on to Laoag in Northern Ilocos. South of Manila, however, hard-paved roads barely extended beyond the environs of Manila. Main roads in Batangas and Laguna, as also in the far south the highway from Naga to Sorsogon, had a layer of asphalt but elsewhere the best surface was traditional macadam, easily damaged by heavy traffic (BPW 1950: Figure 19).

This uneven distribution of road infrastructure was reflected in nationwide traffic data, which first becomes available for the year 1949. The only dense flows were within 100 kilometres of Manila, especially just north of the city in the lower part of the Central Plain. Minor flows extended across the Central Plain to the North Coast. Manila dominated as both origin and destination: there was very little through traffic between northern and southern Luzon or vice versa (PBPW 1950: 56). Elsewhere, including the Outer Islands, traffic was highly localized. Even Luzon did not yet look like an integrated economy.

As late as 1992 the pattern of traffic flows had changed remarkably little, despite a tenfold growth in traffic volume (Figure 4.7). Manila was still the only significant node. The densest traffic flow of more than 60,000 vehicles per day was recorded on the Manila North Expressway, but this fell away very quickly beyond the environs of Manila to less than 10,000 in the middle of the Central Plain. Other roads carrying significant amounts of traffic were southbound from Manila to the neighbouring provinces of Cavite, Batangas and Laguna. Elsewhere, traffic volume was uniformly thin, no more than a few thousand vehicles per day. The only noteworthy change from 1949 was the greater flow of through traffic along the National Highway down the long southeast peninsula to Bicol, previously reliant on rail and sea. Overall the network was well articulated but highly focused, resembling an octopus with many legs but only one head.

Figures 4.7a and 4.7b show more complex patterns for interprovincial passenger and commodity trips per day, towns approximating to point-to-point desire lines without regard to actual roadways. Here the dominance of Manila still shows very clearly but is overlaid with small cross-flows. The pattern is more diffuse for passengers, reflecting circular migration, especially of Ilocano people from the North Coast. Commodity trips are largely concentrated on the region of the lower Central Plain, Metro Manila and the mid-south (Cavite, Batangas, Laguna). What the maps do not clearly show is how minimal is the cross-traffic between North and South Luzon (JICA 1993a).

Given the dominance of Manila, it is logical to scale the road network in terms of concentric radii of travel time from Manila (Figure 4.8). The inner radius of three hours corresponds to the immediate environs of Manila, where travel is fairly slow because of congestion; the outer radius of 12 hours almost reaches the towns of Aparri and Sorsogon in the far north and far south respectively. Nevertheless, because of bad rural roads, some parts of the island are much more remote than measured distance would suggest.

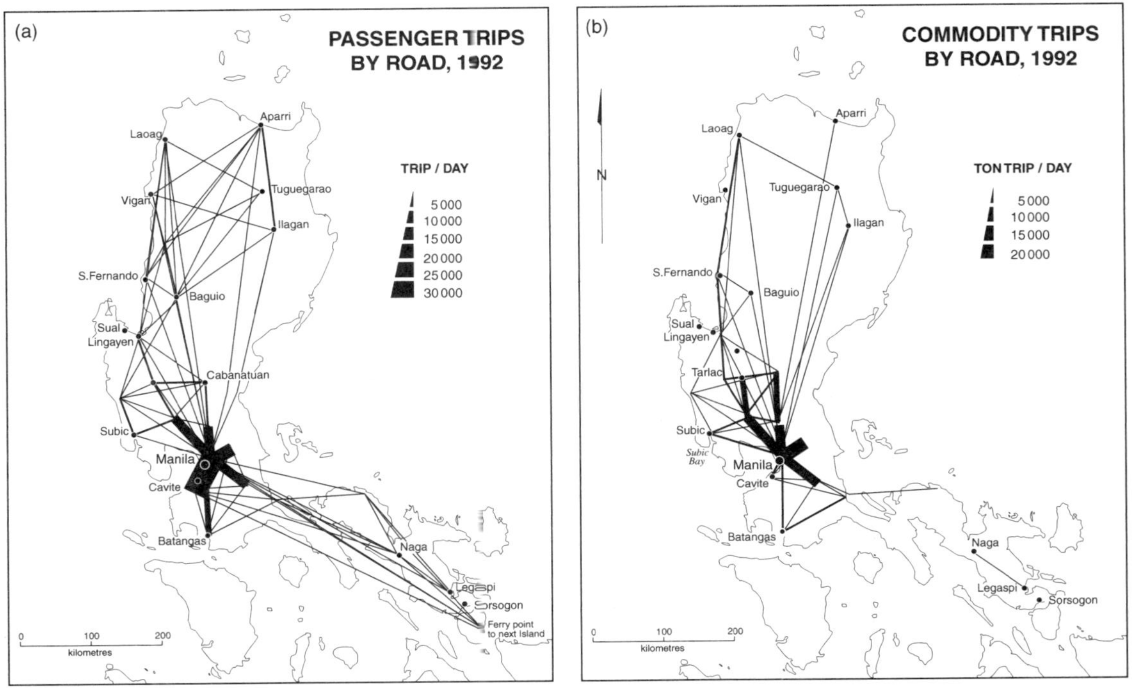

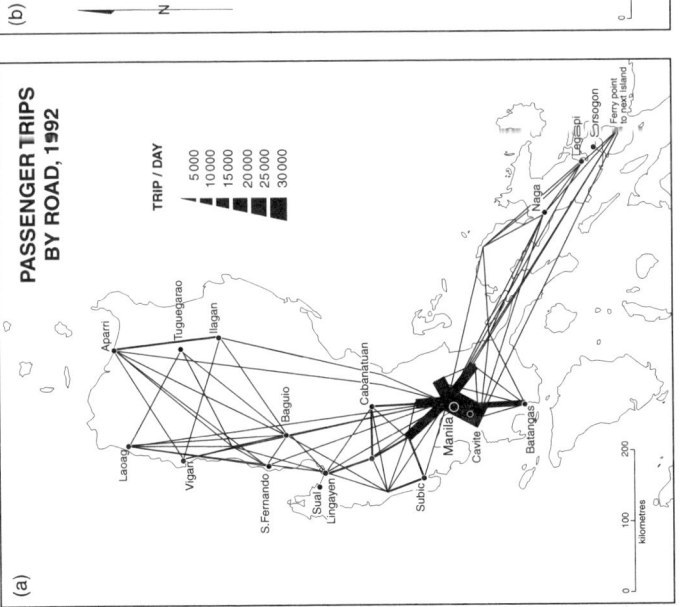

Figure 4.7 Luzon: desire-line chart: (a) passenger trips by road (trip/day), 1992; (b) commodity trips by road (ton-trip/day), 1992 (*Source*: Based on JICA 1993a). Note: A desire line is a directed line from an origin to a destination representing the demand for communication between them; it is not synonymous with traffic flow except perhaps in rural areas where a series of towns are directly connected

Actual road conditions can vary a good deal from what is portrayed in these diagrams. Despite the great advances in road construction, from two-lane macadamized surfaces to multi-lane expressways, nature continues to play havoc with under-engineered roads. For example, in mid-1996 it took several hours to negotiate a short section of the Southern Highway around

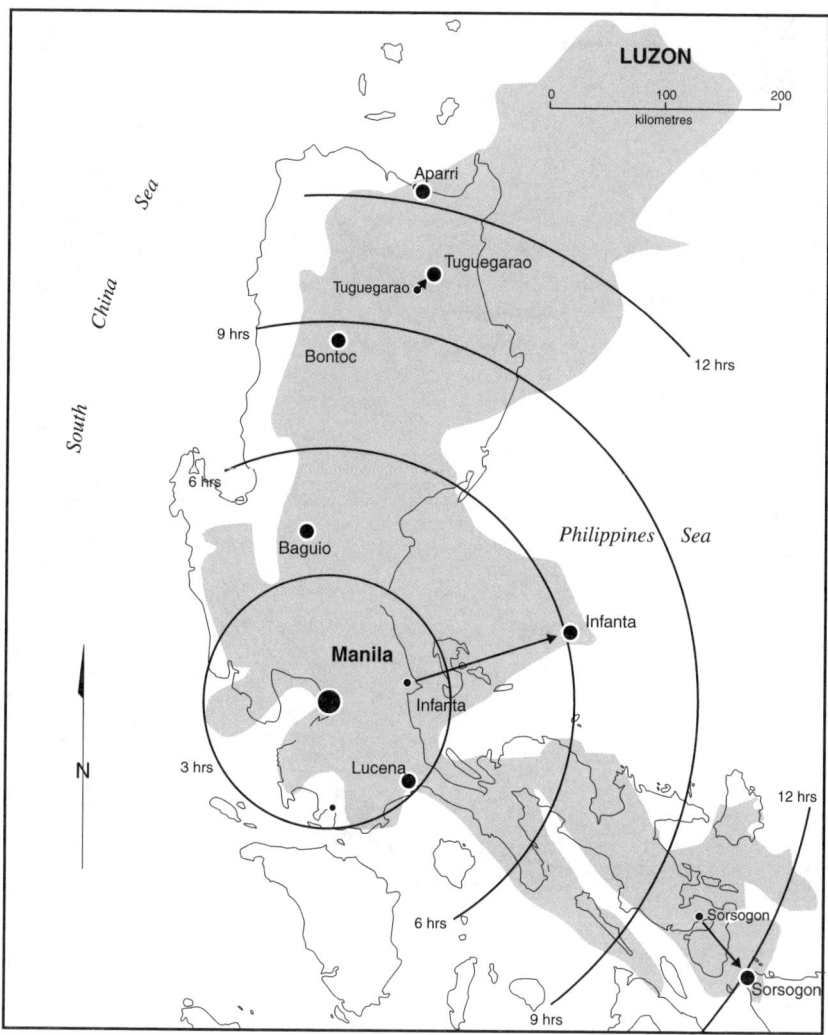

Figure 4.8 Luzon: travel time on the existing road network from Manila, 1992. Poor road conditions in eastern Luzon have reshaped the physical geography. Note the adverse positions of Tuguegarao, Infanta and Sorsogon (*Source*: Based on JICA 1993a)

Pagbilao, where the previous year heavy typhoon rains had caused the concrete pavement to sag and crack. At many other stretches along the road, typhoon repairs delayed traffic for long periods. Sorsogon was certainly not twelve hours from Manila. Similar problems hampered traffic through the mountains of northern Luzon, such as Balete Pass on the main road to the Cagayan Valley. Road conditions also varied greatly from what was shown on official maps. Sealed second-class roads could be so potholed as to be dangerous at more than twenty kilometres per hour, some were closed for repair, and some had been completely carried away by landslides or floods. Third-class macadamized roads could be impassable by normal vehicles, especially in the rainy season. Local knowledge was therefore essential and journey times a matter for discussion. Spanish and American engineers would not have been surprised.

The problem of bad roads can nevertheless be exaggerated. Anyone who travels in the Philippines cannot fail to observe the extraordinary routine mobility. From far northern Ilocos, buses leave for Manila every few minutes. If one is not passing a bus, it is a jeepney bound for the next town, interspersed with trucks, vans, cars and motor cycles. In the Central Plain and Ilocos it seems that every village has families with relatives abroad, as well as scattered throughout the archipelago. Public transport from the village to provincial capital to Manila to Saudi Arabia, Hong Kong or Los Angeles has become a seamless web, that only at the end of the road in remote corners of the island may for a couple of hours be problematic. This is globalization.

Another measure is price differentials. Compared with the Cagayan Valley (9–12 hours north of Manila), around 1992 capital city prices averaged about twice farm-gate prices for agricultural commodities; for manufactures they were about 20 per cent less (JICA 1993a: 29). These are significant differentials but, even in the case of agricultural products, were largely attributable to factors other than direct transport cost. Despite scope for marginal gains in transport efficiency, larger gains would probably flow from improved distribution. Here the main constraint is the modest size of the market in towns and regions beyond Metro Manila.

Railways

The railway system was quickly rehabilitated after the reoccupation but the MRC (now Philippine National Railways, PNR) struggled to compete with road transport. During the 1950s a programme of dieselization helped to rebuild patronage but without quite regaining prewar levels or achieving financial viability (Hartendorp 1958: 396–401). In 1939 passengers and freight had contributed equally to revenue (MRC 1939). By the 1990s freight was insignificant (Table 4.5). Short-trip passengers were mostly captured by local buses and jeepneys, leaving the railway to specialize in long haul, mainly between Manila and Bicol. The northern freight service was 'temporarily' discontinued in 1971 (Mears et al. 1974: 159). In 1985 the

Table 4.5 Manila railroad company passengers (million) and freight ('000 tons), 1904–1991[1]

Year	1904	1913	1929	1938	1952	1960	1991
Passengers	1.3	5.5	10.8	10.2	7	9.4	0.6
Freight	0.15	0.7	1.2	1.5	1	1.4	0.02

Note: 1. Now Philippine National Railways; 1991 excluding Metrotren passengers.
Source: MRC (annual), PSYB (1991), JICA (1993a).

passenger service was cut from four trains to one per day between Manila and Dagupan and soon even this token service was discontinued (Cortes 1990: 218). The southern terminal was temporarily pulled back from Legaspi to Iriga in the mid-1980s and then to the larger town of Naga because there were no longer enough serviceable locomotives to haul the train over the slopes of Mayon volcano (McBeth 1991). PNR thus became all but irrelevant to the main transport task on Luzon. Since 1990 it has been primarily a Greater Manila commuter service (Metrotren) operating along a truncated line between San Fernando, Manila and Laguna.

Conclusion

The past 150 years have seen profound changes in settlement patterns in both Java and Luzon. Scattered hamlets on tidal inlets and along navigable waterways have become nearly continuous ribbons along roads and highways and have enormously increased in density, population and mobility. Their pathways, however, were quite different. Java was the first part of Southeast Asia in which modern industrial revolution technology was systematically applied, in particular to the sugar industry (Dick 1993). Export competitiveness required the scale and unit cost of transport to adjust in line with the scale and unit cost of production. The Dutch achievement in the nineteenth century was to visualize Java as a single productive unit and to coordinate state and private enterprise, railways and public works to that end. Despite large landholdings, Luzon did not develop export plantations on anything like the same scale, so that the demands on infrastructure were also less. In the Spanish time, private investment in railways was just 195 kilometres. The American colonial government, which took over the small, bankrupt system in 1917, gave priority to roads. Whereas the Dutch colonial government in the 1920s and 1930s vigorously defended its railway monopoly against intensifying road and rail competition, the state-owned Manila Railway Company evolved into an intermodal organization. Yet despite these very different paths, by the 1970s the railway system was eclipsed in both Java and Luzon. In the very long term one can envisage three overlapping phases of development: road and cart, railway, and road/motor vehicle.

Java therefore looks to be an anomaly. Until 1900 it was technologically ahead of Japan and until the 1940s still ahead of anywhere else in East Asia. It has been very well articulated by overlapping transport and communications networks: first roads and posts, then telegraphs, railways and tramways, telephones, roads again, airways, tollroads and the Internet. Yet it is still a poor, overpopulated island, again reeling under the impact of economic crisis. Obviously technological advance and good infrastructure are necessary but not sufficient conditions for economic development. The key is the mode of production and the role of the state. The plantation system and its associated state infrastructure had been intrusions of global capitalism. At independence, the nation-state of Indonesia inherited the state railways and in 1958 it nationalized the Dutch-owned plantation system but without any commitment to maintaining a capitalist mode of production. Instead, plantations and railways were absorbed back into an indigenous mode of state patronage that had little regard to efficiency or international competitiveness. Eventually market pressures forced railways to give way to road transport and plantations to smallholder agriculture (Dick 2000).

Yet if Java's railways were a transitional technology, they had a lasting impact on economic structure. In the mid-nineteenth century Java was still birdlike in shape, the body being the populous heartland of central Java and the wings the sparsely populated western and eastern regions. Railways opened up those frontiers to capitalist plantation enterprise and associated settlement. The consequence was a permanent redistribution of economic weight from central Java to eastern and western Java. Boosted by its capital city status as well as proximity to resource-rich Sumatra, by the mid-1990s Jakarta and West Java accounted for 55 per cent of Java's economic activity, East Java, held back by stagnation of its plantation sector, 25 per cent, and Central Java only 20 per cent (ISYB 1997). Recent industrialization has reinforced this pattern.

Here may be the key to another difference between Luzon and Java. Manila, which holds such primacy in Luzon and the rest of the Philippines, has done so since the seventeenth century. Whether under the Spanish, the Americans or since independence, the economic core has been the environs of Manila, namely the Central Plain, around Manila Bay and Laguna de Bay, and the Batangas Peninsula. This primacy has been maintained during recent industrialization (Chapter 8). There has never been a rival centre. By contrast, in Java, the economic primacy of Jakarta is quite recent, hardly predating the 1930s. Until the collapse of sugar exports in the 1930s, Surabaya had been an equal rival as a centre of international and interisland commerce and in the 1990s it re-established itself as a second industrial centre (Dick 2002).

Hence, if the spatial history of nineteenth- and twentieth-century Java and Luzon was of territorial integration, this latest phase appears to be one of redifferentiation of enclaves. Export-oriented manufacturing and its associated global service sector has been heavily urban-biased (Chapter 9). The massive investment in railways was justified by the needs of agro-industrialization. Urban-based, export-oriented industrialization now demands huge investments in toll roads and telecommunications, but only in the immediate environs of the main cities. Java and Luzon do not yet have the tax revenue and purchasing power per capita to justify intercity high-speed railways or toll roads, even though population density would lend itself to such technologies.

5
Rivers: Chao Phraya, Irrawaddy and Mekong

The Chao Phraya has four tributaries:
Ping, Wang, Yom, Nan,
The Chao Phraya flows from Paknampo.

Before the mid-nineteenth century, the political and economic core of mainland Southeast Asia was not the river deltas but the up-country dry farming plateaus and valleys (Spate 1943). The lower Irrawaddy, Chao Phraya and the Mekong were still sparsely populated jungle. Although the British annexed Lower Burma in 1852 this did little more than secure the eastern shores of the Bay of Bengal. The kingdom of Upper Burma, with its capital at Ava (Mandalay), remained independent until 1885. In Siam (Thailand) a mosaic of petty statelets of the northern valleys into southern Yunnan were substantially autonomous. The situation in the Lao-speaking northeast was fluid, as also between Siam and Cambodia. All the northern kingdoms and statelets were linked by a combination of rivers and 'caravan' routes stretching from the Bay of Bengal via northern Siam and Laos to the South China Sea. The trade was not high in volume but gems, textiles and salt were standard items. Wars and later national boundaries shut off official integration (though recent attempts have been made to reintegrate the area through the Mekong project).

This chapter explores the economic and political development of the lower river valleys of the Irrawaddy, Chao Phraya and Mekong since 1850 by focusing on the changing nature of transport and communications patterns. How was the rudimentary river-based economy of the late nineteenth century centred on rice and teak exports influenced by the expansion of the trunk railway and its road-feeder system in Burma, Siam and Cochinchina? How has Thailand's economy and transport pattern been modified by political and economic developments since the 1930s and simultaneous growth of a trunk highway network and extensive feeder road system?

Attention is focused on the Chao Phraya and its tributaries, learned rote-like as above by generations of Thai schoolchildren (Van Beek 1995: 1).

The Irrawaddy and Mekong are referred to for comparative purposes. A key issue is whether the government in Siam (Thailand) has been more successful in consolidating and expanding power over its territory than its counterparts in Burma, Cambodia, Laos and Cochinchina (south Vietnam). Peninsula-like Annam (central Vietnam), which lies outside the Mekong basin, is contrasted with the Malay Peninsula (Chapter 6). Tonkin (north Vietnam), centred on the Red River valley and adjacent to China, is omitted from this study.

The new frontier, 1850s–1930s

The lower valleys of the Irrawaddy, Chao Phraya and Mekong were developed because an emerging regional and world market for rice, combined with increased security and modern transport and communications, gave value to land that hitherto had been worthless and pestilential (Figure 5.1). The consequence was what British-Burmese economist Hla Myint (1971: 120) described as 'vent-for-surplus' expansion: international trade overcame the narrowness of the market and provided an outlet for the surplus product above domestic requirements. Labour was the limiting factor. As the labour force increased through natural increase and migration, jungle was cleared and land brought into rice production with a given pre-modern technology. Plots were larger in the lower Chao Phraya than in the north or northeast because the land had more recently been cleared and blocks had not been fragmented through the natural increase of families. Rice mills were soon erected downriver at the main ports. Land clearing continued until the 1930s. What role did the waterways and railways play in this process? The other major export commodities – tin and, after 1900, rubber – are discussed in the context of the Malay Peninsula (Chapter 6).

Waterways

In 1850 water transport was the rule and land carriage the exception in the rudimentary barter economy of Siam, whose estimated five million people were concentrated on the Chao Phraya and its confluents (Ingram 1955). The principal exports were raw cotton, fish and sugar. The small amount of rice and the absence of teak from this list could be interpreted in terms of their value per unit weight being much less than the other items. Royal monopolies had prohibited rice and teak shipments and sought to maintain Chinese trade privileges, despite the entreaties of envoys from Europe. Western powers threatened gunboat diplomacy to prise open the Siamese market, then encompassing the conquered territories of Cambodia, tributary Laotian territories in the northeast and the tributary states of the Malay peninsula to the south (Chapter 6). A map of the lower Chao Phraya – based on the observations by United States missionaries – showed that the area was imperfectly known and required special investigation (GB 1856).

Rivers: Chao Phraya, Irrawaddy and Mekong 157

Figure 5.1 Chao Phraya, Mekong and Irrawaddy river basins: transport patterns, c.1885 (*Source*: Trading routes from GB 1892: 43)

The Bowring Treaty of 1855 between Siam and Britain allowed the export of rice and teak on payment of appropriate duties. Its provisions were soon extended to the United States and other European powers (Hong 1984; Wilson 1993; Brown 1993). Opening the market to international trade stimulated the more intensive use of land and labour. Between the 1850s and 1890s, economic development was concentrated on the more accessible lower Chao Phraya at the expense of the hitherto more settled upper reaches. Bangkok maintained little contact with the enterprising provincial centre of Chiang Mai, where the Indian rupee rather the Thai baht was still the main currency (HMSO 1901; Ramsay 1976).

By the 1890s the lower Chao Phraya's importance was paramount. 'No river in the world – but certainly no river of equal size – [carried]... so large a number of boats of all descriptions' (E.B.M., 1892). This was confirmed by a five-day survey undertaken at four points on the Chao Phraya in 1898 that revealed an average of 683 vessels at a given spot. In comparison, the 'highest average number of boats passing a given spot in Europe is said to be 200 per day, and the figures for the Rhine where it enters Holland are 160 average per day' (Carter 1904: 230).

Comparisons can be made with the navigability of the Irrawaddy and the Mekong (Table 5.1). Commercial transport on the Irrawaddy could be maintained with one-metre draft for 1,290 kilometres from Bhamo to the sea (Nyoe and Khin 1957: 17; Tinker 1967). On the Chao Phraya 80-ton barges with two-metre draught could be negotiated in the dry season 350 kilometres to Nakhon Sawan (Paknampo) and 700 kilometres to Uttaradit during the flood – this has been reduced to 435 kilometres by dam building. The 4,500-kilometre Mekong River tempted French engineers with the possibility of penetrating from its mouth to the Chinese frontier but this prospect never materialized.

Until the 1890s people and goods in Siam moved easily by boat along the Chao Phraya and associated canals. There were no railways, and roads were restricted to the capital and the vicinity of a few towns where they were constructed as 'meritorious works' or with unpaid corvée labour (Nontawasee 1988). The sparse telegraph network initiated in the 1880s was subject to

Table 5.1 Irrawaddy, Chao Phraya and Mekong rivers: vital statistics, *c.*2000

	Irrawaddy	*Chao Phraya*	*Mekong*
Watershed (sq.km.)	413,710	178,785	805,604
Countries within watershed	3	1	6
Length (km.: incl. longest tributary)	2,150	1,830	4,500
Navigability (km.: flood)	1,288	435	547

Source: Cohen (1998); WRI (2002).

frequent interruptions because of the difficult access without roads to repair the lines (GB 1892, 1895; Tuck 1995). Up-country cattle tracks and elephant paths were the only means of travelling on dry land. During the six-month wet season, movement by people and animals was barely feasible and it was 'quite impossible for carts to make their way over these rough and miry ways' (E.B.M., 1892: 3). In the dry season the trails were dusty and infested by disease-carrying insects. Even in the towns, narrow planks had to be used to afford passage single-file across the marshy soil.

Trade routes

Rivers were combined into northeastern and northern trade routes (Carter 1904; Bowie 1992; Smyth 1898). Wheeled conveyances were used on the northeast routes to Bangkok to carry goods long distances from the river ports of Nong Khai, Nakhon Phanom, and Pakse on the Mekong across the dry plateau to Nakhon Ratchasima (Korat) – a renowned market for elephants and ivory (Figure 5.1). The exception to the ox carts in the northeast was the heavy traffic on the Chi and Mun rivers, which accounted for the importance of Roi Et and Maha Sarakham.

The itinerant northern caravan routes to Bangkok from Chiang Mai, Chiang Saen, Nan and Luang Prabang drew trade from as far north as Yunnan in southern China by combining short porterage legs with navigable rivers, including the Salween, Chao Phraya and Mekong (HMSO 1901: 42). Riverine villages were key trading centres with many foot porters able to carry 25–30 kg per day. Chinese merchants also used pack mules to carry goods of high value per unit of weight such as opium, silk and quilts at speeds of 7 km per day. Phrae was the southern terminus of the Yunnan mule caravans but on occasions they continued to Uttaradit or Moulmein, where 'manchester' goods for the return journey were cheaper.

The journey by river and packhorse from Chiang Mai to Bangkok took between 21 and 35 days. Reputedly, it was cheaper to send a ton of cloth from England to Bangkok than from Bangkok to Chiang Mai (Ingram 1971: 116–17). The return trip could take between three and six weeks for passengers and six weeks to three months for goods. The alternative for Chiang Mai traders was to ship through the port of Moulmein on the Andaman Sea. Profits on imports to Chiang Mai from Bangkok and Moulmein were nearly equal. Yet the cost of transport by the long boats using the river from Bangkok was almost five times cheaper than that charged to Moulmein (GB 1895). Goods from Moulmein transported overland by caravan to Tak (Raheng) for forwarding by the Ping river to Chiang Mai (for distribution to the Shan states and upper Laos) were dearer but of better quality than those despatched from Bangkok. Much effort was invested in extending Chiang Mai's position as a river distribution centre by building a road to Fang, a cart road to Lamphun and a bridge over the Wang. Had there been a railway between Moulmein and Tak, almost every article of import from Bangkok to

Chiang Mai would have been excluded (GB 1896). Without the railway most of Chiang Mai's trade flowed through Bangkok along the Chao Phraya rather than via Moulmein.

Rice and teak

The importance of rivers for transporting rice and logs was emphasized in lectures to the Royal Society of Arts in London by Robert Gordon (1891), a civil engineer, and John Stuart Leckie (1894), a representative of the Borneo Company which had been extracting teak since the 1860s. Foreign interest in river passages stemmed from the availability of surplus rice in lower Siam and teak from forests in the valleys of the Ping, Wang and Yom in northern Siam. The increased demand for these two bulky commodities was stimulated by improvements in ocean transport and the prospects of a local boat-building industry. As the cash-based rice trade and the more speculative timber trade were so dependent on river transport, they are examined in more detail (GB 1896; Furnivall 1931, 1957: 29).

In 1900 the three principal rice markets in the world were Rangoon on the Irrawaddy, Bangkok on the Chao Phraya and Saigon on the fringe of the Mekong Delta (GB 1901). From a low figure in 1850, rice exports from Bangkok by the late 1880s were equivalent to those from Saigon and lower Burma, but less than from the delta plains of the Irrawaddy and Sittang centred on Pegu (Gordon 1891; BRS 1944; Cheng 1968; Morehead 1944; Adas 1974). The Chao Phraya and Mekong deltas were as rich as the plains around Rangoon; water communication equally easy; world markets equally accessible; populations equally industrious; and land tax in Siam one-third to one-half less than in Burma. However, the Pegu area had benefitted from the transformation of 'hopeless jungle' as a result of thirty years of heavy British investment in embankments, canals, roads, bridges and railways (Gordon 1891; Ireland 1907). If the Siamese government had invested more money in additional canals and other public works, perhaps rice production from the Chao Phraya delta could have matched or exceeded that of the Pegu area (Gordon 1891). Good years during the 1890s saw a large rice surplus of over 500,000 tons for export produced from the Chao Phraya delta extending 200 km from its Nakhon Sawan apex to the coast, where the width extended 80 km east and west. Having a lower bulk-to-value ratio, rice was a more profitable cargo for river transport than teak, the second major product.

Stands of teak were located in the dry hills of the upper Chao Phraya and upper Salween river basins north of the 17th parallel. As early as 1835 British subjects of Burmese extraction were working teak from forests around Chiang Mai and sending it by the Salween River to the Burmese port of Moulmein. After the Chiang Mai Treaty of 1883 British merchants were allowed to use the Chao Phraya for moving logs. In 1888 Siam would have been on a par with Burma as an exporter if the teak passing through

Moulmein had been credited to its total (Gordon 1891). By the early 1890s teak extraction in the upper Chao Phraya had come under the control of a few British firms based in Chiang Mai and Lampang (Tate 1979). It took between three and five years to move logs from the rich stands of teak to where they were consolidated into rafts at marshalling points: Lampang on the Wang, Tak on the Ping, Phrae and Sawankhalok on the Yom and Nakhon Sawan on the Chao Phraya. Rafts were moved periodically during the wet season from June to November until they reached the northern outskirts of Bangkok, where they were milled for export.

Other than rice and teak, no major exports flowed along the Chao Phraya. Pepper was brought to the capital by sea from Chanthaburi for shipping to Europe and Hong Kong. Bangkok received little from the Mekong River through the northeast trade routes other than small quantities of gamboge, gum, silk and ivory; the market for imported 'manchester' and 'bombay' textile goods was small. The area was so much shut off from Bangkok that 'trade communication is only kept up by packmen and pedlars, who travel across the hills with their bullock caravans' (Leckie 1894: 658).

River improvements

Attempts were made to improve river transport in Burma, Siam and Cochinchina by clearing the channels of obstructions and constructing arrow-like navigation canals. In Burma river improvements were conspicuously successful. They enabled the Irrawaddy Flotilla Company to maintain a service of fast mail and cargo steamers three times a week each way between Rangoon and Mandalay, and twice a week between Mandalay and Bhamo (Wright 1910; Thet 1989). Indeed, everyone 'of importance used the Flotilla steamers – Viceroys, local officials and, on two occasions members of the [British] Royal family' (Laird 1961: 153).

In Siam engineers went beyond the short cuts and auxiliary canals of the past by diverting the Chao Phraya through Ayutthaya (Van Beek 1995). From the late 1850s the prospect of cultivating sugar cane in the western central plain led royalty to finance the construction of four canals to float the final product to Bangkok. In the 1870s these were complemented by an eastern set of canals linking the Chao Phraya and Bangpakong rivers designed primarily to develop rice cultivation for the price of sugar had since waned (Figure 5.2). In 1889 the Siam Canals, Land and Irrigation Company initiated the large-scale Rangsit Project northeast of Bangkok, though for irrigation rather than transport.

The Mekong was less susceptible to river improvements. Although thought by the French to offer an uninterrupted means of penetrating Cambodia, Laos and western China, an expedition between 1866 and 1873 dispelled any notions of long-distance navigation (Osborne 1975, 2000; Tuck 1995). On the lower Mekong, navigation was 'lively' to the Cambodian port of Kompong Som (Sihanoukville), while the river port of Phnom Penh

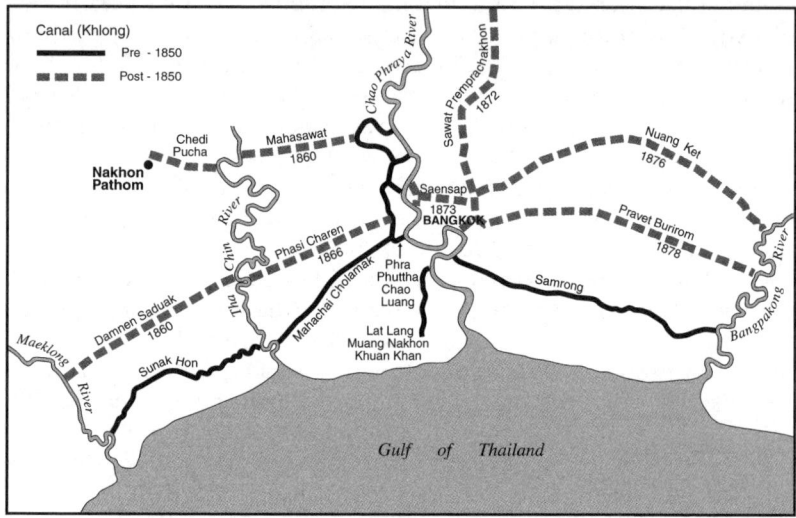

Figure 5.2 Lower Chao Phraya Basin: canals built since 1850 (*Source*: Based on Van Beek 1995)

could be reached by ocean-going vessels, despite obstructions between the river and the Tonle Sap (Robequain 1944). Towards Laos navigation was more difficult, being uncertain between Kratie and Savannakhet and involving transhipment to bypass the Khong (Khone) Falls on the border between Cambodia and Laos. Navigation was again feasible between Savannakhet and Vientiane without breaking bulk, but difficult to maintain throughout the year between the two Laotian capitals of Vientiane and Luang Prabang.

The exploitation of copper, lead, iron ore, tobacco, silk and cotton in the Mekong plain and the then wealthy Siamese provinces of Battambang and Siem Reap awaited the spread of modern overland transport to supplement the rivers and the navigation canals in lower Siam. Rail was a means to extend the river and canal system by opening up a strip of land 10 km either side of the line for cultivation. Progressively, the river lost its pre-eminent role in Siamese commerce and everyday life. The railways enabled people in the north and northeast increasingly to escape the vagaries of river transport.

The railway strategy game

The rice and teak trades were well served by waterways – to all intents and purposes there was no transport constraint. Railways and telegraphs were constructed by the state (British, French or Siamese) primarily for strategic reasons, namely to consolidate central control over up-country statelets and secure the borders of what in the late nineteenth century became territorial states (Figure 5.3). Administrative reforms of the 1880s and 1890s replaced the old revenue system with a new one as the Siamese shifted 'to a much less structured [social order] that was able to assimilate new immigrants, particularly the Chinese, and to meet Western demands for individual autonomy, private property, contracts and wage labor' (Wilson 1993: 143). Without railways, telegraphs and tax reform there would have been no modern state of Thailand (Winichakul 1994). Passengers, especially troops, mail and parcels mattered; bulk freight did not.

Siam was a late starter in railway development compared with British Burma and French Indochina (Table 5.2). As part of its plan of incorporating indigenous Upper Burma, in 1877 the British administration completed Burma's first narrow-gauge 260-km railway from Rangoon to Prome and in 1885 a second from Rangoon to Toungoo. By 1885 the French had opened the first line in Cochinchina from Saigon to Mytho, which was intended as the first stage of a narrow-gauge inland line to Phnom Penh (Bouinais and Paulus 1885). Proposals to extend these colonial railways to claim the fabled wealth of south China threatened Siam's territorial integrity (Chandran 1971, 1977; Lee 1989). Siam's entry into railway development with reliance on foreign capital and expertise was spurred not by concern with the shortcomings of river transport or developing resources but rather by the need to consolidate Bangkok's control over its outlying principalities at a time when the British and French were squeezing the Kingdom. 'The total gain by rail will be perhaps a day and a half, in a country where time is cheap, at a cost of handling [that is] twice more than is necessary' (Smyth 1898: 41). Thus the British and French brought the railway to Asia 'as an agent of imperialism extending the economic and political hinterlands of ports which were always the main centres of Western influence' (Lee 1999: 13). Conversely, the Siamese, like the Chinese, used railways to promote national integration and preserve national sovereignty (Lee 1977; Huenemann 1984).

Table 5.2 Burma, Indochina and Thailand: railway construction, 1869–1940 (route-km)

	Burma	Indochina	Thailand
Pre-1886	536	0	0
1886–1895	867	161	0
1896–1914	1143	1872	1120
1915–1920	39	20	1095
1921–1936	731	1227	885
1937–1940	0	71	30
Total	3316	3361	3130

Source: Nyoe and Khin (1957: 10); Mitchell (1998); RSRS (1947).

Railway routes

During the 1880s five key railway projects were mooted for the north and northeast regions of Siam linking major towns on the Chao Phraya, initially with inland settlements and eventually riverine and coastal ports on or beyond Siam's borders (Figure 5.3). The routes were:

- Bangkok to Nakhon Ratchasima via Ayutthaya and Saraburi (which could be extended to Nong Khai on the Mekong and to Ubon Ratchathani on the Chi-Mun rivers);
- Ayutthaya to Uttaradit (which could be continued to Chiang Mai and Chiang Saen at the Mekong's head of navigation with branch lines to Luang Prabang and to Tak and an extension to Moulmein on the coast);
- Bangkok to Chachoengsao on the Bangpakong River (which could proceed to Cambodia and also link Ubon Ratchathani).
- Moulmein to Szemao (Simao) via Chiang Mai and Chiang Saen (proposed to open up communications with southwest China);
- Mekong above the Khong Falls to the Sea.

Only the first three propositions were feasible. The Siamese government under King Chulalongkorn (1868–1910) and his foreign minister, Prince Devawongse, did not support the Moulmein to Chiang Mai line because a link to Burma would have facilitated British economic penetration and inhibited Bangkok's ability to dominate the north's economy. In 1893 the French takeover of Lao territory east of the Mekong removed the Khong Falls railway from Siam's list.

Siam's first major railway was the 265-km standard-gauge line (1.435 m) between Bangkok and Nakhon Ratchasima and was technically incompatible with the narrow gauge used by the French and British. Its opening on 9 March 1896 was reportedly Chulalongkorn's most auspicious day. The railway showed that 'the extension of Thai political control over the northeast was inexorably connected with the creation of modern communications and transport' (Keyes 1967). The journey from Bangkok to Nakhon Ratchasima was reduced from two-to-three weeks by ox cart to less than one day (Dixon 1978). Although economically unattractive, the line was completed because political pressure from the French in the 'remote' northeast region was considered harmful to Siamese interests.

In selecting foreign bidders the Siamese government was careful to play off rival British and German interests to oppose French designs on Siamese territory. The contractors were British, but Germans exercised executive power in the Royal Railways Department in Bangkok – a recipe for friction and delay. The railway was begun as a joint public–private company with British capital but, after costly litigation with contractor George Murray Campbell, the contract was cancelled and the railway was completed as a

Rivers: Chao Phraya, Irrawaddy and Mekong 165

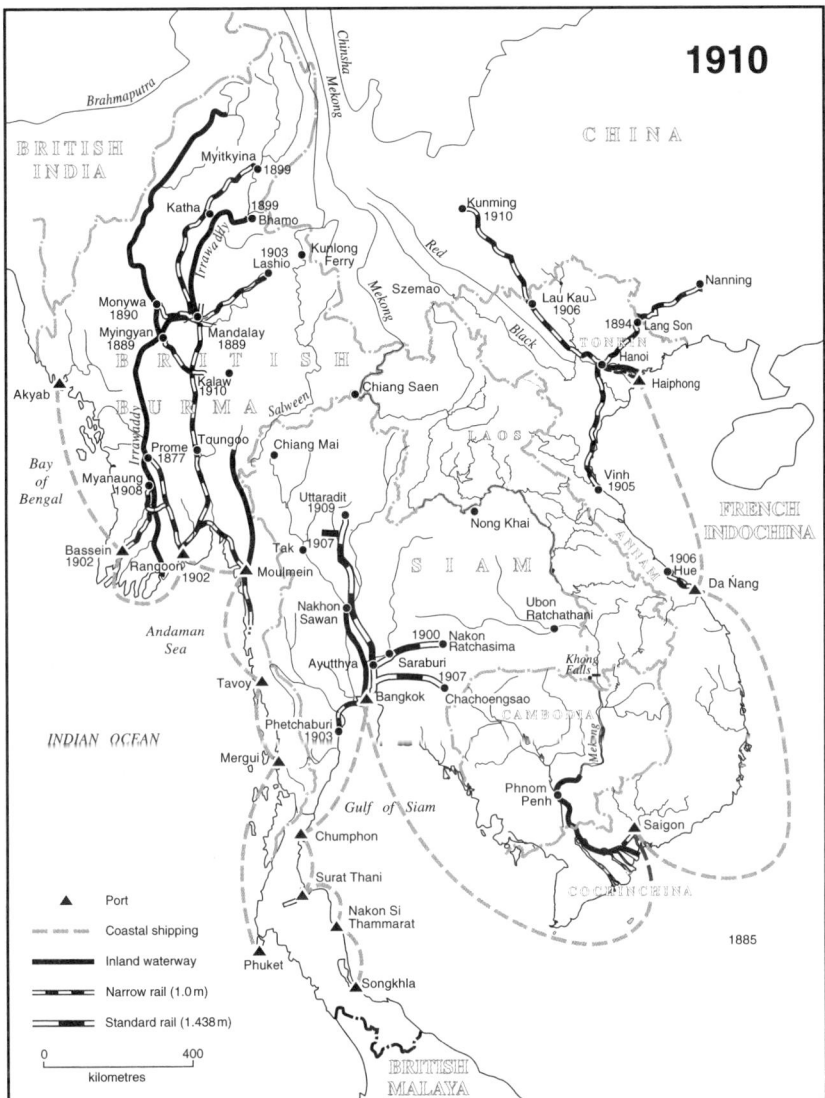

Figure 5.3 Chao Phraya, Mekong and Irrawaddy river basins: transport patterns, 1910 (*Source*: Based on information derived from BRGO 1941; Holm 1978; Théry 1931; Nyoe and Khim 1957). Note: connections within China in Figures 5.3–5.5 are not fully developed.

state enterprise using German technical personnel and Chinese labour. Rates were based on what the traffic would bear. Goods tariffs were divided into four classes according to the value of commodities. Passenger fares were 50 per cent higher in areas without water competition.

The new railway was criticized by the Dutch engineer, J. Homan Van der Heide (1906), who was an adviser to the Royal Irrigation Department between 1902 and 1909. Without favourable conditions within 10 km of the track, he argued that the railway would not automatically generate a surplus product for export. Investment in an irrigation, drainage and flood control scheme for the lower Chao Phraya would have produced a better return than the railways (Feeny 1982; Takaya 1987). While this made good sense, his own irrigation programme was discontinued because the government deemed it too expensive, particularly as farmers could not be expected to pay for the water (Wijeyewardene 1973). Railways were preferred over irrigation schemes because they provided jobs for sons of the ruling elite.

Van der Heide's assessment of the long-term benefits of strategic railways was premature. Admittedly, steam launches competed successfully for passengers between Bangkok and Ayutthaya (Holm 1978). Further, only small quantities of paddy were exported from the dry northeast in the late 1890s, but middlemen were ready to make use of the advantages of improved transport by encouraging Siamese cultivators to shift from subsistence to commercial agriculture (GB 1901). Subsequently, the extension of the state railways, with the aid of foreign loans, engineers and technicians, generated considerable economic benefits. Imported goods were moved by rail to Nakhon Ratchasima for forwarding on to Cambodia and French Laos. After 1900 migrants to the strategic northeast were almost exclusively Siamese (Dixon 1978).

In 1902 the inability of river transport to move troops speedily to a trouble spot triggered the extension of railway links with the north (Brown 1998). When rapprochement between France and Britain in 1904 guaranteed Siam's independence as a buffer state, the government was able to negotiate its first foreign loan for the northern railway (Tuck 1995). While this freed other funds to source the eastern railway from Bangkok to Chachoengsao in 1907, the northern railway proceeded apace, reaching Uttaradit in 1909. 'By 1912 work was under way on the northern railway which would parallel two rivers and cross three, ultimately supplanting them as the prime transport links between north and south' (Van Beek 1995: 127). Cutting through the northern mountains created logistical problems but Chiang Mai (751 kilometres) was reached in 1921 (Figure 5.4) – the journey taking 70 hours with overnight resthouses at Phitsanulok and Lampang (Ramaer, 1994). After the northeast railway had been extended to Ubon Ratchathani in 1930, subsistence farming was transformed into commercial agriculture and pigs and fowls were railed to Bangkok (Pendleton 1943). The increase in commerce led to a programme of double tracking in the Central

Rivers: Chao Phraya, Irrawaddy and Mekong 167

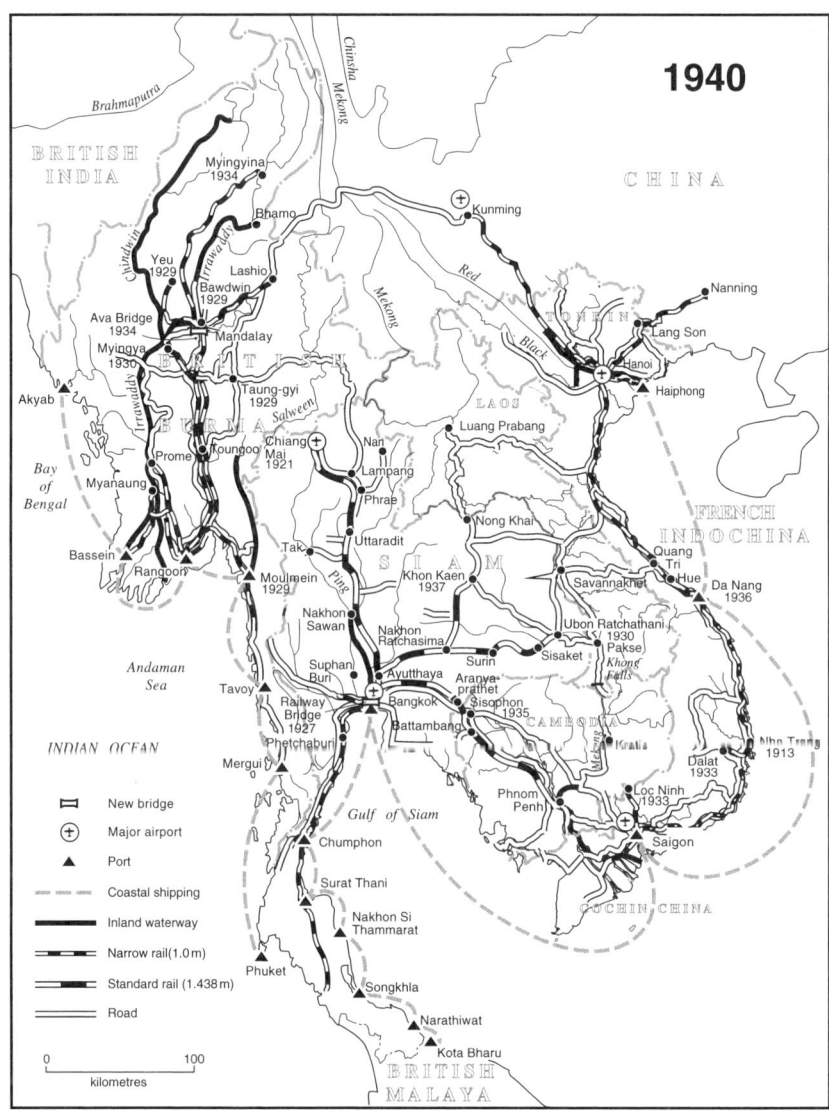

Figure 5.4 Chao Phraya, Mekong and Irrawaddy river basins: transport patterns, 1940 (*Source*: Based on information derived from BRGO 1941; FMS 1939; Holm 1978; Murray 1980; Ramaer 1994; Rimmer 1971a; Robequain 1944; Spate 1943; Tinker, 1967)

Plain, the construction of a second route onto the Korat plateau and the progressive introduction of European diesel-electric locomotives.

Prince Burachat's influence

The Royal State Railways were protected from highway competition. This policy was enunciated by Prince Burachat – the thirty-fifth child of King Chulalongkorn – who had graduated from Cambridge and studied engineering at Chatham and Woolwich (NAT 1919). During the First World War he became Commissioner-General of the Royal State Railways/Royal Railways Department and sought to replace foreign engineers with Siamese by spreading educational and employment opportunities beyond the aristocracy. When he became Minister of Communications (1926–32), the Railways also assumed responsibility for road works and accounted for almost 10 per cent of expenditure (Batson 1984: 91). A pioneer in developing Siam's civil aviation and telecommunications, Prince Burachat preferred building railways to highways because 'in investing money in roads, the State obtains no direct return, whereas in the case of the Railways, a return on the capital is assured' (Holm 1978: 247). Obviously, no thought was given to the potential returns from vehicle licence fees or petrol taxes.

Prince Burachat's rudimentary twenty-year national plan instituted in 1920 saw the railways as being the primary trunk lines supplemented by a system of roads, cart tracks and possibly narrow-gauge railways (1 m) opening up outlying areas. A rule-of-thumb was that dry weather sleds and carts could operate within 50 km of the railhead in the plains area and 15–25 kilometres in hilly districts where trails were restricted to pack mules and bearers. Before any road improvements, traffic censuses were required to determine if the expenditure was justified. Excellent results were achieved under this policy as the substitution of carts for pack animals on the 244 kilometres improved road between Lampang and Chiang Mai led to an 80 per cent reduction in cost per ton-km (NAT 1919). There was an agreement that trunk roads more or less paralleling the railways might be justified in the future where traffic warranted. Although there were criticisms of Prince Burachat's pro-rail policy, he claimed they came from motor car dealers and joy riders (Batson 1984: 101).

There were problems with realizing Prince Burachat's integrated rail–road system to link up places with dense populations and rich land available for cultivation in the northeast and southeast regions. Eventually, the policy led to:

- a route extension from Chachoengsao to Aranyaprathet (255 km from Bangkok) on the Cambodian border built by Chinese labour (completed 1926);
- a rail bridge across the Chao Phraya between Bangkok and Thon Buri to link the northern and southern lines (completed 1927); and

- lines built largely by Lao labour from Nakhon Ratchasima eastwards to Ubon Ratchathani (575 km) and northwards towards Nong Khai (624 km).

Cuts in government expenditure during the mid-1920s precipitated a revision of Prince Burachat's integrated transport plan, leading to the adoption of a uniform narrow-gauge railway throughout Siam by 1930 (NAT 1925; Ramaer 1994). Provinces with good waterways received low priority in the allocation of land transport projects.

As witnessed by the shortage of freight wagons, the railways suffered chronic underinvestment until the late 1920s, even though they absorbed most government investment in infrastructure. The extension of the line to Cambodia was unremunerative and the link through to Saigon was never completed on the French side. By then 3,300 kilometres of narrow-gauge lines of the Royal State Railways linked all outer regions to Bangkok and Siam outstripped both Burma and Indochina in terms of per capita provision (Table 5.3). Freight rates in Siam remained high because there were few economies of scale in using the capital-intensive rail track. These rates couraged the movement of commodities from the northeast and north to Bangkok and held back the spread of cash crop production.

Rail–river competition was varied between the northeast and north. In the northeast, the railway and its local feeder roads did not affect the movements of goods on the river between Roi Et and Ubon Ratchathani. While the railway competed successfully with the slow and small-scale road caravans, it could not rival the cheapness of the paddy boats. In the north, however, the extension of the railways from Bangkok reduced transit times dramatically and caused a shift in traffic. After completion of the railway to Chiang Mai, the freight rate was one-third less by rail than by boat (Manarungsan 1989: 79). This virtually eliminated river transport between Chiang Mai and Bangkok. The timing was fortunate because deforestation was causing river channels to narrow and depths to diminish.

High rail freight rates precluded the movement of bulk commodities from the north. 'With regard to the staple exports, teak and paddy, these will in

Table 5.3 Burma, Indochina, Thailand: access per capita to rail, roads and telegraph lines, late 1930s (kilometres per 1000 people)

	Rail	Roads	Telegraph
Burma	1.96	16.25	7.98
Indochina	1.26	15.50	2.31
Thailand	2.36	2.47	5.79

Source: Tate (1979: 27).

all probability follow the course of the roadways and waterways, for some time to come' rather than the railway (GB 1901: 49). This prescient observation was borne out by a consideration of the effect of railway development on seasonal rice movements on railways in Burma, Siam and Indochina between 1900 and 1940 (Figure 5.4). None of these railways could profit from heavy volumes of freight because the great export commodities – rice and teak – were mostly shipped by river.

Rice and teak again

In 1905 most export rice had come from accessible areas around Bangkok (Ingram 1955). Some 60 per cent of exports came from the districts of Ayutthaya, Rangsit and Suphan Buri where recently cleared holdings exceeded 4-ha. Yet only half of the land in the Central Plain had yet been cleared for cultivation due to the shortage of labour (Manarungsan 1989: 15). Thus the main benefit of the railways to these areas was not in moving their crop to market but in bringing down some 10,000 farm labourers from the northeast. Poor inland transport connections and the resultant high marketing costs limited rice exports from the northeast and north, as also the small size of landholdings (Van der Heide 1906). Even the railway to Nakhon Ratchasima had failed to boost exports because it took six weeks and eight carts to carry one ton of rice there from Siam's frontier (Thompson 1967: 495). Less than 1 per cent of rice exports came from the northeast region and less than 1 per cent were from north of Nakhon Sawan. Extensions of the railways boosted production from the northeast between 1925 and 1935 from 102,000 tons to 276,000 tons compared with 39,000 tons and 78,000 tons for the north (Ingram 1955).

The railways also had a marked effect on the distribution of rice mills. In Burma most mills (including the 'big' European mills) were located outside the ports of Rangoon, Moulmein, Akyab (Sittwe) and Bassein but small mills spread up-country along the main railway lines during the 1910s, particularly as the revision of railway rates favoured rice against paddy (Spate 1958). Similarly, rice mills were decentralized from Bangkok so that more milled rice (including glutinous rice for domestic consumption) was carried in proportion to paddy (Hafner 1973). In the northeast, for example, rice mills were established in Nakhon Ratchasima from 1900 and Ubon Ratchathani from 1928 (Dixon 1999: 44). These mills were isolated from export markets by the railways charging higher freight rates for rice than paddy (Holm 1978: 239–40). In Indochina the principal modern milling centre remained in Saigon-Cholon (Chapter 9), but small mills diffused to all rice areas. No observer has related this shift to the railways (Miller 1947; Murray 1980).

River transport retained its role in the movement of teak from northern Siam because rail freight rates were set too high for bulk commodities. Annual output in this British-directed, capital-intensive endeavour therefore continued to fluctuate erratically according to flood levels (Campbell 1986). As the cost of purchasing an elephant doubled between 1896 and 1912 because of the increased demand for their services, the Borneo Company replaced them in Chiang Mai province with cart roads, light railways and machinery in 1913 (Manarungsan 1989: 130). Further increases in the cost of purchasing elephants during the 1920s led to a shift to two-wheeled carts pulled by buffalo, canals and dams, light railways and timber trucks. The lack of interest of the large teak companies in branch railways and the railways in moving logs from the head of navigation at Nakhon Sawan is attributed to collusive agreements between the two aimed at maintaining the foreign monopoly (Holm 1978: 239). These transport developments in Siam did little to narrow Burma's lead in teak output (BRS 1946; Miller 1947: 406). By the early 1930s Siam's timber trade had begun to decline, but the railways enabled marshalling centres such as Lampang, Phrae and Tak to survive as provincial administrative centres. By then the prime teak market was no longer Europe but East Asia.

Railway era

Unlike Europe, in Southeast Asia railways preceded any mass transit by roads. Focusing narrowly on rice and teak underplays the revolutionary impact of railways between the 1890s and 1930s on agriculture, mining and social mobility in Burma, Indochina and Siam (Desfeuilles 1927; Silverstein 1964). Once administered, reliance on the 'railway drug' was addictive as there was no doubt in the colonial mind that 'the Siamese, like the Burman, is very fond of a jaunt in a railway carriage' (Smyth 1898, I: 41–2). In Burma an early benefit of railway development was that travellers passing through the Myitkyina district were no longer obliged after 1891 to 'light fires around their camps at nights to protect their transport animals from the attacks of wild beasts' (Scott 1921: 312). In Indochina most locals only travelled an average 40 km between the next market place or city during the 1910s and 1920s (Robequain 1944).

In Siam the railways 'changed the ways in which people lived, thought and related to each other' (Lee 1999: 11). The journey time from Bangkok to Chiang Mai in the 1920s was slashed from six weeks to 26 hours, resulting in trains taking precedence over boats in north–south goods and passenger movements. Thousands of up-country people were now able to visit and do business in the capital; high-bulk low-value products of the interior gained access to the world market. By the mid-1920s first-class travellers were able to enjoy an express train with sleeping accommodation. Although freight rates were high, rail traffic to Chiang Mai dwarfed trade by other transport modes and ensured an end to periodic famines in the north. The railway transformed towns in the northeast such as Surin and Sisaket from clusters of buildings into regional market centres. Ubon Ratchathani attracted Pakse's trade from Saigon.

Besides knitting the isolated north and northeast regions to the central region and extending the government's reach, the railway operation boosted Siamese self-confidence about their ability to run efficient technical services without foreign domination. Siam's railways proved to be not only a vital physical asset but also an important psychological one in modernizing the population's social structure and outlook (Thompson 1967). Railways gave birth to a skilled proletariat in the railway workshops, provided initial jobs for immigrant Chinese and trained professional personnel, notably engineers and accountants. Conversely, the railways were claimed to have imposed upon Siam 'an intensified subservience to Western informal imperialism' and to have delayed the construction of irrigation and highway systems (Holm 1978: iii). As in Europe and the United States, railways intensified land speculation and gutted some cottage industries. Although the railways improved Siam's international position and elevated the standard of living, they did much to preserve the power of the ruling elite.

Waterways revisited

Railways did not completely wipe out water transport in Burma, Indochina and Siam. There was evidence that passengers in Burma switched to rail where the journey by country boat was slow and tedious (Rangoon to Toungoo) (Scott 1921: 398). Yet Burma's two principal prewar transport operators – the Irrawaddy Flotilla Company dominating inland waterway movements and the Arakan Flotilla Company monopolizing coastal movements – regularly carried annually more than eight million passengers and more than one million tons of cargo (Nyoe and Khin 1957: 16). Indeed, the Irrawaddy Flotilla Company was reputed have to the world's largest inland waterway fleet with some 270 steamers and 380 engine-less barges (BRS 1946). The double-deck paddle steamers operating between Rangoon and Mandalay were 'the last word for the Express Service' (McCrae and Prentice 1978). If 'contract and private carriers, by country boats and other craft' were added 'the resulting total may be as much as three or four times the amount shown above' (Nyoe and Khin 1957: 16). Indeed, the total volume of river transport was 'not far short of the tonnage carried by rail' (Tinker 1967: 280).

In Cochinchina, 2,500 km of navigable waterways in the Mekong delta – natural and canalized – continued to handle internal traffic. Besides carrying mails on the Mekong to Luang Prabang in Laos, the subsidized Compagnie des Messageries Fluviales offered a service three times a week to Phnom Penh and, during the high-water season, to Battambang. River transport also played a key role in shipping exports of rice and corn from Cambodia and Cochinchina. A railway was built in 1897 to circumvent the Khong Falls and extended in 1920, though little progress was made in carrying goods from Khong Falls to Savannakhet.

Areas beyond the river and rail system in the plains of northeast and north Thailand, Burma and Cambodia remained reliant on the ox cart and the bicycle. In the northern mountains, pack mules and human carriers were also used from the rail and riverheads in Thailand to link Burma, Tibet and Yunnan. An average human carrier travelled 29 km a day with a 27-kilogram load and bartered his goods for salt and *miang* tea (Thompson 1967: 495). Before cars, trucks and buses spread to tracks in the north and northeast, caravans of 40–60 pack animals still carried 2.5 tons to the rivers and railheads. Government-subsidized air services not only attracted crowds to provincial centres but also arrived ahead of vehicles in remote provinces such as Nan. Meanwhile the telephone system remained the 'worst in Asia' (Batson 1984: 99–100).

With the onset of the depression in the early 1930s there was a temporary reversion of traffic in Siam from rail to river but the shift in traffic to trucks and buses was permanent. By undercutting the railway's rates and fares, the waterways forced the abandonment of hill surcharges. This was the 'competitive destruction' that Prince Burachat's transport plan had sought to avoid. Annual net returns from the Royal State Railways varied between 1 and 6 per cent until the depression, when freight rates were reduced and income fell (Manarungsan 1989: 180; Holm 1978). The depression not only stopped the frontier expansion process but also triggered the political changes that led to Prince Burachat, the Minister of Communications, being ousted from government.

Economic development since the 1930s

From the 1930s, economic development in Siam became more democratic. A civil–military coup in 1932 overthrew the absolute monarchy. The timing of technological change of cheap buses and trucks may have been fortuitous, as elsewhere in Southeast Asia, but it coincided with the introduction of more technocratic and populist government in Siam (its name was changed to Thailand in 1939). Although the government promoted the development of partly- or wholly-owned Thai enterprises, roads opened the countryside to Chinese and western commercial development more than the railways. By the late 1930s Japanese entrepreneurs joined the process and consolidated their influence when the two countries became allies during the Pacific War (1941–45). The *Royal State* Railways – note the emphases on both *King* and *bureaucracy* – lost control and never regained it. The involvement of the United States in strategic road-building during the Cold War after the 1950s was the final nail in the railway coffin. Rural Thailand was becoming a smallholder and small enterprise economy.

Roads in the 1930s

When the political turmoil of 1932–33 precipitated the end of the absolute monarchy, the budget for the railways was reduced in a show of populist support for roads by the new government. Prince Burachat left for Singapore in 1933 and died there in 1936. The underfunded Highways Department had built very few passable roads for the increasing number of motor vehicles (GB 1935). During the rainy season the roads quickly became a morass and travelling by motor vehicle an uncertain adventure. Free of any railway competition, the most important highway was between Chiang Mai and Lampang (320 km) (Figure 5.4). It revolutionized the region's economy by facilitating the movement of contraband opium (Thompson 1967). When the Royal Automobile Association of Siam was formed in late 1934, there were just 95 kilometres of first-class road in Thailand (Siam 1935). In 1935, 47 vehicles and 237 small trucks (largely Ford and Chevrolet) were registered in the 'roadless' northeast (Thompson 1967: 510). Motor lorries were moving between the Khon Kaen railhead and Nong Khai (Andrews 1935). In 1936 Thailand only had 1 vehicle per 310 square kilometre compared with 1 to 22 square kilometre in Indochina. The government then introduced a new 14,900-km roadbuilding program to connect principal cities (KOS 1936–37).

Thailand's road development was as slow as Burma's and much slower than Cochinchina (Figure 5.4). In Burma there was no real road system: the short and isolated roads in the Delta were just adjuncts to river transport and ferries and filled the gaps. The only rail–road bridge crossing the Irrawaddy was that built at Ava in 1934 (Nyoe and Khin 1957). 'High-prowed bullock carts' prevailed until the influx of cars, lorries and 'garishly painted' buses were introduced in the 1920s (Tinker 1967: 281). Mule paths provided international connections, as there was no passable road out of the country until the all-weather Burma Road to Yunnan was opened in 1939. For teak extraction, Burma in 1940 still employed 7,000 elephants and 15,000 buffaloes (BRS 1946).

In Indochina road competition since the 1920s had been marked on both rail and waterway transport. A road programme was used to link the various ethnic groups under French jurisdiction. Also a regular public service of motor cars moved overland between Saigon and Hanoi and there was a daily service between Saigon and Phnom Penh (GB 1923). Car travel was also possible on Colonial Road No. 1 from Battambang to Lang Son on the Chinese frontier – a distance of 2584 kilometres. In 1930 an all-weather road linked Savannakhet in southern Laos to the coast (Osborne 2000). When the first railway express travelled between Saigon and Hanoi in 40 hours in 1936 there was already a network of 27,500-km motorable roads in French Indochina and over 17,800 motor cars (Robequain 1944). The diversion of Laotian trade through the railways of northeast Thailand prompted the extension of the Saigon–Kratie road around the Khong Falls to Pakse to divert trade from Bangkok to Saigon. 'In 1900, at low water it took 65 days or more to go from Saigon to Luang Prabang; in 1909 the time was reduced to a maximum of 50 days; in 1937 the upstream trip took 37 days, downstream 27 days; and at highwater, it took 27 or 22 days respectively' (Robequain 1944: 115). Despite this effort 'French Upper Laos was still farther [in time] by water than Saigon is from Marseille' (Robequain 1944: 115). Progressively, the colonial government favoured land transport at the expense of the vessels of the Compagnie des Messageries Fluviales on the Mekong.

The Pacific War and its aftermath

During the Pacific War road transport was boosted by extensive damage to the waterway and railway systems in Burma, Indochina and Thailand. Burma was the most badly affected (Figure 5.5). All of the Irrawaddy Flotilla Company's craft were scuttled, sunk or removed to India and Burma Railways was severely handicapped when the sole bridge across the Irrawaddy River was damaged. Burma's riverine commerce and the railways were immobilized by wartime bombing and insurgency. There was a phenomenal rise in the use of heavy trucks for commercial haul and passenger buses and attempts to integrate roads into a national system. The remnant Irrawaddy Flotilla Company was nationalized under the Inland Water Transport Nationalization Act, 1948. As the Inland Water Transport Board (Irrawaddy Section), its operations were crippled by the insurrection at the time of independence in 1947 and 1948. The insurrection also affected the rehabilitation of Burma's railways, which were singled out for further attack during the civil war. For two years no trains ran for more than 16 kilometres outside Rangoon. Although by 1943 the so-called 'Death Railway' had extended the Thai rail network to the Three Pagodas Pass (415 km) in Burma, the connection was abandoned after the end of the Second World War.

In Indochina the Japanese closed the Khong Falls railway. Bombing also halted railway services elsewhere in Indochina. Two-thirds of the locomotives and rolling stock were destroyed. Conversely, the link between Sisophon in Cambodia and the Thai Railway was completed during the Japanese occupation, but traffic ceased in 1961. Since then part of the track has been removed and what remains is in urgent need of repair (Wickenden 1994). In 1945 the French reopened sections of the Saigon–Hanoi line but they were constantly under attack by guerillas (Joint Development Group 1970). When the French left Vietnam in 1954 the railway system was severed at the 17th parallel and not restored until 1976. By then the Saigon–Hanoi line had only been in service for six years. As traffic on inland waterways was also restricted during this period, a marked shift to truck transport occurred in goods transport and cheap buses in short-distance passenger transport. Increasingly, domestic air transport was available for passengers and mail on longer-distance routes and to inaccessible areas by Air Vietnam, established in 1951 (Davies 1997).

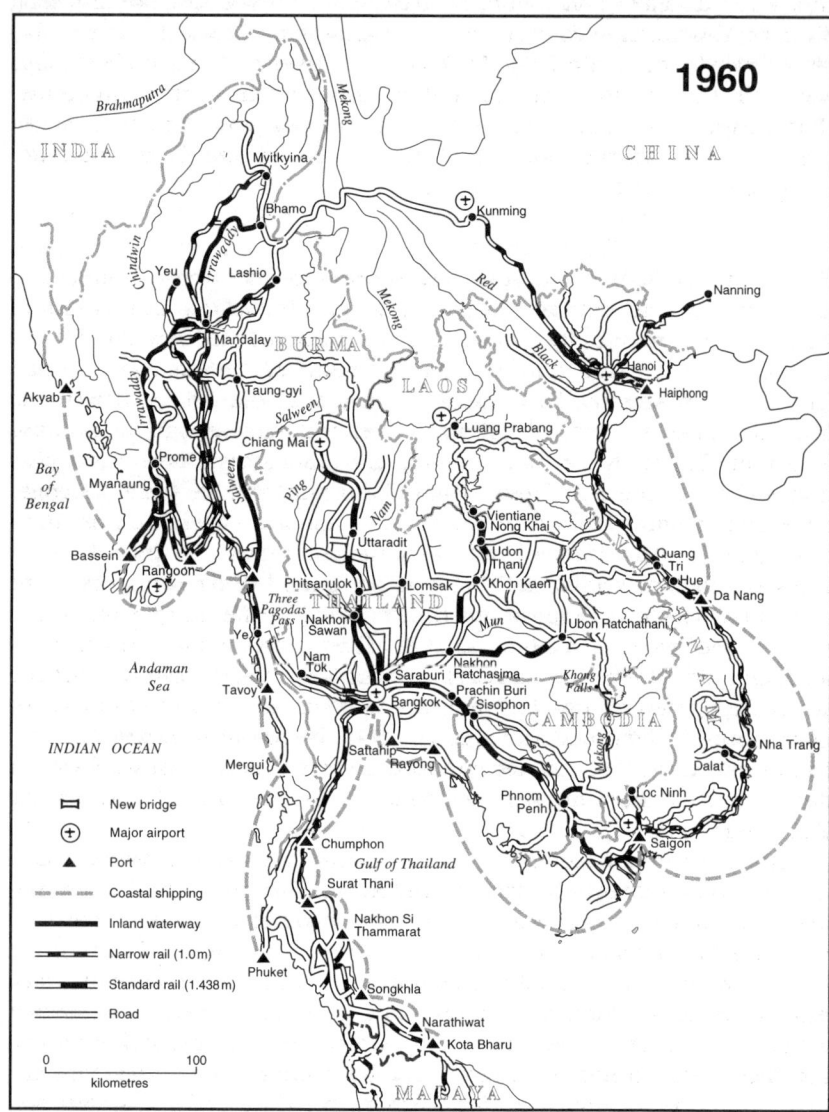

Figure 5.5 Chao Phraya, Mekong and Irrawaddy river basins: transport patterns, 1960 (*Source*: Rimmer 1971a)

In Thailand on the eve of war in 1941 the State Railways were able to carry seven million passengers and two million tons of freight. After repeated Allied bombing of railway 'bridges, principal stations, repair shops, about 50 percent of the rolling stock, signal equipment, etc. were either completely destroyed or unusable' (TSD 1953: A9). Fortunately, road transport could function with little recourse to other modes. However, before it could offer a truly comprehensive service to all parts of the country, an interconnected all-weather road system had to be built and rivers bridged to replace the time-consuming gaps served by ferries. Inter-regional passengers were reliant on the embryonic passenger service provided after 1948 by Siamese Airways (Davies 1997). Even in 1950 roads were an 'unusual feature' in central Thailand and there were many areas of the north and northeast that were essentially subsistence communities (Sharp et al. 1953). Travel in the northeast was particularly arduous on the limited network of 'washboard', laterite roads that became impassable in the rainy season (Muscat 1990). A traveller to the northeast in the early 1950s did not see one car east of Nakhon Ratchasima (Caldwell 1974). Inter-regional goods movements – except those through Bangkok – were non-existent. Although the government received a World Bank loan of US$3 million in 1950 to rehabilitate the railways, this was inadequate to meet the increased transport demand.

Replaying the strategy game, 1950s and 1960s

A modern version of the old strategy game with foreign powers was replayed during the Cold War of the late 1950s and 1960s when the United States moved into northeast Thailand. This led to the construction of a series of US air bases, the military port of Sattahip and the Friendship Highway between Saraburi and Nakhon Ratchasima which was built in 1957 (Figure 5.5). The northeast remained poor (Silcock 1967).

After the 1958 coup by Field Marshall Sarit Thanarat (1958–63), greater efforts were made by the Thai government to expand agricultural production and exports and a policy of import-substituting industrialization was underpinned by an improved road transport system (Dixon 1999). A comprehensive survey by United States consultants in 1958 showed that road transport was still subordinate to river and rail (TCI 1959). Despite increased investment in roads since the mid-1950s, inland waterways dominated freight movements in the Central Plain, while the railway was prominent in traffic to and from the northeast, north and south (Pendleton et al. 1962). The trucking and bus passenger industry did not pose a competitive threat because outside the Central Plain the largely unpaved road system could not provide a reliable all-weather system of interprovincial connections. Telecommunications were skeletal. In the northeast 168,000 square kilometres were served by just 483 kilometres of all-weather highways (ECAFE 1967). Roads were subject to flooding, inadequately maintained and more suitable for ox carts than for motor cars and motor cycles. Bridges were dilapidated. During the four-month rainy season in the northeast access to markets was inadequate as roads were washed out, flooded or boggy. Rather than functioning as a system in their own right, the highways were largely feeders to the railway. With the advantage of comprising the majority of Thailand's administrative and planning personnel, the railways were able in the 1950s to claim a disproportionate share of foreign loans. This included assistance from the United States Operations Mission (USOM) to complete the 60-km railway line from Udon Thani to Nong Khai (Muscat 1990).

In 1961 the Thai government reinforced the government's national economic development policy by devoting most investment to increasing the length and quality of the road system (Rimmer 1971b). This was at the behest of the United States military advisors to combat threats from neighbouring communist countries and the Communist Party of Thailand (Table 5.4). Between 1951 and 1965 US$350 million was spent on road building in Thailand by USOM to provide Thailand with a skeletal trunk network to meet primary strategic objectives of linking military bases in Thailand and Laos (Randolph 1986: 22). This enabled the trucks of the state-owned Express Transport Organization (ETO) to handle all US materials delivered through the new military port of Sattahip on the Gulf of Thailand. Initially, most attention in the road construction program was focused on all-weather highways following the successful completion of the route from Bangkok to Nong Khai, from where a ferry carried traffic across the Mekong to Vientiane, the capital of Laos.

The adoption of a road-oriented policy had much to offer a country with scarce capital resources, which relied on foreign aid donors for assistance, including Australian-funded road projects in the northeast. Roads of a very different type are compatible in the same system and the associated trucking

Table 5.4 Thailand: expansion of the road network, 1950–2000 (kilometres)

Year	State highways		Provincial roads		Total
	Paved	Unpaved	Paved	Unpaved	
1950	809	5,022	—	—	5,831
1955	1,815	5,299	—	—	7,114
1960	2,972	5,474	151	1,967	10,564
1965	5,046	4,436	405	2,388	12,275
1970	8,620	1,781	1,478	4,413	16,292
1975	11,840	819	3,398	4,043	20,100
1980	13,733	160	8,670	5,587	28,150
2000	n.a.	n.a.	n.a.	n.a.	52,960

Source: Hirsch (1987, 137); NSO (2002).

industry did not require the same level of managerial skills as with a centrally managed transport operation. Many of the investments in the feebly developed highway system were largely unquestioned because of the anticipated savings in time and monetary costs on shipping freight and moving people which accrued from a system of all-weather roads (Rimmer 1971a,b, 1973). Speedier, safer, cheaper and more dependable road services have contributed materially to the expansion of economic activity throughout Thailand by lessening the costs of imports upcountry and broadening the base for exports. Roads became synonymous with economic development (Hirsch 1987).

The Friendship Highway illustrates the transport cost savings of new all-weather highways and the accompanying process of spatial reorganization. Its completion reduced the distance by 180 km and journey time by eight hours. Contemporary studies of its effects reported that the reduction in travel costs led to the greater increases in production of upland crops, poultry and pigs than in other areas without the benefit of all-weather roads (Kasiraksa 1963; Jones 1964; Kuvanonda 1969). Contemporary studies were flawed by inattention to world market conditions, government policy, relative population size and geographical area but they were sufficient to show that roads were stimulating economic growth. These changes were instanced in the reduction in stock inventories, an increasing range of goods, and the development of garages, roadside business and trucking firms (Patapanich 1964).

An effective all-weather road system enabled the military and police to respond to recurrent insurgencies in Thailand's border areas with Burma and Laos. The railway system was more susceptible to sabotage. In conjunction with well-conceived and managed investments in power and extension services, highways could literally cement relations with villages isolated from official contact in the past. Roads involve the people more directly in development, but 'the combination of rail and trail holds far less promise for social development than the road net...Trains go by and trucks stop – an essential difference when viewed as carriers of culture not freight' (Haefele cited by Wilson et al. 1966: 201). Given the government's goal of internal integration, the highways have continued to receive the largest share of public investment in the transport sector since the 1960s.

Development economists challenged few of the early highway schemes in Thailand (Hirschmann 1958: 84–5). The extension of the Friendship Highway from Nakhon Ratchasima to Nong Khai in 1961 reduced travelling time from two days to nine hours and in the process created a string of boom towns during the Vietnam War, including Khon Kaen and Udon Thani (Sternstein 1977). Other roads in the northeast linked Nakhon Ratchasima to the US air base at Ubon Ratchathani and the US naval port at Sattahip to Nakhon Rathchasima. The final US major highway project was the East–West Highway that provided an all-weather link between Phitsanoluk in the north and Lomsak in the northeast (Patanapanich 1964).

During the early 1960s there was a switch from the primary road system to developing the provincial road system and farm-to-market roads to accelerate rural development in areas susceptible to Communist influence. An important benefit was the development of new export crops such as maize, kenaf (a jute substitute), sugar cane and cassava. Bus operators also followed the road builders to extend their networks. An examination of the impact of road transport on rail and inland waterways is restricted to rice movements because over-harvesting had reduced teak exports.

Rice

A benchmark study of the costs of transporting rice in Thailand in 1965 illustrated the use of road transport and its relative position compared with river and rail (Usher 1967: 215–19). Fiercely competitive, privately owned trucking companies provided most road transport, but there was also a public trucking company, the Express Transport Organisation (ETO). As reflected in Figure 5.6, truck rates were sensitive to length of haul, road conditions, size of vehicle, demand and type of carrier – ETO's rates were double those of private carriers. Road transport was uncompetitive with river and rail transport over longer distances. Assuming the truck rates for a 100-kg rice sack between Nakhon Sawan and Bangkok to be 100, comparable rates for rail were 80 and river 55.

Trucks had largely superseded the ox cart in carrying paddy from the field to the mills except over very short distances. They had speed but no capacity advantage over river transport because one steamboat could pull between five and ten barges carrying 50–80 tons of rice. Road transport continued to supply cargo to the rivers. Nevertheless, speed gave road transport an edge over the railways, particularly as competition had reduced truck rates to a level where in the off-season they could compete on price with rail. On long-distance routes such as Ubon Ratchathani to Bangkok, rail was 50 per cent cheaper than road during the peak season but its competitive position was eroded because shippers had to pay a bribe of 9 per cent over the official rate to secure a rail wagon (Usher 1967: 218).

The advantages of trucks led rice millers to give a higher priority to road transport in locating new mills. An increasing number of manufacturing activities were located along roads, but a water location still held an attraction for rice millers as it offered 'lower costs, ease in handling bulk rice and paddy, and greater accessibility to producers and distribution centres' (Hafner 1973: 33). In Burma and Vietnam there was no marked shift of rice mills because the development of road construction had stalled.

Accelerated development since the 1970s

Political and economic events in 1973 changed the nature of the Thai economy (Suehiro 1989). The October 14 Revolution and the worldwide oil crisis changed the power structure. Twenty-five years of military rule was

Rivers: Chao Phraya, Irrawaddy and Mekong 181

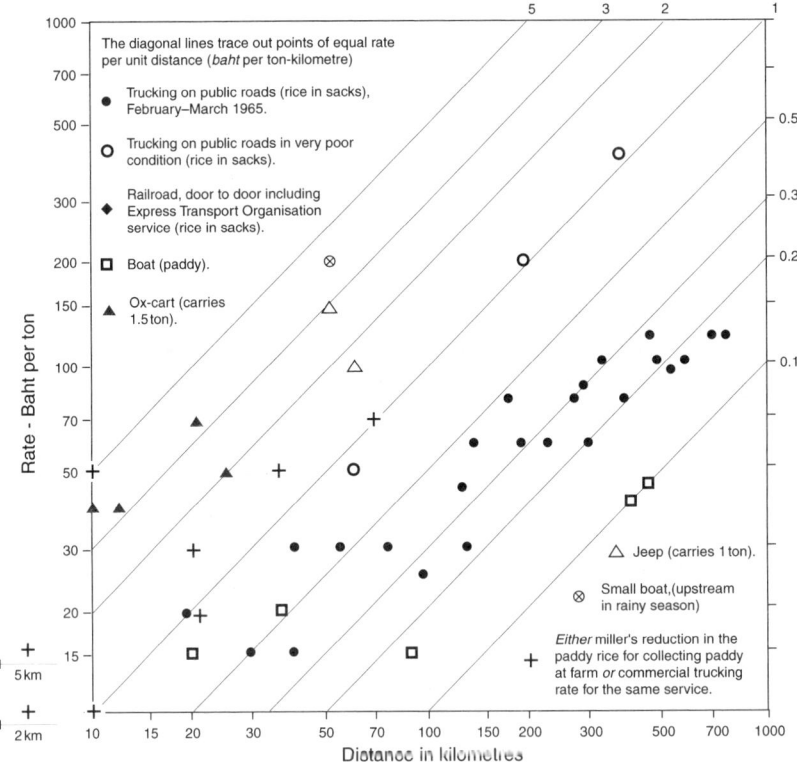

Figure 5.6 Thailand: transport costs, 1965 (*Source*: Usher 1967: 216)

ended and power has been shared with the king, technocrats, capitalist groups and organized citizens. Industrialization accelerated.

Economic development was facilitated by the accessibility of a national highway system (Figure 5.7). The lack of highways could no longer be regarded as a hindrance to economic development, political integration or social cohesion. Low use of up-country roads built instead of alleviating traffic congestion in Bangkok suggest a denial of resources to other modes and sectors of the economy (Chapter 9). Often roads had to be over-designed to offset rampant overloading by truckers and predictable lack of maintenance by Thai authorities. The World Bank suggested that road transport was under-priced compared with the railways but provided US$223 million for extension of the highway network between 1963 and 1978, which suited the needs of the Thai economy. By 1981 the railways had only received US$60 million from the World Bank (Muscat 1990: 95, 106). Despite any misgivings about the impact of misdirected transport investments, the expansion of the highway network – if only to satisfy political motives – was

182 Hinterlands

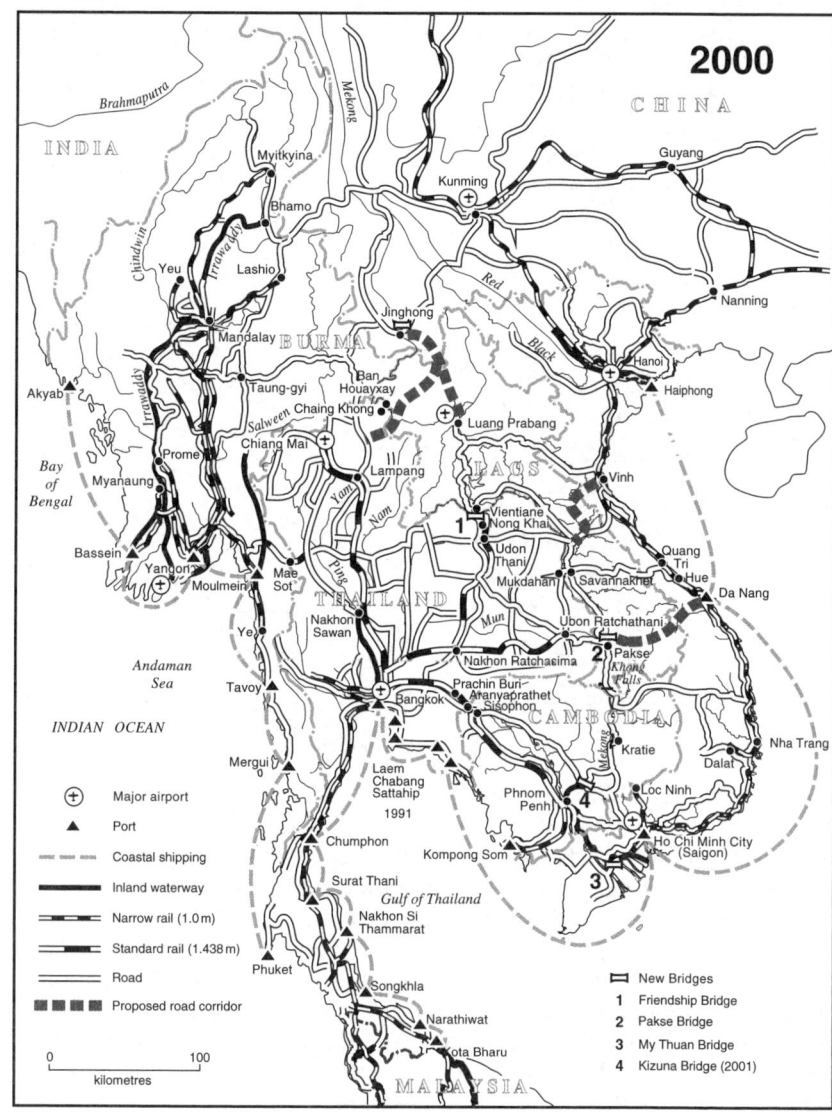

Figure 5.7 Chao Phraya, Mekong and Irrawaddy river basins: transport patterns, 2000

inevitable. Indeed, the only point at issue was the *rate* at which the roads were built. This road-oriented policy strengthened the competitiveness of the trucking industry at the expense of other transport modes.

Since the 1980s Thailand has emerged as a Newly Industrializing Economy (NIE) based on the export of manufactures. Foreign (especially Japanese) investment has underpinned this transformation with leading sectors in textiles, electrical and electronics goods, petroleum and petrochemicals and motor vehicles (Suehiro 1989; Dixon 1999). Most new factories located in and around Bangkok, leaving the hinterland engaged primarily in agricultural production (Chapter 9). Planners sought to decentralize the capital's light export industries into the Extended Bangkok Region (EBR) having a radius of up to 200 kilometres from the city's core (NESDB 2000). A main axis of deconcentration has been the development of the Eastern Seaboard, centred on the new container port of Laem Chabang. Plans to decentralize activities further afield to the fast-growing northeast transport corridor between Nakhon Ratchasima and Nong Khai and to the northern twin cities of Chiang Mai and Lampang are as yet difficult to implement (Kaothien and Webster 2000). Prospects of more even development would be improved by investment in a national, controlled-access highway system like Peninsular Malaysia's North–South expressway (Chapter 6).

Trucks and buses have reduced river transport to little more than movements of sand, gravel and cement for the construction industry in Bangkok. Many river ports have been bypassed and left as backwaters. Rail towns have been absorbed more easily into the new economic fabric. Although truck and bus terminals are now prominent features of the urban landscape, rail retains a loyal clientele for heavy goods and third-class passenger services. Yet, the State Railway has made recurrent losses. Attempts to privatize railway activities in the 1980s and 1990s have made little progress due to strong union resistance and the dominance of road transport.

The Mekong project

Thai transport developments have to be seen in the wider context of supra-national programmes – such as the Mekong project – which perceive possibilities for integration once ruled out by wars and national boundaries. Discussion about the economic potential of the Mekong basin was revived once the colonial powers left Burma (1948), Cambodia (1953), Laos (1953) and Vietnam (1954), but progress has been slow. The emphasis of the Mekong Committee (1957) was on the construction of dams rather than transport. These grand plans were never implemented because Cambodia, Laos and Vietnam became involved in the second Indochina War (1966–75). Passage on the Mekong became hazardous and ceased altogether in January 1975, when a reign of terror swept Cambodia for four years (Osborne 2000).

Meanwhile the final victory of the Communist Pathet Lao in the civil war with the Royal Lao government led land-locked Laos to reorient its trade links away from Thailand towards Vietnam (Jerndal and Rigg 1998). With Vietnamese assistance, roads were improved across the Annamite mountain chain (see Chapter 6).

Economic development of the Mekong basin was in abeyance until Cambodia's problems were resolved, albeit imperfectly, in the 1990s. Rapprochement between Bangkok and Vientiane has also seen a revival of the old axis between Thailand and Laos. To revive cooperation, in 1995 Cambodia, Laos, Thailand and Vietnam established a new body, the Mekong Commission. By then a string of dams in the river's upper reaches in China had proceeded apace, accompanied by large-scale deforestation and erosion – often exaggerated by foreign critics – which threatened downstream locations. Navigation on the Mekong being limited, China sought to improve strategic road connections to expand Yunnan's commercial links with Burma, Laos and Thailand. These connections have been facilitated by completion of the Friendship Bridge between Nong Khai and Tha Deua (completed 1994) and the Pakse Bridge (2000) with the assistance of Australian and Japanese aid respectively, and the upgrading of Route 13, which runs southwards to Phnom Penh and Ho Chi Minh City – another bridge is planned between Mukdahan and Savannakhet using Japanese aid (DOTARS 2002). The My Thuan Bridge (2000) has been built with Australian aid across the Tien Gang, the main branch of the Mekong River as it flows through Vietnam and an international shipping channel for vessels moving to and from Cambodia. The first bridge across the Mekong in Cambodia – the Kiruna bridge – was completed with Japanese aid at Kompong Cham in 2001. Commercial shipping on the Mekong is restricted to ocean steamers that ply to Phnom Penh, passenger ferries to Kratie and large boats operating seasonally above the Khong Falls at high water. River transport on the Mekong upstream to Vientiane, Luang Prabang and from there to Ban Houayxay – close to Chiang Khong in Thailand – has been declining (Jerndal and Rigg 1998). River transport within Laos has decreased with improvements to the country's roads (Rigg 1997).

There is now an Asian Development Bank (ADB) plan for identifying main corridors in the Greater Mekong Sub-region (GMS) to integrate Cambodia, south China, Myanmar, Laos. Thailand and south Vietnam by road transport (PADECO 1994). The plan involves extending the northeast corridor from Nong Khai to Kunming in China's Yunnan province; the Prachinburi–Aranyaprathet corridor – security permitting – to Phnom Penh and Ho Chi Minh City; and the east–west corridor from Mae Sot to Mukdahan to Moulmein in Burma and Da Nang in Vietnam (Webster 1999). More specifically, international development agencies, national governments and chambers of commerce have identified a trans-border Economic Quadrangle in the upper Mekong where arguably regional prosperity would be promoted by improved transport linkages between south China, Laos, Myanmar and Thailand (Jerndal and Rigg 1998). Paradoxically, this arrangement will provide more liberalized trading and transport conditions, but it will also increase the opportunities for state and non-state actors to play an active role in regulating trans-border connections and flows of goods, people and information (Walker 1997). The benefits for the agents of cross-border exchange – transport operators, traders and entrepreneurs – will be unequally distributed (Walker 1999).

Conclusion

Independent states have evolved in all three main river basins: Myanmar in the Irrawaddy, Thailand in the Chao Phraya and Cambodia, Laos and (southern) Vietnam in the Mekong. Of these, the Thai state has been the most successful in using transport and communications linkages to secure its territory and organize its economy. Although a late starter in developing river and rail transport, it has taken the lead in road transport. Burma, Cambodia, Laos and, to a lesser extent, southern Vietnam still have incomplete and unreliable highway systems. Although Thailand's direct road links to Cambodia and Burma are still poor, connections with Laos have been improved by the completion of bridges in the lower Mekong. Notwithstanding the periodic efforts devoted to developing the Mekong river basin as an economic region in its own right, this advantage has prompted the Thai government to seek a 'baht economic zone' between Burma, Cambodia, Laos, Yunnan and northeast Thailand. The Asian Crisis of 1997–98 has not dampened the Thai government's resolve to develop Bangkok as the dominant sub-regional hub (Chapter 9).

6
Peninsulas: Malaya and Annam

The Malay Peninsula is the long, narrow appendage stretching from the Kra Isthmus of southern Thailand to Singapore. Intensive settlement of the jungle-covered and sparsely populated western side facing the Straits of Malacca has been recent compared with Java and the Central Plain of Thailand (Chapters 4 and 5). It was an unfavoured coast that became the favoured coast. Except for Malacca and, later, Penang, it had been less favoured than eastern Sumatra by a rich hinterland and long navigable rivers. The west Malay Peninsula had only a very narrow coastal strip which, except around Kedah, was little suited to wet rice cultivation and could not sustain much more than some trading/fishing communities and swidden agriculture. What made the difference was tin. The west Malay Peninsula became a magnet for tin rushes at around the same time that Australia became a magnet for copper and gold. It was then settled by Chinese immigrants who, with their secret societies, moved into small Malay kingdoms and statelets, leading to conflicts that have been much written about and resulted in British intervention in the 1870s. After some experimentation in the 1880s and 1890s, the rubber boom doubly assured the west coast's prosperity and, though smaller in area than Java, it began to acquire some status as a colonial possession (Drabble 2000). Meanwhile the east coast and the entire Thai section of the peninsula remained underdeveloped.

This uneven capitalist development raises a series of issues about the role of investments in water, rail and road transport. Why was it necessary to create a trunk route along the west Malay Peninsula instead of continuing to rely on a series of port–hinterland links? What requirements did tin and rubber place on transport systems? What is the significance of 'capitalism' (for example, estates versus smallholdings)? How much was infrastructure designed to meet administrative needs? These issues are initially related to diplomatic and administrative developments before the Pacific War.

In 1826 the British had grouped their possessions of Penang (1786), Singapore (1819) and Malacca (1824) into the Colony of the Straits Settlements. Six years later the administrative centre was shifted from Penang to Singapore. In 1867 jurisdiction was transferred from the government of

India to the Colonial Office. British residents successively took charge of the tin-rich Malay states of Perak, Selangor and Sungei Ujong (later part of Negri Sembilan) (1874), Pahang (1888) and the remainder of Negri Sembilan (1895) (Sadka 1968). In 1896 these Protected States were reconstituted as the Federated Malay States (FMS) with the headquarters in Kuala Lumpur. Then in 1909 the Thai government transferred to Britain rights over Perlis, Kedah, Kelantan and Trengganu and a British adviser was appointed. In 1914 Johore became a British protectorate. However, Johore, like Kedah, Kelantan, Perlis and Trengganu, remained part of what the British called the Unfederated Malay States (UMS). Because the Peninsula then had no integrating urban centre, Johore came under the sway of Singapore while Kedah and Perlis became dependencies of Penang. On the east coast, however, Kelantan and Trengganu remained fairly autonomous Malay states behind the natural barrier of the series of high north–south ranges.

Is this pattern of development in the hinterland of the west Malay Peninsula unique in Southeast Asia, or are there parallels with the thin, peninsula-like coastal strip between the Mekong and the Red River basins?[1] In 1850 the strip and the river basins to the north and south were part of the Annam Empire. By 1863 the French had dismembered the Empire with the north becoming the Tonkin Protectorate, the south the Colony of Cochinchina and the middle region the Annam Protectorate.

Water

The sea and rivers were fundamental to life in the Malayan Peninsula (Figure 6.1). Maritime entrepot and river basins on the west and east coasts were the basis of the political units of Malay statelets. Watersheds defined political boundaries; river mouths provided sites for administrative centres; and river patterns defined the nature of the characteristic string-like village settlements located upstream. Sheltered harbours on the west coast, however, offered better prospects for the development of ports than sites on the east coast's lee shore.

In 1850 the hinterland of the west Malay Peninsula was still being carved out of a maritime world from the beachheads of Malacca, Penang and Singapore. Bangkok was remote and administrators lacked detailed knowledge of Thailand's outlying provinces. Sailing ships linked the Straits Settlements ports to the low-lying, coastal settlements engaged in wet rice cultivation and fishing at or near major river mouths on both the west and east coasts (Zaharah 1970). From these transhipment points, navigable rivers allowed shallow-draught vessels and poling boats to penetrate an interior characterized by dense vegetation and steep topography.

Discoveries of huge resources of alluvial tin in Perak and Selangor after the 1840s tested the efficiency and effectiveness of the water-based transport system. The new tin mines attracted tens of thousands of Chinese

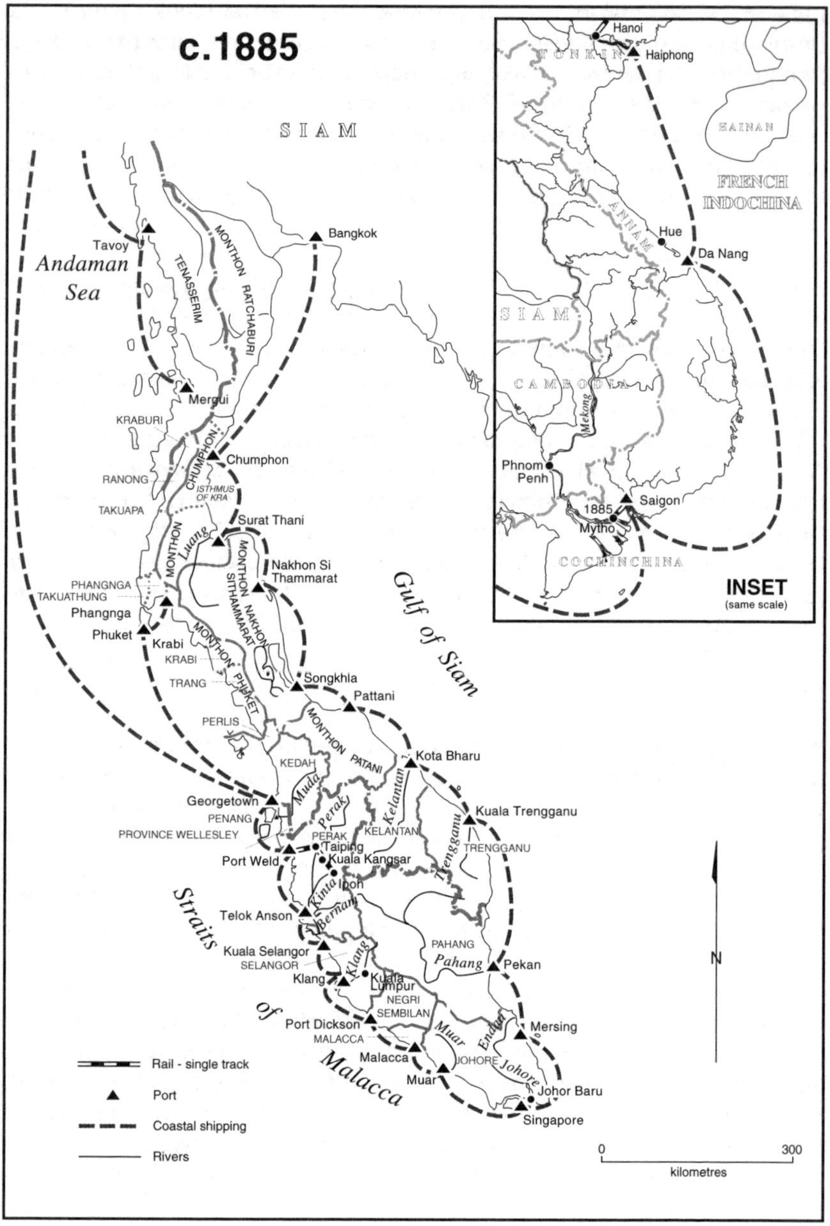

188 Hinterlands

Figure 6.1 Malay Peninsula and Annam: transport patterns, c.1885 (*Source*: based on information derived from Sadka 1968; Marks 1997). *Note*: the coast of Sumatra is not shown on Figures 6.1–6.6).

immigrants to meet the growing demand for the metal. The influx, in turn, led to fresh discoveries and new waves of indentured Chinese labour and capital, their numbers and amounts fluctuating with the price of tin. Malacca and Penang controlled the movements of tin ore from coastal ports within their respective orbits. Steamers from these main ports brought labour and opium to local ports and came back with ore that had been shipped down river for transport by sea. The riverboats returned upstream with Chinese immigrants, mining equipment and provisions. Distances in this waterborne world were measured in sailing or poling days and hours (Lim 1978; Kaur 1985a). Everyone was accustomed to moving about by boat.

This flexible transport system suited Chinese tin miners because little additional infrastructure was required beyond the river. Once the head of navigation was reached and miners pushed their search for tin beyond the riverbanks, the jungle was cut down and elephant tracks and paths were blazed by Chinese miners and Sultans to connect interior mines (Kaur 1985a: 8). The price of tin was high enough to withstand the costs of using elephants. By the late 1860s the crude bridle paths to Malacca's tin mines and outlying agricultural settlements had been upgraded to tracks, enabling the use of bullock carts to convey stores and equipment to the mines and move tin back to the ports (Leinbach 1974: 54).

By 1871, the number of Chinese miners on the Klang River had risen to 12,000 (Parkinson 1964: 61). This influx led to regular steam services between Klang and Malacca and Singapore but not Penang, though there was a regular passenger steamer between Penang and Singapore (Cowan 1961). At that time Klang was the main town and Kuala Lumpur merely a place in the vicinity of the tin mines. In 1872 Frank Swettenham (later British Resident of Selangor) took three days of rowing and poling up river to reach Kuala Lumpur by boat. He returned partly on foot through the virgin jungle – a twelve-hour effort (Parkinson 1964: 67).

As the most accessible mines were worked out, new mines were opened: first in the Larut valley and then in the Kinta valley of Perak which was more remote from the coast and Penang (Wong 1965). Tin from the Larut valley was brought from Taiping to the coast either down river by small flat-bottomed boats or by bullock cart along an earthen road for dispatch by small coaster to Penang (Bird 1883: 264–5; Wayte 1959: 159). As tin could be smelted into ingots this dense but relatively high-value commodity could withstand the costs of transport by slow bullock carts of limited capacity.

When in 1874 the British government brought all three tin-producing states – Perak, Selangor and Negri Sembilan – under its control, the action was justified by the need to quell the anarchy created by rivalries between Malay chiefs over tin deposits and to stifle the activities of Chinese secret societies. Although the British government brought law and order, it did not alter the economy, and the primitive and lawless nature of the mining frontier persisted. Administrators outside town centres still made much use of elephants. On 8 September 1874 the administrator Sir William Jervois proceeded 29 km by elephant between Taiping and Kuala Kangsar (Parkinson 1964: 211).

Even during the late 1870s, sea transport was still a hazardous business. As reported by Isabella Bird (1883: 214–15), a noted woman traveller:

> the quaint little Chinese steamer [the *SS Rainbow* which]…long ago submerged her load line, and is only about ten inches above the water, and still they load…goats and buffaloes, and forty coops of fowls and ducks…the Portugese-Malay captain…is only licensed for one hundred passengers, and the water runs in at the scuppers as she rolls…with about a hundred and fifty souls on board, and not a white man or a Christian among them…

Remarkably, the *SS Rainbow* was still in operation in 1881 as one of six steamers listed as regularly calling at Klang (Sidhu 1965: 7).

During the early 1880s the British encouraged more reliable transport to open up the interior's tin resources to large-scale development. New ports such as Port Weld were established (Wayte 1959). Silted rivers, like the Kinta, were cleared to facilitate navigation, at least temporarily. Existing tracks were expanded into 12-foot roads for bullock carts between inland mining centres and the ports. Although these roads were costly to repair and impassable during the wet season, they allowed Chinese miners and traders to flourish even after mining passed into European hands (Leinbach 1974: 54; Kaur 1985a: 9). Roads altered settlement patterns. They shifted economic activity from riverine and coastal Malay settlements to the towns where shophouses, lodging houses and coffee shops were located along main roads. The roads also facilitated the integration of villages into colonial administrative networks and established the law and order necessary for economic development (Amarjit Kaur, pers. comm.).

In Johore, rivers still served as 'highways of immigration' and later as 'arteries of commerce' (Jackson 1965: 84). In the absence of roads, Chinese agricultural settlers from Singapore moved into Johore by sea and along rivers from the mid-1840s, encouraged by the Sultan (Trocki 1979). By the early 1870s, Chinese brotherhoods had established new communities on 29 estuaries to produce pepper and gambier (used for tanning leather). A Chinese semi-feudal headman governed these communities under a 'river document'. This arrangement spread to 58 rivers over the next decade and extended to Negri Sembilan and Malacca (Jackson 1965: 85, 96).[2] Coincidentally, the large population of Chinese gambier and pepper plantation workers provided a good captive market for Chinese revenue farmers in Singapore, who had purchased the right from British administrators to sell and distribute opium shipped by British merchants from India (Trocki 1990).

Without conservancy measures, the removal of forest for rice farms and plantations often compounded the silting of rivers. The main source of silting stemmed from clearing and sluicing for tin mining, which virtually ended the navigable life of shallow rivers. By the 1880s the Kinta and Perak Rivers in Perak and the Klang River in Selangor were much less navigable than previously. Although the rivers had served the tin industry well, these natural highways did not have the long-term capabilities of the Irrawaddy, Chao Phraya and the Mekong (Chapter 5).

Transport by river, supplemented by cart transport, was becoming time-consuming and costly, as well as being subject to the vagaries of weather. In 1882, for instance, a three-day pole journey was involved up the Klang River to traverse the 29 kilometres between Port Swettenham (later Port Klang) and Damansara, after which the mining centre of Kuala Lumpur was reached by bullock cart. In addition, freight rates between Klang and Damansara, quadrupled during the 'dry season' compared with the 'wet period' (Sidhu 1965: 7).

This state of affairs led to a petition by private interests to build a railway from the coast to Kuala Lumpur, which promised fixed schedules and the opportunity to inculcate the virtue of punctuality among local people (Kaur 1985a: 52). The then Resident of Selangor (Sir Frank Swettenham) did not favour the state's legitimate transport interests being placed in private hands. He was concerned that traffic regulation, timetables, extensions and the location of new stations would be dictated solely by profit. A subsequent inquiry recommended that the state build a one-metre gauge railway to connect with Burma's system at the other end of the peninsula. Without these proposed improvements in inland transport and communications infrastructure, economic activity would probably have remained confined to tin mining.

Annam compared

In 1850 Annam was ruled by a king based in Hue. The Kingdom engaged in trading sugar, salt, oil and rice with Singapore in return for foreign manufactured goods (GB 1856). With no mineral resources the economic development of the Annam Protectorate had stalled after the French takeover in 1863. The French concentrated their infrastructure developments on the navigable channels provided by the Red River and Mekong. They neglected the rough and mountainous intervening area sandwiched between the Annam cordillera and the South China Sea because it had only a few small rivers that 'descend from the mountains and, too quickly to fertilise the narrow region, throw themselves into the sea' (de Corbigny 1878). Seasonal monsoons and the typhoons of the China Sea ravaged Annam's inhospitable coastline. However, like the Malay Peninsula's east coast, the Protectorate was reliant on coastal shipping. Small ships linked the series of populated coastal plains separated by barren spurs from the mountain ranges running down to the shore of the South China Sea (Robequain 1944). Along the coast there was also a narrow 'mandarin road' that had been planned in the early nineteenth century by the Hue emperors and was used by light carriages (or pushcarts) and wealthy travellers in sedan chairs. This 'imperial corridor' was seen as the likely route for subsequent railway development, which would link preferred ports and allow them to expand their hinterlands and capture the trade of weaker rivals (Del Testa 2000).

Railways and ports

Tin was not as valuable as the gold mined in Australia or diamonds in South Africa but during the 'tin rush' it yielded good revenues. The British colonial government invested some of these in infrastructure, so that the west coast of Malaya acquired much better railways and roads than Luzon and other parts of the Philippines (Chapter 4). Although tin ingots could withstand the cost of transport by packhorse, the time taken and the cost of moving exports by river or road from the interior mines to the ports eventually led to the preliminary surveys of railway lines.

These surveys recommended short latitudinal railways using a one-metre gauge to connect inland mining centres with a new set of purpose-built ports that would shorten distances to the shipping place and lessen the expense of road repairs. As a result, the following port-railways were built (Figure 6.2):

- Port Weld to Taiping (opened 1885), then capital of Perak and the centre of the tin industry in the Larut district;
- Klang to Kuala Lumpur (opened 1886);
- Telok Anson wharf to Tapah Road (opened 1895), which was later extended to Ipoh; and
- Prai to Bukit Mertajam (opened 1899), which drew tin from the Taiping area to the smelters.

In 1901 the railways were amalgamated under government control.[3]

These port-railways were substitutes for the river and its feeder tracks on the west coast and became an important source of government income. As reflected by the line between Port Weld and Taiping, tin was shipped out with hides, coffee, sugar and chinchona; food, opium and mining supplies were received (Wayte 1959). Ten trains travelled on the line each day and connected with the daily shipping service to and from Penang. Port Weld's prosperous period was short-lived. Tin production declined in the early 1890s and much of its traffic was diverted once the line to Telok Anson (Telok Intan) was completed. Importing Tamil labour to build the line set a precedent. They did much of the subsequent track work in Malaya and comprised the largest cohort among the railway's 'polyglot' staff (Sandhu 1969; Kaur 1990).

Not all sites for the railway ports were suitable for navigation. The choice of Port Swettenham was attributed to land speculation rather than any special attributes for berthing vessels (Allen 1951). As there was no natural harbour, the port was carved out of a mangrove swamp. Although claims that Sir Frank

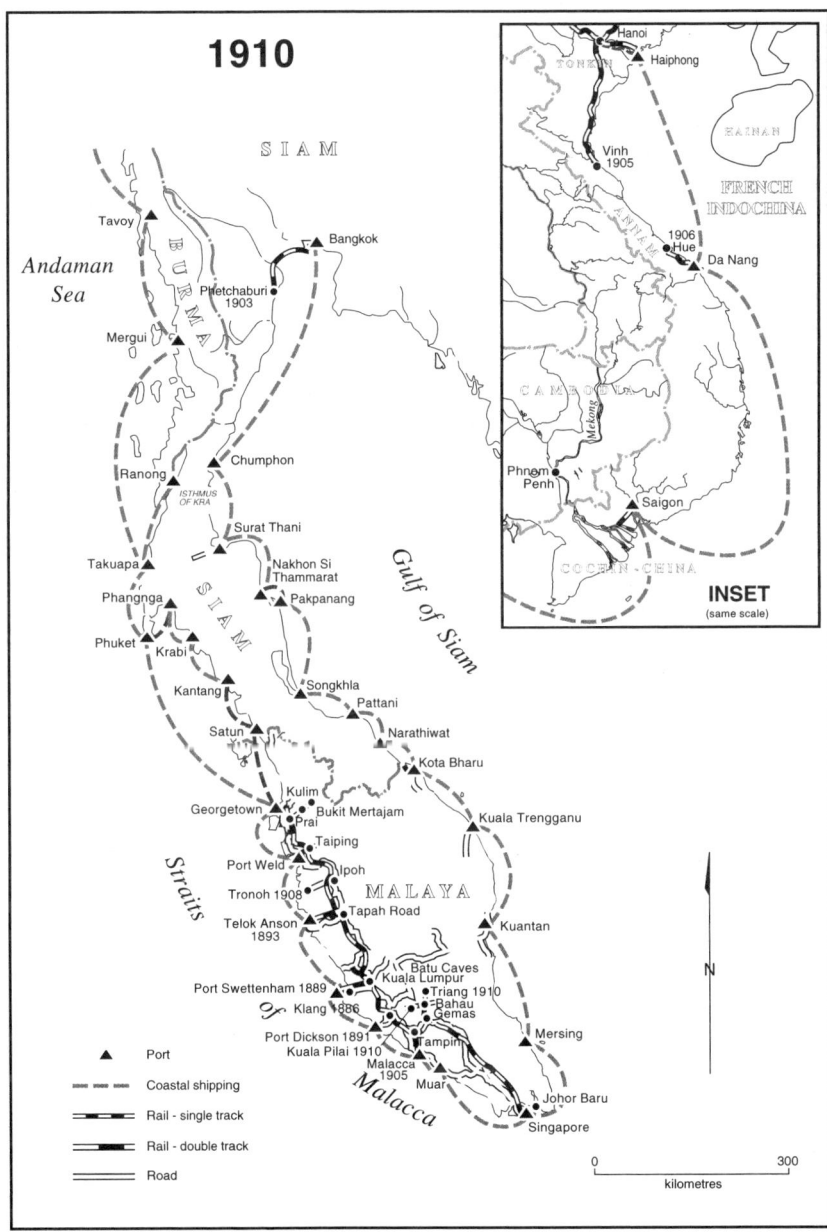

Figure 6.2 Malay Peninsula and Annam: transport patterns, 1910 (*Source*: based on information derived from Kaur 1985a,b; Leinbach 1974)

Swettenham, then British Resident, was involved in land speculation could not be proven, he was implicated in a scandal over the route of the Klang–Kuala Lumpur railway and its extension to Port Swettenham (1889). Once this railway was opened it made a profit of 25 per cent on the investment. A bridge for the railway across the Klang River reduced the scale of the river traffic and ensured Port Swettenham's future as the Peninsula's central port at the expense of Malacca. A deepwater pier was constructed at Port Swettenham in 1903. This close nexus between the shipping companies, ports and railways also led the Federated Malay States (later Malayan) Railway into constructing its own deepwater facilities at Prai wharves opposite Penang.

The railway ports linked up with the sea-lanes to Singapore and Penang. Singapore-based shipping companies provided regional shipping services between Singapore and Penang and the Peninsula's *kampung* and mines (Tregonning 1965). Once exports of tin ore were permitted from the late 1880s, the Straits Trading Company abandoned its Telok Anson smelter and concentrated on bringing tin ore from Selangor, Seremban and Perak to its new, western-style smelter in Singapore that was opened in 1890 and proved more efficient than the primitive Chinese smelters. In 1890 the Straits Steamship Company was formed to carry the ore to Singapore and return with 'coolies' and supplies for the tin mines.

The railways were built primarily to serve the tin mining industry. Differential freight rates favoured the carriage of ore rather than metal and benefitted the large foreign-owned smelters in the Straits Settlements (Jomo 1986: 169). Inbound shipments on the railways comprised equipment, stores, notably rice and opium, and labour. The railways also provided the basic infrastructure for experimenting with other commodities. During the 1900s there was an upsurge in movements of coffee until it was hit by disease (Table 6.1). However, sugar, cassava (tapioca), pineapple and the declining gambier plantations provided little freight. There was a shortage of freight

Table 6.1 Federated Malay States: main commodities carried by rail, 1905–1906

Commodity	1905	1906
Rice (bags)	1,193,710	1,215,494
Tin (slabs)	294,024	286,152
Tin ore (bales)	1,332,991	1,213,093
Opium (chests)	4,346	4,800
Coffee (bags)	25,538	23,630
Kerosene (tins)	598,749	653,900
Poultry (baskets)	33,884	44,635
Pigs (number)	68,182	78,065
Firewood (trucks)	19,148	19,742
Timber (trucks)	5,724	5,383

Source: Wright and Cartwright (1989 [1908]: 178).

until rice, especially from Kedah and Perak, became the principal foodstuff carried by rail.

By 1909 the separate port-railways had been integrated into a north–south trunk route running from Prai to Johor Baru, causing the small railway ports of Port Weld, Telok Anson and Port Dickson to lose their original purpose (Wayte 1959; Khoo 1985). This reliable arterial system was constructed to serve the tin fields, agricultural enterprises and commercial centres. In the process it spurred the growth of inland urban settlements, facilitated the development of administration in the most populous areas, and consolidated the ranking of Prai (Penang) and Port Swettenham ahead of a string of minor ports. Thus tin ore from Taiping was sent through Prai and not through Port Weld or Telok Anson. Feeder lines to the trunk route were also established to enhance the importance of railway junctions at Ipoh, Tampin, Gemas and Bahau. The extension of the line through Johore required the consent of the Sultan. On reaching Johore the railway was connected to Singapore by a train ferry from 1903. Although the train ferry consolidated Singapore's bridgehead, the heaviest traffic was 'always on Sundays: for on that day the gambling farms of Johor pay for return fares of all who come from Singapore to gamble on the premises' (Wright and Cartwright 1908: 184).

In stark contrast, transport and communications networks in southern Thailand continued to be very backward during the 1900s, particularly as plans never materialised to build a canal across the 29-kilometre Kra Isthmus and shorten the sea journey between Calcutta and Hong Kong (Campbell 1904). Chinese tin miners were reliant on boats, elephants and ox carts (Manarungsan 1989: 147). Poor roads and telecommunications led the Sino-Thai Khaw family to base themselves in Penang, the site of their modern smelter (1898), and to use sea transport to access their tin-mining interests on the mainland and Phuket Island (the only area with metalled roads) (Cushman 1991). To open up new mines, British mining and merchant interests in Penang had sought from the late 1880s to build a private railway line from the tin-mining area of Kulim in Kedah to the port of Songkhla in the south of Thailand. The Government of Siam resisted these efforts and appointed the Khaw family as governors, revenue collectors and resource developers. The fear was that the south would be overrun by foreigners and become an enclave economy before Bangkok could assert its control over these remote provinces (Cushman 1991). The first complete map of the Kingdom of Thailand was not finished until 1887 (Falkus 1996).

Bangkok's concern was heightened by a secret agreement in 1897 stipulating that the Thai tributary states of Kedah, Kelantan, Perlis and Trengganu would be under British control (Holm 1978; Marks 1997). Fearing further incursions into the south, the Siamese government responded by giving a concession to Danish interests to build a railway on the west bank of the Chao Phraya from Bangkok via Kra to Ranong and across the Three Pagodas Pass to

Burma. When this was resisted by the British, the Siamese government repurchased the concession and built a railway to Phetchaburi in 1903 using a narrow gauge (1 metre) rather than the standard gauge (1.435 metre) on the east bank of the Chao Phraya river (Chapter 5). Because of competition from canals and coastal shipping, freight rates on the line were set 25–50 per cent below those in other areas of Thailand (Holm 1978).

When Kedah, Kelantan, Perlis and Trengganu were formally transferred from Siam to Malaya in 1909, the Siamese government faced difficulties in financing the extension of the railway from Phetchaburi to the Kra Isthmus, let alone to the border with Malaya. This impasse was resolved through loans from the British government, which stipulated the use of British materials and, where feasible, British engineers (Holm 1978). The main railway line avoided the principal tin regions because it was designed to give locals the opportunity to develop expand smallholder rubber ahead of the international connection with Malaya; it was also intended to give Bangkok traders and edge over their Singapore rivals (Stifel 1973). Spur lines were constructed to the west coast port of Kantang (1914) as an alternative to Penang, and also to the east coast ports of Nakhon Si Thammarat (1914) and Songkhla (1915). In 1918 the main line reached Padang Besar and connected with the railways in Malaya. A second connection was made when a 'branch' railway line was built to Sungai Kolok (1921) on the Kelantan border. Although Hat Yai emerged as a key railway junction, the export of tin and rubber through Penang was reinforced (Stifel 1973). Exports from east coast ports went to Singapore. The boost to tin mining in south Thailand came not from any transport breakthrough but from the introduction of capital-intensive dredging by Australian and British mining interests (Cushman 1991; Falkus 1996).

Acquisition of the east coast Malay states led the colonial government to develop an inland railway route to Kelantan rather than a coastal route, which would have been preferred by commercial interests in Singapore. This gave a fillip to extending the existing line from Gemas Junction to Triang in Pahang, which had been stalled by doubtful revenue prospects. Even with this added incentive, progress on the inland line was slow and work was halted periodically by monsoonal floods and lack of labour.

The pinnacle of railway development was the twice-daily express in each direction between Singapore and Prai (for Penang) (Kaur 1985a: 147). From 1922 these services connected with the international express, providing a 27-hour service between Prai and Bangkok (1,148 km) – a branch of this train was routed to Sungai Kolok on different days of the week. This joint arrangement obviated the 4–6-day voyage between Singapore and Bangkok. In addition, a north–south night goods train service linked important urban centres in Malaya. Once this revenue-generating route was established, British control was extended to all parts of the west coast of Malaya, to which in 1914 Johore had been politically incorporated. Essentially, the trunk rail network, and its supportive feeder roads, defined the pre-rubber economy and enabled Singapore and Penang to expand their functions as the dominant nodes for interior penetration.

The Federated Malay States Railway sought to capitalize on its network by building a deepwater port at Port Swettenham (Mackinder 1931; Allen 1951: 34). Its strategy was to build a central shipment point for Malaya's exports and imports rather than handling everything from the ports of Singapore and Penang at the extremes of the railway system. This doctrine of a single central port would have been logical had not the two existing ports enjoyed so many geographical advantages. As it was, the cramped site was dictated by land ownership rather than its proximity to deep water.

When planters were looking for cheap land for rubber plantations during the 1900s, the port-rail-road infrastructure was already in place in the west coast corridor. The land had been cleared, there was a ready source of immigrant labour from South India, and there was British rule. In few other tropical parts of the world at that time did capitalists have such a free rein to meet the growing demand of the pneumatic tyre industry. Even the east coast of Sumatra could not boast railways that charged such low freight rates or ports so biased in favour of exporters and the interests of foreign investors.

Through accident rather than planning, successive rubber booms transformed the peninsula's economy. By the first decade of the twentieth century the most accessible land to the railway in Perak, Selangor and Negri Sembilan had been alienated for plantations, thereby providing revenue for further rail extensions. Foreign investors used Penang as the springboard to extend rubber plantations into south Kedah by recruiting low-wage Indian labourers. There was also scope for rubber plantations and Indian labour in Johore, especially in the agricultural districts of Muar and Batu Pahat. Isolated from mining areas, these districts had been reliant on irregular shipping connections with Singapore and the 13-kilometre Muar railway opened in 1890. The prosperity brought by the rubber boom enabled a rudimentary network of dirt roads to be built to link with the efficient trunk railway system on the west coast (Khalid 1993: 7). By the 1930s oil palm estates located near rubber plantations were also using the railway. Without this expansion of rubber and oil palm production, the British government would probably have over-invested in railways and feeder roads on the west coast.

A common view is that railways and feeder roads benefited the new export industries and immigrant groups while leaving the Malays on the margins of the fast-changing colonial economy. When the rail and feeder road system on the west coast connected villages to urban administrative centres, the initial beneficiaries were urban dwellers, notably Chinese traders but also the lower ranks of the military and police, schoolteachers, clerks and office boys, railway workers and labourers on local public works. However, opportunities also opened up for rural Malays. Smallholders could now take their rubber sheets to sell in the village shops or towns. Irrigation works increased the number of rice farmers, especially in Kedah. Fish, fruit and vegetables enjoyed a wider market. Malays thereby shared in the prosperity, less so the lowly paid Indian plantation workers. Unable to afford bicycles, they seldom went to town: their life continued to revolve around the rubber estates with their own temples, Tamil schools and shops (Amarjit Kaur, pers. comm.).

The striking disparity in both transport infrastructure and living standards was between the western and eastern sides of the peninsula. The east coast participated only marginally in the rubber boom and, apart from dried fish, produced little else for export. In the absence of railways or highways, trade was totally dependent on small coastal ships, whose safety at the bar-bound river ports and offshore anchorages was precarious during the northeast monsoon (Tregonning 1967: 155–9). The isolation of the predominantly Malay east coast thus persisted along with disadvantages in market access, employment, education and income.

Annam revisited

Annam did not have rubber plantations to warrant railway investment. As rubber was concentrated in Cochinchina's 'red earth' belt, the French colonial government gave priority to the construction of railways and an entrepot in the Mekong delta to match those for coal in Tonkin's Red River delta (GB 1901). The French administration in Indochina argued that Annam lacked the necessary mineral and human resources for priority development (Robequain 1944). It did not pursue the Trans-Indochinese railway (1,729 kilometres) until the 1900s. The Hanoi–Vinh (321-km) section was completed in 1908 and Saigon–Nha Trang (409-km) section in 1915. However, the 300-kilometre rail gap between them was traversed by motor vehicle advertised as a 'pullman autocar' (Madrolle 1930). This meant that Tourane (Da Nang), with the best credentials to be the central port as a convenient outlet to Hue, was only the terminal of a small railway (Desfeuilles 1927). The impetus to bridge the rail gaps stemmed from the demands of southern rubber plantations for northern labourers during the prosperous 1920s.

When the Trans-Indochinese railway was finally opened in 1936 the number of workers travelling south was quite small (Del Testa 2000). Most passengers were locals travelling short distances in fourth-class carriages to and from secondary urban areas (Théry 1931). For long-distance travel, the steamships between Saigon and Haiphong were more comfortable. Major export commodities – rice, rubber, corn and coal – were still carried by river or loaded directly onto ocean-going ships in Saigon or Haiphong (Murray 1980). The principal commodities carried by the railway were not high-value items like tin or rubber but low-value items, notably rice, cattle and lumber. As Annam's coastal plains produced identical crops there was less scope for revenue-generating long-haul, inter-regional movements. Sections of the track ran through 'empty' country and hardly justified the majestic colonial railway stations (Wright 1991: 152). French metropolitan commercial interests had few opportunities to carve up the neglected space to develop cash crop agriculture (Cucherousset 1927).

To break down its peninsula-like character, east–west rail connections were also planned across Annam's mountainous interior. Connections between the Trans-Indochinese railway and the Thai railway system were surveyed but never constructed. The border railway station at Thakhek was finished, but there was little track. The short feeder line to the Dalat hill station was also completed. While the Trans-Indochinese railway offset its passenger expenses from its short-distance traffic, it did not generate enough goods traffic to justify the enormous French investment. In addition, the rail investment had been devalued by parallel construction since 1913 of Colonial Highway 1, which eventually led to competition between the two modes.

Rail–road competition

Changing road densities in Malaya demonstrate how the hinterland was integrated between Penang, Malacca and Singapore (Leinbach 1974). In 1887 the greatest intensity of road development was within the vicinity of these three separate beachheads and was beginning to spread into frontier regions. By 1898 Penang's influence had extended into Kedah and Perak and Malacca's into Negri Sembilan, Selangor and Pahang. The spread of roads into Johore was stoutly resisted by the Sultan. By 1911 the road systems stemming from Penang and Malacca had coalesced with the completion of the metalled north–south trunk road. Development from Singapore into Johore was still stymied. New beachheads were established on the east coast in Kota Bharu and Kuantan. These roads allowed Chinese retailers to set up shops in villages and facilitated the dissemination of agricultural services, particularly the spread of the irrigation and drainage systems associated with rice cultivation. Bicycles also became widespread among villages and immigrant groups and helped policemen to keep the peace (Kaur 1985a).

During the 1920s and 1930s road transport graduated from being a feeder to the railways to becoming a recognized mode in its own right (Figure 6.3). Road transport not only offered flexibility and availability in areas that were not served by rail, but also could handle mixed consignments more easily. Road transport enjoyed a great advantage over railways in that it was under no obligation to carry any but the better classes of cargo. Capital costs were trivial to the road operator, whereas 70 per cent of the railway's costs were overheads. As the direct cost of carriage was small, railways could carry bulky goods at low rates, cutting them where there was competition from road transport or coastal shipping services of the Straits Steamship Company. Such a policy placed an undue charge on Malaya's economy because the west coast was well adapted to road transport.

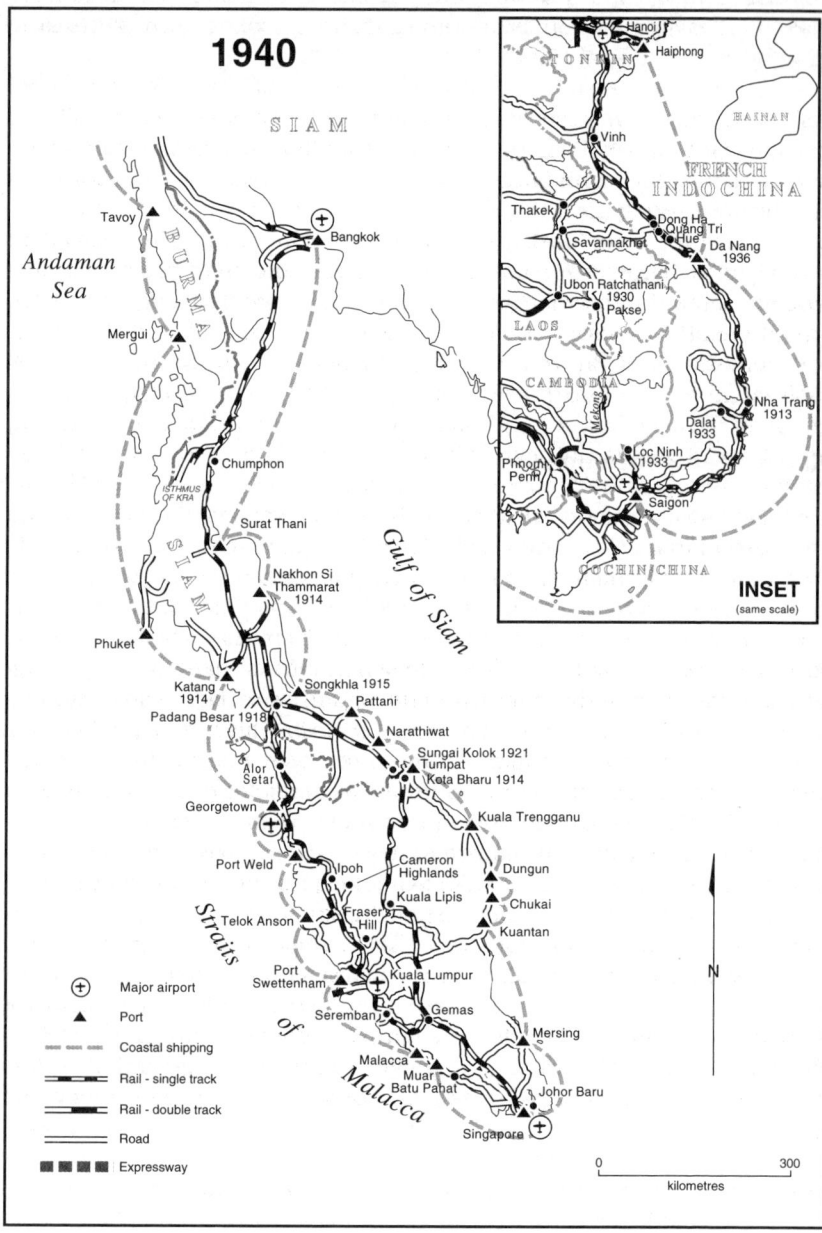

Figure 6.3 Malay Peninsula and Annam: transport patterns, 1940 (*Source*: based on information derived from FMS 1939; Kaur 1985a)

European firms dominated Malaya's economy. Capital-intensive bucket dredges imported from Australia and Alaska to mine deeper tin deposits squeezed out many small-scale, labour-intensive Chinese enterprises (Harrison 1985). Conversely, the owners of large rubber estates had to operate alongside a growing smallholder sector which was mainly Asian (Drabble 1973). The interests of the export sector were also reflected in the transformation of the road and rail networks from a west coast to a national system, albeit at a relatively low level of development (Kaur 1980: 47). In 1915 the completion of a road between Kuala Lipis and Kuantan ended the east coast's total dependence on coastal steamers. A road parallel to the Muar railway in Johore undermined its monopoly; other proposed rail feeder lines were cancelled because of road competition.

Most railway activity during the 1920s was focused on extension of the inland line through Pahang to Kelantan. This line was free of road competition and designed to promote commercial rubber production, tap the trade of Kelantan and south Thailand, and rejuvenate Pahang's declining state capital of Kuala Lipis (Cant 1973). More use was made of the rail connection between Malaya and Thailand to capitalize on the advantage of the railways over road transport in transporting people and goods over longer distances. An east coast link was made between Sungai Kolok and the anchorage of Tumpat in 1921 with a ferry between Tumpat and Kota Bharu. These connections enabled Thai rice to move southwards. Long-distance passenger traffic was also becoming more remunerative than goods traffic as the travel habit became widespread across all classes of people (Kaur 1985a: 152). In 1931 the 535-km inland rail line was finally completed to the east coast.

After the mid-1920s motor cars and heavy lorries captured traffic from the railways within a radius of 50 km, causing the government to regret the expenditure on double-tracking the line between Port Swettenham and Kuala Lumpur (FMS 1924). In Malaya, unlike Thailand, it was government policy to promote unrestricted competition and the railways were forced to reduce their fares and rates. To boost receipts the railways operated motor buses to link their stations with the seaside resort at Port Dickson and popular retreats at Fraser's Hill and the Cameron Highlands. After 1929 the loss of revenue was marked due to the decline in tin and rubber production. Road transport captured higher-rated traffic and left the railway with lower-rated, bulkier goods (Kaur 1980: 57). Because by law the Malayan Railways had to charge the same rate per ton on all routes, they lost their most profitable traffic (Kaur 1980). In 1930 the branch line between Port Dickson and Seremban was closed to passenger traffic (FMS 1930: 42). A proposal for a coastal railway line along the east coast from Mersing to Kota Bharu via Kuantan and Kuala Trengganu was not supported (Kaur 1985a: 110).

Road transport was competing heavily with the single–track railway on the west coast. The main north–south trunk road and 928-kilometre west coast railway between Singapore and Padang Besar were never more than

25 kilometres apart or 80 kilometres from the coast. Sixty small ports complicated this position because there was extensive coastwise traffic parallel to the land routes. Paradoxically, the busiest freight railway line was the short-distance, double-tracked route between Port Swettenham and Kuala Lumpur that carried the bulk of Selangor tin and rubber exports and imports of foodstuffs (principally rice), materials for capital works and the railways, and dredgers for the tin mines.

Prospects for Port Swettenham were limited because the Imperial Shipping Committee recognized that 'it is a railway port pure and simple; its population is confined to shipping agents and railways and port employees and it has no markets' (Mackinder 1931: 9). Perhaps the Committee should have added that Port Swettenham had no road connections. More concerned with Port Swettenham's status as a transhipment port, the Committee noted:

> there are a number of small ports along the west coast ... which are connected by railway to the mainland and which act as outlets for the produce of the estates in their immediate vicinity. The goods exported through these ports are taken down to Penang or Singapore by the coastal steamers and then transhipped. With two such excellent transhipment ports at each end of the Peninsula there is little or no likelihood of transhipment trade developing at Port Swettenham. (Mackinder 1931: 9)

Rubber was the only commodity shipped direct by rail to Port Swettenham in any quantity, accounting for two-fifths of all rail-borne movements of rubber.

The Imperial Shipping Committee recognized that a railway to the east coast would have boosted Port Swettenham's fortunes. Paradoxically, there was no recommendation that Port Swettenham should have road access to its dominant hinterland within a radius of 160 kilometres. Although Port Swettenham had gained at Malacca's expense, it was still subservient to both Penang and Singapore. It had little or no likelihood of becoming a transhipment port, particularly as Singapore's position was enhanced by the completion of a direct road link with the Malay Peninsula.

During the 1920s road construction in Malaya had progressed slowly (Table 6.2). Most roads 'were laid out along the contours of the hills and chiefly followed bullock tracks' (Allen and Mason 1984: 61). By the 1930s motor traction predominated at the expense of the bullock cart and the elephant had all but disappeared. Road traffic counts and motor vehicle registrations had increased markedly. Already, the larger villages had stores that sold both petrol and tyres (Kaur 1985a: 152). The hire car had become the most popular means of travel. Although the private motor car was still in its infancy, it enabled European men and women living in rural areas to reach the social attractions of the larger towns (Butcher 1979).

Truck operations, largely owned by the Chinese, were characterized by irregularity, uncontrolled rates, deplorable conditions of vehicles and

Table 6.2 Federated Malay States: road length, 1922–1927 (kilometres)

	Metalled cart-roads	Unmetalled cart-roads	Bridle roads and paths
1922	3,953	253	2,861
1923	4,014	225	2,887
1924	4,030	193	3,096
1925	4,083	240	2,981
1926	4,127	257	3,138
1927	4,186	182	3,174

Source: FMS (1922–27).

rampant overloading. Because cut-throat competition undermined the revenues long-distance services provided by government railways, the Governor of the Straits Settlements and the Malay States appointed a Committee of senior bureaucrats in 1931: 'to enquire into the conditions affecting the systems of road and rail transport and to make recommendations as to the action necessary to coordinate both systems in the interest of public economy and public convenience' (CIT 1971: 1). Based on British regulations, the Committee recommended the introduction of a licensing system for road vehicles but the legislation was not introduced until 1937. Even these zoning arrangements, designed to 'boost the burdens of the road users in order to bolster up the Railway', were often circumvented, leaving the railways to carry less remunerative cargoes (CIT 1971: 1).

A stock-take of the road and railway networks at the end of the 1930s suggested rather extravagantly that Malaya's transport system was 'unequalled' in any other British colony (SS&FMS 1939). By then the rail system totalled 1,719 kilometres and the road system 12,480 kilometres. Most of the infrastructure was on the prosperous west side of the Peninsula due largely to its mineral and agricultural wealth. The east coast was connected to the west coast at only two places – Mersing in Johore and Kuantan in Pahang: 'To cover the [400-km] of laterite road from KL to Kuantan, it took the best part of a day of hard driving, along "a thin streak of red winding its way through heavy jungle for [kilometre after kilometre]"' (Allen and Mason 1984: 113). Two mountain roads had also been completed: one linking Selangor to Pahang across the central mountain range at 'the Gap' and the other offering access to the Cameron Highlands hill station.

Thus the foundations of the trunk road system were laid prior to the Pacific War (Kaur 1985a,b). The key elements were:

- the main 966-km north–south trunk road from Singapore to the Thai border;
- the east–west Kuala Lumpur–Kuantan road;

- the Kota Bharu–Trengganu–Kuantan road via the minor ports of Chukai and Dungan exporting iron ore to Japan (completed 1931); and
- the east–west Batu Pahat–Mersing road (completed 1934).

As trans-peninsula roads and railways improved, the east coast settlements producing mainly dried fish redirected their focus towards the west coast. Remaining gaps in the east coast's transport network were filled by the coastal shipping services of the Straits Steamship Company, which in 1935 also took over the East Asiatic Company services (Allen 1953: 10; Chapter 3). There were no regular daily flights to the east coast as Wearne's Air Services Ltd operated only between Singapore, Kuala Lumpur and Penang.

Annam, 1910s–1940s

From the 1910s the French built an excellent and extensive colonial highway system. The progressive construction of Colonial Highway No. 1 undermined railway traffic, forcing similar policy changes to those introduced in Malaya. In 1935 competition from motor buses and lorries led to new regulations being introduced to safeguard the railway investments by zoning passenger and freight traffic (Murray 1980). Railway freight rates were designed to favour export crops at the expense of imports. Despite the world depression there was another massive road-building programme which led to two routes being constructed across the forested mountain chain from Colonial Highway No. 1 to Laos linking Vinh to Thakhek and Dong Ha to Savannakhet. Critics have regarded these projects as expensive follies because, although useful for smuggling, they generated little commercial traffic.

Postwar roads

After the outbreak of the Pacific War, the hinterland of the Malay Peninsula underwent a series of political and administrative changes. Between December 1941 and August 1945 the Japanese Army occupied Malaya and Singapore. In September 1945 British troops reoccupied Malaya and Singapore and the British Military Administration (BMA) was established. A return to civil rule in 1946 saw the abolition of the Straits Settlements, the Federated Malaya States and the Unfederated States. The nine Malay states, Penang and Malacca (but not Singapore) were incorporated into the Malayan Union under the direct rule of a British governor. Strong opposition from Malay leaders led the British to confer with the United Malays' National Organisation (UMNO) and the Sultans and to replace the Union by the Federation of Malaya (1948).

Strikes by trade unions against this new constitutional arrangement led the British to declare an 'Emergency' in 1948. Summary arrests forced the mostly Chinese members of the Malayan Communist Party to go underground and begin a guerilla war that spilled into adjacent Thai provinces.

During the twelve-year Emergency, more than 400,000 troops and four air squadrons were deployed against the guerillas (Anon 1987). The British relocated a million people in 'new villages' for squatters and 'regrouping areas' for tin mine and estate workers to cut off intelligence and food supplies from the guerillas. When this programme showed only limited success, the British opted for 'self-rule'. In 1957 the Federation of Malaya became independent, though Singapore remained as a British colony. In 1963 Malaya and Singapore joined together in a Federation of Malaysia that incorporated Sabah and Sarawak as East Malaysia. In 1965 Singapore seceded from Malaysia. This has left the Malay Peninsula divided into Peninsular Malaysia, Peninsular Thailand and Singapore.

These political and administrative changes had their impact on transport networks. During the Japanese occupation road bridges had been destroyed in Kedah but coastal shipping and the railways had suffered most damage. A large part of the Straits Steamship fleet had been lost. The replacement vessels, built during the war, were not well adapted to the peninsula's coastal trade, whose volume dropped precipitously. The Japanese had also removed the track of the Malacca branch line and it was never replaced. In addition, they also partly dismantled 320 km of the inland railway line to the east coast and removed undamaged bridge spans and severed the connection with Thailand railways at Sungai Kolok (FOM 1949: 165).

The Emergency halted railway restoration work in Kelantan and Pahang. In 1950 alone there were 368 attacks on the railways. The track was also interfered with on 101 separate occasions (including 50 derailments), small arms fire occurred 54 times, 19 passengers were killed and 58 injured, and two living quarters and two wayside stations were destroyed by fire. In addition, locomotives were damaged on 47 occasions, coaches on 14 and wagons on 148; and seven coaches were destroyed (FOM 1950). Unrest and sabotage reduced passenger journeys and lowered freight tonnage.

Though some effort was diverted to the construction of 'strategic roads' against counter-insurgency, the railway's difficulties gave a strong impetus to the development of road and air transport and highway and bridge construction. Buses proliferated and an extensive passenger network was developed on the west coast. New motor lorries were introduced and monopolized the distribution of fresh fish. Most 'better class' or short-distance traffic in Malaya was already carried by lorry. This was attributed partly to the advantage of daily door-to-door services and partly to the ease with which lorry traffic could slip through customs (Allen 1951: 87). Petrol rationing was abolished on 1 April 1950 but, despite the multiplication of drivers and garage proprietors, there was still a great dependence on cycling (FOM 1950: 183; Dobby et al. 1957: 37). Not surprisingly, road improvements were concentrated on Route 1 between the Thai border and Singapore, the unsurfaced sections of the east coast road and the two east–west roads across the Peninsula (Ginsburg and Roberts 1958). Large areas of the interior were left with virtually no road (Figure 6.4).

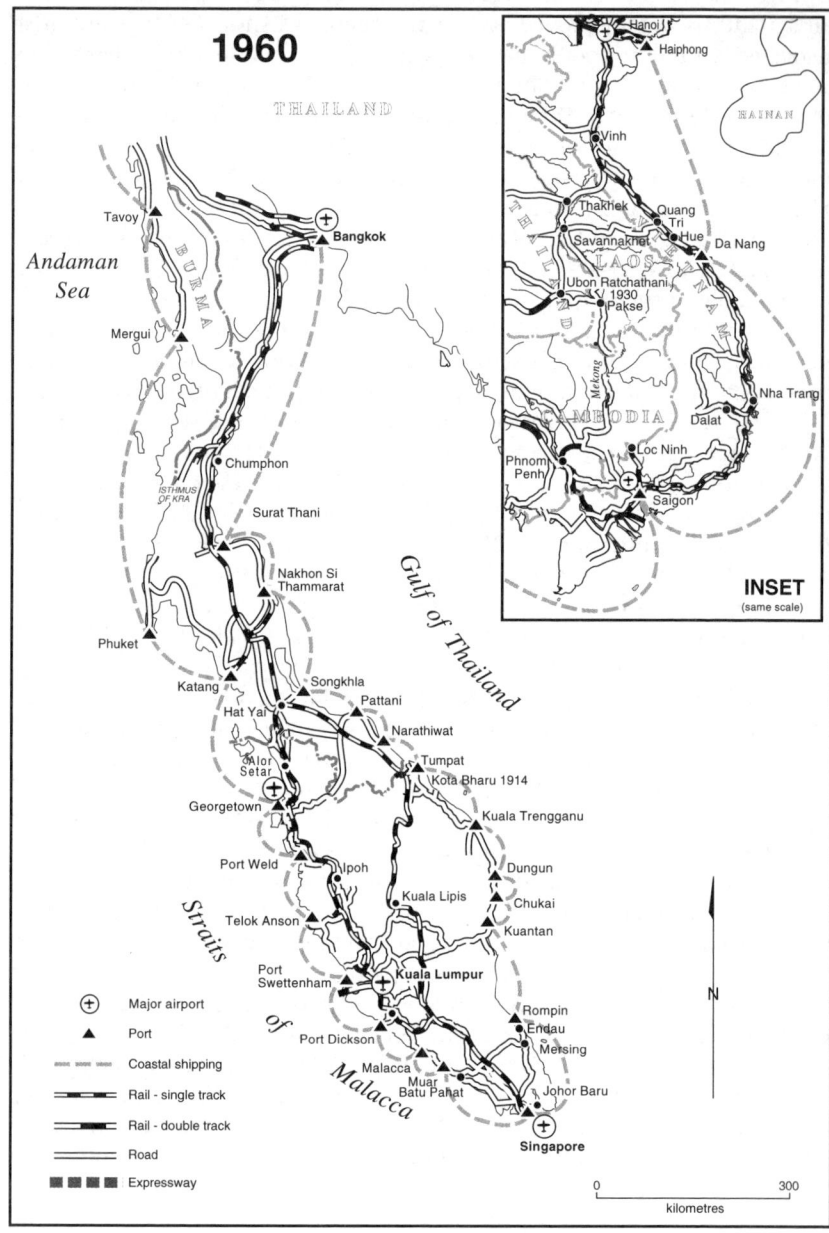

Figure 6.4 Malay Peninsula and central Vietnam: transport patterns, 1960 (*Source*: based on information derived from Ward 1960 and Rimmer 1971a)

A more detailed examination of goods movement in 1950 suggests that road transport had not yet disrupted the fundamental pattern of major commodity movements (Allen 1951). Exports still followed the 'natural routes of exit' by rail. Penang served as far north as southern Thailand and as far south as Ipoh; Port Swettenham from Ipoh southwards to northwest Johore; and Singapore most of Johore and the east coast. This pattern was modified where railway rates were used to influence certain cargoes to bypass minor ports and pass through the railway ports of Prai and Port Swettenham. For example, latex from Kedah was railed to Port Swettenham, bypassing Penang.

Imports exhibited different patterns as around five-sixths were distributed from the efficient and relatively cheap 'free port' of Singapore (Allen 1951). Most imports were high-valued goods, which could withstand the cost of the long haul by rail. Low-cost bulk imports such as cement and fertilizer were taken to the closest port by the Malayan Stevedoring and Transportation Company, which also carried other bulk cargoes such as government rice and sugar. This Singapore-based company had 54 vessels with a capacity of 35,000 net registered tons. It alone provided regular general cargo services with powered ships on the coast. Little general cargo was carried from Penang. Southbound shipments from west coast ports were double those of shipments from Singapore. The trade was more nearly balanced on the east coast because the company carried government rice and sugar.

On the east coast the Malayan Stevedoring and Transportation Company also carried bulk cargoes and specialized in lighter work. In addition, there were a few locally owned ships based in Kuala Trengganu that operated coastal services to Tumpat and Singapore. However, activities at most of the minor ports were limited to 'tongkangs, koleks and junks (many now powered), motor launches, fish carriers and fishing craft in profusion' (Allen 1953: 11). The term *tongkang* refers to the ordinary Singapore sailing trader that carried firewood and timber from minor ports in Johore such as Mersing and Endau.[4]

The progress of road transport reduced the east coast's dependence on sea communications. Shipping was always disrupted by the northeast monsoons (from October to May) which closed the shallow river mouths and turned the east coast into a lee shore treacherous for ships trying to get in or out of harbour. An all-weather coastal port was proposed at Kuantan to open up virgin areas because it had the best natural harbour on the east coast and was linked by road to Kuala Lumpur (Allen 1953). Kuantan was connected to Singapore but the 360-kilometre road via Endau and Mersing was still in a poor condition and the northern section between Kuantan and Kota Bharu involved using ten car ferries. These road conditions led to civil servants using the services of Malayan Airways Limited (established 1947) to travel to six centres on the east coast. However, the most frequented air routes were still on the west coast and involved flights between major urban settlements and Singapore (and, to a lesser extent, Kuala Lumpur).

Singapore remained the peninsula's chief inlet and outlet. Although land connections extended into southern Thailand, the railway provided the only significant international land link. While the Bangkok–Singapore express resumed a twice-weekly service in 1954, the government's major effort during the Emergency was to link smaller towns and 'new villages' built along the trunk road network.

Independence

After Malaya's independence in 1957 the west coast of Peninsular Malaysia underwent significant economic diversification (Figure 6.4). Large-scale oil palm plantations and import-substitution industrialization occurred during the 1960s and petroleum production and export-led manufacturing appeared in the 1970s (Jomo 1986). These were supported by a new road transport infrastructure, despite Malaysian and Thai national plans emphasizing the need to develop east–west roads and to expand and capitalize on the spread effects of potential growth poles such as Kuantan and Hat Yai-Songkhla on the east coast.

After 1960 motor vehicle registration and the paving of roads in Peninsular Malaysia proceeded apace (Table 6.3). Roadworks included the East–West Highway (155 km) between Penang and Kota Bharu and the Kuantan–Segamat Highway (149 km) (KKRM 1984; Siew 1987). These assisted the movement of Malays to urban areas, particularly to Kuala Lumpur and the Klang Valley and close to Singapore in Johor Baru (Chapters 7 and 10). Malays took up occupational niches in the administrative and protective services, taxi driving and trishaw pedalling, hawking and domestic service without disturbing the predominance of non-Malays in manufacturing and commerce (Rimmer and Cho 1981). A greater shaking

Table 6.3 Peninsular Malaysia: private vehicle registrations and length of roads by surface type, 1965–1990

Year	Motor vehicles (thousand)				Road length (kilometres)			
	Motor cycles	Cars	Other	Total	Paved	Gravel	Earth	Total
1965	175	154	51	380	12,464	2,107	785	15,356
1970	350	232	68	650	14,761	1,991	665	17,417
1975	722	398	110	1,230	16,465	1,741	932	19,138
1980	1,286	669	169	2,124	19,676	6,982	1,041	27,699
1985	2,290	1,152	269	3,711	20,086	5,147	982	26,215
1990	2,678	1,367	485	4,530	32,029	2,715	4,970	39,714

Note: For 1975 road length was unavailable (1974 data are used); for 1990 motor vehicle data are unavailable (1989 data is used).
Sources: MOC (1975), MOT (1979), MOT (1985–90).

loose of Malay migrants was stalled, at least temporarily, by the building in remote areas of earth and gravel roads to service large-scale agricultural and land development projects such as the Pahang Tenggara programme (Higgins and Savoie 1995).

Because employment opportunities did not expand quickly enough to absorb unskilled Malays, after the race riots of 1969 the National Economic Plan (NEP) of 1970 aimed to draw them not only into urban areas but also to bring them into the modern sectors of the economy. The New Economic Policy (NEP) was detailed in the Second Malaysia Plan (Malaysia, 1971). Its main objectives were summarized as: 'Comprising two main prongs, the NEP seeks to eradicate poverty among all Malaysians and to restructure Malaysian society so that the identification of race with economic function and geographical location is reduced and eventually eliminated, both objectives being realized through rapid expansion over time' (Malaysia 1971: 7). To ensure that race would no longer be identified with occupation and geographical location, further attention was given to mitigating the disparity in transport infrastructure, particularly between the west and east coasts, though to no immediate effect. Between 1970 and 1980 biennial traffic censuses recorded an annual 10 per cent increase in traffic volume on Federal Route 1 along the west coast. As a single-carriage highway it needed upgrading, particularly as the number of registered motor vehicles over the period in Peninsular Malaysia increased at annual rate of 13 per cent from 670,000 to 2.4 million (Siew, 1987).

During the 1980s the Malaysian government, now led by Dr Mahathir Mohamad, vigorously promoted export-oriented industrialization centred on industrial parks and free-trade zones. On the west coast, the relative importance of rubber and tin declined. The transformation from an agriculture- and mineral-based economy to an industrial-based economy boosted road traffic, which in turn occasioned congestion, longer travelling times and high accident rates near major cities and towns. This led to the progressive construction of ring roads around the city centres of Kuala Lumpur, Johor Baru, Penang, Malacca and Kuantan. Subsequently, accelerated economic growth and rising real incomes led to the near doubling of the road network in Peninsular Malaysia from 30,400 km in 1980 to 57,000 km in 1990. Express intercity bus services gradually replaced shared taxis services.

After the mid-1980s the Malaysian government continued to build rural roads to support smallholder rubber and oil palm activities but the main focus of attention shifted to the construction of inter-urban toll expressways on the west coast (Figure 6.5). The construction of the dual, two-way carriageway of the North–South Expressway was progressively opened beginning with the Kuala Lumpur–Seremban segment in 1977 (Olszewski and Tay 1996). In 1980 the Malaysian Highway Authority was established to construct the entire 772-kilometre expressway and collect tolls. Due to the recession of the early 1980s only 324 km had been finished by its planned

210 Hinterlands

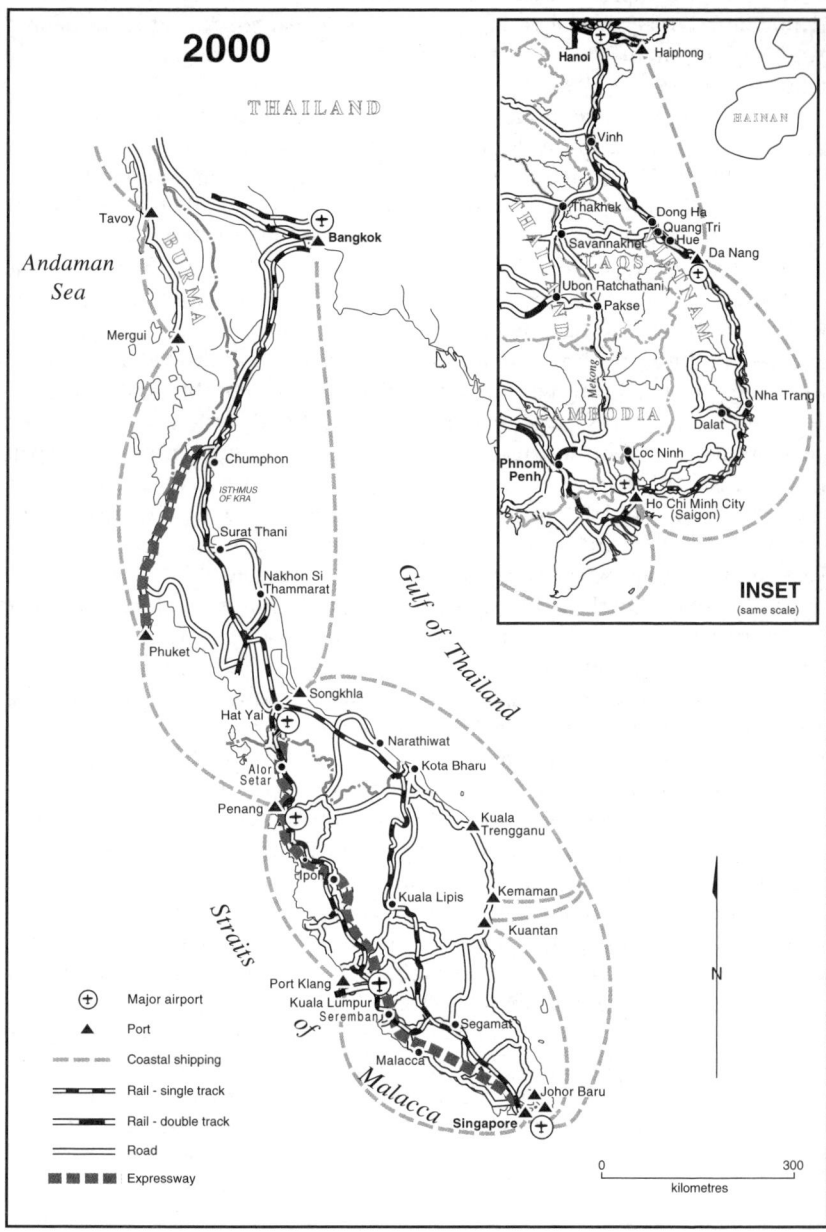

Figure 6.5 Malay Peninsula and central Vietnam: transport patterns, 2000

completion date in 1988. The government decided to privatize the construction and operation of the North–South Expressway in a programme to encourage private sector participation in construction and operation. In 1988 a 30-year concession was awarded to United Engineers of Malaysia (UEM) to complete the remaining 513 kilometres. The expressway was finished in 1994. Travel time between major west-coast cities was reduced by 20–30 per cent, leading the Malaysian Airlines System to cut competing air services by one-third. In a supply-driven approach of building infrastructure ahead of demand, the government initiated two other privatized projects – the East Coast Expressway and the West Coast Expressway. These new highways, together with the post-1980 offshore oil boom, finally ended the isolation of the backward east coast and began to break down its long-standing disadvantages in market access, education and income.

In 1993 these projects in Peninsular Malaysia were incorporated into the *Highway Network Development Plan* (HNDP) intended to develop the inter-urban network as the main transport system by 2010 (JICA 1993b). This plan sought to create a 10,580-km network to ensure high transport accessibility and reliability between the national capital and regional centres in the urban hierarchy and principal growth areas of transport and industrial development. National planners have focused the proposed highway network on three designated north–south corridors (west coast, east coast and central spine) and five east–west corridors (Figure 6.6). A parallel plan for Peninsular Thailand envisaged increased capacity on the central spine and existing east–west road links. Even when these new roads are completed, service levels will still be relatively low compared with advanced capitalist countries.

The transport corridors in Peninsular Malaysia were designed to underpin the proposed urban hierarchy centred on the existing west coast metropolitan conurbations of Penang, Kuala Lumpur, Johor Baru and potentially Ipoh and their outer ring of small and intermediate towns (Figure 6.7). The outer rings on the west coast were expected to merge into a continuous urban strip by 2020. An emerging east coast metropolitan axis was also identified. Similarly, the 'dynamic urban regions' of Hat-Yai-Songkhla and Phuket were recognized in Peninsular Thailand but no merger of urban areas was expected without the implementation of the Southern Seaboard Development Program – the latest in a series of proposals for bridging the Kra Isthmus (NESDB 1991a, 1992). Concerns about ecological damage have led to the canal concept being replaced by a landbridge and an emphasis on moving containers by rail or road between ports on the two coasts.

Clearly, roads have displaced railways in shaping urban development. Modernization plans for the railways were outlined in 1955 but the expected impetus from dieselization never materialized and the railways on both sides of the Malaysia–Thailand border have struggled to pay their way (Rimmer 1971a; Kandiah 1972). While the railways retained their role as a goods haulier, their position as a passenger carrier was seriously undermined

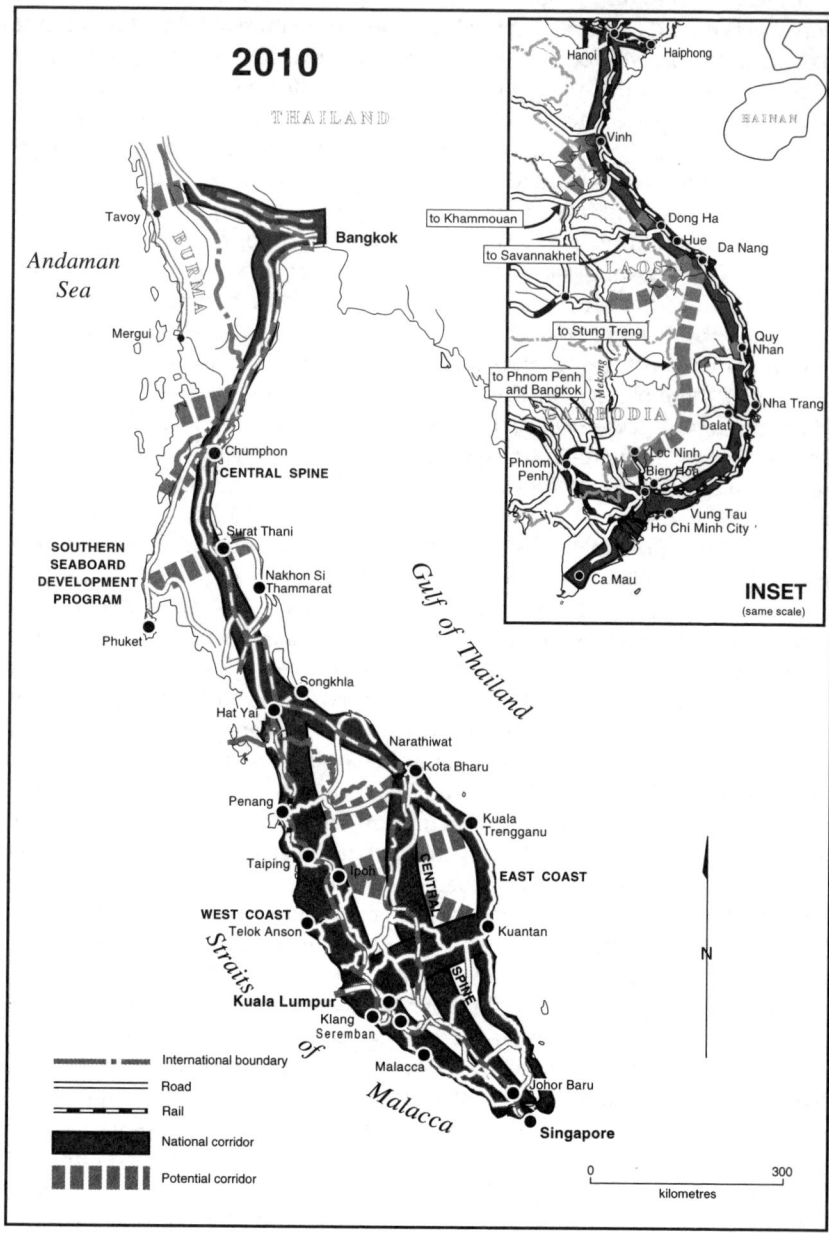

Figure 6.6 Malay Peninsula and central Vietnam: proposed transport corridors, 2010 (*Source*: Data drawn from JICA 1991; MDS 1996; NESDB 2000)

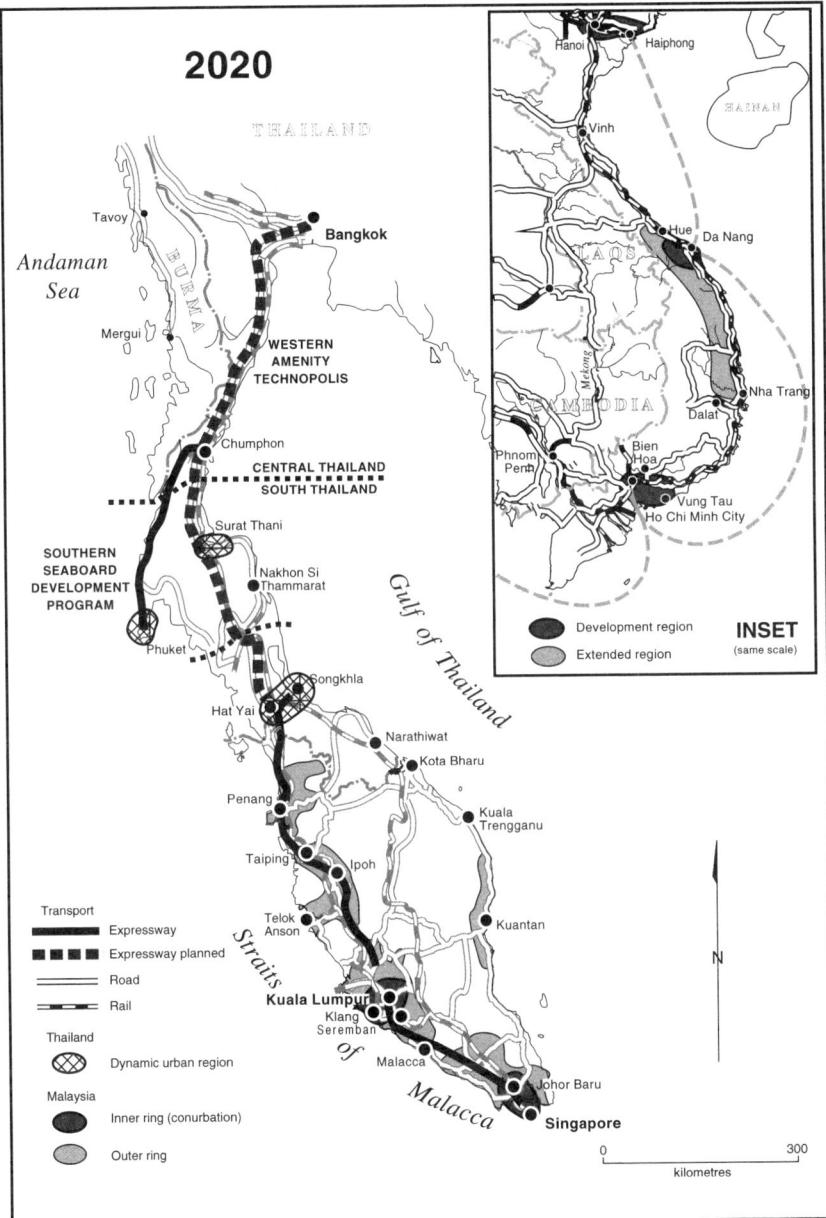

Figure 6.7 Malay Peninsula and central Vietnam: projected urban developments and expressways, 2020 (*Source*: Data drawn from JICA 1991; MDS 1996; NESDB 2000)

by competition from both road and air transport. In 1992 the Malayan Railway (*Keretapi Tanah Melaya*) was corporatized, but the rail unions in Thailand resisted commercialization and privatization. By the early 1990s protracted delays in railway modernization and the lack of links to new townships and new ports handicapped the railways in competition with road transport, particularly as intercity distances are relatively short. Even if the existing track on the west coast of Peninsular Malaysia was doubled and electrified, and the proposed new east–west lines were built on both sides of the border, the railways would fulfil only a minor share of traffic demand.

The railways have been forced to contract and to concentrate their efforts in areas where their services are still in demand. A marked increase in average train speeds is required to retain passenger traffic. In 1993 the Orient Express was introduced between Singapore and Bangkok, but there has been no progress on a high-speed train between Singapore and Bangkok via Kuala Lumpur. Most effort has been directed to the carriage of bulky goods (cement, coal, petroleum products, chemicals and logs). Plans for moving containers between Port Klang and Bangkok to bypass Singapore have not materialized but they are being railed between Port Klang and inland clearance depots within Peninsular Malaysia.

Central Vietnam compared

In central Vietnam rail transport has also suffered during both the Pacific War and the divided Vietnam economy (1955–75). Although repaired by 1959, the railway was subject to repeated sabotage by Ho Chi Minh's guerillas and there was a gap in the section between Saigon and Dong Ha on the 17th parallel (JDG 1970). When the railway was reunified after the end of the Vietnam War in 1976, it was not competitive with coastal shipping for military cargoes, or truck transport for bulk shipments under 140 kilometres or for small shipments under 650 kilometres. However, little had been done to improve highways in central Vietnam since American troops developed Cam Ranh Bay as the region's central port and upgraded roads and bridges on Highway No. 1 (ADB 1972).

Reconstruction of the deteriorated highway network came to a halt in the late 1970s following Vietnam's military engagement in Cambodia and did not resume until after the government's commitment to an economic reform programme in 1986. Much remains to be done to lower the overall cost of transport by rehabilitating and modernizing Highway No. 1 (BFTB 1996; World Bank 1999). Sections of Highway No. 1 offer divided highways in the vicinity of Saigon (Ho Chi Minh City) but there are no expressways (Figure 6.5). A north–south superhighway has been proposed along the alignment of the wartime Ho Chi Minh trail but this initiative has been dogged by financial problems. These difficulties have prompted international agencies to emphasize the importance of enlisting private participation in justifiable infrastructure construction and maintenance projects.

Paralleling developments in the Malay Peninsula, most attention by Vietnamese planners is being focused on Nha Trang-Hue as the main development corridor, with Da Nang as the core region (Figures 6.6 and 6.7). This will complement the northern focal economic area centred on Hanoi-Haiphong and the southern focal economic area of Ho Chi Minh City, Bien Hoa and Vung Tau (McGee 1995). Regional planners have also highlighted the need for improved east–west road connections between Da Nang port and Thailand (PADECO 1994). If these eventuate central Vietnam will lose its peninsula-like qualities.

Conclusion

The hinterland of the west Malay Peninsula has become the most prosperous urbanized area in Southeast Asia. This uneven capitalist development par excellence initiated by the British has some parallels with the clearing and planting of virgin territory by the Dutch on the east coast of Sumatra, but the latter never attained the same dense settlement pattern. Why was this very thin coastal strip so favoured for capitalist development when it had to be conjured out of sea and jungle from the strategic havens of Malacca, Penang and Singapore? There was cheap land in Sumatra or Mindanao, labour had to be imported and a huge foreign investment was necessary. Without tin and rubber and the development of the transport and communications infrastructure economic activities would have been arrested and economic activities limited to Singapore and Penang and their trading functions. The slower development of the remainder of the Malay Peninsula is attributed to the absence of rich mineral resources, the inhibiting effect of inaccessibility and geographical isolation on commercial agriculture, and lack of immigrants.

A comparison between the Malay Peninsula and Vietnam's peninsula-like area adjacent to the South China Sea has been instructive. During the colonial period the transport policies pursued by the British in the Malay Peninsula seemed identical to those followed by the French in Indochina. Both administrations duplicated the cheaper sea route by building parallel railways. Then both administrations devalued the railways by constructing a matching highway system. However, the outcomes were not identical because tin and rubber gave the Malay Peninsula a development edge over Annam. Timing was also important: infrastructure development was much slower in Annam and the outcomes were much closer to those observed on the east coast of the Malay Peninsula rather than to the developments of the west.

After the Pacific War the fortunes of the Malay Peninsula and central Vietnam diverged. Construction of the road network proceeded apace in the Malay Peninsula under both Malaysian and Thai governments to accommodate the demands of their Newly Industrializing Economies (NIEs). By contrast, industrialization and infrastructure development were delayed in central Vietnam until the 1990s. Since then there has been a belated convergence in national planning policies in Malaysia, Thailand and Vietnam because new and upgraded road networks have allowed planners to identify transport corridors to link their metropolitan enclaves.

Notes

1. This strip is 1,200 kilometres in length and 150 kilometres wide, compared with the 1,207-kilometres length of the Malay Peninsula, which widens to 306 kilometres at its greatest extent before tapering down to a point at Johor Baru.
2. These agricultural settlements were well suited to the maritime world but the transport difficulties in Kedah and Perak led to rice being shipped from small ports along the coast until the opening of the Alor Star-Kedah Peak Canal helped consolidate shipments (Short and Jackson 1971).
3. The sole exception was the rail link between Port Dickson and Seremban, which was built in 1901 and operated by a private company until it too came under government control in 1908. Presumably, Port Dickson was preferred because silting necessitated the lightering of cargo offshore at Malacca for all but the small coasters (Allen 1951: 50).
4. Confusingly, the term *tongkang* was also given to the Penang sailing lighters, which operated as far south as Port Swettenham (Gibson-Hill 1949, 1952).

Part III
Cities

Nation – region – city. As the smallest of these units, the city is invariably seen as local, and therefore the least interesting. However, this says more about a curiously aspatial contemporary way of thinking than it does about the true importance of the city. Relentless nationalist ideology propagated through television, the press, and the classroom has constructed the 'imagined community' of the nation as the frame of reference in place of the *experienced community* of the city. As the political, economic, cultural and symbolic centres of the nation, capital cities are hardly 'local'. Here intersect most of the connections between nation and the world and between nation and sub-national regions. Cities are the focal points in the circulation of information, people and goods. To study the capital city is therefore to study the nation, but without the baggage of a preconceived uniform national space. The intellectual challenge is to do this explicitly with a methodology that reveals the relations between centre and nation.

Part III compares and contrasts the individual histories of each of the major cities of Southeast Asia: Singapore, Jakarta and Manila, Bangkok, Rangoon and Saigon, and Kuala Lumpur – as in Part II Hanoi is excluded along with the Red River. First, by way of a framework, the historical pattern of their incorporation into a regional (and global) system of cities and the comparative size of their economies is considered. Then a long-term perspective of urban development in Southeast Asia is provided as a yardstick for studying the dynamics of individual city systems.

A system of cities

In Southeast Asia, as elsewhere, industrialization has changed the relationship between cities and hinterlands. When the main cities of Southeast Asia were still primarily trading cities, their populations were numbered in the hundreds of thousands and links with their hinterlands were attenuated. During the late nineteenth century, land-extensive, export-oriented agricultural development integrated cities and hinterlands (Part II). Urban population was boosted by the processing and shipment of agricultural commodities and service functions of management and finance, but without much rural–urban migration.

By 1900 the largest city in Southeast Asia, Bangkok, had an estimated population of 600,000 (Figure III.1a). As the one capital not under colonial administration, it may be seen as having preserved a pre-colonial pattern into the modern era. Even Singapore, the commercial capital of Southeast Asia, had a population of not quite 250,000, about the same as Manila and Rangoon. By 1930 Hong Kong and Singapore had caught up with Bangkok, but no other cities exceeded 400,000 (Figure III.1b). The nearest cities of more than one million were outside the region, notably Calcutta, Canton, Shanghai and Sydney.

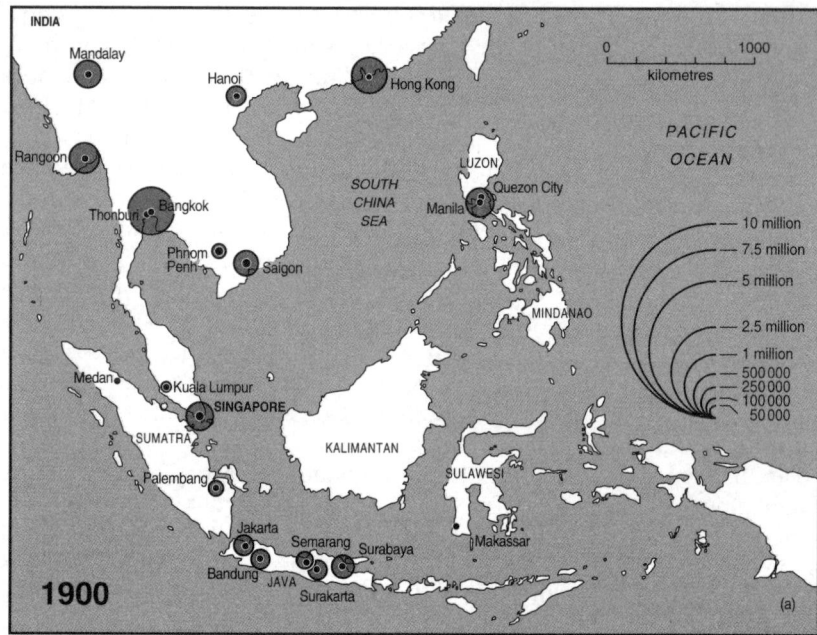
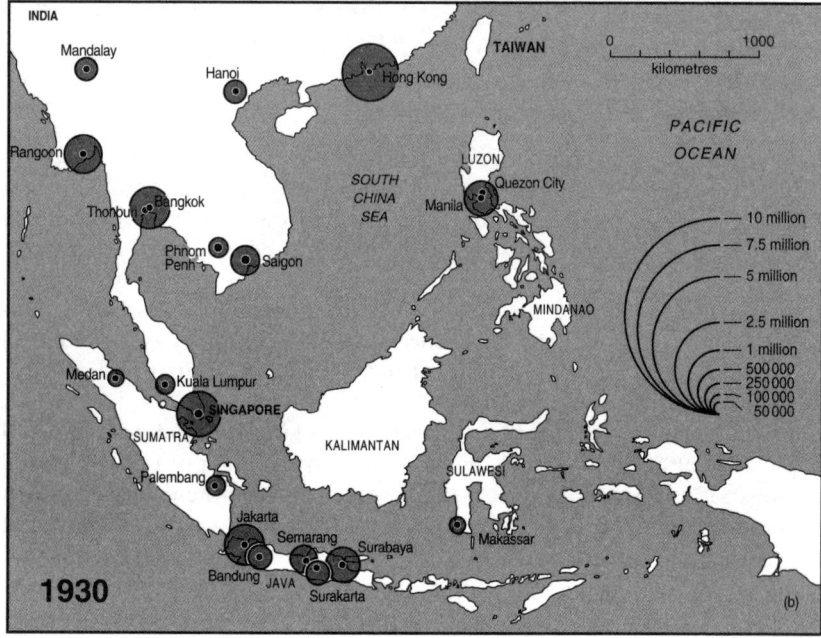

Figure III.1 Southeast Asia: urban population, 1900 and 1930 (*Source*: Based on data from Mitchell 1998)

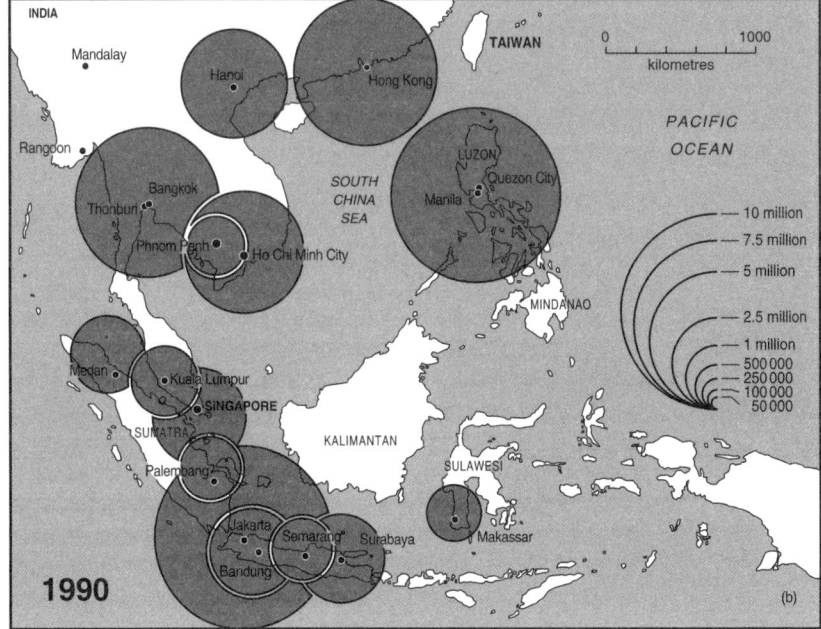

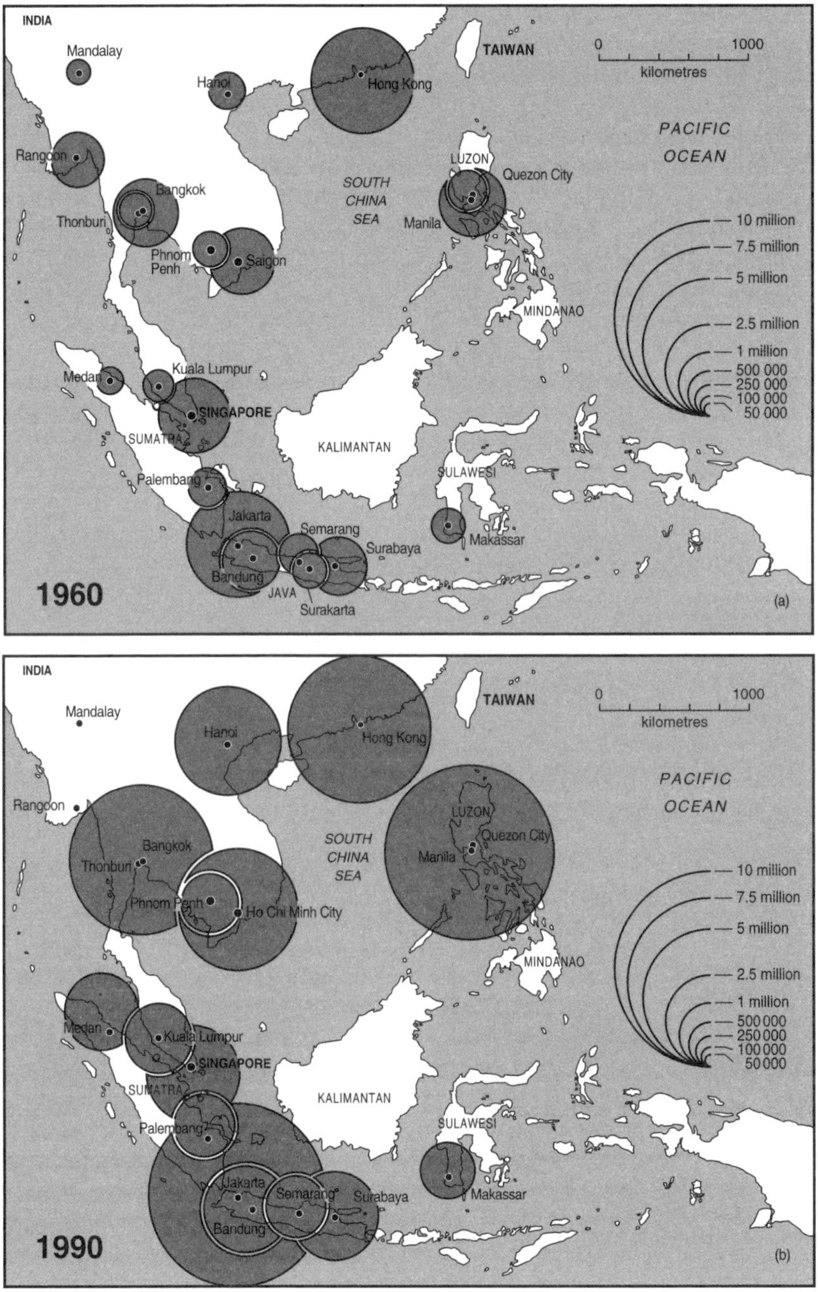

Figure III.2 Southeast Asia: urban population, 1960 and 1990 (*Source*: Based on data from Mitchell 1998)

The Japanese occupation and subsequent decolonization broke down the barriers to rural–urban migration. Since the end of the Pacific War Southeast Asia's urbanization has been spectacular. By 1950 Hong Kong, Jakarta, Manila and Saigon had all reached one million. Ten years later Singapore, Bangkok and Surabaya had also qualified (Figure III.2a). Since then industrialization of the immediate hinterland and associated economies of agglomeration have sustained rapid migration. From the mid-1980s, foreign investment and exports have led to manufacturing becoming the leading sector in the extended metropolitan regions of Jakarta, Manila, Bangkok and Kuala Lumpur. These megacities are once again closely linked with the world economy but now much less dependent upon their agricultural hinterlands. This would seem to be a new enclave pattern.

By 1990 Jakarta and Metro Manila were approaching ten million and even provincial cities in Indonesia like Medan, Palembang, Bandung, Semarang and Makassar were achieving one million status (Figure III.2b). Second cities such as Surabaya (Indonesia), Cebu (Philippines), Penang/Butterworth (Malaysia) and Nakhon Ratchasima (Thailand) have enjoyed steady growth from industrial development but not of a scale to challenge the primacy of their capital city's population and economies.

Urban economies

Capital cities are concentrations not only of population but also of economic activity. For the city-state of Singapore this is self-evident, since the urban economy is the national economy (though functionally the industrial town of Johor Baru and offshore Batam Island in Indonesia are part of a Greater Singapore economy). Greater Jakarta, Greater Manila and Greater Bangkok may also be regarded hypothetically as island economies. Table III.1 ranks the

Table III.1 ASEAN capital cities by population and economic size, 1995[a]

City	Population (million)	GRDP (market prices, US$ b.)	GRDP (PPP, US$ b.)
Singapore	3.0	84	84
Bangkok	9.4	65	171
Jakarta (Jabotabek)	16.7	40	148
Manila (NCR)	9.5	24	63
Kuala Lumpur	1.3	28	60

Note: a. Note inconsistencies between demographic and economic areas. Greater Jakarta is calculated as the Capital City Region (DKI) plus the adjacent districts of *Bo*gor, *Ta*ngerang and *Be*kasi (Botabek) with exclusion of rural population; Greater Bangkok is the Bangkok Metropolitan Region (BMR) for population but GRDP includes Central Plain and Eastern Region; Kuala Lumpur is the Capital City Region (population) plus the surrounding state of Selangor (GRDP); Manila is National Capital Region (NCR) only.
Source: National statistical yearbooks; World Bank (1997).

five core ASEAN capital cities in terms of their absolute economic size as measured by gross regional domestic product (GRDP) at market prices. For the sake of comparability, these figures are also converted at purchasing power parity (PPP) in terms of US dollars in proportion to World Bank estimates of national GDP. These figures are approximations but they show relativities. In terms of current prices, Greater Jakarta is about half the size of Singapore and somewhat smaller than Bangkok. If 1995 exchange rates are adjusted by World Bank estimates of PPP however, Greater Jakarta becomes almost twice as large as Singapore, though not quite as large as Bangkok. Greater Jakarta and Greater Bangkok therefore look to be substantial economies in their own right, irrespective of the size of their encompassing national economies.

The importance of the capital-city economies can be highlighted visually by disaggregating the economies of Southeast Asia into sub-national components and showing their capital-city economies as separate components (see Figure 1.12). These blocks are reshuffled here to show the metropolitan components as a consolidated central urban core (Figure III.3). Taken together, this urban core represents between one-quarter and one-third of the combined five original ASEAN economies. If more accurate allowance were made for urban overspill beyond capital-city boundaries, the capital-city proportion could approach 40 per cent. Such a broader definition would be functionally appropriate in terms of immediate market size and labour pool.

Southeast Asia's capital cities are not just arbitrary segments of national economic activity but constitute substantive economies in their own right. Their economic primacy is unchallenged because no second city in Southeast Asia, with the possible exception of Surabaya in Indonesia, has yet achieved a minimum efficient size. Capital cities provide domestic and international enterprises with the most efficient locations. They offer the most valuable component of the national market (large populations and concentration of the middle class); the largest market for skilled labour, professionals and managers; the most frequent national and international transport connections; and the most accessible information and cheapest search costs. All these factors generate externalities leading to increasing returns to urban size (Krugman 1996b). Non-capital cities are therefore likely to impose higher overall unit costs and place firms at a competitive disadvantage. Being the location for national governments and the locational preference of corporate managers and their families for big city lifestyles reinforce the dominant position of capital cities in Southeast Asia.

City systems

When research into specific cases becomes separated from knowledge of the general, bizarre hypotheses gain currency. However, discussing the city in Southeast Asia as a 'port city' has little theoretical utility because the city outgrows the dominance of the port and acquires a host of other functions

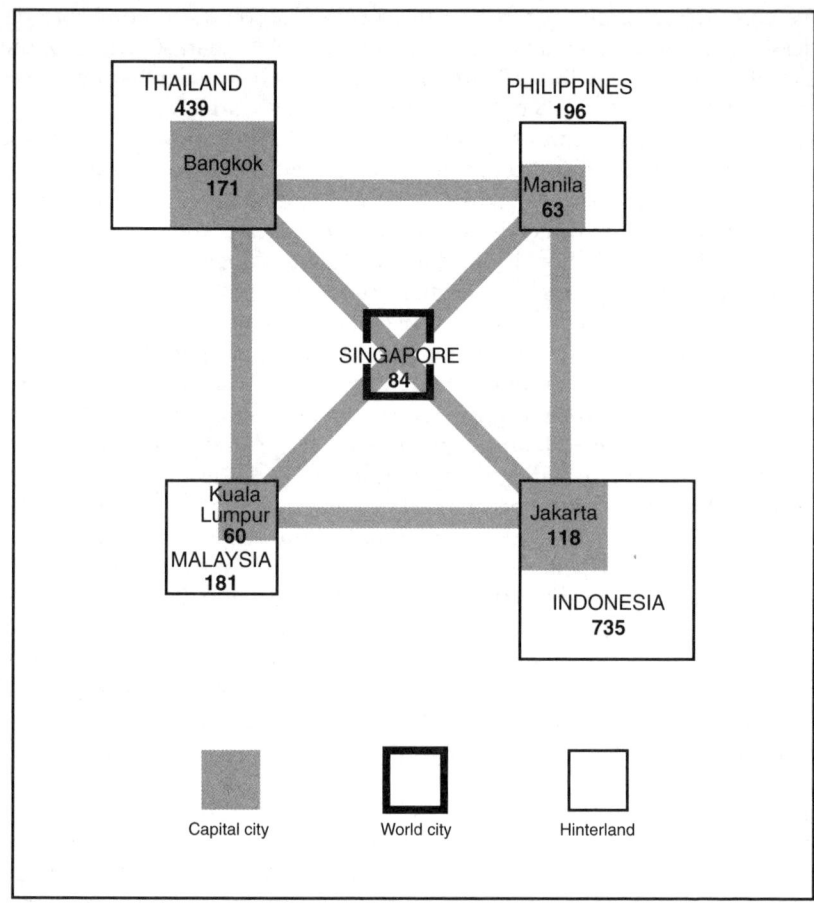

Figure III.3 Southeast Asia: schematic diagram of Singapore's interurban connections, 1995. Figures show relative size of the capital city and national economies in gross domestic Product (US$ billion) (*Source*: based on Table III.1; World Bank 1997)

(Broeze 1989, 1997). Also untenable is the continuing description of the city in Southeast Asia as a 'Third World city'. Even invoking the hybrid rural–urban (*desa–kota*) model to account for the distinctive spatial morphology of cities occurring in the wet rice areas of Southeast Asia is fraught with problems (McGee 1989, 1991, 1998). Unlike the definitive western experience of an expanding city encroaching on the countryside, the model seeks to demonstrate that the countryside in Southeast Asia has itself been transformed into an urban landscape by *in situ* urbanization. This is an observable phenomenon as hybrid settlement patterns in urban hinterlands were seen in Southeast Asia as long ago as the nineteenth century (Chapter 8).

However, the model is essentially a misguided attempt to extend the Third World city (albeit with unique Southeast Asian characteristics) to the phase of late industrialization (Dick and Rimmer 1998a). The 'Third World city' was in fact a transitional phase characteristic of the early decades after the Pacific War. In the late colonial period, Southeast Asian cities were becoming more like western cities, particularly after the opening of the Suez Canal in 1869. Since the mid-1980s in the era of globalization, this process of convergence has re-emerged leading to the development of megacities, which has prompted the new economic growth and trade theories to emphasize geography, space and agglomeration economies (Krugman 1997).

What is needed is a long-term perspective which shows that, because of extended periods of convergence and divergence, the historical pattern of urban development has not always run parallel between metropolitan cities and cities in Southeast Asia. Figure III.4 shows the phases of convergence and divergence between cities in Southeast Asia and metropolitan cities (Dick and Rimmer 1998a). Taking the latter as the yardstick the figure shows a time scale down the vertical axis with city size scaled by population in orders of magnitude. The horizontal dimension shows the nature and intensity of interaction between metropolitan cities and cities in Southeast Asia. Periods of strong interaction are denoted by bold lines, periods of weak interaction by dotted lines.

This figure is distinguished by three separate phases defined by the intensity of technological transfer and adaptation moderated by power relations and resources. New technologies flow readily between metropolitan cities and those in Southeast Asia, but the investment needed to embody them in urban infrastructure is sensitive to both political and economic conditions. Under favourable conditions, as under late colonialism or recent globalization, investment is high and cities in Southeast Asia appear to converge with metropolitan models. When countries were disengaging from the world economy, as in the 1940s to 1960s, investment was low and the cities appeared to diverge from metropolitan models.

No capital city in Southeast Asia has experienced all phases outlined in Figure III.4 – pre-colonial, late colonial, Third World and global – and their different trajectories may fruitfully be compared and contrasted. Singapore, lying at the centre of Southeast Asia, is logically considered first as a truly 'global city' that best resembles Hong Kong (Chapter 7). Jakarta and Manila follow as a pair, being broadly similar as both 'early and late colonial' model cities and, more recently, as sprawling Third World cities (Chapter 8). Bangkok has come to resemble Jakarta and Manila with First World technology and global interactions and their associated middle-class elite superimposed upon a sprawling, populous city with only the most basic public amenities – the newly industrializing city. However, its history as a non-colonial capital was quite different from Jakarta and Manila and, as a river port, it is better compared with Saigon and Rangoon (Chapter 9). Kuala Lumpur stands on its

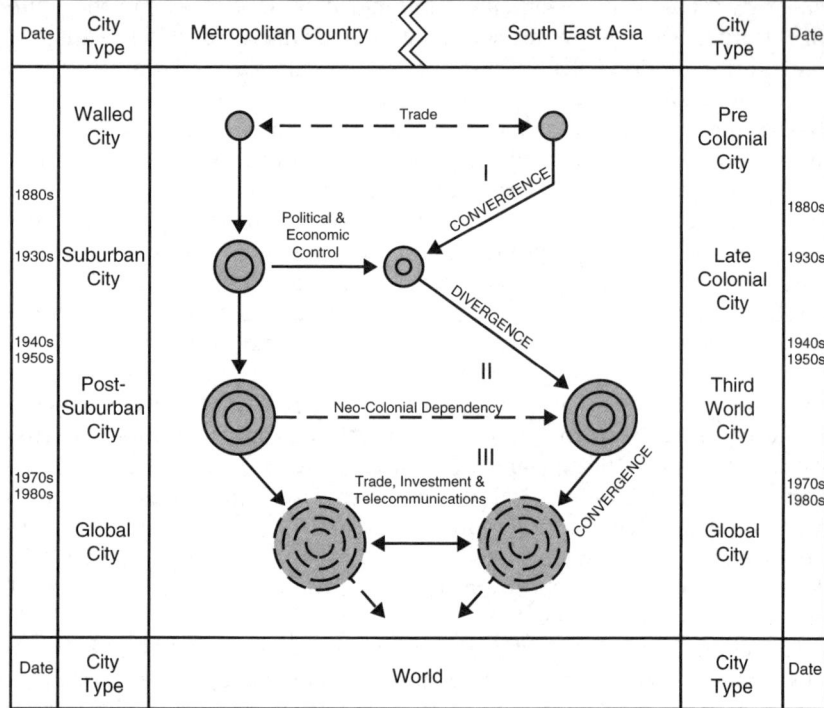

Figure III.4 A model showing phases of convergence and divergence and the associated economic processes in the historical development of city types in Southeast Asia against the yardstick of city types in metropolitan countries (*Source*: Dick and Rimmer 1998: 2306)

own as the one capital city in Southeast Asia, which has been built almost from the beginning around the technology of the motor car – here is the Asian Los Angeles with 'global city' aspirations (Chapter 10).

Urban dynamics

Each chapter traces major economic, political and social shifts and their reflection in changes in urban transport and communications networks. Studying these networks identifies the circulation of people, goods and information both within the city, between the city and its hinterland and with the world beyond the nation.

The studies of individual cities show the coexistence of different socio-economic levels and circuits, which provide the structure and movement, texture and rhythm to urban life. New transport and communications technologies tend to be taken up more readily through the globalized, educated, high-income, elite circuit. In colonial times the hegemonic elite was the European elite; now it is an indigenous middle-class elite. By contrast, the localized, low-income circuit of the mass of the population lacks the resources to acquire much in the way of consumer durables and has little political leverage to gain access to the networks. This socioeconomic contrast shows up vividly in access to the air-conditioned built environment and its transport links (Dick and Rimmer 1999).

A recurrent theme of all chapters is the way transport and communications networks complement urban planning (or the lack thereof) by municipal, provincial and national governments (Rimmer and Dick 1998b). To a considerable extent cities 'happen'. Like the growth of coral reefs, they are the accretion of many individual land-use decisions by households and firms – that is, market forces. However, governments and utilities must decide when and where to provide infrastructure and locate public buildings. Deliberately they may try to realize a vision of what the city should be. Planners laid out pre-colonial royal capitals like Bangkok to align ruler, kingdom and cosmos. Colonial planners imposed a European stamp, whether in the walled cities of Dutch Batavia and Spanish Manila or the garden cities of American Manila and French Saigon. Since independence, national planners have tried, more successfully in Singapore and Kuala Lumpur than elsewhere, to modify these urban forms to reflect new national aspirations.

Another theme is the way transport and communications networks interact with municipal administration (Rimmer and Dick 2000). Colonial rule provided a channel to transfer best practice western technology, so that the main Southeast Asian cities received the modern benefits of transport and communication and other utilities – gas, electricity and water – not much later than metropolitan countries (though administrative restrictions ensured that they flowed predominantly to colonial elites). After 1942 the Japanese occupation and independence changed this dynamic as massive rural–urban migration was accompanied by a loss of local government control and stagnation in infrastructure spending, giving rise to the features of a Third World city. In recent decades political elites have sought to improve their urban lifestyles by renewed spending on infrastructure to produce an updated version of the colonial city – First World enclaves in Third World settings.

7
World City: Singapore

> You will be happy to hear that the Settlement I had the satisfaction to form in this very centrical and commanding station has had every success, and that our Port is already crowded with Shipping from all Native Ports in the Archipelago. (Raffles, 1819)

Singapore is a pivotal component of the international network of cities, ranking with Hong Kong only one step below the topmost level comprising New York, London and Tokyo. As outlined in Chapter 1, this city at the centre of Southeast Asia acts as a transit point for airline flights, a transhipment port, a business hub, a field for investment, a shopping emporium, a tourist stopover and a meeting place. As an anomalous city-state in a world of nation-states, a common response to Singapore is to regard it as no longer a real part of Southeast Asia. However, just as Hong Kong's prosperity is inconceivable without China, Singapore has become what it is by virtue of its regional trading role in Southeast Asia (Chapters 3 and 6).

This chapter highlights the evolution of Singapore's urban functions at the interface between the world economy and Southeast Asia and examines the relationship of this with the size and structure of the city. As seen in subsequent chapters, governments of neighbouring countries have sought at times to emulate Singapore in the planning and management of their capital cities, but have been overwhelmed by forces beyond their control (see Chapters 8, 9 and 10). Logically, Singapore should therefore lead the several chapters on the capital cities of Southeast Asia.

Singapore's urban development is the outcome of two very different modes of production separated by a brief hiatus during which Singapore escaped becoming a Third World city (Ooi 1969). The first system reflects the colonial government's creation of Singapore in 1819. This involves examining the initial British town plan and tracing the city's structural development, dictated by its staple port activities between 1820 and 1940 (Figure 7.1). The second concerns the virtual destruction of the colonial city and its remaking under the mandate of the dominant People's Action Party,

230 *Cities*

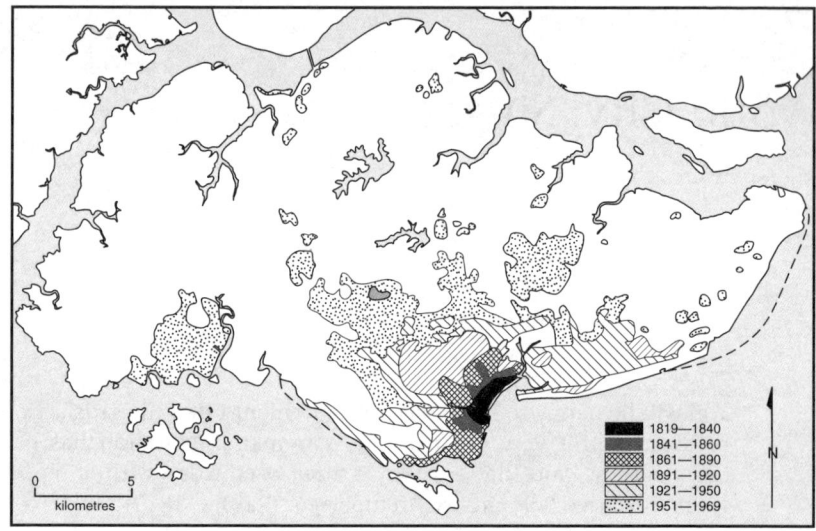

Figure 7.1 Singapore: urban growth, 1819–1969 (*Source*: CMPS 1971, Vol. II Map 1)

which since the late 1960s has single-mindedly focused on restructuring Singapore as a modern, western-based, global metropolis. Attention is focused not only on physical structure and transport patterns but also on the equally important creation of institutions that have dictated the pace and direction of urban growth.

Raffles Town

Singapore's structure was largely in place by 1850. To understand its subsequent development, we have to return to its inception and the original town plan. From the outset, Singapore was a British East India Company trading post, commanding approaches to the Indian and Pacific oceans. While other regional ports imposed levies and restrictions on trade, an emissary of the British crown, Sir Stamford Raffles, sought to capitalize on Singapore's geographical advantage – in 1819 he declared it a free port and adopted a laissez-faire policy. This new emporium offered an attractive alternative to the monopolistic trading practices of rival Dutch ports in archipelagic Southeast Asia. By 1822 a standard gridiron town plan of Singapore with its rectangular plots had been drawn according to instructions given by Raffles to the newly appointed town committee (Figure 7.2). This plan was designed not only to facilitate government and 'anchor' the mercantile community but also to generate revenue from orderly land-use development and to segregate ethnic and occupational groups (Buckley 1867; Bristow 1992).

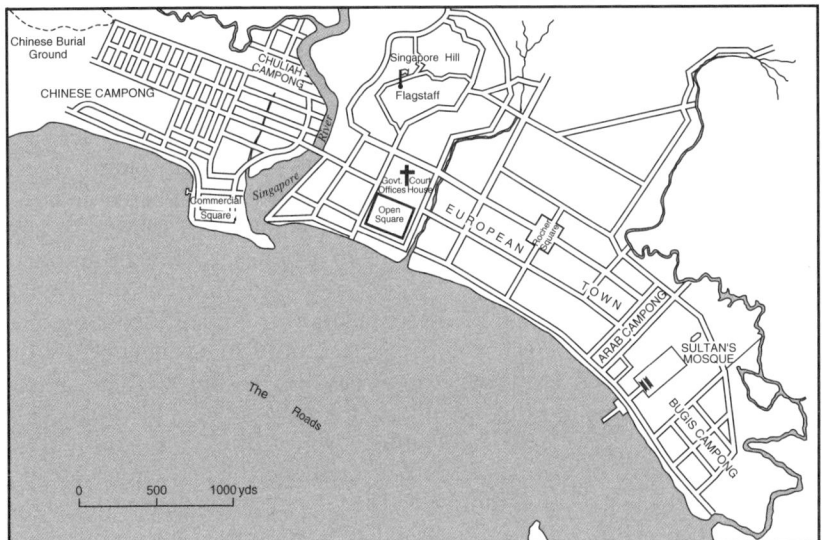

Figure 7.2 Singapore: the 1822 Town Plan drawn to Raffles' specifications by Lt Phillip Jackson (*Source*: Based on Crawfurd 1828: facing 529)

Raffles established the town's military and civic components in relation to the Singapore and Kallang Rivers. Once this distinction had been made, a central administrative area, comprising courthouse, church and government offices, was established on three sides of an open square east of the Singapore River. Adjacent to the administrative quarter was the 'European Town'. This stretched inland from the esplanade and recreation area on the seafront, to the fort on Singapore Hill. As its name implies, this land was allocated as the residential quarter for European administrators, merchants and bankers. Beyond the European Town were the Arab *kampung*, the Malay *kampung* surrounding the Sultan's palace and mosque at Tanjong Glam, and the Bugis *kampung* located near the mouth of the Kallang River, which housed traders from the east and southwest coasts of Borneo, the Celebes and Bali. A commercial area comprising a line of wharves and warehouses to facilitate trade was designated Commercial Square and developed as the principal business district on drained land west of the Singapore River. It was land previously occupied by Chinese traders and raft houses. Also established west of the river were a small *kampung* for the Chulias (south Indians) and a tightly packed *kampung* for the immigrant Chinese. Telok Ayer was merely a fishing village.

The original map is misleading because it does not indicate critical maritime features such as water depths and the distribution of sand and mud. Without this additional information, as shown on Admiralty charts and Coleman's 1838 map, the rationale for the location of the Bugis settlement is not immediately apparent. It was the sand and mud at the mouth of the Kallang River that provided protection for sailing ships during the monsoons. The greater importance attached by Raffles to the Bugis rather than the Chinese stemmed from their potential geopolitical role. The Dutch in particular sought to control key sea-lanes, notably the Straits of Malacca and the Sunda Strait. The Bugis were ideally placed to play a fifth-column role in informing the British about the activities of the rival Dutch based in Java.

The structure of Raffles' plan has continued to influence Singapore's subsequent development. The 400-hectare town area has remained the core of Singapore's Central Area. In 1901, 91 per cent of Singaporeans still lived in Raffles Town and 63 per cent of the population still lived there as late as 1957 (Savage 1992: 18). Segregation persisted until Independence. By the 1840s the designated European Town was left to the Chinese as the Europeans headed first for the 'suburbs' in the surrounding hills and later to the detached European villas and bungalows within compounds in the Tanglin area. Both the Garden City movement and the desire to escape the town area's endemic diseases motivated the second shift (Chua and Edwards 1992; Edwards 1992). Only the wealthy Chinese followed the Europeans in the search for social status in the garden suburbs. But there was an overspill of Chinese and their distinctive two-storey shophouses westwards from their original *kampung* into the new port area developed in Tanjong Pagar. The designated 'Chulliah' compound remained a focus for Indian migrants and a base for Chettiar money lenders and spice merchants. Subsequently, the non-mercantile Indians moved from Chulia, Malacca and Market Streets into other clusters, but they retained their affiliations with Chettiar. In particular, they moved to the Serangoon Road area ('Little India') which was the centre of the cattle industry and an Indian convict jail. Some Indians also took up residence in the dock area when the port expanded.

Colonial city

Between 1850 and the late 1950s, Singapore's outward expansion from the planned town was centred on the commercial heart that developed around Commercial Square (renamed Raffles Place in 1853). Much of the expansion of the original 1824 planned town, designed to house 10,000 people, was to accommodate the unanticipated influx of traders and migrants from China, India and other parts of Southeast Asia. By 1849, Singapore's population had quintupled to almost 53,000, with a decline in the proportion of Malays that initially had been the dominant group. Five decades later, in 1901, Singapore's population had increased to nearly 227,000 with the Chinese exceeding 72 per cent (reflecting an influx of male immigrants from southern China), followed by Malays with almost 16 per cent, Indians 8 per cent and others around 4 per cent (Table 7.1). By 1931 the population had doubled to close to 558,000 with little significant change in the ethnic

Table 7.1 Singapore: distribution of population by ethnic groups, 1824–1957

Year	Chinese No.	%	Malays No.	%	Indians No.	%	Other No.	%	Total No.
1824	3,317	31.0	6,431	60.2	756	7.1	179	1.7	10,683
1849	27,988	52.9	17,039	32.2	6,284	11.9	1,580	3.0	52,891
1901	164,041	72.3	35,986	15.9	17,047	7.5	9,768	4.3	226,842
1931	418,640	75.1	65,014	11.7	50,811	9.1	23,280	4.2	557,745
1947	729,473	77.8	113,803	12.1	68,967	7.4	25,901	2.8	938,144
1957	1,090,595	75.4	197,060	13.6	124,084	8.6	34,190	2.4	1,445,929

Note: Rounding errors.
Source: Perry *et al*. (1997: 31).

composition of the population, as migration continued to compensate for negative or low fertility rates stemming from a gender imbalance in favour of males. In 1947 Singapore's population exceeded 938,000. Chinese accounted for almost 78 per cent, Malays around 12 per cent, a slight increase, but there was a decline in the relative shares of Indians and other ethnic groups. By 1957 Singapore's population was almost 1,446,000 due primarily to natural increase as the volume of in-migration had decreased because of the uncertain economic conditions.

Urban form

The transformation of Singapore's physical fabric to accommodate this influx of population was strongly influenced by the location of navigable estuaries, natural anchorages and harbours (Figure 7.3). Progressively, jungle, marsh and seashore were replaced by shophouses, detached colonial houses and their compounds, roads and port infrastructure as reclamation pushed out the foreshore, and filled back swamps between the city and Mt Faber. Until 1860 the city was still within a 3-kilometre radius of the Singapore River and beset with a lack of funds to overcome unwholesome sanitary conditions and lack of policing (Savage and Yeoh 1993). By 1890 ribbon development had occurred along main roads. Over the next thirty years as residential suburbs for the wealthy were established west of the city, the working classes stayed behind in the high-density inner districts. By 1950 the town had grown by accretion with pronounced ribbon development occurring in the west and more consolidated expansion in the east.

Rhythms in Singapore's spatial development until the 1960s closely reflected variations in its fortunes as a port, market and financial centre. These activities 'put regional traders and producers [in the Malayan archipelago] in contact with world markets, stimulated enterprise and capital accumulation and furnished the necessary "inducement goods" of manufacturing from the West and food from Southeast Asia' (Huff 1994:

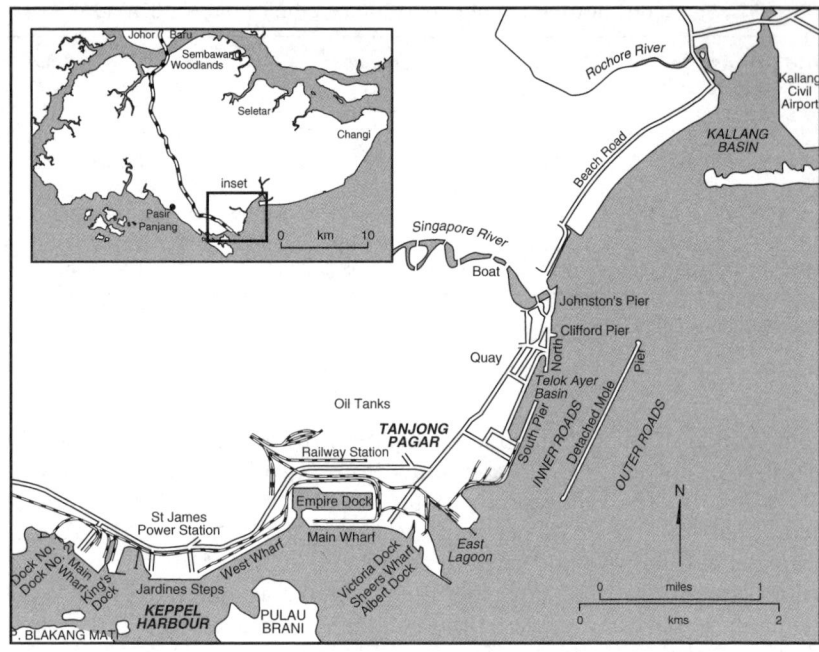

Figure 7.3 Port of Singapore, 1939 (*Source*: Various)

274). Of critical importance was the maintenance of Singapore as a 'free port', which gave both Asian and European traders the unique opportunity to operate unfettered by monopolies, duties or other trade restrictions.

Until 1900 Singapore expanded as the premier entrepot port in East Asia, redistributing imported manufactured goods, foodstuffs and Straits produce (agricultural and mineral products from the Malayan Archipelago and more generally Southeast Asia). Singapore's exports were limited to smelted tin, polished rice and tropical produce such as the short-lived gambier industry, canned pineapples and sago products. Between 1900 and 1942, Singapore's exports centred largely on the new staples, rubber and petroleum; tin was less influential, rice declined and tropical produce stagnated and changed composition (copra, sago and areca nuts replacing gambier and tapioca). Their growth was uneven with a marked decline during the depression of the 1930s (Huff 1994). Redistribution of manufactured imports from Britain, Europe, America and Japan continued with a greater emphasis on producer goods and consumer durables at the expense of the old trade in textiles. After the Second World War, Singapore's trade expanded substantially, particularly following the declaration of independence from Malaysia in

1965. As Singapore's fortunes were so tied up with its port, it is often seen among cities in Southeast Asia as the quintessential port-city, at least until the 1960s.

Not surprisingly, key land-use changes were focused on the port. The island's sheltered and commanding position at the narrow gateway to East Asia accounted almost entirely for the settlement's early prosperity and importance. Although the seafront between the Singapore and Kallang Rivers retained its importance, the port expanded beyond the confines of Raffles' planned area into New Harbour, south of Chinatown. This in effect created three distinct functional areas: the New Harbour (later called Keppel Harbour), the Roadstead and the Kallang Basin (Figure 7.3). From the beginning, Keppel Harbour's deep water accommodated the mail ships, opium ships and deepsea freighters that connected Europe and India with China, Indochina and Thailand. The Roadstead provided a protected anchorage for small inter-Asian ships to discharge and await return cargoes; lighters operated to and from Boat Quay in the Singapore River, where British and Chinese shipping agencies were located. As noted, the Kallang Basin provided shelter for the Bugis prahu carrying low-cost commodities such as timber and firewood.

Boat Quay and the Roadstead were the dominant foci of activity until steamships ended the dominance of sailing ships from the 1870s. When the first scheduled steamship service operated by the Peninsular and Oriental (P&O) Steam Navigation passed through Singapore in 1845 the facilities in New (Keppel) Harbour were rudimentary (Gibson-Hill 1958: 157). In 1864 the first docks were built in Tanjong Pagar. Over the next fifty years a deeper harbour was developed by private enterprise, particularly after the opening of the Suez Canal in 1869. This comprised deepwater wharves, especially for coaling, docks and an extensive reclamation at Telok Ayer. By 1887 a specialized berth had been added at Pulau Brani for handling Malayan tin ore which was collected, smelted and exported by the European-owned and -funded Straits Trading Company (Chapter 6). In 1891 the first storage facilities for kerosene were established by Shell on Pulau Bukom and Singapore developed as the storage, transhipment and distribution centre for the Asiatic Petroleum Company and Standard Vacuum Oil Company in East Asia.

By 1899 the pioneering Tanjong Pagar Dock Company had gained almost complete control over the five dry docks that repaired oil tankers and the 2000-metre wharf at Keppel Harbour. In 1903, the Harbour was linked to the Malayan Railway system and wharf facilities were expanded. These facilities were appropriated by the government in 1905 because the Dock Company was unwilling to undertake the necessary upgrading of its existing wooden wharves to accommodate the expanding trade. By 1907 the publicly owned Tanjong Pagar Dock Company had finished the necessary harbour works to provide shelter in the Roadstead and three years later completed a second reclamation at Telok Ayer. In 1913 the newly established Singapore Harbour Board became responsible for the development and management of Keppel Harbour, the Singapore River and Telok Ayer Basin. During the 1920s the British Government made Singapore its principal installation in East Asia and established military bases at Pasir Panjang, close to the urban core and on the northern peripheries at Sembawang, Tengah, Selatar and Changi (Phang 1992).

The large-scale bulk movement of products from plantation agriculture was still accommodated in the roadstead (including rubber until the mid-1920s). To accommodate the increase in rubber exports, in 1932 the Singapore Harbour Board completed the Telok Ayer reclamation to provide berthing for local feeder vessels (Huff 1994). By then Singapore had become the pivot of the regional transport system operated for rubber and passengers by the Straits Steamship Company and the Dutch-owned Koninklijke Paketvaart Mij (KPM) (Chapter 3). An associated development was the construction of shipyards in the Kallang Basin. This was also the site of the first commercial airport opened in 1937. Kallang Airport replaced the Selatar Air Base and, like its predecessor, permitted seaplane and airplane landings by KLM, Imperial Airways and Qantas.

Following the development of Keppel Harbour, Raffles Place, Battery Road and Chulia Street became the core of the Central Business District. Warehouse-cum-offices belonging to European trading houses and companies disappeared. They were replaced by the major banks, shipping offices, large retailers and the General Post Office (Chua 1989). In 1890, a warehouse-office was demolished to accommodate the Hong Kong and Shanghai Bank, which in turn was replaced by an even more imposing building in 1919 (Chua 1997). By 1930, few of the original two-storey buildings occupied by European traders survived. Boat Quay's shallow waters were used for the intra-Asian trade in Straits produce, controlled by Chinese entrepreneurs.

Cathedrals, European hotels and English-language schools augmented the administrative quarter across the Singapore River. This quarter contrasted markedly with two densely populated areas hemming in the Central Area – Chinatown adjacent to the commercial area south of the Singapore River and an area to the north of the administrative area. Both areas were characterized by back-to-back, two- or three-storey shophouses, subdivided tenements, cubicles, eateries and pleasure quarters. The enforced provision of back lanes to shophouses improved rear access but the establishment of the Singapore Improvement Trust had only a limited impact on housing conditions (Ho and Lim 1992; Chua 1997). Street congestion in these crowded areas made the smooth flow of traffic difficult.

Urban transport

The transport task in colonial Singapore was to connect the port, the Central Business District focused on Raffles Place, the government quarter and the inner and outer residential districts (Rimmer 1986a, 1990). Bullock carts were used for moving cargo along dirt roads to and from port. Horses and carriages had been imported from the United Kingdom for the well-to-do as early as the 1820s. By the 1860s there were two-wheeled hackney carriages (*gharries*) owned and operated by 'Kling coolies', for hire, in addition to sedan chairs and the *redi* for high-class ladies. Most people walked until the first two-seater rickshaws (*jinrikisha*) were imported from Japan via Shanghai in 1880. The rickshaws were operated from public stands controlled by different clan groups in a way peculiar to Singapore (Table 7.2). They offered taxi services linking home, office, club and the Malay Street brothels (Rimmer 1986b; Warren 1995).

The unreliability of animal and man-powered transport systems led to demands for steam and electric-powered replacements. In 1885 a steam tram was introduced to bridge the three-kilometre gap between the growth points at Singapore River and Keppel Harbour. The tram carried goods and passengers until it ceased operations in 1894, unable to compete against the rates offered by the entrenched interests of rickshaw pullers, porters and carriers (York and Phillips 1996). In 1905 electric trams were introduced on a 25-kilometre track but, despite predictions, they did not run the rickshaw off the streets, because the travelling public disliked having to board at fixed stops. Although the trams reached a peak of 16 million passengers in 1913, they could not provide adequate services or run profitably because of antagonism between the South Indian drivers and the Chinese conductors, the cheap fares charged and systematic defrauding of the company by its 'Chinese inspectors' (Rimmer 1986a). The tram did little to promote suburban development (Huff 1994).

Gradually, Singapore progressed from electric systems to motorized transport for moving people and goods. Although the first motor car was introduced in 1896, not until after the First World War did numbers become significant. In 1920 17-seater buses were introduced by the Singapore Municipality to supplement the trams, but these only lasted one year. Their function was taken over by Chinese bus companies operating seven-seater

Table 7.2 Singapore: division of public jinrikisha stands into clan districts, late 1910s

Clan	Location
Cheow Ann	Bukit Timah Road
Chin Kang	Adelphi Hotel
Heng Hua	Europe Hotel (Inside), Orchard Road
Hok Chia	Europe Hotel (Outside), Raffles Hotel, River Valley Road, Tank Road Railway Station
Hui Wah	Johnston's Pier, Tanjong Pagar and Kreta Ayer

Source: Rimmer (1990: 146).

'mosquito' buses. The rapid growth of these bus companies prevented the Singapore Municipality rejuvenating the ailing electric tramway company. Eventually, it passed into the hands of the receiver in 1922 (Straits Settlements, Tramways Commission 1921). Efforts were made to restrict the rickshaws and mosquito buses in a failed attempt to allow the Shanghai Electric Construction Co. Ltd the opportunity to revitalize the electric tramway. In 1926 the tramway was replaced by a 31-kilometre trolley bus system, later supplemented by large petrol-driven buses. During the 1920s lorries also took over the functions of bullock carts and after the Johore Causeway was completed in 1923 they were able to provide services to the Malayan Peninsula (Chapter 6).

The effects of these changes are reflected in the traffic censuses which list the array of vehicles crossing Anderson's Bridge and Institution Bridge in 1917, 1923 and 1930 (Table 7.3). In goods transport there was a sharp decline in bullock carts and handcarts and an increase in the number of light and heavy lorries. Similarly, in passenger transport there was a reduction in the number of gharries, rickshaws and bicycles, and a marked rise in both public transport (trolley buses and heavy and light motor buses) and private transport (motor cars and, to a lesser extent, motor cycles).

Table 7.3 Singapore: vehicle types passing over selected bridges during traffic censuses, 1917, 1923 and 1930

Mode	Anderson's Bridge			Institution Bridge		
	1917	1923	1930	1917	1923	1930
Goods vehicles						
Bullock carts	358	135	25	1,006	312	25
Hand carts	44	25	14	58	49	32
Heavy lorries	46	242	520	41	208	405
Light lorries	46	215	1,044	42	126	880
Trailers	0	0	13	0	0	4
Sub-total	494	617	1,616	1,147	695	1,346
Passenger vehicles						
Jinrikisha	2,741	2,101	1,666	8,599	6,950	3,225
Cycles (ordinary)	398	n.a.	n.a.	651	n.a.	n.a.
Gharries etc.	527	183	27	645	236	42
Tramcars/trolleys	181	178	1,909	0	0	0
Heavy motor buses	0	10	278	0	15	6
Light motor buses	0	167	228	0	190	238
Motor cars	2,067	5,491	11,805	924	2,435	4,172
Motor cycles	221	420	315	115	89	108
Sub-total	6,135	8,550	16,228	10,934	9,915	7,791

Note: n.a. not available (cycles were not counted in 1923 and 1930).
Source: Singapore Municipality, Vehicle Registration Department (1931: 28F–29F).

In 1935 the Municipality successfully amalgamated the mosquito bus operators into eleven Chinese bus companies capable of operating larger buses. Notwithstanding these changes the Committee appointed to inquire into Singapore traffic in 1938 still had concerns about the traffic congestion in narrow streets. These arose from the ongoing mix of pedestrians, slow-moving vehicles such as bullocks, rickshaws, bicycles and tricycles, private cars, lorries and public service vehicles (Straits Settlements 1938). Emphasis was placed on removing street vendors to allow the free flow of motor traffic. Little was achieved before Singapore was drawn into the Pacific War, which resulted in a hiatus in its development.

Hiatus

Between 1942 and 1945, Singapore was occupied by the Japanese. This extended the moratorium imposed on waterfront development by economic pressures stemming from the decline in rubber and tin exports during the Great Depression. The war interrupted both shipping and air transport. Public transport equipment deteriorated rapidly during the occupation. Rickshaws were eliminated in 1947 and replaced by 4,000 trishaws (bicycles with passenger-carrying sidecars). By 1949 trolley bus services had been re-established and shipping and air transport had regained their prewar eminence (Singapore 1956). In the same year Singapore handled 71 per cent of Malaya's imports and 63 per cent of its exports (Cheng 1991). The staple port had revived.

Local politicians, intent on highlighting their post-colonial achievements, overplayed the downturn in Singapore's fortunes during the 1950s. Singapore had the fundamentals for rapid postwar recovery. Besides its commercial banks and pool of skilled labourers, Singapore retained its international markets in rubber and tin, possessed a ship repair industry and rubber-processing activities, and continued to be a major oil distribution centre. There was no marked decline in the port's traditional entrepot trade, particularly as the rubber trade remained buoyant. Petroleum exports quadrupled after the Royal Dutch Shell Group located their East Asia regional offices to Singapore and Caltex opened its marketing operations (Huff 1994). Singapore was still the region's key maritime centre for Indonesia and Malaya and had rapidly developing air and telecommunications connections. In 1956 Queen's Dock was added to the existing complement of five dry docks and new wharves were completed in the East Lagoon. Because its grass runway could not cope with the larger aircraft and heavier traffic, in 1955 the airport was relocated from Kallang to a site occupied by squatters at Payar Lebar.

Declining trade returns per capita underlined the need for a more diversified economy. Although Singapore had a manufacturing base associated with the rubber industry and traditional woodworking, it was never an industrial city like Bombay or Shanghai (Huff 1994). Singapore's population continued to increase, despite the end of mass migration in the 1930s. This led to unemployment and an acute housing shortage. In Chinatown's Upper Nankin Street most of the population lived in shophouses divided into cubicles; some households even shared cubicles (Kaye 1960). Trading spilled over onto the sheltered pathways, forcing pedestrians and itinerant hawkers to compete with motorized and non-motorized vehicles on the streets. Squatter settlements appeared on the city's periphery. The deterioration of Singapore's physical conditions prompted the colonial government in 1958 to produce a master plan. Echoing Abercrombie's Greater London Plan (1944), it recommended deconcentration with the development of new towns and a green belt to stop the expansion of the Central Area. The master plan was never implemented because it seriously underestimated population and vehicular growth. Consequently, there was no systematic transport planning.

Industrial unrest in Singapore led in 1956 to the 136-day 'Great Strike' by Singapore Traction Company's workers (Singapore, Legislative Assembly 1956). As competition from the 11 Chinese bus companies intensified, the trolley buses ran at a loss (Singapore 1956). Public transport was left largely untouched throughout the 1960s apart from the replacement of trolley buses by motor buses in 1962 and an inquiry into unlicensed 'pirate' taxis (Singapore 1966). Consequently, inadequate bus services had to compete with pirate taxis providing superior services to squatter slum dwellers who had been transferred to public housing estates on the city's periphery (Buchanan 1972). This 'unsatisfactory' state of affairs led not only to the creation of the Ministry of Communications in 1968, but also to a comprehensive attempt to streamline Singapore's land use and transport system as part of the government's wider bid to become a global city.

Global metropolis

Within the three decades since the late 1960s, the low-rise, colonial city with its distinctive shophouses has been virtually reconstructed as a new high-density, high-rise 'global city'. Singapore's fortunes are now inextricably linked with London, New York and Tokyo. The city's built-up area has more than doubled to accommodate the needs of the international economy. Housing units have trebled, commercial floor space and industrial land have quintupled and farmland, forests and tidal wastes are much reduced. Slums and squatter areas have been virtually eliminated and most of the population lives in dwellings in public housing provided originally by the Housing and Development Board in new towns (Table 7.4). As occupiers were able to purchase their flats, over 90 per cent of Singaporeans now own their dwellings (Yuen 1998). These developments have necessitated restructuring urban land use and replanning the traffic system.

Table 7.4 Singapore: economic and social indicators, 1960–2000

	1960	1970	1980	1990	2000
Population (million)	1.6	2.1	2.4	3.0	4.0
GNP per capita (current US$)	450	920	4,420	11,160	24,740
GNP per capita (world ranking)	33	27	21	19	12
Life expectancy	65	68	72	74	78
Infant mortality (aged 0–1 per thousand live births)	35	14	12	7	4
Population per physician	2,360	1,400	1,150	840	n.a.
Literacy rate (10 years and over)	52.3	72.2	83.0	90.1	92.0

Sources: Huff (1994; 352); World Bank (2002).

Land-use changes

State ownership of almost four-fifths of the island facilitated the transformation of Singapore's space economy to accommodate rapid economic development. Control of land use through inherited zoning and density regulations gave the government the opportunity to guide urban renewal and plan additions to infrastructure. They were able to achieve this objective without inflationary pressures by drawing upon employee pension funds held in the Central Provident Fund and deposits in the Post Office Savings Bank. This funding enabled the Singapore government to concentrate on the comprehensive, long-range planning of the island's physical development according to guidelines established by the United Nations Development Programme and implemented by State and City Planning Project (1967–71) (CMPS 1971). The plan's aim was to optimize the distribution of jobs and residences to reduce the need to travel by public and private transport.

The most dramatic land-use change stemmed from the development of manufacturing, which until the 1960s had been restricted to river and port locations (Figure 7.4). New projects to address massive unemployment and underemployment occurred outside the existing port area following recommendations by the Canadian, F.J. Lyle, under the Colombo Plan in 1959 and Dr Albert Winsemius, under the auspices of the United Nations in 1961 (Cheng 1991). In 1963 the Jurong Industrial Estate was created on swampland in the southwest corner of Singapore, as the site for an industrial port, factories and petroleum and petrochemical activities. By 1968 Jurong had been developed as a self-contained satellite town on 1,447 hectares of state-acquired swampland (Yuen 1998). The Jurong Town Corporation was established to administer its development. Although it was not immediately successful in triggering industrial production and exports, it did boost the construction sector and attract firms requiring large sites and special facilities such as the National Iron and Steel Mills, the National Grain Board, Singapore Sugar Industries and Jurong Shipyard,

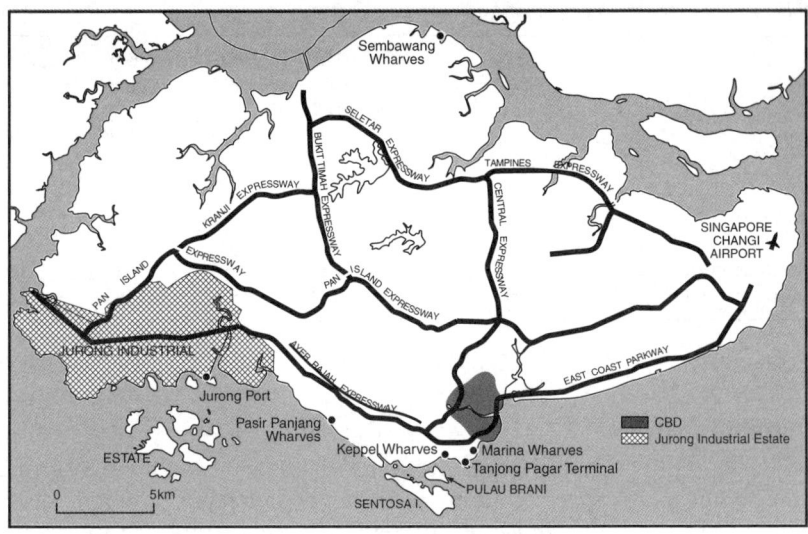

Figure 7.4 Port of Singapore and expressway system, 2000 (*Source*: Various)

which in 1972 launched the island's first ocean-going ship and was soon joined by the Japanese-owned shipbuilder Hitachi Zosen. Ship repair also became a major industry.

A large oil-refining and petrochemical complex was established on small islands offshore following the opening of the Shell oil refinery at Pulau Bukom in 1960. Subsequently four other refineries were developed (by Esso, Mobil, BP and BP/Caltex/Singapore government) and an offshore supply terminal was developed in 1971 (Pinder 1997). These activities had important spin-offs because they enabled Singapore – the 'Houston of the East' – both to attract large tankers and to offer other ships cheap bunkering services (Huff 1994). Indeed, the petroleum refining industry linking Middle East crude suppliers and East Asian markets has remained an important contributor to Singapore's manufacturing sector.

In 1962 the *M.V. President Buchanan* brought the first container to Singapore. This event heralded that machinery and capital goods would eventually supersede the staple exports of tin and rubber. To meet the fresh challenge, the existing port area of the Keppel and Tanjong Pagar wharves was redeveloped by the Port of Singapore Authority (PSA) established in 1964. Specialized roles were allocated to individual terminals as the port moved from labour-intensive operations to highly mechanized systems of handling to accommodate the needs of container operators such as the government-owned Neptune Orient Lines. In 1967 Tanjong Pagar was selected as the location of a purpose-built container terminal which was completed with World Bank assistance five years later. Since 1968, Telok Ayer Basin, adjacent to the Keppel and Tanjong Pagar wharves, had been used for the coastal and lighterage trade. In 1972 Pasir Panjang was created as a new centre for regional trade and lighterage. In 1973 Sembawang in the northwest of the island was added to the list of port sites when the British withdrew from Southeast Asia and the Royal Navy transferred ownership of the terminal to the PSA. It now handles timber products.

By 1983 lighterage had been moved from the Telok Ayer Basin back to the Singapore River to allow for the foreshore reclamation and construction of the Marina Centre as Singapore's second Central Business District. In 1990 the Keppel wharves were converted to container use and in 1992 the Pulau Brani container terminal became operational. Jurong Port, like the Pasir Panjang and Sembawang terminals, has been used mainly for non-containerized and bulk cargoes under the management of the Port of Singapore Authority.

Subsequently, the Jurong Town Corporation (JTC) became the government's leading developer and manager of industrial facilities with the Economic Development Board (EDB) attracting investors and its offshoot, the Development Bank of Singapore (DBS), providing the necessary capital. The Corporation has built thirty industrial estates on 6,000 hectares dispersed throughout Singapore and still has two-thirds of the land under its control. Initially, the Corporation's industrial estates were designed to attract labour-intensive, low technology firms to multi-storeyed factories. During the late 1950s the colonial government had encouraged the development of apparel and textiles but the major transformation came with the government's adoption of an industrial programme offering companies incentives and subsidies. Workers were to work longer hours and take fewer public holidays and the power of unions was eroded to offset Singapore's reputation as a high-cost Asian producer. Although this industrial programme paralleled developments in the other Newly Industrializing Economies of Hong Kong, South Korea and Taiwan, Singapore was ahead of anywhere else in Southeast Asia.

After a very brief period of import substitution, Singapore gave state-owned enterprises an opportunity to engage in entrepreneurial activities. The government also sought to attract multinationals from Europe, Japan and the United States to use Singapore as an export platform. This was especially the case for multinational enterprises engaged in 'footloose' consumer electrical and electronic industries (including semi-conductors) and to a lesser extent textiles and garments. By 1970 this export-oriented policy, dependent on increased female labour force participation, was so successful that manufacturing became the leading sector of the economy. Within three years the labour surplus had been eliminated despite the withdrawal of British forces from east of Suez between 1968 and 1971 (Huff 1994).

The preference for large multinational corporations has continued, probably to the detriment of truly indigenous technological research and development. Similarly, the government has persevered with efficient state-owned enterprises administered through holding companies and statutory boards, as exemplified by Singapore Airlines and Singapore Telecom (SingTel). However, the agency houses, which underpinned Singapore's pre-1970 development, have virtually disappeared. While the local Chinese entrepreneurs still dominate primary commodity exports, they are no longer so important (Huff 1994). Between 1970 and 1980 manufacturing jobs increased from 25,000 to 287,000, delivering full employment (Cervero 1998).

A conscious effort was also made to integrate labour-intensive industries into the high-density, high-rise public housing estates, which also incorporated schools, recreational facilities and shopping centres. The Housing and Development Board has thereby been the principal agent influencing Singapore's physical structure and transport demand. This has occurred by influencing the vertical and horizontal expansion of residential land to redistribute the population away from the Central Area. In 1960 the statutory board replaced the colonial Singapore Improvement Trust, which had managed 23,000 public flats at a time when there were an estimated 250,000 living in slums and 300,000 living in squatter areas around the city's periphery (Chua 1985; Wong and Yeh 1985). The Board's initial housing estates, typified by Kallang, Whampoa and the more comprehensive satellite towns of Queenstown and Toa Payoh, were within 6–8 kilometres of the city centre. They were designed to accommodate residents uprooted by urban renewal and to give former migrants a stake in the new country. A special effort was made to disperse concentrations of Malays in the Geylang Serai/Eunos area across the city to integrate them into the general population (Van Grunsven 2000).

By the 1970s the Housing and Development Board's role had switched from flat builder to developer of new towns at greater distances from the Central Area. Following advice from United Nations consultants in 1971 – which reflected Dutch experience – self-sufficient new towns such as Ang Mo Kio, Bedok, Clementi and Telok Blangah (each housing 250,000–300,000 people) were located in a ring around the central catchment area inter-linked by expressways or mass transit. When regulations were relaxed in the 1980s, more attention in the new towns was paid to visual design and the distinction between town centre, neighbourhood and precinct. As exemplified in Bukit Batok, Tampines and Yishun, precincts incorporated low walls to demarcate the resident's territorial space and multi-storey carparks. Although the Board's powers over urban renewal were assumed in 1974 by its offshoot, the Urban Redevelopment Authority (URA), housing production peaked in the mid-1980s. Increasingly, attention has been given to upgrading old estates. Since its inception the Board has built 750,000 homes in twenty new towns throughout Singapore (Table 7.5). While much reference has been made to the creation of Singapore as a 'Garden City', its post-colonial manifestation, however, has less to do with Ebenezer Howard's vision and more to do with planting trees to give the city a green mantle.

Table 7.5 Singapore: housing and development, residential properties by town, 1931–1999

Town	Date first block taken over by management	Dwelling units under management, 1998	Estimated total dwelling units
Queenstown	1931	29,084	47,000
Bukit Merah	1937	46,976	68,000
Kallang/Whampoa	1947	34,271	43,000
Other estates	1952	23,752	25,000
Geylang	1958	30,980	46,000
Bedok	1963	57,238	72,000
Toa Payoh	1966	35,500	45,000
Jurong West	1970	49,659	95,000
Bishan	1972	19,396	29,000
Woodlands	1973	46,826	88,000
Hougang	1974	44,133	65,000
Sembawang	1974	3,881	64,000
Ang Mo Kio	1975	47,988	56,000
Pasir Ris	1975	27,516	44,000
Serangoon	1975	21,497	28,000
Jurong East	1976	22,796	27,000
Bukit Batok	1977	31,050	47,000
Choa Chu Kang	1977	32,019	55,000
Clementi	1977	23,928	34,000
Yishun	1977	45,735	84,000
Sengkang	1978	5,985	95,000
Tampines	1981	61,853	83,000
Bukit Panjang	1986	21,598	40,000
Punggol	—	—	86,000
Total	1931–1999	763,661	1,366,000

Note: Other estates are Bukit Timah, Central Area, Lim Chu Kang and Marine Parade.
Source: Chief Corporate Development Office, Housing and Development Board, Singapore (pers. comm., 5 March 1999).

A conservation programme to attract the tourist's gaze was also instituted by the Urban Redevelopment's Authority to recapture Singapore's 'Asian identity', 'traditional values' and 'local culture'. These had been lost in the bland urban redevelopment of the 1960s and early 1970s, characterized by demolition and rebuilding rather than renewal. The development of the firm-centred economy and the 'industrialization of everyday life' had been at the expense of the bazaar economy (Chua 1998). Since the mid-1980s there has been a belated program to conserve civic and cultural areas that figured prominently in the life of the old colonial city including Fort Canning and the Singapore River (Chua 1997). Also involved are the ethnic areas of Chinatown, Little India, Kampong Glam, Bugis Street and the Malaccan-style terraces in Emerald Hill (Savage 1992; Kong and Yeoh

1994; Perry et al. 1997; Chang 2000). Symptomatic of the changes, the Raffles Hotel built in 1887 and made famous by the visits of writers Joseph Conrad, Somerset Maugham and Rudyard Kipling has been 'upgraded' into the high-rise Raffles City development. This has bundled the old hotel together with a new hotel and a retail and office complex.

Transport changes

In the late 1960s the Singapore government ended the laissez-faire growth of the country's transport system. A new system was required to meet the spread of industry, dispersal of residential areas and population, the diffusion of port facilities over five sites and the proposed relocation of the international airport to Changi from Payar Lebar. Following the recommendations of the State and City Planning Project, inner and outer ring roads were constructed around the Central Area. Since 1968 development has been concentrated on pan-island transport corridors with 140 kilometres of grade-separated expressways running along the island's southern shore linking key transport nodes between Jurong and Changi and in a ring around the central catchment area (Table 7.6). As noted, major satellite housing areas associated with this 'ring concept' have been Ang Mo Kio, Bukit Bukok, Bishan, Woodlands and Yishun.

The reliance on private buses, private taxis and pirate taxis to service the new towns was undermined by the government's development of a comprehensive and affordable public transport system. The standard of bus

Table 7.6 Singapore: motor vehicle registrations and road length, 1965–2000

	1965	1975	1985	1995	2000
Vehicles (no.)					
Private motor car	104,729	142,045	221,279	342,245	388,702
Public motor car	3,621	6,492	14,971	21,661	25,578
Buses	1,617	4,935	8,717	10,723	11,938
Goods vehicles	21,617	43,761	109,596	137,913	128,324
Motor cycles and scooters	60,838	83,145	127,564	129,587	134,745
Total	192,422	280,378	482,127	642,129	689,287
Roads (km)					
Expressways	—	—	73	132	150
Major arterials	222	257	435	567	569
Collector road	109	122	202	326	360
Local roads	1,114	1,788	1,935	2,031	1,992
Total	1,445	2,167	2,645	3,056	3,071

Note: The record on motor vehicles was computerized in 1974. Earlier numbers of private vehicles were overestimated. Public motor cars include taxis and private hire cars.
Sources: DOS (1965–1998) and Land Transport Authority (pers. comm.).

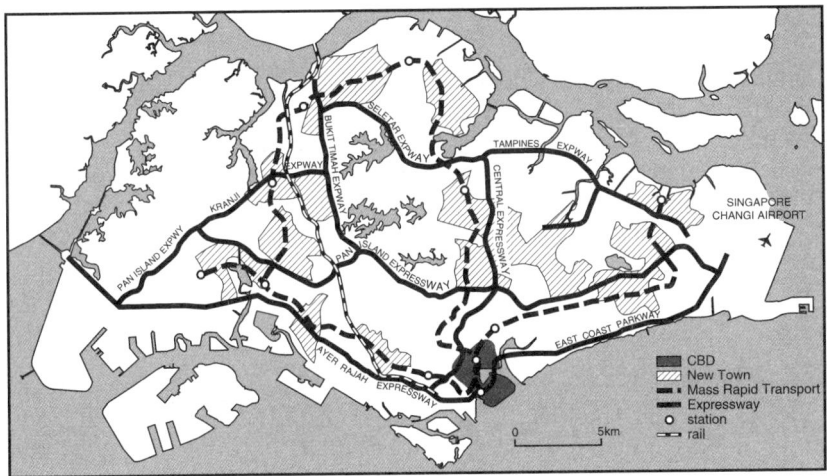

Figure 7.5 Singapore: new towns and the Mass Rapid Transit System, c.2000 (*Source*: Various)

services was poor, speeds were low, schedules unreliable and comfort and convenience minimal (Registry of Vehicles 1970; Singapore 1974; Seah 1975). Between 1971 and 1978 bus companies were merged, services reorganized and rescheduled, routes reallocated and the bus fleet modernized to get workers to work on time. As an all-bus system could not cope with the anticipated demand for public transport, the State and City Planning Project proposed a rail-based Mass Rapid Transit (MRT) system in 1971 as the backbone of the public transport system. A decision was made in 1976 to build the MRT. After much debate, construction began in 1983 financed by the sale of reclaimed land (MRT Review Team 1980). The first train ran in 1987 on a US$2.2 billion network covering 67 kilometres and with 42 stations linking the new towns (Figure 7.5). A 16-kilometre link and six stations were added in 1996. By improving the accessibility of both workers and shoppers, the system has made the Central Area and the Orchard Road area more attractive office and retail locations. Although the system and its feeder bus services have received accolades, the less glamorous bus system run by Singapore Bus Services and Trans-Island Bus Services still carries more than three million passengers per day – three times more than the MRT (Cervero 1998).

Complementing the emphasis on public transport was a private car control policy. Piecemeal road improvements and initiatives such as staggered working hours and carpooling had done little to alleviate congestion (Rimmer 1986a). A decision was made in 1975 to restrict the growth in private car ownership by introducing high taxes on vehicle purchase and vehicle quotas, and restricting car use (Area License Scheme) in the six-square-kilometre Extended Central

Area. Initially introduced during peak hours, the scheme was later extended throughout the working day, with a lower entry fee being charged during midday hours. While the scheme did not reduce the car population, with the aid of a quota system on new vehicles it helped cut the rate of growth from 6 per cent in the late 1980s to less than 3 per cent in the late 1990s.

The designated restriction zone included Orchard Road and areas either side of it. This zone has been redeveloped since the 1970s to accommodate the tastes of international tourists attracted by the development of modern high-rise hotels, shopping complexes, fashion boutiques, ubiquitous fast-food chains and international-chain cafés. In contrast, the old Central Area – the Singapore of Raffles – still had 155,000 jobs in 1968 and 230,000 residents, most living on the second floor of shophouses lining narrow streets in which motor vehicles increasingly conflicted with pedestrians and hawkers. After the late 1960s Raffles Place ceased to be the premier shopping area – Robinson's store was destroyed by fire in 1972. Orchard Road then became the new 'High Street' (Yeung and Savage 1996).

Restructuring land use

The State and City Planning Project's scheme for increasing jobs and population in the Extended Central Area was overtaken by external economic trends, which accelerated the further diversification of Singapore's economy and redevelopment. The prime force was the development of producer services during the 1970s. This required higher levels of education for employment in advertising, engineering, finance, insurance and real estate, legal, and transport and communications management services. In turn, this led to changes in land use within the Central Area. Of pivotal importance was the emergence of Singapore as a global financial centre, seizing the opportunity offered by its positioning between the London/New York and Tokyo/Hong Kong time zones following the initiation of 24-hour trading. In addition, Singapore was able to act as an honest financial broker, transferring surplus funds, for example, from Bahrain to meet the demands of borrowers drawing loans from Hong Kong (Huff 1994). The Central Area's skyscrapers housing Singapore's leading Chinese banks were developed on the west bank of the Singapore River using the architectural skills of I.M. Pei and Kenzo Tange.

The old administrative area on the opposite bank of the Singapore River was left intact. Because the narrow central business district focused on Raffles Place could not accommodate the demand for space by corporate business services, from 1968 the Shenton Way financial district was created on the south waterfront by facilitating the repossession of commercial premises and reclamation (Chua 1989). This 'Golden Shoe' district (so-called because it was shaped like an upturned shoe) attracted banks that desired to capitalize on the Asian Dollar Market created in 1968 and the regional headquarters of multinationals. Furthermore, the government's key financial agencies such as the Monetary Authority of Singapore, the Treasury, the Ministry of Finance and the Ministry of Trade and Industry were relocated to this 'Wall Street of the East'.

After 1978 producer services were the fastest-growing sector of Singapore's economy (Huff 1994). During the early 1980s there was also a shift in government policy away from the previous emphasis on labour-intensive, export-oriented industries by deliberately raising wages to persuade firms to relocate their activities in low-cost sites outside Singapore. In their place the government sought to use incentive schemes and programs to advance labour skills. As part of Singapore's 'second industrial revolution', these were intended to attract technology-intensive industries such as disk drives, computer peripherals, semi-conductors and audiovisual equipment. High technology parks and the Singapore Science Park were developed to house them.

This initiative stalled during the 1985–86 recession, leading to the repatriation of some 200,000 foreign workers and conspicuously unsuccessful attempts by local firms to invest in high-tech European and North American firms. The revolution was revived by the renewed demand in North America for electronic goods and by direct Japanese investment following the appreciation of the yen. Much of this development was underpinned by increased female participation in the manufacturing workforce. Women also augmented the number of professional and skilled workers in producer services. This was made possible by the recruitment of temporary female domestic servants from Indonesia, the Philippines and Sri Lanka and construction workers from the Philippines and South Asia (Yeoh and Huang 1998). Social polarization and segregation on ethnic lines has virtually disappeared in Singapore by distributing Chinese, Indians and Malays throughout the new housing estates. However, the 'divided' city has appeared in another guise with attempts to proscribe – not always successfully – predominantly Filipino and Indonesian migrant workers from urban public space (Baum 1999; Chang 2000).

By the late 1980s the Singapore government had shifted emphasis from developing an export platform economy towards creating regional control and logistic functions (Perry et al. 1997). Overseas companies were encouraged to locate their operational headquarters (OHQ) or business headquarters (BHQ) in Singapore to develop it as a 'total business centre' (later reformulated as 'international business hub'). Assistance was also given to overseas firms with already established offices. Although the services sector offers fewer jobs than manufacturing, it almost matches its contribution to Gross Domestic Product. Key additions to Singapore's 'financial supermarket' role included foreign exchange, financial futures and fund management (with participants in the government's Central Provident Fund being able invest in approved funds) (Huff 1994). This emphasis on producer services has led to the further urban redevelopment of the Central Area and general recognition that Singapore has joined the pantheon of world cities by becoming an increasingly sophisticated financial centre, albeit in the second tier below London, New York and Tokyo.

By the 1990s the cumulative effects of the accelerated development of the producer services sector had triggered the creation of knowledge-intensive industries. High-technology exports of advanced materials, biotechnology and opto-electronics exceeded the number of those shipped from Hong Kong, South Korea and Taiwan. Unlike industrial estates, new towns and transport facilities the indivisible effects of information technology were not place specific. Consequently, there has been a strong stress on ensuring that information technology diffuses throughout the island and permeates all sectors of society (including air and sea transport). The aim is to strengthen Singapore's position as a global hub to attract operational headquarters and service export firms, notably international freight forwarders such as DHL, Worldwide Express and Federal Express. These sophisticated business services suggest that Singapore's entrepot function was only temporarily eclipsed by manufacturing and has returned in a more modern and expanded guise, as reflected in the expansion of telecommunications and transport services (Krause et al. 1987).

A revised concept plan was published in 1991 by the Urban Redevelopment Authority to enhance Singapore's status as a global hub for the Asian-Pacific region in trade processing, banking, insurance, repackaging and marketing (URA 1991). The new focus was reflected in a change in the authority's remit from a preoccupation with slum eradication to creating visionary urban designs for a 'tropical city of excellence'. The new national structural plan replaces the ring concept with a constellation plan embodying a hierarchical pattern of urban centres interconnected by the MRT. The Central Area covering the long-standing government quarter, the Golden Shoe District and the newly reclaimed Marina Bay will maintain its status as the island's commercial and financial hub. However, a new downtown business, entertainment, cultural and residential district tied together by elevated and mid-air pedestrian links was recommended around Marina Bay to inject new life into the area at night.

Under the constellation plan four regional centres – Jurong East, Woodlands, Tampines and Selatar – will 'orbit' the expanded Central Area and provide satellite retail, office and recreational centres serving populations of 800,000 each. Within this orbit there will be mini-CBDs with smaller rings of sub-regional centres and fringe centres offering commercial and service activities. More varied housing styles are to be offered together with an expanded and improved transport system with extensions to the MRT system (the number of stations increasing from 48 to 130). New light-rail transit branch lines are also envisaged at Bukit Panjang, Punggol and Sengkang. Business parks to attract new generation information-based industries are to be located within two technology corridors and close to transport nodes. The southwestern corridor stretches from Singapore University in the east and Nanyang Technological University in the west; while the northeastern corridor extends from Changi Airport in the east to a planned regional centre in the west. Tampines New Town in the northeastern corridor has become a test-bed for information technology developments. They include Telepark, the first business park in Asia designed for telecommunications tenants, and Tampines Webtown, an Internet project for interconnecting residents with each other and local authorities (Corey 2000).

The 1991 constellation plan is intended to offer visions for 2000 and 2010 and to serve until 'year X', when Singapore will have an optimal population of four million within an urban area comprising 70 per cent of the island. It is a space-consuming strategy because land required for airport, port, telecommunications and domestic transport exceeds that occupied for housing. This allocation is a tacit recognition that transport and communications, particularly air transport and telecommunications, are now the overriding factors underpinning Singapore's global city status.

Transport and communications

By reducing business expenses, transport (and communications) infrastructure, in effect, subsidizes local firms. This was underlined in 1996 by the Land Transport Authority's White Paper, which detailed four ways to realize a world-class transport system within 15 years (LTA 1996; Chin 1998). The first way was to integrate land use, town and transport planning by developing high-rise sites adjacent to the railway lines and a tunnel system through the Central Business District. The second was to expand the road network and maximize its capacity by adopting green waves and automatic traffic monitoring systems. The third way was to manage demand for land use by using an electronic road pricing (ERP) system (adopted in September 1998) and phasing out the Area License Scheme. The final way was to provide quality public transport by expanding the MRT system to 160 kilometres and developing Light Rail Transit (LRT), as buses were not a solution for compact Singapore. Characteristic of these planning improvements is the new town of Punggol 21 – the model for the twenty-first century – which will be served by both MRT and LRT systems. Residents will not have to walk more than 300 metres to the nearest station and fares will be paid using a 'smart card'.

Since 1981 the Singapore government has given special emphasis to developing Singapore as an international telecommunications centre through the efforts of its National Computer Board. Subsequently, Singapore was established to serve as a pivot in optic fibre cable networks connecting it with Malaysia, Hong Kong, Taiwan and Japan, as well as a host to three satellite earth stations. The thrust of its National Information Technology Plan, published in 1986, was to enhance communications technology to boost Singapore's role as a passenger, warehousing and distribution hub by computerizing and streamlining air and sea data and documentation systems. TradeNet was the world's first electronic data interchange system. A subsequent plan *IT2000* published in 1992 was intended to develop Singapore as a regional multimedia hub (NCB 1992). The plan was designed to advance Singapore as an 'intelligent island' to ensure that information technology and computer networking pervades both business and home activities. In 1996 the *Restructuring for IT2000* initiative permitted public–private partnerships and strategic alliances in information technology. Since then the National Computer Board has been reorganized. Singapore One has been launched as a nationwide broadband system which can be accessed by the digitized telephone cable network. Singapore Inc. has been created to coordinate and market government support for information technology. Access to the latest in information technology and more liberal rules over Internet use are intended to encourage professionals to stay in Singapore by offsetting restrictions on car use and the constraints of apartment living.

Changi International Airport, opened in 1981, has been able to take advantage of these developments in information technology in order to enhance Singapore's status as a regional logistics hub. A second runway began operations in 1984, when control of Changi passed to the Civil Aviation Authority of Singapore. Since then Changi has been developed into an 'airtropolis' handling passengers and cargo, aircraft servicing and repairs, refuelling, aircraft meals and duty-free shopping. A second terminal was added in 1990. Plans have been made for a third runway and terminal that will increase annual passenger capacity from 24 million to 32 million as part of the government's deliberate policy to build facilities ahead of demand. Changi Airfreight Centre works round-the-clock and is designed to fit into computerized just-in-time logistics systems by acting as a transhipment point for multinational freight forwarders and as a central warehouse for manufacturing companies (Raguraman 1997).

Singapore Airlines accounted for 60 per cent of all Changi's flights in 1998 and derived maximum benefits from its hub information on passengers, car hire and hotels. Other carriers, such as Qantas, also make use of Singapore's hub location between Europe and Australia. Although Bangkok is equally well placed, Singapore has been more successful in encouraging passengers to break their journey. The Singapore government has pursued an 'open skies' policy allowing regional airlines to compete with Singapore Airlines for passengers. Sea–air transhipment opportunities – in by sea and out by air – have been pursued with mixed success.

The Port of Singapore Authority's facilities and organization has also been progressively redeveloped in response to changing needs, technologies and preferences (Ho 1996). During the 1980s and 1990s, the Authority's emphasis was on investing in information technology to optimize port operations and management of facilities. These initiatives have included the Computer Integrated Terminal Operation System (CITOS) and PortNet, which provides access to vessel, cargo, container and shipping information. This was part of a bundle of policy measures designed to capitalize on the global hub's high potential for container traffic, including appropriated terminal service agreements for specific shipping lines (Ho 1996). In 1998 a fourth container terminal was opened on reclaimed land at Pasir Panjang. These developments allowed Singapore to capture 30 per cent of the world's transhipment cargo in 2000 because it served not only Southeast Asia but increasingly also Australia and South Asia on mainstream services linking Europe and North America. The Port of Singapore was connected by 250 shipping lines to some 600 ports in 123 countries (PSA 2000: 32–33). There were two sailings daily to the United States, five each to Europe and Japan, seven to South Asia, ten to China, Hong Kong and Taiwan, and 72 to Southeast Asia. Singapore was the world's largest port in shipping tonnage, vessel calls and bunkers and was second only to Hong Kong in containers and cargo throughput. Over 2,600 ships are registered in Singapore, although one-fifth of these belong to the locally based Neptune Orient Lines.

In 1997, as part of the government's partial divestment programme, the Port of Singapore Corporation superseded the Port of Singapore Authority. Two divisions of the Corporation focused on container terminals and warehousing and logistics while the third concentrated on international business as part of its plan to offer its management services worldwide. Since the early 1990s these activities have been promoted by the Singapore government's 'go-regional' program to encourage local firms to prepackage their accumulated skills in urban planning and transport and communications and export them to create offshore space as part of their policy of developing 'Singapore Unlimited' (Willis and Yeoh 1998).

The most striking export of the Singapore model of urban management and national economic development is in Suzhou industrial town near Shanghai. There are also investments in industrial estates in the Yangtze River Delta region in China, India and Vietnam and developments in Burma, Cambodia and Thailand. SingTel has invested in Indonesia. Singapore International Airlines has taken shares in Air New Zealand and in Virgin Airlines of the United Kingdom and has become a member of the global Star Alliance. Neptune Orient Line has taken over American President Lines. The Port of Singapore Corporation has contracts to transform China's Dalian and Fuzhou ports into container hubs. Apart from providing communications and transport infrastructure, there is an emphasis on transferring administrative skills, particularly to Chinese cities. This process of creating offshore space has been facilitated by the revision of regulations on foreign investment, the offer of taxation incentives, and the provision of educational support for the children of expatriate Singaporeans and employment assistance for wives (Perry et al. 1997). It is also complemented by the spillover of activities from Singapore to neighbouring areas of Indonesia and Malaysia.

Conclusion: Greater Singapore

Since the early 1990s there has been growing recognition that the reach of Singapore's functions has outgrown the island (Figure 7.6). The biggest barrier to Singapore maintaining its status as a world city is a shortage of land, labour and water. Already, some 25,000 people commute daily from Johor Baru to Singapore. Between 1960 and 1990 Singapore's land area has been expanded by 44 square kilometres from 582 square kilometres at a cost of $4.24 billion (Savage and Yeoh 1993). Reclamation is so expensive that Singapore cannot be self-sufficient and has to rely on its neighbours for water supplies and waste disposal facilities. To permit the cross-border spillover of its economic and social activities there is a pressing need to improve Singapore's connections with adjacent areas of Johor Baru in Peninsular Malaysia and the nearby Riau islands in Indonesia.

In these circumstances urban affairs are no longer an internal issue but have become a matter of foreign policy. This is reflected in the 1994 scheme for a 'growth triangle' with Johore and Riau. Johore is reached by a major road link and Riau by a 30-minute fast ferry. In this informal Singapore–Johore–Riau (Sijori) arrangement, Singapore provides the producer services and draws separately on Johore's skilled labour and Riau's land and less skilled labour (there is no Johore–Riau link). Since 1980, there has been increased Singapore investment in manufacturing activities in Johore, especially in electronics and information technology. Also Singaporean capital has been used to build industrial estates and leisure facilities on Riau's two larger islands, Batam and Bintan. Most notable is the Batamindo Industrial Park developed by Indonesia's Salim Group, the Singapore government-owned Singapore Technologies, and Jurong Environmental Engineers. Sites on the Karimun islands have been designated for heavy marine engineering and oil storage. Other sites have been reserved for animal husbandry, though one limitation is the scarcity of water on the Riau islands (Lee Tsao Yuan 1995).

There are many rivalries between Indonesia, Malaysia and Singapore. The second causeway link between Singapore and Johore was not opened until 1998 – 55 years after the first road link. The Malaysian government also has an active policy of trying to persuade shippers to use its national ports as an alternative to Singapore (Chapter 3). During the 1970s the Malaysian government built Pasir Gudang in south Johore to capture Singapore's traffic by placing restrictions on freight moving across the causeway. Another attempt to attract mainline calls was made in 1999, when a new port was opened at Tanjung Pelapas in Johore with a 7-kilometre link to the

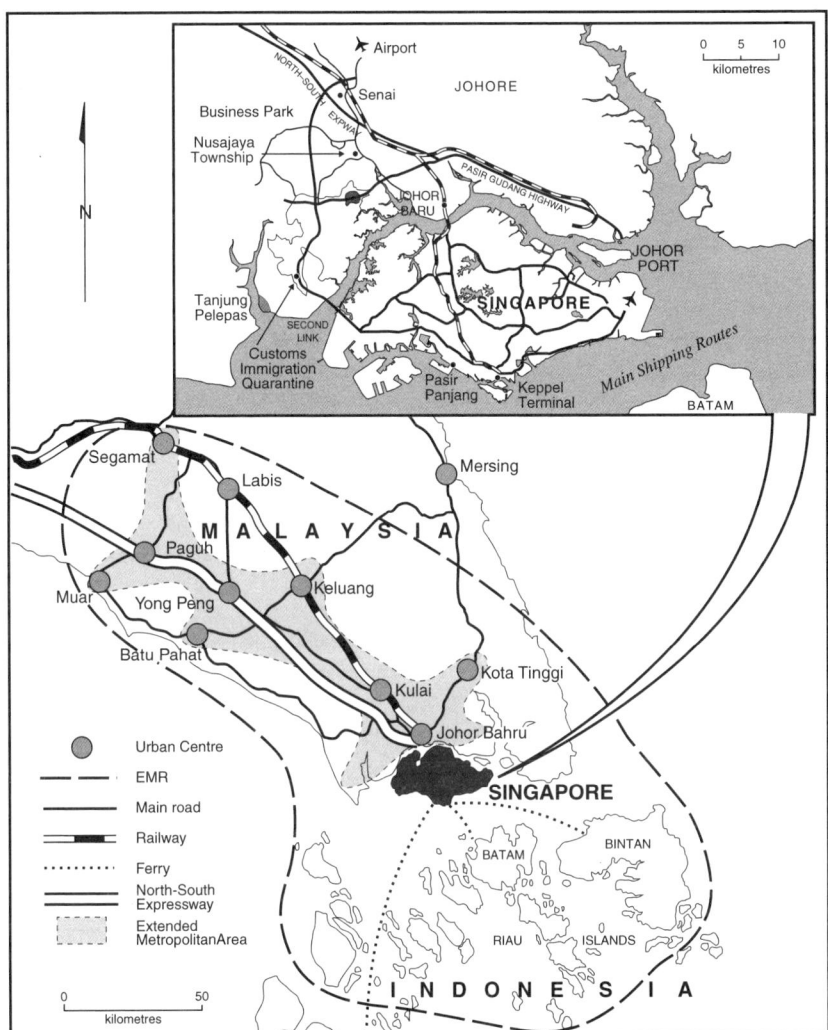

Figure 7.6 Greater Singapore: spatial structure, c.2000 (*Source*: Based on Rodrigue 1994 and other sources)

Malaysia–Singapore Second Expressway in Johore. To date Singapore has been able to withstand local competition by concentrating on its global investments.

Thus the concept of Greater Singapore marks a return to the city's roots. Singapore began as an urban enclave. Then it became a significant component of the international network of cities servicing the colonial development of its hinterland in the Malay archipelagoes (Chapters 1, 3 and 6). In the post-Independence process of industrialization based on foreign direct investment and the state's highly regulated approach to integrating land use and transport, connections with the adjacent resource-based hinterland weakened as global links strengthened. In the guise of Greater Singapore the settlement has returned to being an urban enclave, though its 'footprint' as a First World city-region is much larger than its initial mark.

8
Archipelagic Cities: Manila and Jakarta

For all their obvious differences, Manila and Jakarta are in many respects the two most similar capital cities of Southeast Asia. Established as river ports by local rulers, they were conquered and rebuilt as model European cities but relied heavily upon Chinese trade networks for regional trade (Blussé 1986). During the twentieth century, their populations increased by two orders of magnitude (Tables 8.1 and 8.2). Around 1900 they were still comparable with Malacca four centuries earlier (Reid 1993: 69). In the 1940s their populations leapt to over one million. By the 1990s with expanded boundaries the capital city provinces of Manila and Jakarta had both reached almost ten million, qualifying them for megacity status. Another ten million people lived in their peripheral industrial and residential zones. Either of these two extended metropolitan regions thus equates to the entire population of Malaysia or Australia. By the 1990s the middle classes of Manila and Jakarta, which may be equated with car-owning families, were larger than their respective total populations at the end of the colonial period.

The most marked difference between the two cities is the degree of dominance over their respective national economies. In 1995 Manila accounted for 32 per cent of national GDP, compared with 16 per cent for Jakarta (Figure III.3); Greater Manila had 40 per cent of large- and medium-scale manufacturing employment, compared with 27 per cent for Greater Jakarta (PSYB 2001; BPS, Statistik Industri). Whereas Greater Manila (NCR) was nine times larger than the leading provincial cities (Cebu and Davao), Jakarta was only about four times larger than Java's other large industrial city of Surabaya, closely followed by Bandung (see also Chapters 3 and 4). Nevertheless, the dynamics of capital-city growth have been much the same for both Manila and Jakarta.

With some differences in timing, the spatial evolution of Jakarta and Manila shows the same alternating phases of divergence from and convergence to metropolitan cities (Figure III.4). These phases can be related to broader trends of globalization and localization, for which capital inflow and rural–urban migration may serve as proxies (Table 8.3). During the early colonial phase,

Table 8.1 Manila and Metro Manila: population, 1903–2000

	1903	1939	1948	1960	1975	1995	2000
Manila	0.2	0.6	1.0	1.1	1.5	1.7	1.6
Metro	0.3	1.0	1.6	2.5	5.0	9.5	9.9

Source: Philippines, Census (1992), PSYB (1997, 2001).

Table 8.2 Jakarta:[1] population, 1905–2000

	1905	1941	1949	1961	1980	1995	2000
Jakarta	0.14	0.5	1.3	3.0	6.1	9.1	8.4

1. Municipality 1905–1949 figures; Province (DKI) 1961–2000 figures.
Source: Hugo et al. (1987), DKI (1972: 44); BPS, Supas (1995a); BPS (2001).

the foreign presence was very small, allowing the local elite a good deal of autonomy in its urban lifestyle. These colonial enclaves adopted the customs of the local-born European and Eurasian community and acquired little in common with distant metropolitan cities. After the mid-nineteenth century, local autonomy came under pressure from increased communications with the rest of the world, growth in international trade, rising capital inflow, and a larger European presence. In the early decades of the twentieth century, aggressively modernizing colonial governments began investing in urban infrastructure to create at least the façade of a model tropical city. Sleepy backwaters were transformed into model colonial cities with a privileged, garden suburb, elite lifestyle protected by local government restrictions on rural–urban migration, squatting and the informal sector.

This artificial situation was upset by the Japanese occupation. Subsequent nationalist governments discouraged capital inflow and squeezed the expatriate community, while allowing massive rural–urban migration. This was the era of the sprawling, overcrowded Third World city. Yet it was only a phase. From the 1980s, as the Philippines and Indonesia accelerated their processes

Table 8.3 City types, phases and trends of globalization and localization

City type	Period (approx.)	Spatial form	Relation to metropolis	Capital inflow	Rural–urban migration
Early colonial	pre-1900	compact	divergent	small	slow
Late-colonial	1900–1942	suburban	convergent	large	moderate
Third World	1942–1980s	sprawling	divergent	small	rapid
Post-colonial	post-mid-1980s	nodal	convergent	large	moderate

of industrialization with more liberal trade and investment regimes, urban elites literally reclaimed lost ground, often by driving off squatters or cheaply buying out poor communities to clear large tracts for redevelopment. The aim in this fourth phase was not, as in colonial times, to transform the whole sprawling city, for that task was now too great, but rather to create enclaves, gated communities, commercial precincts, and entire satellite towns where the middle class would be comfortable and safe. This chapter compares the development of Manila and Jakarta through these four phases with particular attention to town planning, infrastructure and public transport.

Manila

The Spanish time: a mestizo town

The Spanish occupied Manila in 1571 as a matter of convenience. Denied a trading post on the China coast, they found in Manila Bay the closest safe transhipment point for the Acapulco galleon trade, a monopoly of the Spanish Crown. Conceived and built on new foundations as a model colonial city, its layout reflected the imperatives of trade, security and religio-ethnic segregation (Reed 1977). At the mouth of the Pasig River, the citadel of Fort Santiago guarded the Spanish walled city (Intramuros), whose drawbridges were raised each night until 1852. Adjacent was the Chinese quarter of Parian, moated but without walls. Across the Pasig lay Binondo, settlement of the christianized Chinese and Spanish mestizos. Behind Binondo and along the tidal creeks (*esteros*) were the villages (*barangay*) of the local people.

This urban morphology began to change during the nineteenth century in response to growth in private foreign trade. The Parian withered away and the site eventually became the Botanic Gardens. Business concentrated in Binondo on the other side of the river, which by 1855 was already three times more populous and was the location for all the new foreign trading houses (Bowring 1859: 24–5). Even Spaniards began to move away from the claustrophobic Intramuros, the world of officialdom and the church. Bowring (1859: 11) observed of Intramuros that 'few people are seen in the streets, and the general character of the place is dull and monotonous'. Binondo, by contrast, was bustling.

Despite the prominence of Europeans and Chinese, the population was predominantly Filipino,[1] both indigenous and mestizo (Bowring 1859: 24–5). The main group was the urban poor, mostly villagers who had migrated from elsewhere in Luzon in search of work and opportunity. Their largest and most crowded settlement was Tondo, the labour pool for nearby Binondo. The main form of employment, especially for young women, was in the cigar factories, while men worked as waterfront and warehouse labour (Doeppers 1984). As household staff and labourers, Filipinos were also numerous within Intramuros itself. There was also a small, but growing educated Filipino elite.

Although Spanish Manila was a very compact city, walking was regarded as undignified:

> No-one in Manila ever thinks of moving about in the daytime without a carriage; even the lower class of 'mestizos' keep them, and they are termed by the Spaniards 'les zapatos del pais', literally the shoes of the country.... They may be seen standing at the door of every mercantile house and most private ones, and on account of the heat of the country they are used if it is only to cross the street. (Ellis 1859: 48)

In late afternoon it was the custom for Europeans to dress in their finery and parade in their carriages around the outer walls, enjoying the evening sea breeze. After reclamation of the foreshore, this promenade was extended along the seafront (Malecon) to the central park (Luneta).

Mid-nineteenth-century Manila nevertheless extended beyond the twin towns of Intramuros and Binondo (Figure 8.1). Bowring (1859: 27) reported that the population of Manila *and its suburbs* was around 150,000. These 'suburbs' are no more than glimpsed in contemporary accounts but some of them had a long history. During the seventeenth century the religious orders had founded parishes around Manila. As landed estates were consolidated, the churches became the nucleus for small settlements (Reed 1977). Some estates later passed into the hands of mestizo Chinese, who leased land for private dwellings. As early as the seventeenth century rich Spaniards had begun to build summer houses, of which one survives as Malacañang Palace (Reed 1977). By the 1870s there were also grand mansions to the south of Intramuros along the sea front of (H)ermita and Malate (Campbell 1993). Like Chinese, the first foreign merchants lived over their offices in Binondo but by mid-century their successors had moved into bungalows up river from the town. Junior staff lived nearby in shared 'messes', commuting each day by river or carriage. Educated Filipinos joined the trend.

Despite the tree-lined streets and spacious bungalows along the river, most of the land beyond the city of Manila was still rural with 'suburban villages' supplying rice, fruit and vegetables, eggs and chicken to the urban populace. Some villages also specialised in handicrafts (Bernad 1974: 4). A network of small waterways (*esteros*) facilitated the movement of goods into the city (Campbell 1993: 23). Bowring (1859: 27) observed that many Filipina women walked 'a considerable distance' each day to and from the cigar factories in Binondo. This passing remark of the 1850s seems already to point to what McGee (1991) claimed to be a modern Southeast Asian phenomenon of *desa-kota* (village-town) or quasi-urban settlement. In this early case the young daughters of farming families provided a pool of cheap labour for urban manufacturing. Bowring (1859: 24) also refers to the 'washermen and women' of Sampaloc. Washing was collected in the city and

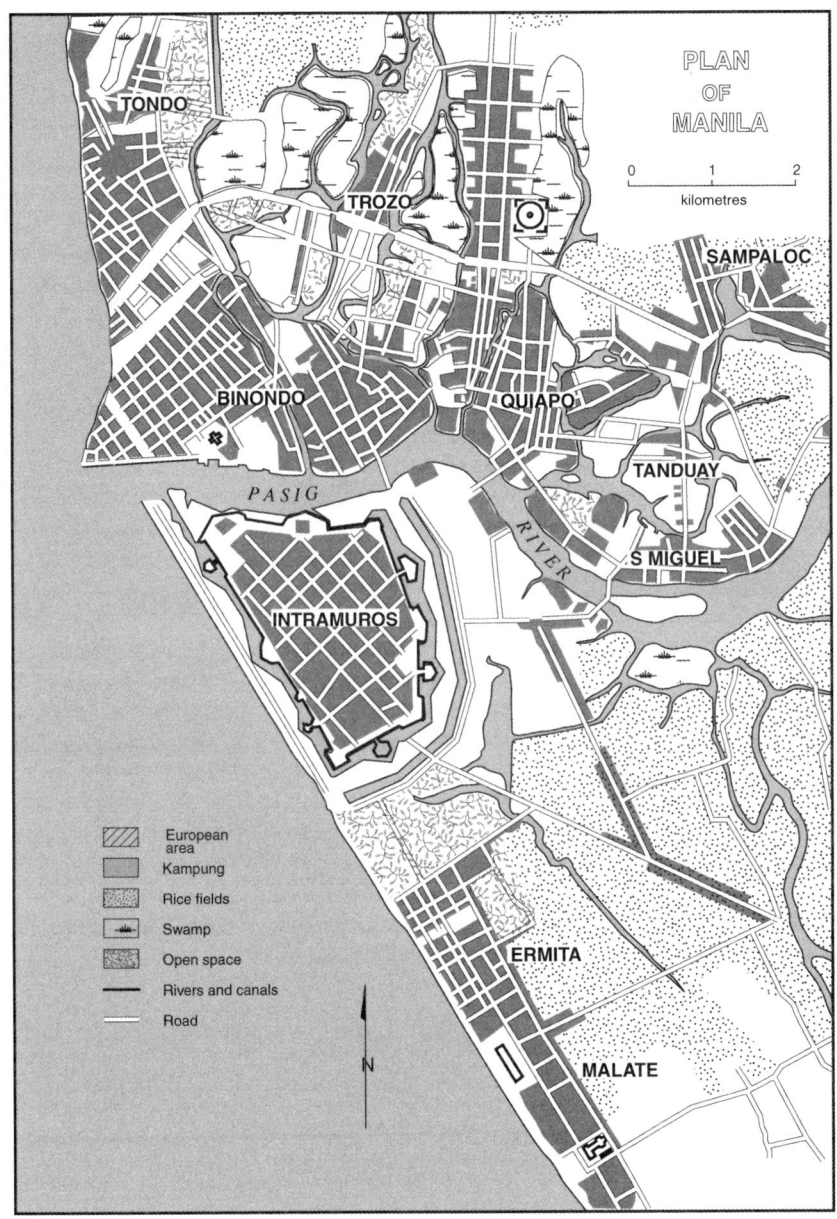

Figure 8.1 Manila: land use, c.1895 (*Source*: Guia 1898)

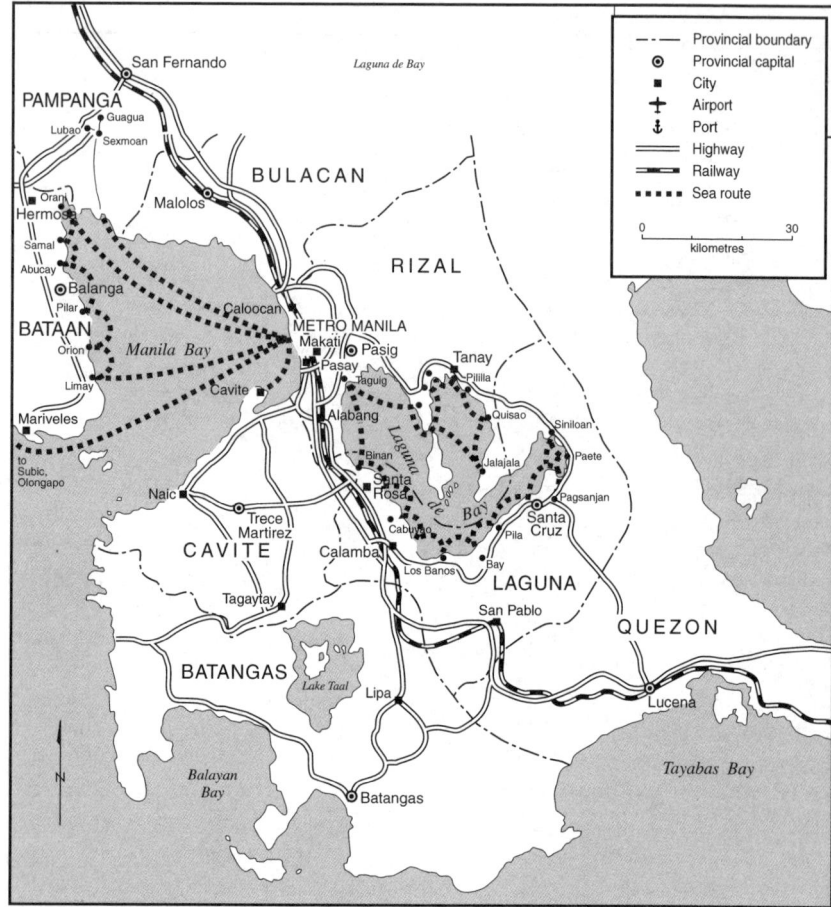

Figure 8.2 Manila: local hinterland and transport routes, 1900s (*Source: Atlas of the Philippine Islands* 1900; MT, transport schedules)

taken by boat to this northeastern barrio, where the water would have been cleaner. By the late-nineteenth century, more accessible parts of the district were becoming genuine suburbs, poorer streets with local thatch (*nipa*) houses, others with more elaborate bungalows (Campbell 1993: 46).

The first modern land-based urban public transport was the horsetram. The concession was awarded in April 1881 to Senors Zobel and Bremon, who vested it in the Compania de los Tranvias de Filipinas (Philippine Tramway Company) (Philcom, Act 1112, 11/4/04). Its cars were like an omnibus on rails, seating 12 inside and eight standing on either end, and drawn by a pony driven hard and often recklessly. They ran from the main shopping street of Escolta through the business district of Binondo to the working-class quarter of Tondo. In October 1884 Zobel was granted a further concession for a connecting steamtram from Tondo via Caloocan to the outlying fishing port of Malabon, the site of a sugar refinery. The one-class cars, which around 1900 provided an hourly service (Philcom 1901, I: 50), were crowded and naturally regarded as disreputable by Europeans (Devins 1905: 52).

Beyond the paved city streets, the main form of transport was by water (Figure 8.2). By the late 1860s small steamers were providing daily sailings from the commercial district of Binondo to the naval base and dockyard at Cavite, to landing stages along the Pasig River and around Laguna de Bay, and to towns in the delta of Bulacan and Pampanga. Most of this ferry business was later consolidated in the hands of the Yangco Steamship Company. Slower and cheaper, especially for those travelling with goods, were covered dug-out canoes (*banca*) that were rowed or poled along the river and *esteros*. Villagers living beside an *estero* used these *banca* as household utility vehicles. Larger cascoes were used to lighter goods from the offshore anchorage or the Pasig River wharves to merchant warehouses around the commercial district.

The American era (1898–1935): beautiful Manila

The all-conquering Americans were not impressed by Manila and determined as a matter of pride to modernize it in the shortest possible time. Atkinson, former General Superintendent of Education, was blunt:

> Three years ago the lack of luxuriant tropical vegetation; the dirty, narrow, poorly paved streets; the inferior artificial light; the miserably equipped (tram)car line, patronized only by natives of the poorer classes; the groups of nipa shacks, more primitive as a dwelling than the worst American shanty; the narrow Escolta, the main business street; the sultry climate; the mixed currency; the inability to speak the language...all combined to make the first impressions of the newcomer hardly of the most pleasant sort. But conditions have changed much, even within this short time, and the city is becoming more and more American or, in other words, up to date. (Atkinson 1905: 214)

The impulse was imperial, the manifestation technological.

The creation of a model colonial capital was also a matter of aesthetics and prestige. In 1904 the Philippine Commission called upon the services of leading Chicago architect Daniel H. Burnham, whose 1905 report became the vehicle for the 'Manila Beautiful' movement (PHYB 1934: 81–3). Since Manila was not a greenfield site, Burnham sought to improve and enhance rather than to rebuild. After 1910 his approach was embraced by the new City Government with ongoing funding from the Insular Government. Manila residents began to regard their city with pride as the most beautiful in the Far East, reviving the Spanish appellation of 'Pearl of the Orient', echoing Havana's fame as the 'Pearl of the Antilles'.

Having established law and order by suppressing Philippine nationalism, the colonial government embarked with great energy upon a massive programme of public works. This affected every province, city and town, but nowhere was more money spent or greater care lavished than in Manila. By about 1915 Manila had in a very short time acquired most of the infrastructure characteristic of a modern city in the United States. The initial investment was huge but, once installed, the networks could be extended at incremental cost.

The most expensive item, ultimately costing several million dollars, was the construction of a safe, deepwater port to accommodate the largest ships in transpacific trade (Chapter 3). This involved construction of a breakwater to protect against typhoons, dredging to a depth of over thirty feet, using the spoil to reclaim a large area in front of the Intramuros, and from there building out long finger piers for alongside cargo-handling. Completion took about a decade. The city's stinking *esteros* were also dredged, restoring shallow-draft navigation between the Pasig, Malabon and the delta of the Central Plain (Philcom 1914: 185).

These port works allowed for reclamation around the old city. Although backed by gentle hills, the Spanish town of Manila had been built in marshes at the mouth of the Pasig River, which provided ideal breeding grounds for malarial mosquitoes. The American government began to fill these marshes for parks and housing. South of the river and fort, the harbour spoil was used to extend the old seafront boulevard (Malecon) and the Luneta park, around which land was provided for the luxury Manila Hotel, the Elks' and the Army and Navy clubs. This showcase area still retains the layout of that time. Dewey (now Roxas) Boulevard, named after the conqueror of Spanish Manila, was laid out around the foreshore from the Luneta past the elite suburbs of Ermita and Malate to the naval base at Cavite. To the north, a swampy margin remained as the preserve of poor urban migrants (Figure 8.1).

Drainage and reclamation were complemented by sewerage and garbage disposal systems. In the Spanish period sewerage and garbage had flowed into the *esteros* or, in the old stone dwellings of Intramuros, piled up in deep pits. The American administration introduced a collection (pail) system and from 1909 began to connect houses to underground pipes (Philcom 1908: 108–10; 1909: 67). In 1913 garbage collection was let out to a private contractor using motor trucks; refuse was either burned or used as fill (Philcom 1913: 44). Daily street cleaning was introduced. Water supply was improved by new reservoirs and extension of the reticulation system (Philcom 1908: 107–8). Poorer districts were provided with public hydrants, bath houses and laundries, while the washing of clothes in rivers or *esteros* was prohibited

(Philcom 1914: 43). By 1914 better-off households and commercial premises also enjoyed reticulated gas.

The Americans were also dismayed at the outmoded forms of public transport. Part of the problem was that capacity had not expanded in line with growth in demand. The horsetram was infrequent, unreliable and invariably full. By 1902 some 4000 horsecabs (*carromata, quilez*) had been licensed for hire, but their number did not suffice for peak hours, when it was said to be 'practically impossible' to secure public transport (USNA, BIA: 6300/5, 22/12/02). Rickshaws were tried with little success. The better-off kept their own carriages; others waited or walked.

Looking to American cities, the Philippine Commission planned an electric tramway network, then the ultimate in modern urban infrastructure. A franchise was awarded in March 1903 to a Detroit tramway entrepreneur, who proceeded to float the Manila Electric Railway and Light Company (later Meralco). Taking over the Compania de los Tranvias, Meralco rebuilt the entire system and in April 1905 began operations along a network of 40 kilometres, later connected to the new deepsea piers (Philcom 1907, II: 365–7). Under separate franchise, the subsidiary Manila Suburban Railways Company in 1908 completed a 12-kilometre extension from Santa Ana to the then outlying market towns of Makati and Pasig via Fort McKinley (later Bonifacio) (Philcom 1908, I: 336–7).

The tramway was a great improvement in terms of both efficiency and equity. To bid for the patronage of the better-off, including Europeans, two-class tramcars were introduced but the mainstay was popular custom. For those without their own carriage, the best public transport option had been to hire a *carromata* that – if available – cost 15–20 cents for the journey (USNA, BIA 6300/5, 22/12/02). Now it was possible to travel anywhere by tram within the municipality at the flat-rate fare of only 10 cents second class (12 cents first class). As noted in the original press release, 'Manila is in a country where coolies ride' (USNA, BIA 6300/5, 22/12/02). In the year ending 31 August 1907, the tramways carried 10 million passengers (Philcom 1907, II: 367). Cheaper public transport in turn allowed people to become more mobile and to live in better or cheaper housing further away from their place of work. The tramway system thereby encouraged the spread of what had been a very compact city.

Ironically, just as Manila acquired a modern and comprehensive public transport system, the automobile began to alter the demand for infrastructure. To improve vehicle access, old city streets were widened, straightened and extended. Rapid growth in traffic caused much expense in maintenance. As of 1908 most of the city's 150 kilometres of streets were still macadam, which quickly deteriorated in the wet season (Philcom 1908: 105–6). After much experimentation with paved and hardened surfaces, bitumen was found most practical (Philcom 1914: 43). New bridges were built over the Pasig River to improve connections to new garden suburbs on higher ground.

In terms of social dynamics, the automobile was reactionary. As in America and Europe, the tramway had the potential to integrate a city hitherto stratified by class and ethnicity. In Manila it scarcely happened. Instead of abandoning their inefficient private carriages for efficient public transport, elite families could now substitute an automobile for a carriage without altering their pattern of socialization. By January 1911 some 440 vehicles were reported to be registered in Manila (USNA, BIA 18343/22). As numbers soared, the mix of traffic began to alter and with it the very nature of the street as shared public space. Being faster than horse-drawn carriages, cars were a much greater hazard to pedestrians and animal-powered vehicles. The new vehicles were expected to respect the needs of other road users but it was not long before all other traffic was being reorganized to allow cars to travel faster without impediment.

Because decision-makers themselves were among the first to drive – or be driven – it is hardly surprising that their needs so quickly took precedence. Formation of the Manila Automobile Club in January 1911 provided a lobby group of the rich and powerful (FER 1912). The elite now applied the force of law to marginalize modes of transport other than the automobile. By 1940 the city's 15,000 automobiles and 5,000 trucks were squeezing out other modes of transport (YPS 1940). There were even complaints about traffic congestion, especially on the bridges over the Pasig River. Meralco remained the dominant transport operator but after 1927 relied upon motor buses to extend the network. In 1941 the tram still carried 28 million passengers but another 31 million were carried by Meralco's buses (Torre 1981: 42). Supplementary transport was provided by taxis, including three-wheeled mini-taxis, ferry services on the bay and river, and the perennial *carromata*. As colonial Manila caught up with America and Europe, it drew further apart from the rest of the country. The only other city or town which enjoyed an equivalent or perhaps even greater investment per capita was the summer capital of Baguio, which was really an outpost of Manila, or rather elite Manila (Chapter 3). Other cities and towns, especially provincial capitals, benefited according to their size and function from electric power, fresh water, better streets, regular garbage collection, telephone exchanges and improved port facilities, but these were in the nature of add-ons rather than networked systems. Westerners travelling outside Manila often observed how primitive were the amenities.

The improvements brought about by American rule met with the approval of the Filipino elite, who were quick to adjust their lifestyles and expectations. Their involvement in the formation of the Automobile Club and the enthusiasm with which they embraced Baguio as a summer resort were only two manifestations of this. Manila increasingly attracted the leading provincial families to set up mansions, where they could enjoy a gay social life while advancing their interests through inter-family alliances and access to government. This trend was given great impetus by the introduction in 1907 of the Philippine Assembly, which from the beginning was dominated – as it still is – by provincial families. After 1916 the Filipino elite was directly involved in shaping the city through a Municipal Board of ten elected councillors appointed by the governor-general with consent of the Senate. Compared with colonial Indonesia, the Filipino elite was larger, more urbanised and more involved in both business and government.

The role of women, especially the matriarchs, should also be recognized. One innovation of the Manila Beautiful movement was an annual House Beautiful contest, intended 'to lead in all that means modern comfort, beauty and elegance in citizen's homes and residences', both 'hygienic and artistic' (PHYB 1934: 83). This quintessentially middle-class display was more than a transplanting of middle-class America. It was a new and modern field of contest for elite Filipino families to compete with the Americans and with each other. Much later Imelda Marcos would turn this into politics on a grand scale.

Self-government and independence: urban sprawl and spatial segregation

Inauguration of the Commonwealth under President Manuel Quezon in November 1935 led to new initiatives in urban planning. One was a high-profile campaign against poor and unsanitary housing. Former Governor-General Murphy had already declared 'an end to slums' and earmarked funds for low-cost rehousing, the focus of concern being the crowded settlement of Tondo, just north of the Pasig River. In 1938 the Commonwealth purchased almost 1600 hectares of the Tuason family's Diliman estate, of which the large part was allocated to the People's Homesite Corporation and 500 hectares for a new campus of the University of the Philippines (Duldulao 1995: 35–9). Plans for low-cost housing were soon overtaken by more grandiose ideas. In October 1939 Quezon City was designated as the site for a new administrative capital with a landbank from adjacent estates increased to over 5000 hectares. Quezon Boulevard was laid out but little construction took place before the Japanese occupation on 1 January 1942.

The case for relocating the seat of government was strengthened by the terrible destruction of Old Manila at Japanese and American hands during the liberation of February 1945 (Aluit 1995; PHC 1947: 21). No Asian city outside Japan suffered worse physical devastation. On full independence in July 1946 it seemed much easier to rebuild the capital on a spacious new site than to attempt to restore the layout of the Old City. Act 333 of July 1948 thus confirmed Quezon City as the seat of government (Hartendorp 1958: 411–14). A Capital City (later National Urban) Planning Commission was formed to draw up a masterplan, which in 1954 finally emerged as a physical plan. Boulevards were laid out and a few government offices relocated but the seat of government remained in downtown Manila. The reasons for the failure of the Quezon City initiative are complex, but the most salient is the conflict between public planning and private development. At the

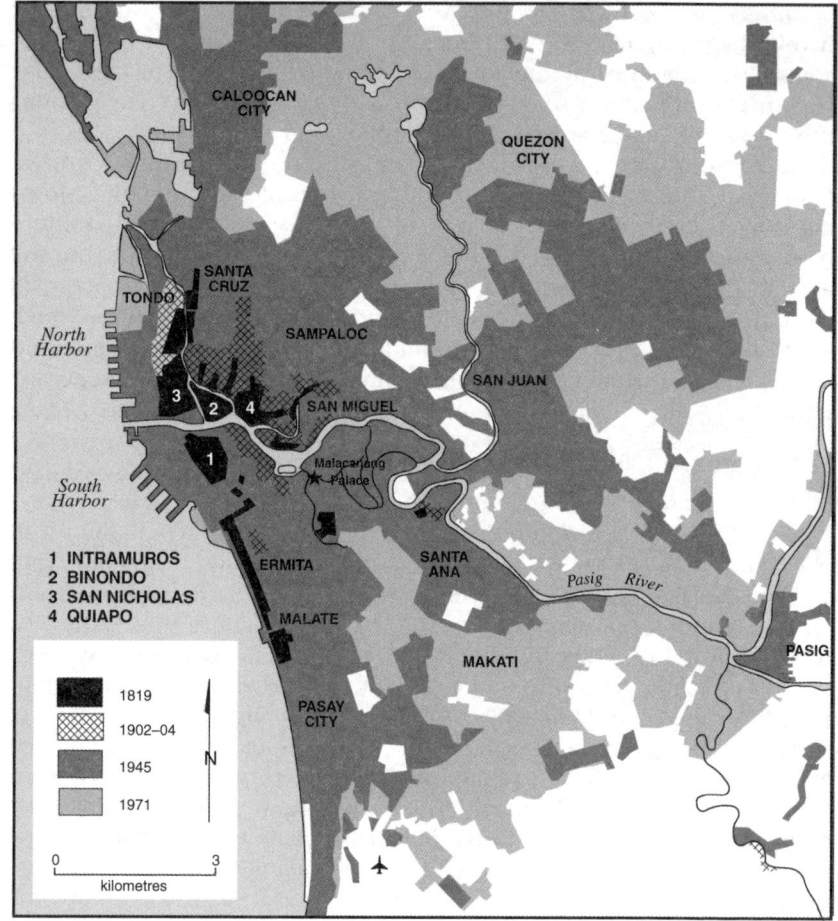

Figure 8.3 Manila: growth map, 1819–1971 (*Source*: Kolb 1978)

national level, Congress and the Senate were both factionalized by rivalry between the leading families. Despite the Local Autonomy Act of 1959, Manila's city government was weak. Manila therefore rebuilt itself spontaneously and almost without regard to urban planning (Figure 8.3).

While Quezon City was starved of public funds, the private sector invested in a 900-hectare satellite town at (San Pedro de) Makati, part of an old hacienda to the southeast of the city beside the American Army base of Fort McKinley (later Fort Bonifacio). The new town was the prewar brainchild of Joseph McMicking, a former United States army officer who married into the landowning Ayala-Roxas-Zobel families (Goss 1990: 117–20). After the surrender he was instrumental in setting up the Makati Development Corporation as a subsidiary of Ayala Corporation. At first the emphasis was on elite housing in a country club atmosphere. The Polo Club and the Manila Golf Club had been encouraged to relocate there in the late-1940s (Gleeck 1977: 286). In the 1950s Forbes Park, the first gated community in Southeast Asia, was promoted as a secure and prestigious suburb for the expatriate elite. With Hukbalahap guerilla attacks on the outskirts of Manila, the security threat was very real (Gleeck 1977: 316). The Manila elite soon followed the expatriates. Later in the 1950s a second subdivision of San Lorenzo was promoted for middle-class expatriates and Filipinos (Gleeck 1977: 316). In the early 1960s big business began to move away from the crowded central business district of Binondo to modern high-rise offices in Makati, which were adjacent to the new housing estates and offered good road access and parking for car-owning expatriates and the middle class. By the end of the decade Makati had become the New Manila, an off-centre CBD and elite residential enclave only loosely attached to the rest of an increasingly amorphous and non-functional metropolis. Makati was a triumph of planning but it was exclusive private-sector planning that rested on monopoly control of land (Goss 1990: 120–4). The Ayala Corporation profited fabulously.

The rest of a more and more crowded, low-income Manila was left to fend for itself. Increasing insecurity in the countryside together with the inability of either large landholders or the state to enforce land rights in and around Manila made the city increasingly attractive and porous to rural–urban migration from both the Central Plain of Luzon and the typhoon-prone islands of the Visayas (Juppenlatz 1970). Despite an official policy of slum clearance, squatters quickly became de facto tenants and in those democratic times found protection at law and from political bosses (Stone 1973). By the 1970s a third of Manila's population was estimated to be either squatters or slum dwellers (Yeh and Laquian 1979: 53–4). The worst 'slum' was the Tondo Foreshore, a 184-hectare reclamation that before the intended port development had been taken over by squatters (MHS 1982: 63–7). By 1974 it contained 170,000 residents, overwhelmingly first-generation migrants and having incomes below the poverty line. Most houses were impermanent, having walls of wood, bamboo, galvanized iron and other scavenged materials, and most families occupied only one room (Calinao 1990: 41–5). Nevertheless, as in other poor communities, most families owned their own dwelling and had some kind of income-earning employment, usually in fairly close proximity. Poverty it was, but also survival.

Manila's huge expansion in population was accommodated with minimal investment in physical infrastructure. Only electricity was widely distributed. Water supply met only 73 per cent of needs and garbage collection was grossly inadequate (Yeh and Laquian 1979: 38–40). By 1969 the prewar sewerage system reached only 12 per cent of the population. The formal public transport system was overwhelmed. Amidst the rubble of

Old Manila, Meralco decided against costly rehabilitation of the tramway system in favour of a bus fleet but in May 1948 even the buses were sold as unprofitable. Monopoly capital thereby surrendered the streets of Manila to small-scale entrepreneurs. It was now the Age of the Jeepney. At first merely a stopgap, the jeepney was an American Army jeep converted to carry around ten and later up to sixteen passengers. They were franchised to ply fixed routes but, unlike buses, would pick up and set down at any point, which suited passengers who mostly travelled only short distances. Their jaunty style also suited the local character (Grava 1972). How many jeepneys came onto the streets of Manila was unknown. Around 1980 it was estimated that only 25–30 per cent of jeepneys were legally registered and franchised, giving rise to an estimate of some 60,000 vehicles (Diaz 1984). In terms of passengers, jeepneys dominated all but longer journeys with 57 per cent of person-trips compared with car (including taxi and truck) 25 per cent, and other modes including buses and motor tricycle 18 per cent (NTPP 1982: 36–8). On many roads traffic comprised a stream of gaudily painted jeepneys.

For a sprawling, low-income city the jeepney was the ideal mode of mass public transport because of its adaptability to any land-use pattern. In the 1950s, when the city was still fairly compact, traffic still flowed quite freely. As population and the number of vehicles grew relentlessly, however, congestion became a serious problem. Construction in the late 1960s of the ring road Epifanio de Los Santos Avenue (EDSA) facilitated movement between the emerging centres of Caloocan, Quezon City/Cubao, a nascent Ortigas, Makati and Pasay, without solving congestion on feeder roads to and from the ringroad. By 1981 on main radial roads the average travel speed of bus and jeepney was less than 10 kilometre per hour, falling to 5 kilometres per hour in peak periods (MMUTIP 1981: 18).

In the 1970s the pendulum swung back towards big government. Having declared martial law in 1972 in order to avoid the need to step down at the end of his second term, in 1974 President Marcos turned his attention to gaining control and reforming the administration of Manila. The City of Greater Manila had been declared on 1 January 1941 to include the seven surrounding districts of Caloocan, Quezon City, Pasay, San Juan, Mandaluyong, Makati and Paranaque. After prolonged negotiations, in November 1975 the component cities of Manila, Caloocan, Quezon City, Pasay were amalgamated with 13 surrounding municipalities into the Metro Manila Commission (MMC), which in 1978 was placed under the Ministry of Human Settlements. The president's wife, Imelda Marcos, was appointed governor and general manager of MMC, serving concurrently as Minister of Human Settlements (Naerssen et al. 1986: 172–4).

Martial law was justified to tackle mass poverty and undermine support for the Communist Party. In 1973 the World Bank's first Urban Survey mission identified Tondo as its high-priority area and proposed rehousing on site in new apartment blocks (TFHS 1975: 23). In 1975 a National Housing Authority was established, followed in 1979 by the 'Bliss' low-cost apartment program (Yeh and Laquian 1979: 22–3; Keyes 1982: 43). In 1974 the Metro Manila Transit Corporation was formed to amalgamate 132 private franchised operators and renew the bus fleet (Rimmer 1986a: 169–89). Ten years later a 16-kilometre, elevated light-rail transit system (LRT) was opened to run north–south through the central business district. Funds were therefore poured into Metro Manila but the achievements were disappointing. Most large projects – even those funded by the World Bank – were a means of massive appropriation by the Marcos family and cronies (Manapat 1991). The poor found their de facto land rights eroded by officially sponsored projects and well-connected speculators backed by a brutal repressive apparatus. Protest leaders were subject to beatings or arrest (Pinches 1984: 145–7). Despite the rhetoric of poverty alleviation, the poor were even more marginalized by relocation and loss of informal sector jobs. The little public housing actually built was beyond their means.

The downfall of Marcos in February 1986 revived the confidence of private investors. Developers bought up cheap land at strategic locations. Large-scale retailing was the catalyst. The city's first modest shopping mall, Harrison Plaza, had opened in Pasay in the mid-1970s. Much larger in scale, in 1985 SM (Shoe Mart) City mall opened at Cubao near Quezon City and began a trend towards location at hubs around the EDSA ring road. High-rise office construction was stimulated by a 1991 law allowing foreign investors to take up to 40 per cent of the value of new buildings. Ayala Corporation began to rebuild Makati, while a new business district rose quickly on EDSA between Makati and Cubao at the crossroads of Ortigas, where the Asian Development Bank occupied a new complex in 1991. Office space was built almost as fast at Ortigas as at Makati. Investors also renewed interest in industrial estates around the urban periphery (McAndrew 1990: 190–1).

In the 1990s rival real estate consortia invested in competing satellite towns. The Gotianun group's Filinvest Land in 1991 acquired the 244-hectare former government stock farm beside the southern expressway at Alabang as the site for Filinvest Corporate City (Tiglao 1994a,b). In 1995 an 'anti-Ayala' consortium led by First Pacific group (Metro Pacific Corporation) won the tender for the 214-hectare former Fort Bonifacio (Fort McKinley) site east of Makati that was promoted as Bonifacio Global City (Patterson 1999). At the end of the decade long-standing plans were revived for massive reclamation along Manila Bay (Naerssen et al. 1996: 190–7). These took form as Centennial City, a 750-hectare project led by former Marcos crony Eduardo Cojuangco, who had resumed control of the San Miguel group. The development was to be the main component of a planned 1,450-hectare Boulevard 2000 waterfront project. On a smaller scale, Megaworld Properties revitalized Quezon City with Eastwood City Cyberpark and prestige high-rise offices for IBM and Citibank (MT 15 June 1999). There has been some corporate investment in low-cost housing on the urban fringe, but for the most part real estate interests have responded to the demand for offices, housing and amenities by those who have the most money to spend, namely expatriates and the Filipino elite and middle class.

Driven by the interaction of elite demand, real estate interests, 'crony capitalism' and 'money politics', Manila has therefore coalesced from an amorphous sprawl with an off-centre CBD (Makati) to a series of hubs around the EDSA ring road and an extended north–south axis (Figure 8.4). This morphology has been consolidated by the long-delayed expansion of the LRT network like a bow-string around the EDSA ring road. The first ten-kilometre section of elevated light-rail line between Quezon City and Makati opened in December 1999, followed by the rest of the loop around to the ends of the original LRT axis at Pasay (2000) and Caloocan (2002). The loop was intended not only to provide a fast link between the satellite nodes but also by transfer to allow low-income workers faster access from cheap housing areas on the western fringe, as does the Metrotren commuter rail service along the north–south San Fernando–Laguna axis.

Since the 1940s Manila has become a megacity heavily dependent upon the motor vehicle but, unlike large American cities, with a much higher population density and much higher use of public transport. Road space or accessibility is the link between place of residence and place of work. Historically, it has been the middle class that has commuted, mostly by private vehicle. By 2000 Manila accounted for 75 per cent of automobile registrations on the main island of Luzon, compared with 50 per cent of public transport vehicles (Table 8.4). Cars not only occupy much more road space but also carry fewer passengers. A traffic count on the EDSA ring road in 1980 showed that cars (excluding taxis and jeepneys) made only one-third of person trips but represented almost three-quarters of vehicle flow (NTPP 1982: 38).

The bias in transport funding towards road and freeway construction has neither relieved congestion nor improved accessibility. In 1980 pedestrians accounted for one-third of all trips yet 42 per cent of primary links were

Table 8.4 Manila (NCR) and Luzon, Jakarta (DKI) and Java: passenger vehicle registrations by type, December 2000 ('000)

	Manila	Luzon	%	Jakarta	Java	%
Cars	492	659	75	1,238	2,246	55
Jitneys	534	1,083	49	312	448	70
Buses	12	26	47	397	1,038	38
Motor cycles	168	734	23	2,213	8,365	26
Total	1,206	2,502	48	4,160	12,097	34

Sources: PSYB (2001), ISYB (2001).

Archipelagic Cities: Manila and Jakarta 273

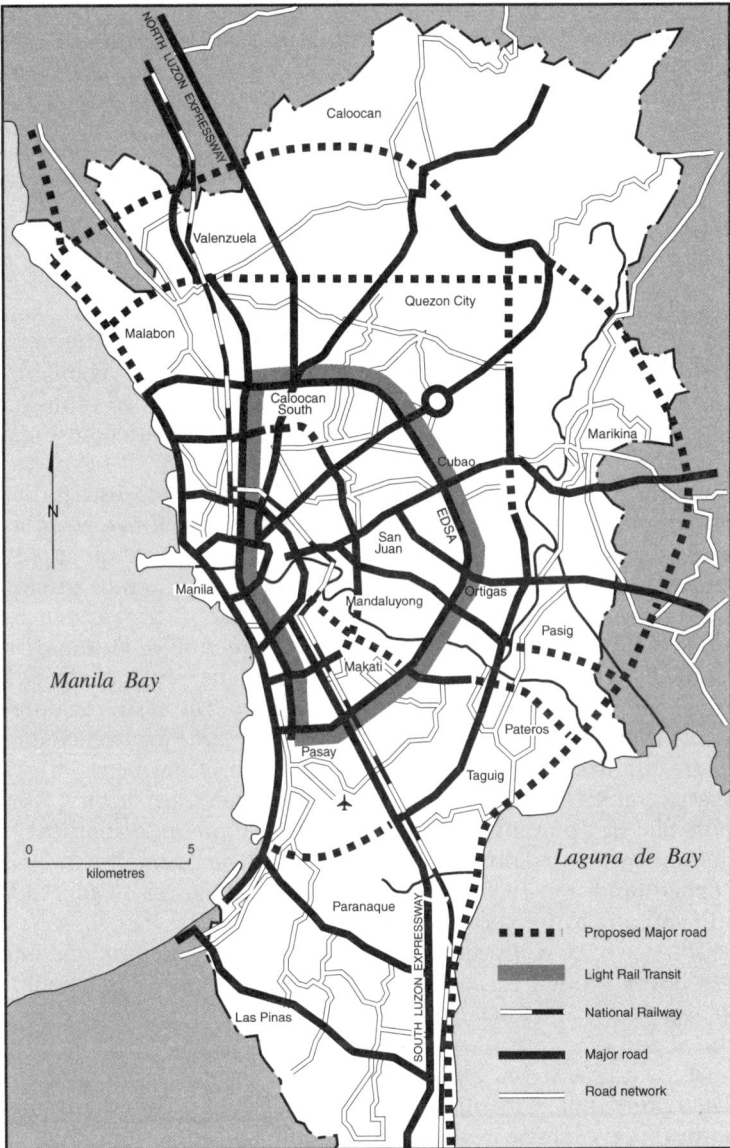

Figure 8.4 Manila: transport network, 2000 (*Source:* Anon 1996b: 27 updated from various sources)

found to have 'poor' or no footpaths (NTPP 1982: 8). Urban sprawl and relocation have meant that increasingly the poor also face long trips over congested roads. In 1999 it was reported that in ten years the average vehicle speed had fallen from 18 kph to 12.6 kph (MT 9 July 1999). Mass public transport such as the LRT cannot solve the dilemma unless it is a well-articulated and affordable mode. In the meantime, Manila has almost ceased to be an integrated city, either spatially or socially.

Jakarta

Before 1905 the main cities of Java had as yet no clear identity. The city existed as a built environment and a set of functions but not as a common society or political entity. Administratively they were part of the surrounding residency, equivalent in size to a Philippine province. The Residents' prime concern was peace and order (*rust en orde*), not urban infrastructure and planning, and they were kept busy attending to matters in the countryside. Urban populations were fragmented. Constrained by law to live within the Chinese quarter (*kamp*) beside the old business district (Kota) or the outlying markets of Tanah Abang and Senen, the Chinese were administratively autonomous under their own officers appointed on approval of the Resident. The *kampung* of the various Indonesian ethnic groups, the local Betawi plus Javanese, Sundanese, Ambonese, Buginese and others, were also self-governing under their own headmen and customary (*adat*) law, subject to oversight by the 'native' bureaucracy, whose upper levels coexisted with the colonial European bureaucracy. The small but dominant European community organized itself by a variety of means, including the bureaucracy, Army and Navy, the Chamber of Commerce, the Trade Association, clubs and societies, the churches, and Freemasonry. Sectional interests had no common voice. If there were common forums, they were the new invention of daily newspapers. By 1850 there were Dutch-language newspapers and by 1900 a flourishing vernacular press for educated Indonesians and assimilated Chinese (Adam 1995).

The absence of civic institutions and civic identity had profound implications. Indonesia inherited almost nothing of the vigorous Dutch tradition of urban autonomy. The inhabitants of Jakarta (Batavia) had a sense of neighbourhood and place but no sense of a common society. Civic identity and national identity emerged together in the early twentieth century as part of the same process of modernization and globalization. Faster transport and communications, newspapers, electricity, the spread of education and a more paternalistic government all played a role in shaping this new awareness. Enjoying legal privilege and wealth, Europeans and Chinese participated more actively in the new civic institutions, whereas educated Indonesians were more oriented towards the national challenge. On Independence, civic institutions were firmly subordinated to the national and central government. Nevertheless, as the capital city, Jakarta became the focus of the nation. For the Indonesian elite, Jakarta society became almost indistinguishable from national society.

Beyond the early colonial city

Like Spanish Manila, Dutch Jakarta (Batavia) was established as a compact, walled city but much sooner burst its container (Figure 8.5).[2] By the eighteenth century the model town at the mouth of the Ciliwung had become notoriously pestilential (Blussé 1986). Land clearance in the hinterland had increased the rate of siltation, causing the river mouth to prograde and the grid of urban canals to clog up, creating ideal breeding conditions for malarial mosquitoes. High officials withdrew to country villas. In 1808 the reformist Governor-General Daendels decided to move the seat of government several kilometres upstream to the former country estate of Weltevreden. The musty old fortress was demolished and the stone used for construction of an 'Empire'-style palace, completed in 1828 as offices for the growing colonial bureaucracy (now the Department of Finance) (Heuken 1983: 144–8). Better-off Europeans gradually moved out of the Old Town (Kota) into the exclusive new districts of Rijswijk and Noordwijk along the road between the social club (Harmoni) and the new government offices. By the end of the nineteenth century the growing European population had spread south to the area around Konigsplein (King's Square), now the National Monument (Monas) (Figure 8.6). Only the Chinese remained behind in the Old Town, with small outlying settlements around the markets of Tanah Abang, Senen and Meester Cornelis (Jatinegara).

A parsimonious colonial government provided only the bare minimum of urban infrastructure. Transport access, as well as sewerage and garbage disposal, was by river. Water was drawn from wells. Road surfaces were hardened dirt, muddy or dusty according to the season. Hospitals were places to die. Schools were few. Death rates and morbidity were high, especially from malaria and intestinal diseases. Family life adapted to local conditions. In the European and Chinese communities, women were much under-represented, so that formal or informal habitation with local women was the norm (Taylor 1983). Eurasian children were often accepted by their fathers and could achieve position in local society, which had many mestizo elements. The wealthiest men were Chinese revenue farmers, agents for the government's tax collection. Society was therefore more integrated than ethnic boundaries would suggest.

All this began to change in the late nineteenth century. Opening of the telegraph from Europe via Singapore was followed by the inauguration of a direct steamship line from Amsterdam via the new Suez Canal (Chapters 2 and 3). European society in the main cities of Java was suddenly brought into much closer contact with the mother country. More women began to travel out and to bring up respectable, God-fearing families. European society began to measure itself against metropolitan norms. Mestizo elements

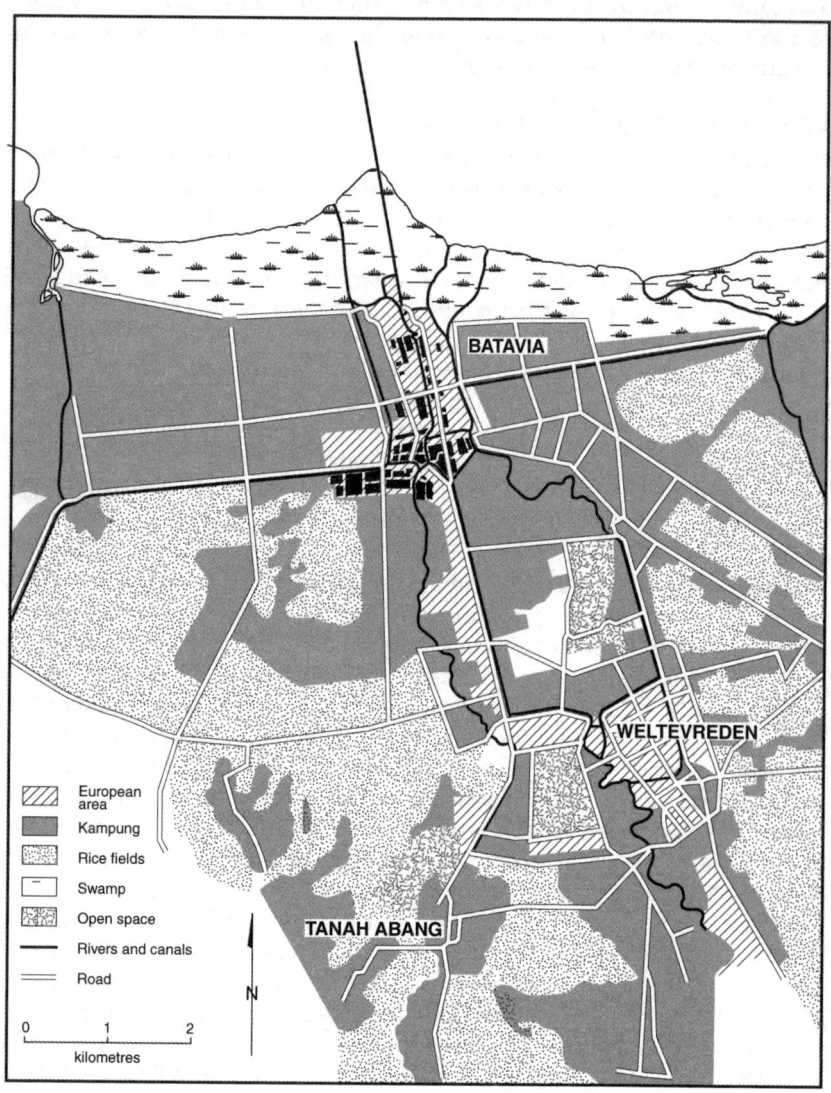

Figure 8.5 Jakarta: land use, 1858 (*Source*: Weitzel 1860)

Archipelagic Cities: Manila and Jakarta 277

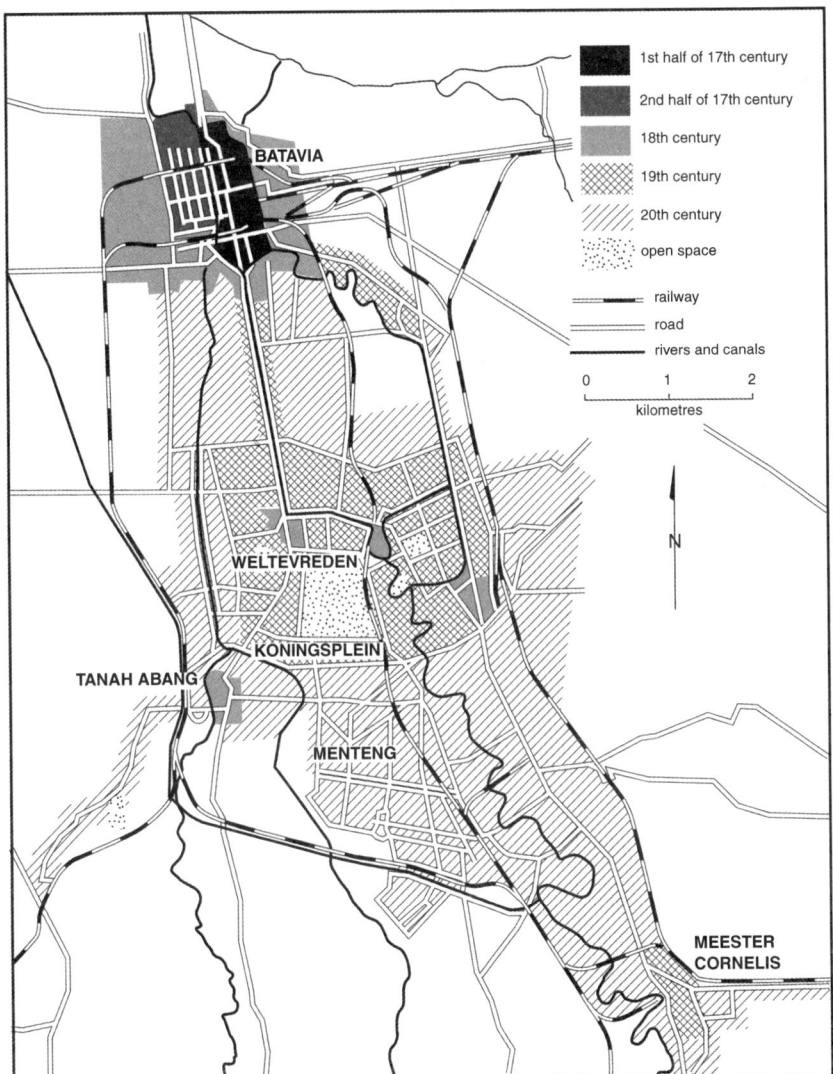

Figure 8.6 Jakarta: growth map, c.1600–mid-1930s (*Source*: Stadsgemeente Batavia 1937)

fell into disrepute and were practised more furtively. One small example was the decline among Dutch women of the custom of wearing a sarong outside the home in the morning (Weitzel 1860: 127). Modern technology also intruded in the form of steam launches, horse- and steam-tram lines, town gas, electricity, telephones and daily newspapers.

As in Manila, the most expensive item of urban public infrastructure in the late nineteenth century was a deepsea harbour at Tanjung Priok, previously a fishing village east of old Batavia. Since the seventeenth century, larger ships had anchored off the mouth of the Ciliwung and relied on tenders and lighters to bring passengers and cargo ashore. After Jakarta had become the terminus for the deepsea mail steamers, as well as home port for the interisland fleet of the NISN, something less primitive was called for. Work began in 1877 and by 1886 a harbour basin had been fitted out with wharves and cargo sheds (ENI 1921: 69–70). A canal, road and later railway connected to the nearby town. Jakarta was thereby established as a modern port two decades before Manila and three decades before Surabaya.

Because of its elongated form, Jakarta was also the first commuter city in Southeast Asia. Although Europeans now lived in the new residential districts around Weltevreden, commercial employment continued to be in Kota (Figure 8.6). For households of means, a carriage was therefore essential. Each day around eight, merchants set off for the Old Town, returning around three for a meal and siesta (Weitzel 1860). At dusk the Old Town was devoid of all but pedestrian traffic, while Weltevreden bustled as families went out in their carriages to take the air, go visiting or seek entertainment.

Fixed route public transport therefore developed early in Jakarta and kept up with developments in technology. In 1869 Hr Dummler introduced a horse-drawn tram, which at that time was still modern technology in European and American cities (Duparc 1972). This was a convenience for households without their own carriage but in the tropical climate there proved to be a high death rate among the ponies. In 1881 the Nederlandsch-Indische Stoomtram Maatschappij (NISTM) was floated to replace the horsetram with a steamtram, comprising two or three carriages pulled by a little boiler on wheels. A 14-kilometre line between Kota and Meester Cornelis was completed three years later. In 1899 the Bataviasche Electrische Tram-Maatschappij introduced a supplementary network of ultra-modern electric lines.

In 1905 the Decentralization Law brought into being a new form of self-governing urban municipality (*gemeente*). First to be set up in April 1905 were Batavia (Jakarta), the adjacent town of Meester Cornelis, and the summer capital of Buitenzorg (Bogor) (Eggink 1930: 22). Chinese, Arab and 'native' communities continued to be self-governing and were excluded from municipal authority. The municipalities at first represented only the European community and that part of the urban infrastructure for which it was prepared to accept tax-paying responsibility. Since their financial position was at first extremely weak, that meant little more than sanitation. As ways were found to increase revenues, the role of the municipalities increased. Divisions of tasks were negotiated with central government departments, such as with Public Works for roads, bridges and drainage, Public Health for hospitals and Education for schools. The much-vaunted autonomy of the other communities came more and more to be ignored whenever the municipality deemed that interference was necessary on grounds of health and sanitation, which gradually embraced the type of building materials, drainage, water supply, garbage collection, and eventually *kampung* improvement (Abeyasekere 1987: 119–25).

The spread of the town was constrained by a ring of private estates (*particuliere landerijen*), which had been sold to raise revenue early in the nineteenth century (Abeyasekere 1987). These were akin to manors, in which the inhabitants worked as serfs. On those closest to the town cultivation gradually gave way to *kampung*, whose occupants paid rent. Some parcels were also rented to wealthy Chinese or Europeans to build mansions or country houses. After the turn of the century, the high demand for suburban-style housing made it profitable to sell estates for urban development. In 1908 the municipality took the initiative of buying up the Menteng estate for subdivision and by 1920 it had acquired a landbank of 800 hectares (Eggink 1930: 107). By the 1930s settlement extended all the way along the main highway to Meester Cornelis, which in 1935 lost its separate status as a municipality (Figure 8.6).

Jakarta's infrastructure boom of the 1910s and 1920s was a decade or so later than in Manila but the broad programmes were almost identical. As also in Singapore (Chapter 7), the aim was to make the colonial city a worthy showpiece, whose tropical environment and exotic elements were civilized to metropolitan comfort and convenience. Modernization included public transport. In 1927 the State Railways completed electrification of its urban loop, giving a connection every 15 minutes to the port of Tanjung Priok (KRN 1930: 77). The electric rail loop more or less defined the city's area until the 1940s (Figure 8.6). This new competition in 1930 prompted a merger of the electric- and steamtram companies into the Bataviasche Verkeers Maatschappij (BVM), which in March 1934 opened an all-electric network of 35 kilometres. Frequency was excellent: every six minutes in the peak (7.5 minutes off-peak) on the main line and 7.5 (10) minutes on other lines (Duparc 1972).

The Municipality responded to public pressure to improve the roads, which were still mostly hardened earth. To control dust in the dry season, the Municipality sent out workers with a shoulder pole and two pails to spray water several times a day (Eggink 1930). In the wet season, however, there was no remedy for the mud, which was spattered by every passing car. After 1920 roads were progressively overlaid with bitumen, a by-product readily available from local oil refineries. This was of great benefit to public health, by reducing dust-related illnesses in the dry season, as well as improving the comfort of motorists and even more so cyclists and pedestrians.

Better roads increased competitive pressure on the tramways. From a peak of 21 million passengers in 1920, tramway patronage halved to 11 million in 1930 and then again to a trough of under 6 million in 1936 (KV/IV annual). This sharp decline occurred despite growth in the population, from about 300,000 in 1920 to over 500,000 in 1930. Part of the explanation was the decline in prosperity from the postwar boom of 1919/20 to the depths of the depression and the company's failure to reduce fares accordingly. The other part was competition from both the electrified rail loop and rapid growth in the number of road vehicles. At the end of February 1937 the city had registered 6,500 passenger cars, including 1,300 taxis. A phenomenon of the 1930s depression was that some small automobiles were rebuilt as jitneys with a wooden rear cabin to carry about six passengers. Known as autolettes – later more commonly *opelet*(tes) – and similar in concept to Manila's postwar jeepneys, they ran in relay along the tram routes, offering a cheaper fare and stopping on demand. There also survived about 2,400 two-wheeled pony carts (*sado* and *de(e)l(e)man*) in the role later taken over by pedicabs (*becak*) (Stadsgemeente Batavia 1937: 181–2).

Compared with Manila, in 1940 Jakarta had only half the number of automobiles but many more bicycles (YPS 1940). This reflected both social attitudes in the Netherlands, where the bicycle was popular among all classes, and in the 1930s the cheapness of Japanese imports. Because few Indonesians could afford private cars, the bicycle was their revolution in personal mobility (see also Chapter 4). By 1937 some 70,000 bicycles were registered in Jakarta, making up 60 per cent of all vehicles (Stadsgemeente Batavia 1937: 167). Contemporary photographs attest to their general use. Offices, factories and stores were equipped with large cycle racks. Bicycles would explain a good deal of the tramways' loss of patronage, especially among third-class passengers who made up around 85 per cent of the total (IV 1941: 410–11).

Because land and labour were so much cheaper, Europeans enjoyed a much higher standard of living in their colonial cities. Once disease was banished, obvious signs of poverty cleared from the streets, and ice, refrigeration and electrical ventilation made available to all households, Europeans could live in garden suburbs in large bungalows staffed by servants, with a car in the garage and a driver on standby, work in offices under electric fans, enjoy the amenities of the cinema, the dance hall, the club or the municipal swimming pool, sip coffee or eat ice cream on the terrace of a local café or shop for the latest fashions. Here was tropical urban paradise. Offstage there were less happy voices, but the triumph of colonialism was that they were not heard.

Independence: Third World city

Unlike Manila, Jakarta suffered very little physical damage during the Japanese occupation. During this period, change was more social and political. Living standards markedly declined, but requisition and rationing meant that food was actually more scarce in the countryside, leading to rural–urban migration (Sato 1994). Dutch citizens were interned and Indonesians rose to senior positions in the bureaucracy. Batavia was renamed Jakarta. There in August 1945, a few days after the Japanese surrender in Tokyo, Sukarno and Hatta proclaimed the independent nation of Indonesia. Six weeks later British forces arrived to supervise the Japanese surrender and release prisoners of war. Dutch military and civilians returned in their wake and in July 1947 seized full control of the city (Abeyasekere 1987).

The Dutch intended that their reoccupation would be accompanied by rigorous postwar town planning. In mid-1948 the Town Planning Ordinance (*Stadsvormingsordonnantie*) was proclaimed with the aim of 'ensuring a well-considered urban development and in particular a rapid and purposeful rebuilding of war-affected areas'. Since 1941 population had doubled and in the first half of 1949 it exceeded one million (DKI 1972: 44). In this context, the satellite town of Kebayoran was planned and the first stone laid on 18 March 1949 (DKI 1972: 14). The 730-hectare site was on higher ground several kilometres southwest of the then outer suburb of Menteng. To the northeast, the ricefields and *kampung* of Kemayoran were also to be redeveloped as the international airport. However, little was accomplished before December 1949, when the Dutch were forced by a combination of guerilla warfare and international pressure to recognize Indonesia's independence.

The new Republican government did not reject Dutch concepts of town planning. On the contrary, the approach was seen as modern and practical by energetic and idealistic nationalist bureaucrats, especially those who had graduated as engineers from the Bandung Institute of Technology (ITB). But there were now more constraints. One was a lack of funds. Another was the greater difficulty of enforcing municipal by-laws. It was much harder for Indonesian officials to demolish illegal dwellings, evict squatters, remove streetsellers, enforce road laws, insist on public cleanliness, and clamp down on illegal electricity and water connections than when acting under orders from colonial authority. Indonesians felt that they had now reclaimed their city and were reluctant to obey regulations whose logic and justice they had never accepted. Since the prime beneficiaries had been the urban elite, they had good reason for this attitude.

The greatest single influence on Jakarta's planning in the two decades after Independence was President Sukarno, a graduate in Civil Engineering from ITB. He understood the symbolic power of town planning and harnessed it to Indonesian nationalism, applying his influence beyond any formal powers. The Asian Games of August 1962 were an opportunity to show off Jakarta to the rest of the world and central government and aid funds were used for various prestige projects, including the stadiums at Senayan, the luxury Hotel Indonesia, ring roads, and new public transport vehicles to replace the electric tramway system (DKI 1977: 98). Jalan Thamrin, as it emerged in the 1960s, Jakarta's answer to Makati, was also Sukarno's vision, a new central boulevard lined with modern buildings, from the National Monument and Bank Indonesia at one end to Hotel Indonesia at the other, in between the offices of international organizations, embassies, and the luxury state-owned Sarinah department store. The site opposite Hotel Indonesia was allocated to Jakarta's first genuine skyscraper, the 29-storey Nusantara Building finally completed in 1972. Although derided at the time as 'prestige projects' which the country could ill afford, decades later they are all still part of the structure, fabric and character of the city.

Sukarno was also instrumental in having the trams abolished in 1960–61. Unlike in Manila, the system had survived the Pacific War intact and was nationalized in 1954 but no new investment was forthcoming to expand the network to cope with a population which soared from around 0.5 million in 1941 to almost three million in 1961 (Table 8.3). Trams became the cheapest form of transport, because they were so overcrowded that payment of fare was optional. Those willing to pay a fare in return for a seat could ride by *opelet*, which now flowed in a steady stream along the main thoroughfares. When the tramway was shut down, the *opelet* easily took over their customers. The state-owned (D)jakarta Transport or PPD (formerly BVM) was allocated a very inadequate fleet of buses but could offer no real competition. In the 1960s the *opelet* were supplemented on new routes by *bemo*, smaller three-wheeled Daihatsu Midget vans with a two-stroke engine and room in the cabin behind for six passengers (Dick 1981a: 68–9).[3] This was more or less the jeepney solution to public transport and it worked very well.

The other remarkable feature of Jakarta's public transport after Independence was the *becak*, a three-wheeled, man-powered pedicab or trishaw that served as the poor person's taxi. Clumsy prototypes had appeared in the late 1930s but when motor vehicles were requisitioned during the Japanese occupation and petrol and tyres became unavailable, a lighter model suddenly became a transport solution (Robinson 1953). In 1952 Jakarta was a two-wheeled city with 26,500 motor vehicles but 123,000 bicycles and 25,000 *becak*, whose drivers formed a union 50,000 strong (Republik Indonesia 1952: 272). Loathed by traffic engineers and car drivers, the ungainly *becak* became a popular symbol of the new urban freedom, just as the tough, proud drivers, for the most part migrants from rural Java, became the people's mobile strike force in the event of any clash or riot (Critchfield 1970). Hampered by more complicated logistics and less able to jostle for road space, pony carts retired to the outskirts to serve the local markets. By 1970 *becak* were estimated to number around 150,000, most unregistered and each providing employment for two or three men (Rimmer 1986a: 163). Drivers, owners and repairers together represented almost 400,000 people, a substantial component of the informal sector work force. This combination of *opelet*, *bemo* and *becak* explains how in early 1969

Jakarta's public transport system was able to function with a fleet of only 550 buses, many of them unserviceable, to replace the former trams (DKI 1972).

One of Sukarno's last acts before being forced to surrender power in March 1966 was to appoint a young and dynamic marine commander, Ali Sadikin, as Mayor of Jakarta (Abeyasekere 1987: 215–42). His aim was to make it a more functional city for the rich, but also a better city for the poor. Up to two-thirds of the city could become flooded after heavy rain, only 20 per cent of the population had access to electricity and 15 per cent to reticulated water, and there was a huge housing deficit (DKI 1972). Sadikin tackled these immense problems with the verve of Imelda Marcos and to more effect. Needing revenue, he took the controversial step of legalizing and taxing gambling and steambaths (de facto prostitution), then used the funds to build infrastructure.

The Sadikin era and the acceleration of economic growth drew the line more sharply between the city of the rich and the city of the poor. As in Manila, most of the population lived in dense, low-rise *kampung*, once village-like hamlets that with rapid population growth had been overrun by the city and fragmented into minute parcels for each family (Krausse 1978: 12). Until the 1960s, poor migrants could still occupy land as squatters and find work in construction or in informal sector activities such as hawking, *becak*-driving or small-scale industry (Jellinek 1991). Informal sector incomes were high relative to the cost of basic necessities and there was little harassment from government. President Sukarno encouraged them to believe that Jakarta was their city. The New Order assisted the urban poor with *kampung* improvement, flood control, education and family planning and health clinics, but took away informal sector jobs by crackdowns on street traders and *becak*, and made migration more difficult by tightening up on residence cards and deporting those without valid cards. Sadikin (Bang Ali) nevertheless remained a popular figure. Only after his departure in 1977 did the poor become convinced that government and big business threatened their survival.

The New Order: renewed modernization

Jakarta's New Order transformation owed little to urban planning and, as in Manila, a great deal to market forces. The driving force was national industrialization, which was highly concentrated around the outskirts of Jakarta (Kuncoro 2001). However, contrary to the *desa-kota* hypothesis of McGee (1991), factories did not diffuse across the hinterland like the former agricultural estates but located in very distinct ribbons along the east–west and north–south highways and freeways that gave access to the CBD, port and airport (Gardiner 1997: Map 5; Figure 8.7). Workers migrated to nearby villages, which quickly became densely settled urban *kampung*. Real estate developers sought prime land further afield to build complexes of middle-class and luxury housing with facilities such as shopping malls, entertainment centres and golf courses. Thus in the 1980s and 1990s most of Jakarta's population growth occurred beyond the adminis-

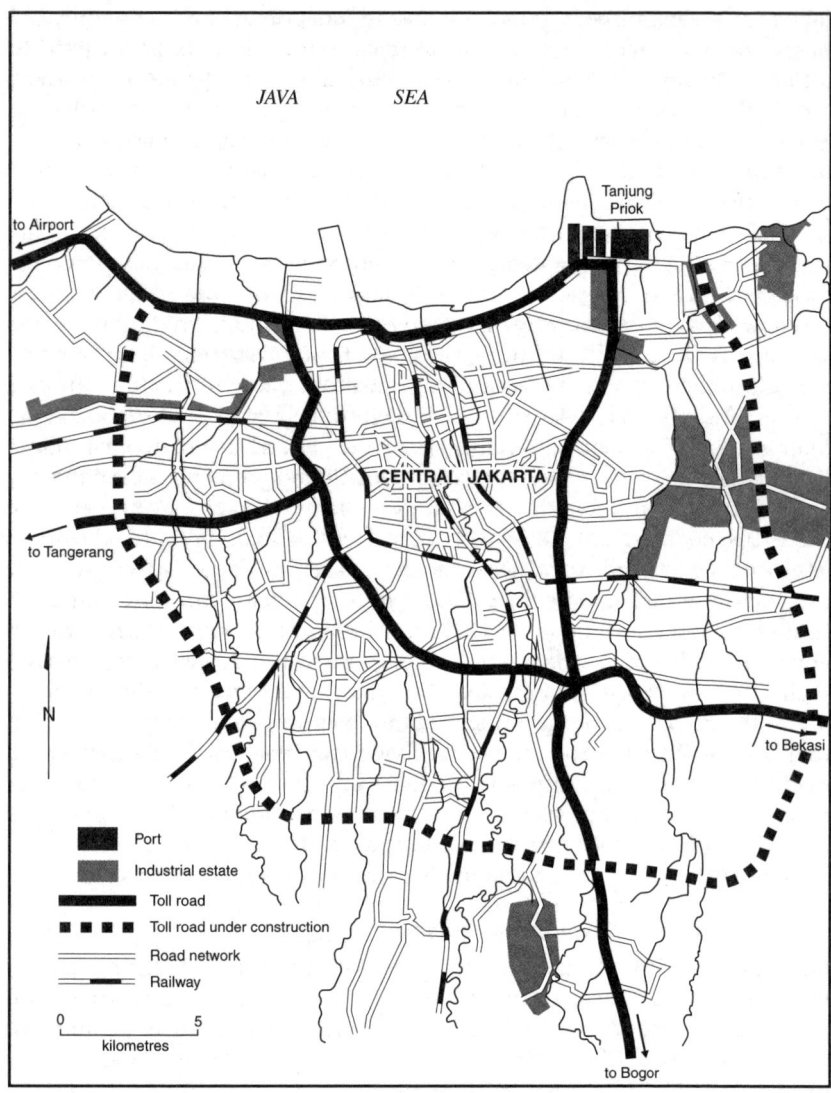

Figure 8.7 Jakarta: expressways, 1995 (*Source*: Anon 1996a: 19)

trative boundaries of the capital city province and was distributed fairly equally between reclassification of rural villages to urban *kampung*, natural increase and net migration (Gardiner 1997: Table 7.2, Maps 3–4). In the 1990s the population of the official capital city province barely grew at all (BPS 2000; Firman 1997: 112).

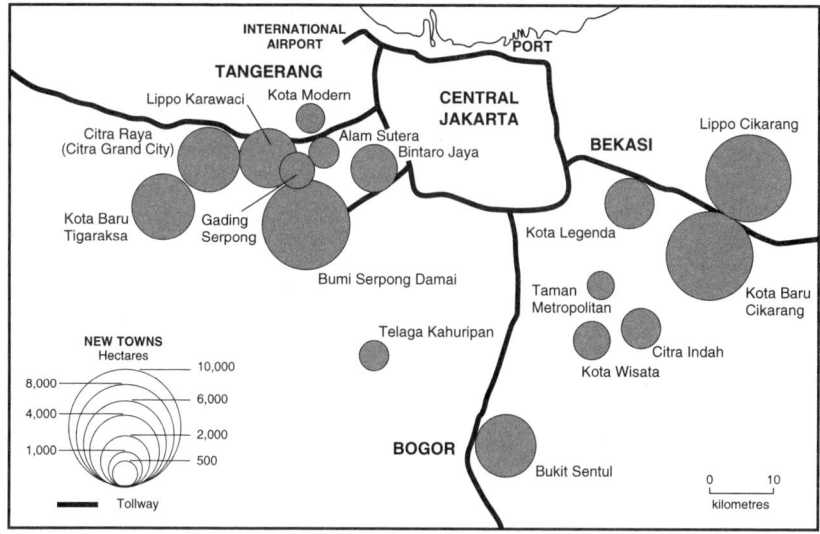

Figure 8.8 Jakarta: new towns, 1997 (*Source*: Dick and Rimmer 1998a: 2315)

The leading force in new town development, like the Ayala Corporation in Manila, was Ir Ciputra, an engineer of ethnic Chinese background, who after 1961 was the main local contractor for the city's large infrastructure projects. In the 1970s with other partners he began to launch elite housing projects, while consolidating land around the edge of the city for whole satellite towns, the first planned development since Kebayoran. In 1985 his Metropolitan Group launched the 461-hectare Pondok Indah ('Beautiful Village') as the prime elite housing complex, including a golf course and international school and targeted at expatriates. The shopping centre grew into a mall, to which an office tower, apartment blocks, hotel, entertainment centre and theme park were to be added (Ciputra *c*.1996). Even more ambitious was the 6000-hectare new town of Bumi Serpong Damai to Jakarta's southwest (Figure 8.8). The banking and finance Lippo Group's new towns of Lippo Kawaraci (2630 hectares) and Lippo Cikarang (5500 hectares) were provided with a full range of facilities at the outset: golf course and country club, international school, international hospital, mega-mall, business district, apartment towers, roads and landscaped gardens and themed housing complexes with controlled gateways and security patrols. These self-contained new towns were designed by American consultants and, with some adaptations to local taste, applied an American approach (Dick and Rimmer 1998a).

In the 1990s, spurred on by strong capital inflow and low-cost international finance, big business scrambled to acquire strategic sites for redevelopment as new central business districts and satellite towns. The desirability of sites was determined in the first instance by their proximity to actual or projected freeways, secondarily by access to the airport, major industrial estates and recreation. By October 1996 about 90,000 hectares in and around Jakarta was estimated to have been acquired for redevelopment, far exceeding the 66,000 hectares of the entire Capital City province, though only 13,000 hectares had been developed by the time of the Asian crisis (Kompas, 21 October 1996). In inner Jakarta, land was cleared for multiple new CBDs with massive 'superblock' complexes. The overall scale of development of office towers, shopping malls and luxury apartments between Jakarta, Manila and also Bangkok was broadly comparable (Table 8.5). Unhedged foreign currency loans for urban real estate were a contributing cause of the Southeast Asian economic crisis of 1997/98 (McLeod and Garnaut 1998).

To achieve such redevelopment, densely settled *kampung* had first to be demolished. Between 1980 and 2000 the population of the Central Jakarta district fell by 350,000 (BPS 2000). Compensation was but a fraction of what the land was worth after consolidation, and even some of this was bled off by officialdom. Most *kampung* dwellers had to relocate to the urban periphery, either losing their income-earning activities or having to commute long distances (Jellinek 1991). All this bred tremendous resentment. Here was the front line between the modern air-conditioned city of the elite and middle class and the vibrant, but highly insecure and increasingly alienated urban village of the populace.

Becak bore the brunt of urban restructuring. In the early 1970s they were identified as the cause of worsening congestion and in 1974 amidst much controversy were banned from operating during daylight hours in favour of small vehicles such as *bajaj* powered by motor cycle engines (Dick 1981; Rimmer 1986a). By municipal by-law of 1988, *becak* were prohibited altogether. Those intruding into the closed area were impounded by police and destroyed or dumped at sea. Tens of thousands of drivers lost their livelihood.[4] However, traffic flow did speed up, which necessitated overhead bridges for pedestrians to cross in safety. Yet vehicle numbers continued to soar. By 2000

Table 8.5 Manila, Jakarta, Bangkok: modern building stock, mid-1999

	Office (million m^2)	Retail (million m^2)	Condominiums (units)
Manila	2.9	2.9	8,700
Jakarta	2.9	1.5[1]	14,300
Bangkok	2.6	3.3	7,800

Note: 1. June 1998 before riots. Includes 0.3m. m^2 in outer metropolitan ring.
Source: JLL (2000).

Jakarta had 1.2 million cars, 0.8 million commercial vehicles and 2.2 million motor cycles (Table 8.4). As people were forced to live further away from place of work, bottlenecks and congestion reappeared on wider, multi-lane roads.

The late-New Order 'solution' to congestion was to expand the high-speed road system (Figure 8.7). In the 1990s the extra-urban links to the south, west and east along with those to the airport and Tanjong Priok were connected by a continuous urban ring road and by 1997, on the eve of the economic and political crisis, work was well advanced on an outer ring road. These urban toll roads greatly enhanced the mobility of cars and trucks and long-distance buses that could afford to pay the tolls and save time and petrol. Even more effectively than in Manila, it knitted together the satellite new towns, industrial estates, shopping malls, airport, seaport with expanding, high-rise central business districts into a bustling, air-conditioned, quasi-First-World city. Beneath and around it was the slow-speed, heavily polluted, Third World city of the large majority of the population.

Conclusion

Despite differences of timing and degree, since the late nineteenth century there has been a continual and almost parallel increase in the scale and complexity of both Manila and Jakarta. Modest trading centres have become national capitals and been transformed into sprawling megacities. This chapter has followed this evolution through the spatial systems of transport and communications and set it in the context of closer integration with the respective hinterlands of Java and Luzon and the fluctuating intensity of relations with the global economy.

The most recent phase of export-oriented industrialization has tied Manila and Jakarta more closely to the world economy while reducing dependence upon their agricultural hinterlands. The capital-city province of Jakarta now merges like a bird with outstretched wings into the planning region of Greater Jakarta (Jabotabek), including the adjacent districts of *Bog*or, *Ta*ngerang and *Bek*asi. More constrained by mountainous terrain, Greater Manila is elongating into a long corridor, southwards into the adjacent planning region of Calabarzon (*Ca*vite, *La*guna, *Ba*tangas, *R*izal and, nominally, more remote Que*zon* provinces) and northwards through Bulacan and Pampanga provinces into the Central Plain. Both cities are now huge urban agglomerations of some 20 million people and large enough to be nations in their own right (Figures III.2–3). Increasingly they appear to be becoming economic and social enclaves.

These extended metropolitan regions have continued to expand and flourish despite huge gaps in public infrastructure. Congestion and pollution are just the most obvious symptoms. Budget constraints make it impossible to close the gaps in utilities, housing, health, education and transport for the whole vast area and populace. Instead, a high-speed, air-conditioned city for the elite is being superimposed upon the sprawling, slow-speed, non-airconditioned, slum city. The comfort zones of the middle-class elite coexists uneasily with the amorphous world of the working poor, who struggle for survival on narrow margins and without tenure of land or employment. Yet despite all the obvious inequities and discomforts, Jakarta and Manila are still the economic frontier, places where ordinary people can better find or supplement a livelihood than in the more equitable but poorer rural villages.

Notes

1. In the contemporary sense, rather than the nineteenth-century meaning of locally born Spaniards.
2. Batavia was the name bestowed by Coen upon the fortified town laid out after the defeat of the local ruler of Jayakarta in 1619 and used by the Dutch until the transfer of sovereignty in 1949. Jakarta (after Jayakarta) was a nationalist renaming that sought to restore continuity with the precolonial era.
3. These *be*cak *mo*tor were originally intended to replace the man-powered *becak* (see below).
4. In the aftermath of the Asian crisis the ban was relaxed in June 1999 and an estimated 7000 *becak* returned to the city before the ban was reimposed in April 2000 (JP 25/2/00).

9
River Cities: Bangkok, Rangoon and Saigon

Once known as the 'Venice of the East', Bangkok is the most water-based city in Southeast Asia. The flow of barges on the Chao Phraya River from the hinterland through the city and the many ferries along and across the river are the pulse that connects Bangkok back to Ayutthaya and the 'age of commerce' between the fifteenth and seventeenth centuries (Reid 1993). Bangkok, like Ayutthaya, has become both a great trading centre and a symbolic religious, political and administrative heart of the kingdom. However, when founded in 1782 it was an enclave rather than the centre of a productive hinterland. The hinterland developed only after the Bowring Treaty of 1855 and the emergence of the rice trade was dominated, like most commerce and crafts, by the immigrant Chinese (Chapter 5). At the same time, the Bangkok-based Siamese (Thai) state aggressively extended its control over what would become the modern Thai nation. Thus the city grew into a nation and the nation was encapsulated in the city.

While Siam escaped colonization, this is only subtly revealed in Bangkok. Between the 1850s and the 1950s the original enclave settlement developed into a fairly typical cosmopolitan Southeast Asian commercial city. After the 1932 *coup d'état* the transition from absolute monarchy to military-bureaucratic alliance and more aggressive nationalism led to changes in the land use-transport system. From the 1960s to the mid-1980s Bangkok exhibited many of the characteristics of a Third World city: squatters, a large informal sector and extreme traffic congestion (Table 9.1). The economic boom of the late 1980s and early 1990s overlaid an untidy array of First World elements, giving rise to a dual or hybrid city which, like Jakarta and Manila, may be described as a newly industrializing or post-colonial city-region.

Rangoon and Saigon, by contrast, were not royal capitals. Despite pre-existing settlements, they evolved from the outset as colonial port cities and administrative centres. After the Pacific War, Saigon developed into a Third World city but, following the Communist takeover in 1975, it too stagnated like Rangoon. These stagnating cities probably warranted a new category – the Fourth World city. In the late 1980s the governments of Vietnam and Burma

Table 9.1 Bangkok Metropolitan Administration (BMA) and Bangkok Metropolitan Region (BMR): area (sq. km) and population (million), 1913–2000

	1913	1936	1950	1960	1970	1980	1990	2000
BMA (built-up area)	13	43	67	96	184	344	637	n.a.
BMA population	0.6	0.9	n.a.	2.2	3.1	4.7	5.9	7.1
BMR peri-urban ring	n.a.	n.a.	n.a.	1.1	1.4	1.9	2.7	3.7
BMR total population	n.a.	n.a.	n.a.	3.3	4.5	6.6	8.6	10.8
THAILAND	(8.9)	(11.7)	19.6	26.3	34.4	44.8	56.3	61.9
Percentage	n.a.	n.a.	n.a.	12.5	13.1	14.7	15.3	17.4

Note: BMR (7800 sq. km) comprises Krung Thep Maha Nakhon, Samut Prakan, Nonthaburi, Nakhon Pathom, Pathum Thani, Samut Sakhon. Built-up area figures for 1958, 1971 and 1988. Population for 1910 and 1930.
Sources: Sternstein (1982, 1995), Punpuing (1996), Alpha Research (2001).

(Myanmar) relaxed foreign direct investment controls so that Saigon (Ho Chi Minh City) and Rangoon (Yangon) could vie to join the ranks of the newly industrializing city-regions. Few elements of the First World city had materialized before the onset of the Asian crisis in 1997–98. Recurrent parallels are drawn with Rangoon and Saigon to illuminate the nature of Bangkok's development.

Southeast Asian city, 1850s–1950s

Bangkok's original layout was modelled upon cosmopolitan Ayutthaya, the capital of Siam for 417 years until its sack by the Burmese in 1767 (Sternstein 1964, 1976). A few years later the Chinese merchant-warrior, Phya Tak, founded a modest new capital at Thon Buri on the west bank of the river from Bangkok and became the self-appointed King Taksin. An uprising led to Taksin being deposed, executed and replaced by the highest-ranking general Chao Phraya Chakri, who later became King Rama I and moved the capital across to what was then Rattanakosin Island on the east bank of the river. There was a conscious attempt to recreate the Ayutthayan way of life, culture and administration within the new royal capital of Bangkok, which housed both the Grand Palace and Temple of the Emerald Buddha.

By 1850 this moated, gated and fortified site had been developed to accommodate the king's household, nobility and their servants (Figure 9.1). Outside the town walls, the commoners lived in floating dwellings along the river or tributary canals. As reflected in the large number of temples, the capital also functioned as a religious centre. Above all, it was an international seaport that attracted people from the countryside, migrants from southern China and ex-prisoners-of-war to give the town its distinctive character as a cosmopolitan commercial city (Viraphol 1977; Cushman 1993).

River Cities: Bangkok, Rangoon and Saigon 291

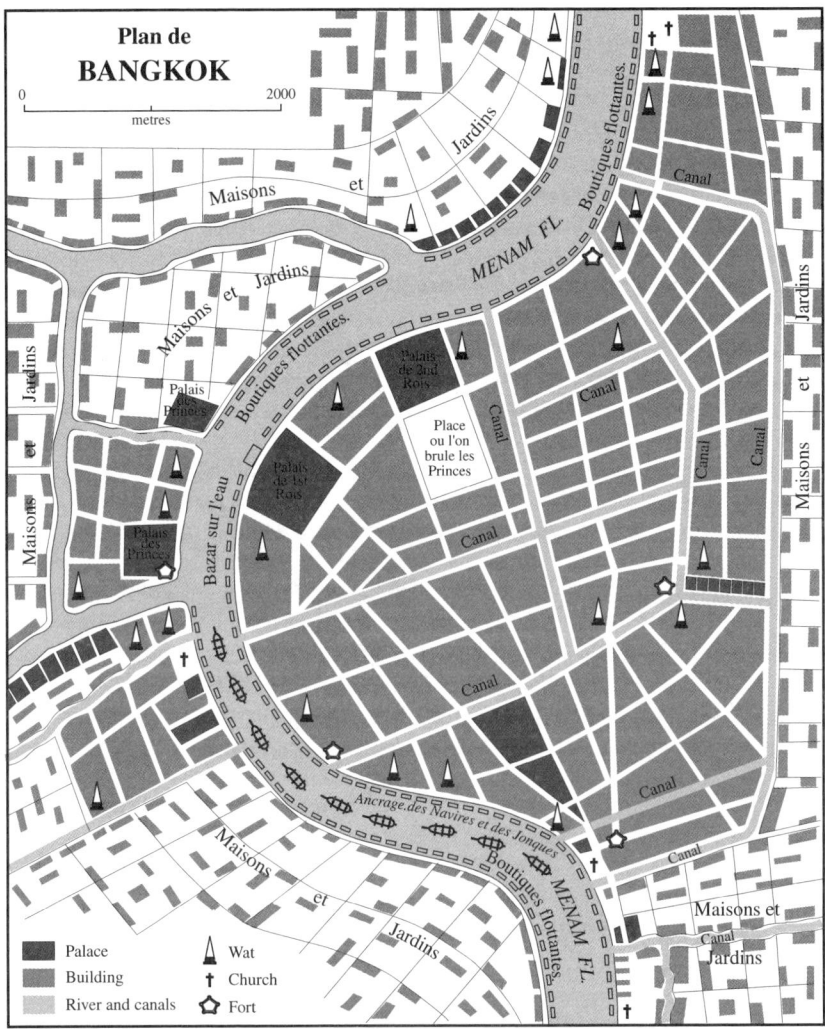

Figure 9.1 Bangkok, 1850 (*Source*: from map appended to Pallegoix 1854, vol. 1). The regular street pattern in this stylized plan should be disregarded and 'replaced by a maze of meandering footpaths and innumerable creeks, canals and ditches' (Sternstein 1982: 15)

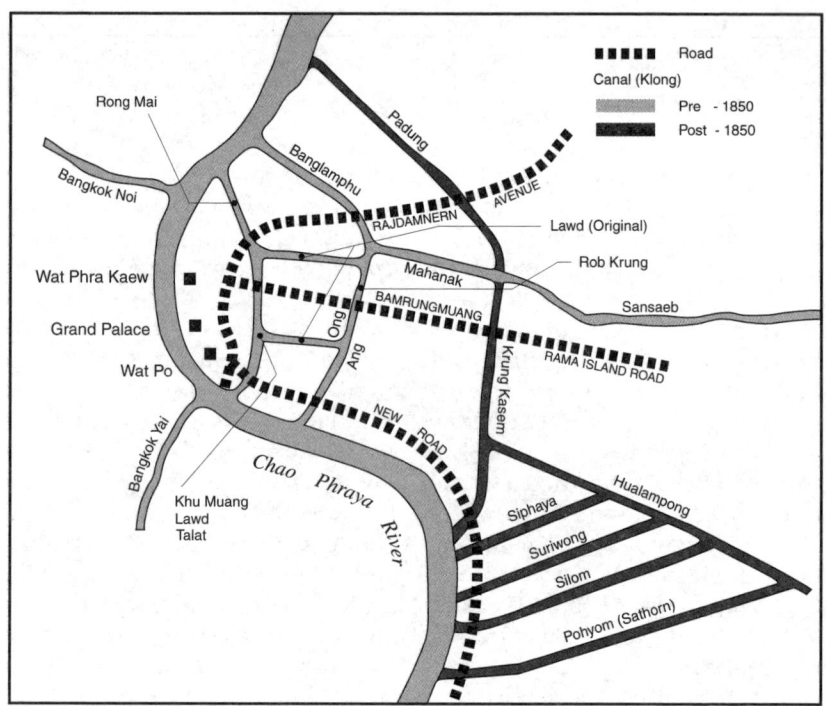

Figure 9.2 Bangkok: Canals and roads, c.1860s onwards (*Source*: Based on Van Beek 1999)

Waterways dominated the everyday life of the settlement with a population in Bangkok's built-up area of around 50,000 (Sternstein 1982). Apart from the river, the main transport system comprised three rings of canals encircling the Royal Palace with radial canals leading out to beyond the city fringe (Figure 9.2). Within the compact area bounded by the outermost Krung Kasem Canal (built 1851–52), roadways were primitive and unconnected by bridges. Roads were not needed for commerce. Sampans offered door-to-door transport services for passengers, hawkers and food sellers.

Creating the new Bangkok, 1850s–1920s

The need for at least a rudimentary land transport system became apparent following the Bowring Treaty of 1855. Westerners wished to live on land in brick and stucco dwellings in garden settings, a style soon copied by the nobility. Foreign consuls then complained that since 'coming to live in Bangkok they had found there were no roads to go riding in carriages and on horseback for pleasure' (Smithies 1986: 37).

Successive rulers of Thailand and their expanding bureaucracy sought to refashion Bangkok by applying western transport technologies. Their aim was both to establish Bangkok's status as a modern capital and to consolidate the city's position as the kingdom's transport and communications hub and sole commercial centre (Korff 1986). An early initiative was the construction of New Road (Charoen Krung) on a former elephant track behind the churches, consulates, dockyards, mills and warehouses fronting the river. Roads were added for horse-drawn carriages but their use was restricted by the absence of bridges over canals and floods during the southwest monsoon. For business or pleasure one still had to go by water.

During the 1870s and 1880s, King Chulalongkorn (1868–1910) promoted road transport by inviting foreign advisors and sending the princes overseas to study western technology. One hundred and ten new roads complemented the introduction of telecommunications (1881) and postal services (1882) to provide for an estimated population of 170,000 (based on the 1882 postal census), which was still largely tied to low-cost water transport (Sternstein 1982). By the late 1880s, 67 bridges traversed the canals – 22 were wooden and 24 were in need of repair. Good roads were restricted to areas adjacent to the palace and houses of the many Siamese princes (Ouyyanont 1999). Along the roads moved an array of imported vehicles laden with passengers or goods (Fournerau 1984; Carter 1904). They included gharries and ox carts brought from India, rickshaws from Japan, modern bicycles with European designs and, after 1888, unsprung 'native' omnibuses based on a British prototype. Worsening road congestion was a consequence of Bangkok's rising population, which in turn flowed from the flourishing international rice trade and the continuing influx of Chinese migrants.

After the 1890s, shortcomings in the road system were addressed through the creation of the Ministry of the Capital (1892), which improved royal supervision over transport (Ouyyanont 1997). Working closely with the Crown Property Bureau, the ministry expended a major share of its budget on roads and bridges. In 1894 foreign engineers, mostly Italian, were recruited to design a series of more substantial marble and iron bridges and over the next 15 years constructed over 200 kilometres of carriageways illuminated by electric lights. Conspicuous among these new carriageways was the three-kilometre Rajdamnern Avenue, built with road metal from the demolished city wall. Styled after the Champs Elysées, the broad royal avenue connected the Grand Palace and the Chitlada Palace and princely villas north of the Krung Kasem Canal. This once-forested area became the location for main government offices. Rising land prices also attracted royal investments into housing projects. Feeder roads to the railways and in the Sampeng commercial district served the Crown Property Bureau's shophouses. The Bureau was also responsible for ensuring road development met the needs of royal activities in other parts of Bangkok.

In 1893 Bangkok became the first Asian city to introduce electric trams. These replaced the recent but unsatisfactory horse-drawn trams and served the Hualampong terminus of the first railway opened also in 1893. Although designed to ferry troops to protect Bangkok's riverine gateway at Samut Prakan, the 23-kilometre railway line also became patronized by foreign passengers travelling to the beach (Van Beek 1999). As the railway system extended to the north and northeast, Hualampong became the terminus for up-country passengers and also the focal point of the urban tramway system (Chapter 5).

During the 1900s the Danish-owned tramway company shaped the spread of the city by offering traders and bureaucrats frequent suburban services on four routes totalling 17 kilometres in length – additional lines were added in the following decade. By 1908 the number of motor cars had grown in eleven years to 300, but when some fine vehicles from Europe took part in motorcades the following year, the occasion was still rare enough to be advertised in local newspapers (Smithies 1986: 41; Van Beek 1999: 27).

By 1910 Bangkok's ill-connected water- and land-based transport systems served a population of 360,000, still heavily concentrated around the palace and the river's commercial areas (Carter 1904; Ouyyanont 1997). The traditional water-borne system remained centred on the port of Bangkok, where steamers anchored mid-stream in the Chao Phraya River. Although steam launches were prominent, sails and oars were still used to steer vessels handling rice and teak downstream. Flat-bottomed boats towed by steam launches were used to top-load or lighten steamers at Koh Si Chang off the bar at the mouth of the Chao Phraya River. The river's dawn floating market offered a strong contrast to the day and night-long commercial activities in Chinatown's bustling Sampeng district, where merchants used the street and pavements to display their wares (Wright and Breakspear 1903/1908). Most personal transport was still by water or on foot and virtually 'all goods reached the market dangling from the end of bamboo poles' (Sternstein 1982: 21).

The city's dual water- and land-based transport systems continued to operate more or less independently into the 1920s. Motor vehicles had much less impact than in other Southeast Asian cities. Some new main roads were upgraded (Silom, Sri Phraya and Suriwong) but the government's financial constraints precluded any large program of road construction and little was done to extend the network eastwards (Ouyyanont 1999). Trams were still the backbone of public transport, supplemented by rickshaws and horse carriages (Seidenfaden 1928). Motor vehicles found their niche as taxis and hire cars. Later Nai Lert, a Sino-Thai entrepreneur, introduced buses. Construction of a road network on the poorer and mainly Thai western side of the river was inhibited by the absence of any bridges across the Chao Phraya River. Here most transport was still by canal.

Land use, 1930

By 1930 these transport developments had helped transform the original concentric settlement into an elongated pattern (Korff 1986: 86). The royal town remained at the centre of the two legs but the administrative function had moved to the northeast, which had acquired the elite residential areas of Samsen, Dusit and Phayathai (Figure 9.3). The Chinese quarter of Sampeng maintained its renown for daytime shopping and night-time revelry in brothels, opium dens and gambling houses. Other prominent ethnic concentrations included the Indians of Pathurat (cloth merchants), the Mons of Ban Tanao and the Vietnamese of Ban Yuan (Askew 1994). However, the original European consular area along the river had been transformed into a district for international banks, insurance and trading houses; the embassies and consulates moved downstream along the riverbank. Parts of 'Old Bangkok' had been demolished to accommodate modern commercial premises but few new buildings challenged the height of the old temples (Batson 1984).

Around 1930 the port was handling over one million tonnes of cargo, mainly rice exports and imports of textiles, small manufactured goods and petroleum (Falkus 1993). Serving this trade was a string of rice mills, saw mills and petroleum depots on both sides of the river towards the Gulf. Other large factories manufactured cement, soap, cigarettes and leather products. A very lively intra-regional trade within Southeast Asia had spurred the growth of these portside activities. Commercial aviation was handled at Don Muang airport, which had been established in 1914 some 20-km north of the city to avoid marshy ground (Davies 1997).

The military-bureaucratic era, 1930s–1950s

In June 1932 amidst the Great Depression, the People's Party led a bloodless *coup d'état* against King Prajadhipok (Rama VII). The bureaucracy was then freed from the power of the nobles. A ruling alliance of the bureaucracy and the military was to reshape Bangkok over the next thirty years (Batson 1984). Already in 1922 the royalist Ministry of the Capital had been abolished and Bangkok's administration transferred to the Ministry of the Interior. Under the 1933 Municipal Act, Bangkok (*Krung Thep*) became a province in its own right and in 1937 the Bangkok Municipality and the Thon Buri Municipality were created. These organizations maintained stability through numerous coups and changes in national government.

The Minister of the Interior had taken initiatives to integrate land transport systems on both sides of the river. In 1926 the first railway bridge had been completed across the Chao Phraya River to carry trains from the southern line through to the main terminus of Hualampong (Van Beek 1999). Then in 1932 the opening of the Memorial Bridge (Phra Bhudda Yodfai Bridge) allowed vehicular access and stimulated the western extension

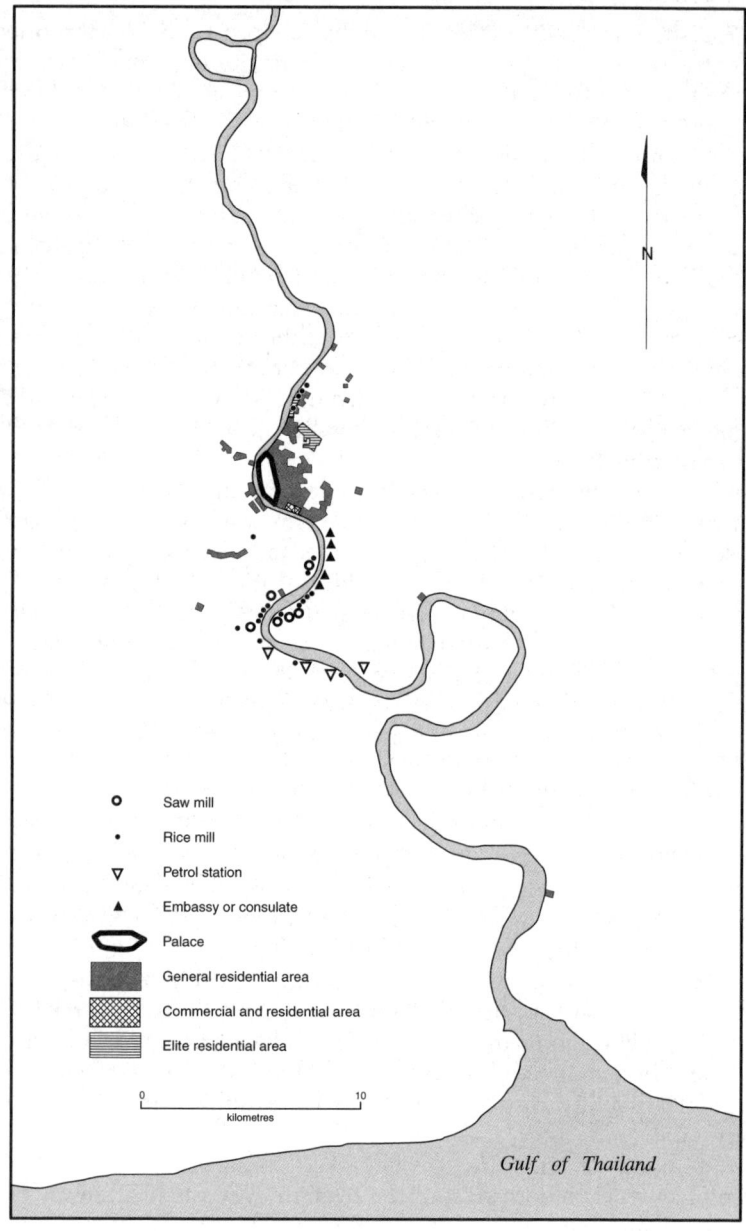

Figure 9.3 Bangkok: land use, *c*.1930 (*Source*: Dilokwanich 1995)

of the road network at the expense of the inland waterway system. By 1940 the epithet 'Venice of the East' was no longer applicable.

During the Pacific War the Japanese did not occupy Thailand like the rest of Southeast Asia. Collaboration between the Japanese and Thai governments allowed small Bangkok firms to expand their activities and a number of major businesses were established, including banking. During 1944–45 Allied bombing caused some damage to Bangkok's port facilities, bridges and railways. After the Pacific War spontaneous rural migration to Bangkok accelerated the city's population growth but this was not matched by investment in housing, transport and other public utilities.

The military-bureaucratic alliance gave priority to the development of the road network and allowed the urban waterway system to decay. This policy was defended as modernization but it also satisfied military demands for a faster response time during frequent coups. At the behest of Prime Minister Field Marshall Phibun Songkhram (1948–57), much effort was given to constructing roads to connect military camps in Bangkok with Don Muang airport, hitherto remote from the city. The process was accelerated under Field Marshall Sarit Thanarat (1958–63), who overthrew Phibun. Canals were reduced to stagnant pools. Waterways were buried beneath avenues. Streets were widened by filling in the parallel ditches and canals (Sternstein 1982). Despite improvement of the port at Klong Toey in the late 1950s, even the arterial Chao Phraya River lost a large measure of its importance. Waterways ceased to play a central role in the functioning of the city. Their absence as natural drainage systems has been keenly felt in the increased frequency of flooding.

During the late 1950s Prime Minister Sarit sought to accelerate the modernization of Bangkok by also banning pedicabs and introducing land-use planning techniques (Textor 1961). In 1958 the American consultants Litchfield Whiting Browne & Associates were appointed to produce Bangkok's first masterplan. The *Greater Bangkok Plan 2533* published in 1960 provided a land-use framework within which specific physical plans could be developed for the next thirty years (Sternstein 1982). Perceiving Bangkok's primacy as a problem, the plan recommended population control and decentralization to other cities, while identifying the need for a mass transit system. The plan was never adopted and an early opportunity to control Bangkok's infrastructure growth was lost. Land-use change became driven by the 'demand for urban space by competing urban groups and institutions, and controlled by landowners' (Askew 1994: 38).

Land use, 1960

By 1960 the old pattern of land use along the river could no longer accommodate the enlarged population of 2.2 million, an increase of 150 per cent since the mid-1930s. The northern and southern legs of the once-compact residential area had expanded eastwards across the Central Plain by integrating villages, orchards and paddy fields into the urban fabric (Dilokwanich 1995). Military bases were dispersed but there was an important concentration north of the palace area, which was left intact (Figure 9.4). Some government departments and utilities followed the military to new locations in the north (Dusit, Samsen, Phayathai and Bangsue); others moved to the south (Bangkapi, Phrakhanong and Bangna). Embassies and consulates had moved away from the river. New high-class residential areas had been established in both the northern and southern parts of the city and suburban housing had spread outside the confines of the earlier settled districts. 'Chinatown' survived as a commercial and financial centre, but a new commercial, industrial and residential area was developed along the riverbank, which included warehouses, low-income housing and small industries. Beyond this area were conglomerations of saw mills and rice mills and a major petroleum terminal along the Chao Phraya River. A new port was developed below the city at Klong Toey.

The most striking land-use changes stemmed from the adoption in the late 1930s of a policy of economic nationalism intended to overcome the domination of manufacturing industries by Europeans and Chinese entrepreneurs. This pro-nationalist policy had led to the location of large, state-owned secondary industries on extensive greenfield sites typified by the Royal Ordnance Factory at Bangsue north of the city and the Thai Tobacco Monopoly at Bangkapi in the south. Public funds were also used to establish rice mills, textile plants, tanneries and pharmaceutical factories along the river.

Rangoon and Saigon

Like Bangkok, between the 1850s and the 1950s the colonial creations of Rangoon and Saigon were also transformed from raw settlements into modern commercial towns that were the focus of river, rail, road, air and telephone systems. Rangoon's population increased from 56,000 in 1860 to 501,000 in 1940 and Saigon from 6,000 in 1860 to 400,000 with the adjacent Chinese settlement of Cholon in 1940 (Nguyen 1998). During the boom years of the 1920s Rangoon became the world's largest rice and teak port and a base for exploration of Burma's oil and gems (Pearn 1939; Spate and Trueblood 1942). Meanwhile Saigon-Cholon became a major rice port linked by barges and junks through extensive canal systems to the Mekong Delta (Murray 1980). Whereas the well-planned colonial seaport of Rangoon was 'provincial and, except in patches, rather shabby', Saigon (but not Cholon) was developed as a model colonial town of boulevards, gardens and terraces, recreating French culture in a tropical guise (Spate 1991: 74; Wright 1991). Rangoon was ruled from India but Saigon was more truly than Hanoi the colonial capital of Indochina. The Governor-General's palace and the

River Cities: Bangkok, Rangoon and Saigon 299

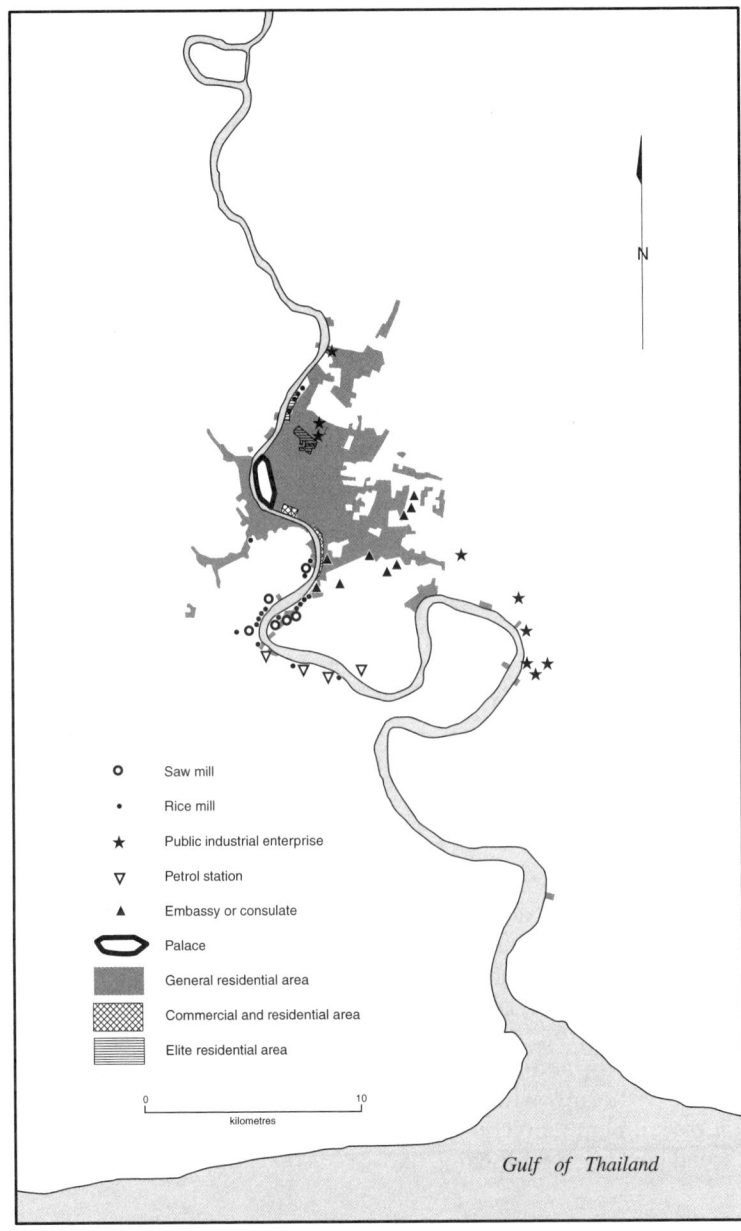

Figure 9.4 Bangkok: land use, c.1960 (*Source*: Dilokwanich 1995)

Banque de l'Indo-Chine symbolized French domination of traditional culture and the strength of the colonizer's commercial interests (Wright 1991).

Land transport services in both cities were also notoriously bad. Handcarts were used for moving goods while rickshaws provided short-distance trips for expatriates. Trams were first introduced in Rangoon in 1880 (electrified 1906) and Saigon in 1889. First brought to Saigon in 1903 and Rangoon in 1905, cars were soon disruptive – the first stoplight was installed on Saigon's busy Rue Catinat in 1909. Suburban expansion was brought about by the introduction of regular bus services in Rangoon from 1926 and increasing car registrations in Saigon, which resulted in the citadel's moat and canals being filled and asphalted and expatriates retreating to more salubrious locations. Without peripheral ring roads, vehicles travelled through both town centres and congestion ensued as they circumnavigated Rangoon's octagonal Shwedagon pagoda and Saigon's opera house and cathedral. Rangoon's powerful Municipal Commissioner with direct access to the Governor was responsible for improving traffic flows but in the late 1920s the French brought in an urban designer, Ernest Hébrard, to develop land-use zoning and transport plans for the commercial-industrial region of Saigon-Cholon (Wright 1991).

After the end of the Pacific War and some wartime damage, Saigon-Cholon matched Bangkok's development but Rangoon failed to keep pace. The French reoccupied Saigon-Cholon in 1945 but after years of bitter warfare were obliged to withdraw from Vietnam under the Geneva Accords of 1954 (Figure 9.5). There was little Vietnamization because Americans replaced the French in the southern part of the partitioned country. Under American influence the puppet South Vietnamese government led by Ngo Dinh Diem transformed Saigon-Cholon into a vibrant capital, belying the apparent decline in its population from 1.9 million in 1955 to 1.4 million in 1960 (Nguyen 1998).

Rangoon did not match the vibrancy of Saigon. The British returned only briefly to their model colonial capital before ceding independence in 1948. Then the Burmese reclaimed the city from the British, Indians and Chinese, kept the wharves, hospitals and pagodas and virtually let the rest decay (Spate 1991; Figure 9.6). During the brief period of democratic government between 1948 and 1962 Rangoon was bloated by an influx of rural refugees fleeing communist and ethnic insurgencies (Nwe 1998). Although the city expanded northwards with the creation of three satellite towns under the central government's direction, there was a pervasive sense of decay, particularly when compared with Bangkok.

Third World city, 1960s to mid-1980s

During the 1960s Bangkok maintained its economic supremacy. A widening range of agricultural exports and tourism continued to drive the city's

River Cities: Bangkok, Rangoon and Saigon 301

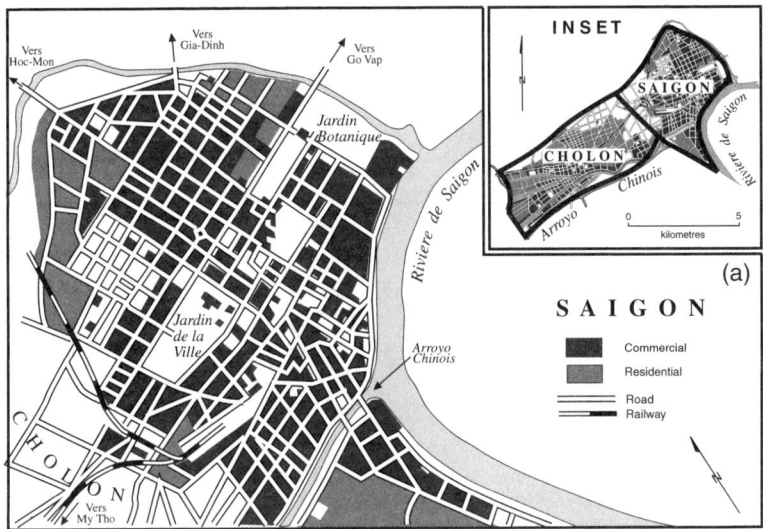

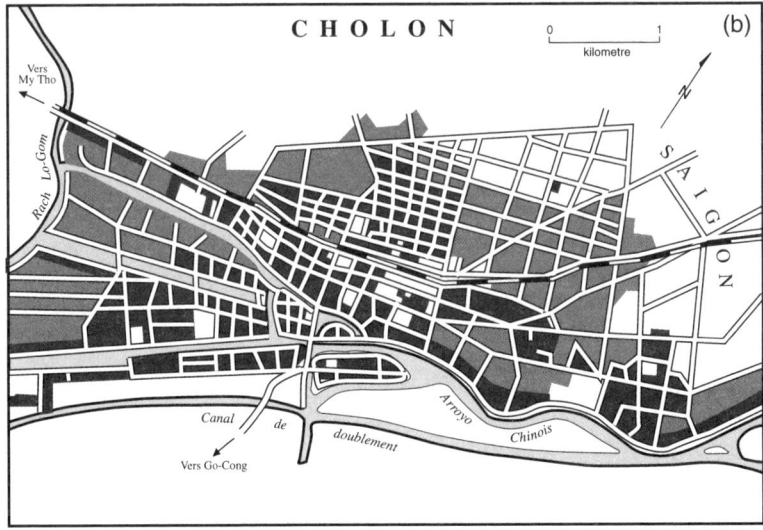

Figure 9.5(a) Saigon: land use *c*.1930s: The city was constructed by the French around a citadel surrounded by the Saigon River and two creeks close to the Chinese village of Cholon, established in 1778 some 72 kilometres from the open sea; (b) Cholon: land use *c*.1930s (*Source*: Not identified)

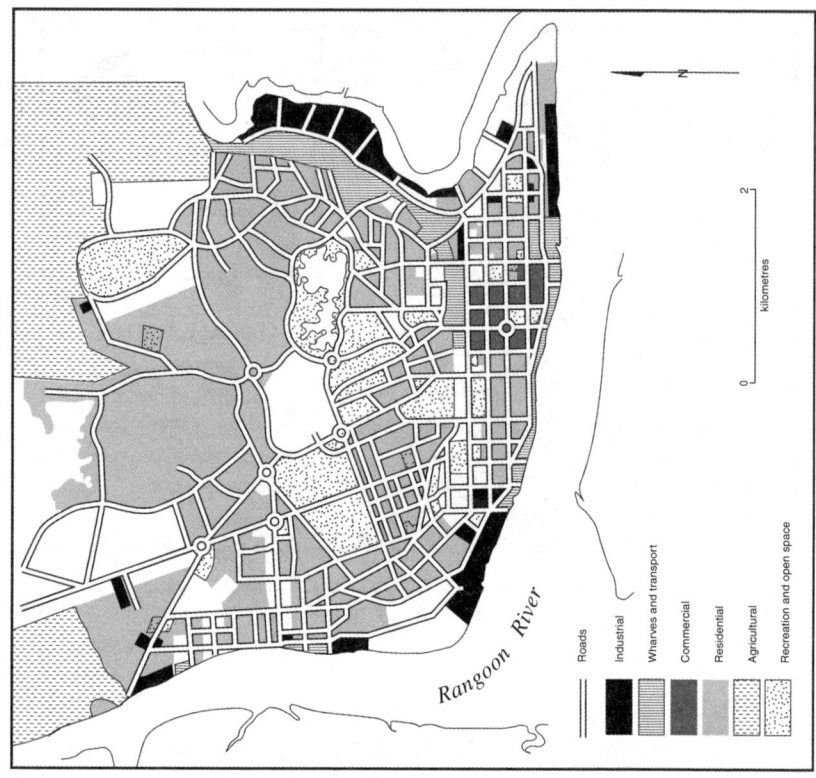

Figure 9.6 Rangoon: land use c.1940: the British built Rangoon on the site of an important religious shrine – the Shwedagon pagoda – and palace town (*Source*: Spate and Trueblood 1942)

economic growth and consolidate it as a service-cum-industrial centre. Bangkok was the prime beneficiary of investments by state enterprises and multinational enterprises, initially from the United States but later from Japan (Suehiro 1989). Most of all, the city benefitted from the activities of Sino-Thai business groups engaged in import-substituting industries, agribusiness and the financial sector, typified by Siam Motors, Charoen Phokphand and the Bangkok Bank Group respectively. Bangkok's economy was further buoyed by the expenditures of United States bases and American soldiers on rest and recreation during the Vietnam War. However, the city's population expansion outstripped the rate of job creation in the formal sector, generating a large informal sector and squatter settlements, hallmarks of a classic Third World city (McGee 1971). The military-bureaucracy was unable to control these developments. Its authority over Bangkok was diminished by civil unrest in both 1968 and 1973, oil shocks in 1973 and 1978 and the loss of revenue from the departure of American servicemen (Warr 1993; Phongpaichit and Baker 1995). The uneasy transition from authoritarian military government to democracy also undermined Bangkok's control over national politics, which was now challenged by powerful provincial barons enriched by agribusinesses (Ockey 1992; Girling 1966; Phongpaichit et al. 1998).

The epitome of traffic congestion

During the 1960s bus operators struggled to take over Bangkok's growing transport task from the trams (which were eliminated in 1965) and the waterways. By the late 1960s there was growing congestion and lengthening commuting times between home and work, particularly for travellers crossing the Memorial Bridge between Bangkok and Thon Buri (Table 9.2). Traffic congestion was compounded by the poor articulation of road networks. Many narrow streets had once been private lanes leading to residences. Traffic signals were controlled manually by police officers and inappropriate one-way restrictions created unnecessary detours and U-turns. Progressively, traffic was reduced to a crawl. Travelling any distance began to take more and more time as Bangkok became the epitome for congestion. This situation was aggravated in the late 1960s by the Thai government's preoccupation with 'goldplating' up-country highways at the expense of extending the city's road network (Rimmer 1971).

Progress in tackling these problems was slow because of the division of responsibility between an array of government departments (Table 9.3). Administration of the city was vested primarily in the national government but city streets were the concern of the Bangkok Municipality (later Bangkok Metropolitan Administration); national and provincial roads were under the Department of Highways and bridges over the Chao Phraya River under the

Table 9.2 Bangkok: motor vehicle registrations by categories, 1950–2000 (thousands)

Type	1950	1960	1970	1980	1985	1990	1995	2000
Passenger cars	n.a.	50	151	221	420	598	940	1,317
Microbuses and passenger pick-ups	n.a.	n.a.	n.a.	78	143	301	321	295
Vans and pick-ups	n.a.	n.a.	n.a.	55	78	268	403	737
Motorcycles	n.a.	n.a.	n.a.	172	486	729	1,617	1,965
Total	6	70	275	526	1,127	1,896	3,281	4,314

Sources: Various.

Table 9.3 Division of responsibilities in Thailand's urban transport administration

Acronym	Organization	Task	Department
NESDB	National Economic and Social Development Board	National planning, formulating and coordinating plans for infrastructure development	Office of the Prime Minister (OPM)
OCMRT	Office of the Commission for the Management of Road Traffic	Formulating transport-related plans, coordinating government organisations, assessing projects	OPM
MRTA	Metropolitan Road Transit Authority	The Bangkok subway system	OPM
PD	Police Department	Traffic management as part of nationwide Thai police operations	Ministry of the Interior (MOI)
DTCP	Department of Town & Country Planning	Urban planning	MOI
PWD	Public Works Department	Regional roads and Chao Phraya River bridges in Bangkok's suburbs	MOI
ERTA	Expressway & Rapid Transit Authority	Urban expressways in Bangkok and other cities	MOI
DOH	Department of Highways	Thai national highways throughout country and provincial highways	Ministry of Transport & Communications (MOTC)
DLT	Department of Land Transport	Compulsory motor vehicle testing, driver licences, truck terminals and other land transport administration	MOTC
SRT	State Railway of Thailand	Construction and operation of railways nationwide	MOTC
BMTA	Bangkok Mass Transit Authority	Operating bus services in Bangkok	MOTC
	Bangkok Metropolitan Administration	Responsible for road and railway construction and urban planning as part of Bangkok's local government system	—

Source: Based on Watanabe (1996: 9).

Department of the Interior's Public Works Bureau (Tarvisin and Suwarnarat 1996). The prime strategy was to restrict truck access to the city. Meanwhile, without an inner-city-railway system to take over from the defunct tramway system there was a shortage of public transport facilities. Bangkok had to rely on buses run by 22 private and two public companies, which had thwarted an attempt in 1963 to amalgamate them into a single company.

During the 1970s the Thai government refocused its attention on Bangkok's growing traffic congestion. In 1971 the government entered into a technical cooperation agreement with the Federal Republic of Germany to conduct the *Bangkok Transportation Study* (KRR 1975). This comprehensive plan provided for expressways, rail mass transit systems and truck terminals but, because of the huge costs involved, was implemented in a partial and piecemeal way. An immediate outcome was the creation of a series of instrumentalities to tackle specific transport tasks. The Expressway and Rapid Transit Authority was established in 1972 to organize the building of the proposed expressways and heavy-rail mass transit system. While the proposed mega-projects were being investigated, some flyovers were built at key intersections and the public and private bus companies in Bangkok and neighbouring towns were amalgamated into the government-owned Bangkok Mass Transit Authority.

Almost from the outset, the authority had difficulty in coping with demand. The regular services patronized by city workers and students were inexpensive but the buses were notorious for being slow, overcrowded and unreliable. The authority augmented its fleet by franchising 2,500 private minibuses. Next were introduced 700 air-conditioned buses that charged a premium fare and provided a more comfortable ride. Experiments in moving students from public transport to school buses and park-and-ride schemes were unsuccessful. The authority had problems operating profitably allegedly due to management inefficiency and corruption. After the Democracy Movement of 1969–73, the Thai government was reluctant to grant fare increases to the authority because of the anticipated political backlash. Between 5,000 and 9,000 illegal minibuses were allowed to operate in competition with the authority's fleet. A scheme for privatizing the authority in the mid-1980s came to naught. There was even a scheme to make more use of the waterways. River buses were introduced on the Chao Phraya River and proved so popular that there were concerns over boat crowding and safety.

When the reorganization of the buses failed to resolve Bangkok's transport problems, the Thai government sought assistance from the World Bank (Rimmer 1986a). In 1978 the Bank established the Bangkok Traffic Management Project, which proposed road pricing, parking fees, staggered work hours and traffic noise abatement. This led to the creation of the Office of the Commission for the Management of Road Traffic (OCMRT) to coordinate these proposed transport developments. In 1980 the Commission integrated traffic signals to improve intersection control; 95 kilometres of bus lanes were built and another 50 kilometres added later. The World Bank's scheme for charging for road space was not accepted. Computerized area traffic control, reversible lanes and compulsory vehicle inspection were implemented but – in marked contrast with Singapore – parking fees have remained ridiculously low.

Bangkok had European levels of public transit patronage through extensive use of buses and minibuses supported by para-transit comprising taxis (with and without meters), three-wheel motorized vehicles, four-wheel small open vans and motorcycle taxis. This situation was changed by the rapid increase in the number of motor vehicles to 530,000 by the mid-1980s, a 670 per cent increase in 25 years. Bangkok did not have North American levels of parking provision to accommodate them because the area devoted to road space in its inner core was one of the lowest in the world, just 8 per cent compared with New York's 25 per cent (Poboon et al. 1995).

Land use, 1985

By the mid-1980s Bangkok had spread in an uncontrolled way at the expense of paddy fields, orchards, vegetable gardens, outlying settlements and former canals (Figure 9.7). This spread was stimulated by widespread land speculation following the nomination of Nong Nga Hau as the site for Bangkok's second airport (Korff 1986; Dilokwanich 1995). The city's northeastern leg was extended to Rangsit, the southeastern leg to Samut Prakan and the middle infilled to Minburi and Bang Shan. In the west this advance was accommodated by Bangkok's annexation of Thon Buri in 1972. Even with this extension, the general residential area had expanded well beyond the 290 square kilometres controlled by the Bangkok Metropolitan Administration (though this area was three times greater than in 1950). Only the palace area, the government quarter and Chinatown survived more or less intact.

A major change in land use had stemmed from the concentration of a large number of industrial areas devoted to import-substituting industries in Bangkok such as chemicals, electronics, motor car assembly and synthetic textiles. Medium-scale manufacturing plants were attracted to industrial parks that offered cheaper land on which to expand (textiles, gunny bags, sugar and agribusiness). Most small-scale industries had continued to operate from shophouses. Older shophouses were still located in the inner city, but the postwar model of four or five-storeys characterized Bangkok's peripheral expansion and ensured that mixed rather than homogeneous land-use development was the norm.

Another new landscape feature was the development of a massive commercial and residential complex on the east bank of the river and its extension to the west bank. The European business precinct on New Road expanded to incorporate Silom, Suriwong and Rama IV Roads to form the city's most expensive real estate wedge. Silom Road was fast becoming an international financial precinct and mini-business districts appeared on Ploenchit Road (Prathunam and Siam Square), Phetchaburi Road and Asoke Road. Smaller commercial and residential areas with markets and modern shopping centres were also scattered in the extended areas. Some elite residential areas had been consolidated and new ones had appeared in the

River Cities: Bangkok, Rangoon and Saigon 307

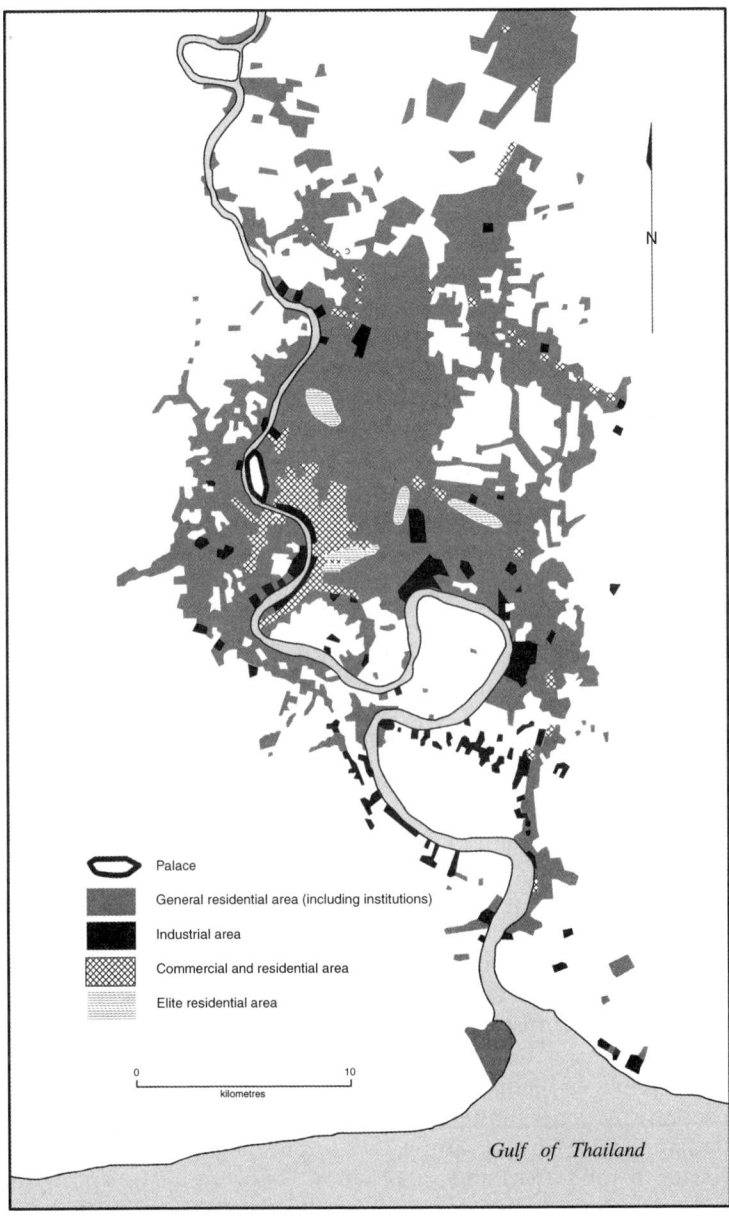

Figure 9.7 Bangkok: land use *c*.1985 (*Source*: Dilokwanich 1995)

southeast as middle-class families began to desert the city for suburban bungalows such as Bangkapi. The narrow lanes (*soi*) off Sukumvit Road had been redeveloped but settlement was still relatively sparse.

Missing from the generalized land-use map for 1985 were several inner-city 'slums' in old neighbourhoods and squatter settlements on vacant land near canals and adjacent to the port of Klong Toey. These offered first- and second-generation rural migrants the opportunity to survive in the city. The general land-use map was also too generalized to show low-income housing developments in Din Daeng, Khlong Prapa and Khlong Chan. Also omitted were the neighbourhood 'strips' that were created to serve the American servicemen stationed in Thailand or on leave from Vietnam – areas 'of bars and jewelry stores, massage parlors and curio shops, ill-concealed brothels and "Hong Kong tailors"' (London 1980: 30).

Governments had paid little attention to coordinating land-use and transport developments. The promulgation of the *Town and Country Planning Act* in 1975 had provided the Department of Town and Country Planning with an opportunity to implement the earlier *Greater Bangkok Plan 2433* [AD 1990] (Sternstein 1982). As noted, this land-use plan incorporated a green belt to control the conversion of agricultural land into urban land but its provisions were easily circumvented. The planning process was also hindered by conflicts between the Department and the newly formed Bangkok Metropolitan Administration (BMA): the 1975 BMA Act gave the governor of Bangkok greater control over planning land use and transport but few extra financial resources.

Rangoon and Saigon revisited: Fourth World Cities?

Saigon-Cholon remained vibrant under American patronage and acquired the characteristics of a Third World city like Bangkok. As the South Vietnamese government and its American allies were unable to control the countryside, the city experienced marked population growth. By the late 1960s Saigon-Cholon's population was estimated at five million. American troops withdrew in 1974. The following year the South Vietnamese government collapsed and the Democratic Republic of Vietnam from the north and the National Liberation Front from the south entered Saigon-Cholon. To celebrate the Communist takeover, the city was renamed Ho Chi Minh City and conservative street names were changed to radical ones. However, the Socialist Republic of Vietnam made little investment in the urban fabric, not least because its very limited resources were drained by its invasion of Cambodia in 1978. Ho Chi Minh City's surplus population was encouraged to relocate in rural areas and small cities. Following de-urbanization, the city's population fluctuated around 3.5 million until the mid-1980s.

Rangoon experienced further decay. After the Burmese junta had cut its ties with the West in 1962, urban infrastructure gradually deteriorated. Crumbling colonial façades, particularly in old Rangoon's rectangular core, testified to a once-modern colonial city. Development of suburban commercial nodes boosted Rangoon's census population from 2.0 million in 1973 to 2.5 million in 1983 but the decline of the vital rice trade and slackening economic conditions resulted in some incipient 'hollowing out' of the city centre. Presumably, if Thailand had not enjoyed the intervention of the United States and lapsed into political isolation Bangkok's city centre would have experienced similar stagnation.

The newly industrializing city-region since 1985

Between the mid-1980s and the Asian Crisis of 1997–98, Thailand experienced an economic boom which allowed market forces to reshape Bangkok's development (Phongpaichit and Baker 1996). This boom was initiated by the migration of Japanese manufacturing to low-cost sites in Thailand after the Plaza Accord of 1984 (Warr 1993; Phongpaichit 1990). Labour-intensive firms from Taiwan and Hong Kong followed to take advantage of the devaluation of the baht and the availability of female workers. The boom was sustained by Sino-Thai 'shophouse tycoons' boosting their overseas connections and moving their capital into export-oriented manufacture such as textiles, electronics, jewellery, leather goods, wood products, processed foods, computer components and motor vehicle parts. These new industrial developments were concentrated in 'Bangkok and Vicinity' comprising the Bangkok Metropolitan Administration and the five neighbouring provinces of Nakhon Pathom, Nonthaburi, Pathum Thani, Samut Prakan and Samut Sakhon – all this was later referred to as the Bangkok Metropolitan Region. Because the boom was based on manufacturing and not agricultural exports, the Bangkok Metropolitan Region's connections with the rural economy atrophied. The urban economy has become an industrial enclave, whose links are stronger with the international economy than with the hinterland.

First World elements

Bangkok's expansion sparked a secondary boom in domestic producer services (finance, insurance and real estate). This was reflected in the transformation of the city's skyline by the addition of many high-rise buildings, including the 94-storey, 309-metre Bayoke II Tower. These First World buildings with names like Centerpoint, Wall Street Tower and World Trade Center, symbolized that city business had come to dominate metropolitan development at the expense of the bureaucracy. This development was not matched by complementary investment in First World transport infrastructure, encapsulating Galbraith's contrast of private affluence and public squalor (Friedman 1977: 14).

By 1990 the economic boom had brought Bangkok's traffic to a virtual standstill. The cost in travel time, health care, lost productivity and foregone tourist investment was estimated at US$2.3 billion per year (Yordphol 1997).

Little relief was afforded by the creation of the Bangkok Mass Transit Authority to run the buses. There was anecdotal evidence of students leaving before dawn for a two-hour commute to school. Traffic jams were so long that drivers equipped their cars with portable toilets (Fairclough 1995). Pedestrians were knocked down on footpaths as motor cycle taxis sought to circumvent the traffic jams. Children, traffic police and road workers were reported with high lead levels in their blood (Tasker 1990). This dilemma led the National Economic and Social Development Board (1991) to highlight urban transport in its plans for the 1990s. An additional US$200 million in financial assistance and technical cooperation was provided by the World Bank and Japan for additional transport studies (HFA 1985, 1991; JICA 1990). Japanese planners concluded that the full programme for 2006 would require the injection of First-World transport elements: 19 expressways and eight mass rail transit lines.

When these recommendations were made an expressway network was already being constructed by the Expressway and Rapid Transit Authority to link proposed sub-city centres within a 50–75-kilometre radius of Bangkok (Figure 9.8a). Between 1990 and 1995 average traffic speeds had increased from 11 kph in 1990 to 16 kph in 1995 (ERTA 1991; Kubota 1996; Fairclough 1996). By 1998, 76 kilometres of the planned 200–300-kilometre network had been completed including the First Expressway, the Second Expressway and the first stages of the Bangna–Chon Buri Expressway and the Bangpa-in-Pakkred Expressway (Table 9.4). Delays in completion of the network were due primarily to problems of land acquisition because of opposition from affected groups and disputes with foreign contractors (Askew 1994; Daniere 1995; Rüland and Ladavalya 1996). Other road construction included the completion of the Middle Ring Road and the rationalization of

Table 9.4 Bangkok: number of trips by expressways, 1985–2000 (millions)

Expressway	First stage	Second stage		Bagna-Chon Buri	Bangpa-in-Pakkred	Total
Official name	Chalerm Mahanakorn	Sri Rat	Chalong Rat	Burapha Withi	Udon Rattaya	
Inception	1981	1993		1998	1998	
1985	45	—	—	—	—	45
1990	104	—	—	—	—	104
1995	125	43	—	—	—	168
1996	130	51	2	—	—	183
1997	135	94	19	—	—	248
1998	129	93	18	1	—	241
1999	117	87	14	6	3	227
2000	121	112	14	8	6	261

Note: Chalong Rat refers to Ram Indra-At Narong section of the Second Stage Expressway.
Source: Alpha Research (2001: 227).

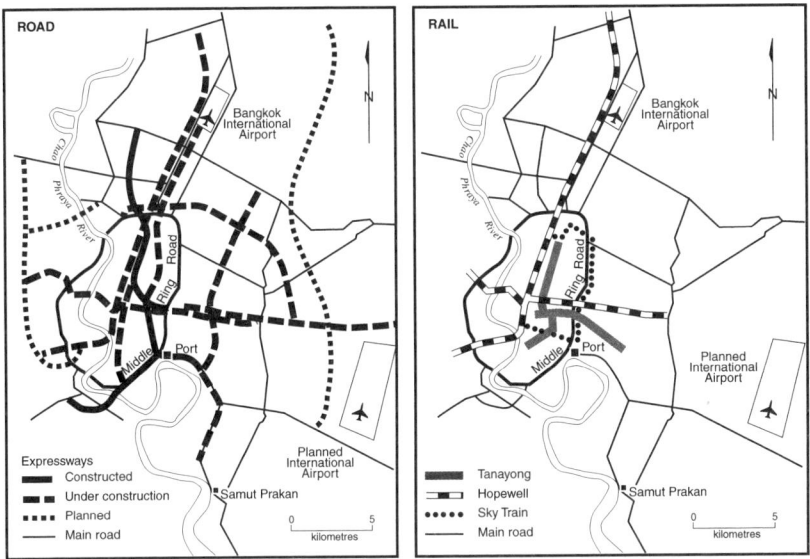

Figure 9.8 Bangkok: urban transport, 1990s (a) expressways (b) mass transit railway
(*Source*: Based on ERTA 1991)

inner-city road networks. An urban freight distribution terminal has been completed on the periphery of Bangkok, so that goods can now be transferred from larger line-haul trucks to and from smaller pick-up and delivery vehicles for city distribution.

Three mass transit railways were also to be constructed by privately sponsored concessionaires on a thirty-year 'build, operate and transfer' basis (Figure 9.8b). Rights were given to three systems promising 100 kilometres of track and capable of carrying three million people per day, about 40 per cent of all commuters (Suwat 1994). They were sponsored by three different government agencies, used three separate technologies, intersected at thirty locations and conflicted with existing expressway systems necessitating cantilevered arrangements (Rimmer 1995). Shifts in project priorities occurred after each of the eight national elections between 1988 and 1996. One project was changed from an elevated railway to a subway; another was scrapped because of the cost of articulating the US$3.2 billion Bangkok Elevated Road and Train System; and a third was opened only to find that patronage levels fell short of expectations because fares were set so high and cheaper buses ran below (Table 9.5). Although there was a plan for reorganizing the buses to act as feeders to the new system, the Bangkok Metropolitan Transport Authority did not respond. Reportedly, local politicians control the Authority's bus routes as part of an elaborate patronage system.

Table 9.5 Bangkok: proposed mass transit systems

Project	Date	Project	Length	Counterpart	Status
1A SNC-Lavalin (Canada)	1989	Skytrain	36 km	Expressway and Rapid Transit Authority	Original project rescinded 1993
1B Metropolitan Rapid Transit (Bangkok Land)	1992	Subway	20 km	Expressway and Rapid Transit Authority	Under construction
2. Hopewell Holdings (Hong Kong)	1990	Bangkok Elevated Road and Train System (BERTS)	60 km	State Railway of Thailand	One-fifth built when scrapped in 1997
3. Bangkok Transit System Corporation (Tanayong)	1992	Skytrain monorail	24 km	Bangkok Metropolitan Transport Authority	Completed in 1999

Source: Various.

These large elevated expressway and mass rapid transit projects proved difficult for the Thai government to manage. As they were not divisible into small fragments to satisfy the myriad of competing political factions, there was much indecision and backtracking on agreed decisions by a succession of governments (Vatikiotis 1995; Rüland and Ladavalya 1996; Unger 1998). Retrofitting these First World elements has also created accessibility problems. During the construction phase the projects have disrupted pedestrian and vehicle movements. On completion these expensive high-order transport systems have displaced the earlier network and denied mobility to those who do not qualify to use them. Social polarisation has been the inevitable result.

Much blame for Bangkok's continuing traffic impasse has been attributed to the business elite (Daniere 1995). The elite has promoted economic growth and job creation but there is no comparable capacity for dealing with its repercussions on the travel needs of ordinary citizens. This is reflected in a reduction of the tax rate on imported vehicles from 150 to 50 per cent in 1991 without compensating improvements in bus services. Bangkok has one of the highest per capita ownerships of Mercedes cars in the world (Mehrtens 1998). The power of the elite has prevented any impingement on middle-class property rights such as restrictions on car use in downtown areas.

Land use, 1995

Reflecting Thailand emergence as one of Asia's Newly Industrializing Countries, by 1995 industrial areas dominated Bangkok Metropolitan Region's land-use pattern (Figure 9.9). Transnational corporations from Japan and, to a lesser extent, Taiwan and other Newly Industrializing Economies had established large factories to manufacture plastics, electronics and electrical goods and petrochemicals on industrial estates in Minburi, Samut Prakan and Chachoengsao (Dilokwanich 1995). In addition, these estates had attracted Sino-Thai factories from inner areas of Bangkok wanting cheaper land and labour to expand their activities. These activities have changed the intended heavy and chemical character of the Eastern Seaboard project (commenced 1981) centred on the ports of Laem Chabang and Rayong.

Bangkok also became a great consumption centre to meet the demand by the new urban middle class for the newest and best (Phongpaichit and Baker 1996). The major downtown commercial area with its new world-class stores has expanded horizontally and vertically. Marked by high-rise buildings, the Silom area emerged as an international financial centre and several sub-financial districts developed along Sukumvit Road, New Petchburi Road, Wireless Road and Ratchadaphisek Road. Apartments, offices, hotels and boutiques invaded the lanes off Sukumvit Road, though many pockets of Sino-Thai shophouses still persisted. The head offices of large businesses and retail functions dispersed away from the downtown area (Kidokoro 1992). Freestanding, high-rise commercial sub-centres, typified by Petkasem Road, Lad Prao and Wipawadi-Rangsithad, were built in the major traffic corridors for the growing number of mobile middle-class consumers; they featured modern shopping malls, entertainment complexes, fast-food outlets and parking. Activities were also 'bundled' into precincts comprising offices, condominiums, malls and sports centres such as Century Park and Tana City. To attract tourists and local visitors, theme parks appeared to the north and east of Bangkok, including Ancient City and King Kong Island.

Suburbanization has led to increased land-use specialization. New areas for government institutions were designated in areas north and west of the city. A range of conspicuous self-contained, outer-suburban private housing estates were designed at Phayathai, Bangkapi, Bangkhen and Bangkok Noi for the middle class and wealthy. These gated communities were characterized by ornamental lakes, winding streets and mock Roman façades that mimicked the giant shopping complexes supplying the consumer goods to the resident families (Askew 1994: 174). The National Housing Authority built public housing estates at Amphoe Pakkret for low-income people but increasingly the task of providing low-cost units, such as terraced shophouses and high-rise housing was undertaken by the private sector. Squatter settlements appeared near industrial estates to house workers from the provinces. The persistence of informal sector trading and the spread of squatter settlements to every part of the city prevented the development of

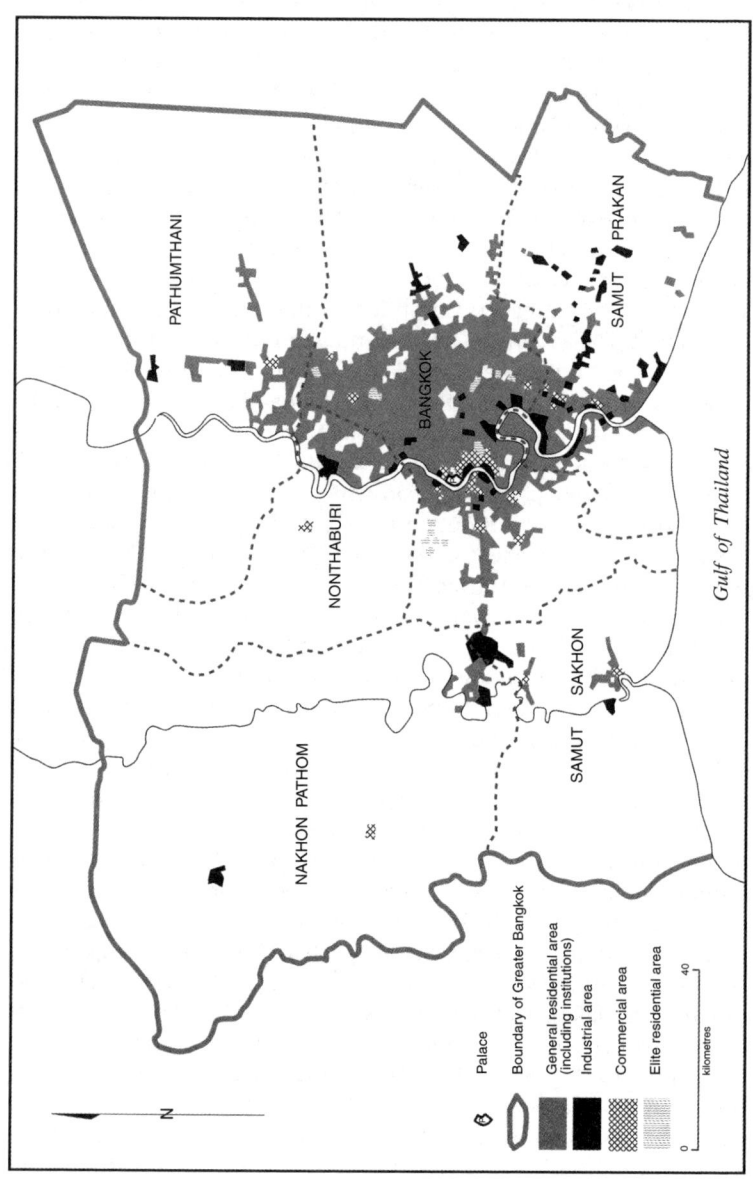

Figure 9.9 Bangkok and vicinity: land use, c.1995 (*Source:* Dilokwanich 1995)

uniform land use areas. One-third of all households still lived in shophouses with the family business on the ground level and higher levels providing living quarters (Van Beek 1999).

In the absence of effective controls during the boom, ribbon development spread along highways to Ayutthaya, Chachoengsao, Nakhon Pathom and Samut Sakhon. Development of the blocks between these highways funnelled traffic to the main roads (HFA 1991; NESDB 1991b). Less accessible and ill-drained sites were left for the urban poor. In response, the Bangkok Metropolitan Administration commissioned a planning vision with development corridors, a range of sub-centres and height controls for the high-rise buildings springing up along one-lane roads (MIT 1995). The plan covered only a small area of the Extended Bangkok Metropolitan Region (EBMR), comprising the Bangkok Metropolitan Region and the six adjacent provinces of Ayutthaya, Chachoengsao, Chonburi, Ratchaburi, Rayong and Saraburi. Within 50 kilometres of Bangkok there was almost total gridlock as the number of motor vehicles had risen to 1.6 million in 1995, an increase of more than 165 per cent within 15 years.

Looking ahead

After enjoying a manic boom, Bangkok bore the brunt of the Asian crisis of 1997–98. Thailand's vulnerability stemmed from the build-up of the private sector's foreign debt and the flight of capital. Many of the new office towers and apartment blocks were left empty and unfinished. Severe unemployment arose from lay-offs in Bangkok's construction, financial and manufacturing sectors (Phongpaichit and Baker 2000). This was cushioned for the first eighteen months by reliance on savings, informal sector employment and job sharing before migrants returned to their rural relatives for support. Expressway traffic declined. Recovery from the crisis reasserted the centralization of power, prestige and media controls in Bangkok because the power exercised by provincial barons over national politics had been weakened by a coalition of business and the new middle class.

By 2000 there was renewed support for the long-term objective of challenging Singapore as an economic pivot for Southeast Asia (Chapter 8). However, without efficient infrastructure and logistics Bangkok cannot compete as a regional production base for multinational capital. Its aspirations be the 'Center of Air Transportation' in the Asia/Pacific Region in competition with Singapore will also be sabotaged by traffic congestion. Thus, the persistence of Bangkok's fragmented land-use and transport system will hamper the transition of both capital and nation into a postindustrial society that can attract or retain associated transnational activities and feature domestically owned high-tech industries (Dixon 1999).

The National Economic and Social Development Board (NESDB) had already included the Extended Bangkok Metropolitan Region (population 9.4 million in 2000) into its plan for the spatial development of Central

Thailand (Figure 9.10). Its proposed Chao Phraya Multipolis will cover 18 provinces within a 200-km radius of Bangkok with the capital and Saraburi envisaged as the northern and southern pivots of the area's inter-urban motorway network. By 2010 the metropolis would have an estimated population of 32 million – 44 per cent of the country's total (Kaothien and Webster 1995; Kaothien 1995). As national planners have no power to enforce their recommendations, an expansion of governance to coincide with the effective urban area is long overdue.

Ho Chi Minh City and Yangon

Belated opening to foreign investment brought some revival to Ho Chi Minh City and Yangon without closing the gap with Bangkok. Since 1986 Vietnam's policy of renovation (*doi moi*) and the encouragement of market relations has allowed Ho Chi Minh City to renew links with the global economy. The city's boundaries were expanded and a core city-region was defined as the Southern Economic Focal Zone (the Ho Chi Min City–Bien Hoa – Vung Tau triangle). In 1994 the United States lifted its trade embargo and the following year Vietnam joined ASEAN. Foreign investment, mainly from East Asia, financed some consolidation of retailing and commerce into new high-rise buildings, while factories were located in special economic processing zones and industrial parks. The 2,600-hectare Saigon South new city project was planned and a tunnel under the river was investigated to link the city to the undeveloped opposite bank. Following the Asian crisis these projects have stalled and Ho Chi Minh City (population 4.6 million in 2000) has remained a low-rise city characterized by short trips and movements of commodities in small packages (UN 2001). Heavy container trucks have nonetheless exposed the inadequacy of narrow roads (10–15 metres wide) and intensified conflicts with local traffic, mainly motor cycles. This has led authorities in Ho Chi Minh City to impose curfews on truck movements, to construct new roads in industrial areas and to develop a transport master plan favouring affordable public transport.

Yangon (population 4.2 million in 2000) also revived with some influx of foreign investment and a flourishing illegal drug trade (UN 2001). Liberalization began in 1988 after the State Law and Order Restoration Council (SLORC) had suppressed opposition. The junta sought funds to revitalize the sluggish state sector but most foreign investment went into high-rise buildings in the city and industrial estates on the fringe. Since then Yangon has transformed its elongated north–south dimension into a diamond-shape with the development of new satellite towns and townships to house relocated squatters (Than and Rajah 1996; Cangi 1997; Nwe 1998). In 1997 Myanmar entered ASEAN and the military government was renamed the State Peace and Development Council (SPDC). This led Singaporean investors to commence building Yangon's large cross-river Thanlyin-Kyauktan industrial zone to take advantage of the new bridge. The scheme was still in

River Cities: Bangkok, Rangoon and Saigon 317

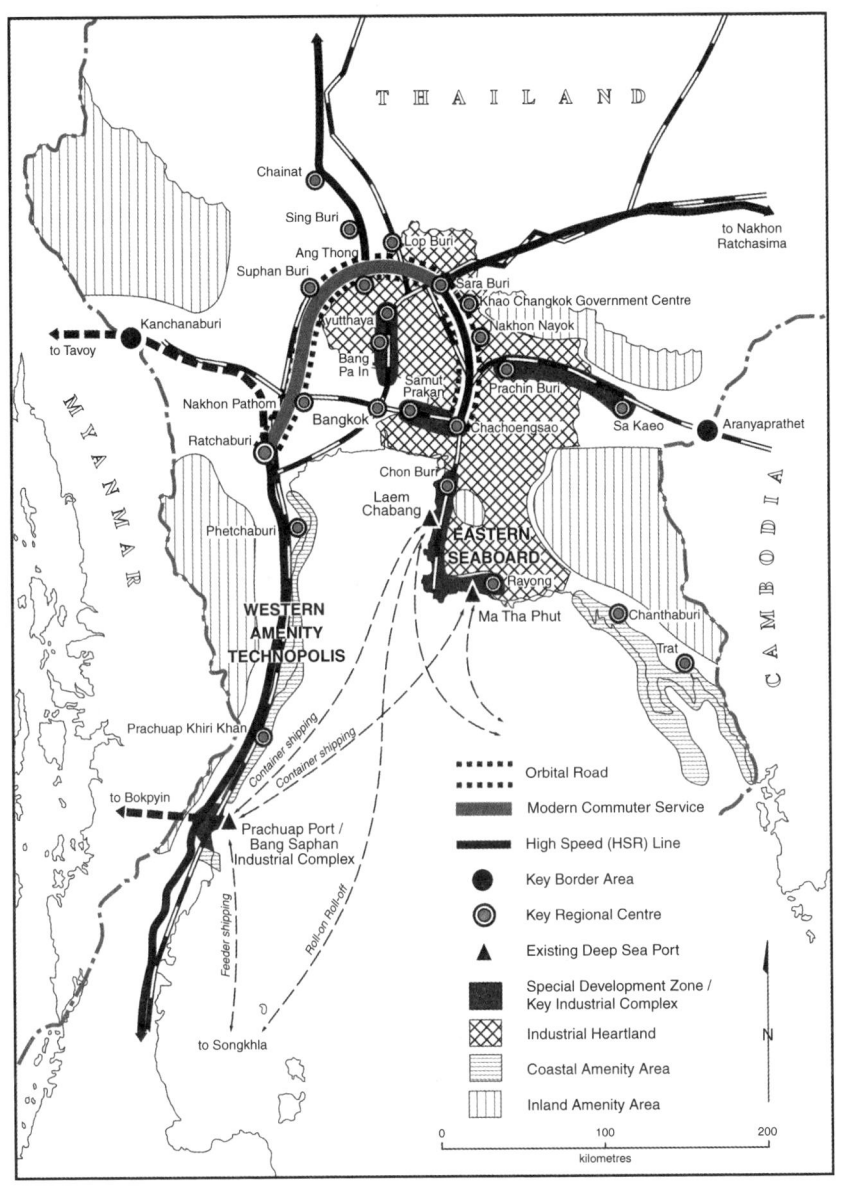

Figure 9.10 Central Thailand: Spatial Development Framework, *c*.2010 (*Source*: Based on NESDB 2000)

the early stages of implementation when it was stalled by the Asian economic crisis in 1997–98. High-rise developments under construction in the inner city were also halted until there was a return of confidence.

Nevertheless, Ho Chi Minh City and Yangon do have some advantages over Bangkok. Most of the causes of Bangkok's congestion either do not occur or are less influential in Ho Cho Minh City and Yangon (UNDP 1997). Land is owned by the state and private sector developers do not enjoy the permissiveness extended to their counterparts in Bangkok. There are fewer problems in obtaining land for large developments compared with Bangkok, where the primacy of private property rights has restricted the scale of urban development.

Conclusion

After 150 years of development, Bangkok has become a huge but dysfunctional city. Without a good system of transport infrastructure like Singapore and Kuala Lumpur, or even a poor system like Jakarta and Manila, Bangkok's uncontrolled land-use development has been overwhelmed by population growth (Chapters 7, 8 and 10). As in Jakarta and Manila, however, social polarization has created a dual or hybrid city-region of rich and poor.

These issues are not easily resolved because Bangkok has not enjoyed the single-minded style of planning offered by the Communist Party in Ho Chi Minh City, the junta in Yangon, the People's Action Party in Singapore or the United Malays' National Organisation in Kuala Lumpur (Chapters 7 and 10). Since the late 1960s the Thai political system has been contested. This has had two consequences: intense competition between political factions and the manipulation of land-use and transport projects to generate benefits for the party in office. The powerful rural lobby has undermined the government's emphasis on large transport projects in Bangkok. Consequently, ribbon development along the major highways has been characteristic of the newly industrializing city.

This spread of the Bangkok Metropolitan Region has created complex cross-movements which have made the transport task more difficult. Because it is massively expensive to retrofit infrastructure such as a mass transit railway, on a marginal basis there are no low-cost solutions for combatting Bangkok's sprawl. The envisaged Chao Phraya Multipolis will merely reflect existing trends and mirror Kuala Lumpur's extensive style of development without being able to achieve any similar level of efficiency.

10
First World City: Kuala Lumpur

Not until the 1960s was Kuala Lumpur a significant international city (Chapter 1). Georgetown (Penang) was the more important service centre on the western side of the Malay Peninsula (Chapter 6). However, after 1896, as the capital of the new Federated Malay States, Kuala Lumpur gradually became the hub of the Peninsula's land transport networks and in population moved ahead of Ipoh (Table 10.1). Despite population growth during the 1920s, 1930s and the Emergency (1948–60), the administrative centre of Kuala Lumpur and its outlet of Port Swettenham (later Port Klang) did not challenge the commercial dominance of Singapore (Chapter 7).

At Independence in 1957 the small colonial town of Kuala Lumpur was upgraded to the status of capital of Malaya and then in 1963 of the Federation of Malaysia. It was seen as the likely 'Washington, DC of the region with Singapore City serving as an important entrepôt similar to New York City' (Hamzah 1965–66: 138). However, the growing rivalry and tension between rival politicians in Kuala Lumpur and Singapore culminated in the latter separating from the Malaysian Federation in 1965 (Tregonning 1966). This separation made Kuala Lumpur a political centre in its own right. With the consequent expansion of both administration and commerce, Kuala Lumpur outstripped Penang in population and acquired some of the adverse characteristics of a Third World city. Meanwhile Kuala Lumpur was forming a much larger conurbation with the satellite town of Petaling Jaya, which was later incorporated into the Klang Valley transport corridor.

Since 1981 the Malaysian government, under the direction of Prime Minister Mahathir Mohamad, has sought to raise Kuala Lumpur to the first rank of Asian cities. Much effort has been directed at building a series of image-enhancing mega-projects, including the Twin Petronas Towers (452 metres) – the world's first and second tallest buildings. Because Kuala Lumpur is still rather small in population compared with other major Southeast Asian cities, there were no extensive pre-existing physical and functional urban forms as in 'Old Singapore' on which to superimpose the elements of the modern First World city. High levels of vehicle ownership

Table 10.1 Ipoh, Georgetown and Kuala Lumpur: population, 1896–1957 (thousands)

Population	1911	1921	1931	1947	1957
Ipoh	24.0	36.9	53.2	80.9	125.8
Georgetown	101.2	123.1	149.4	189.1	234.9
Kuala Lumpur	46.7	80.4	111.4	176.0	316.2
Area (sq. km)	21.0	44.0	44.0	93.0	93.0

Source: Tsou (1967).

and an extensive expressway system have resulted in a sprawling low-density city region. Before examining the urban planning implications of the transformation of Kuala Lumpur into a mini-Los Angeles, the small colonial town and Third World city, conurbation and corridor are revisited (Figure 10.1).

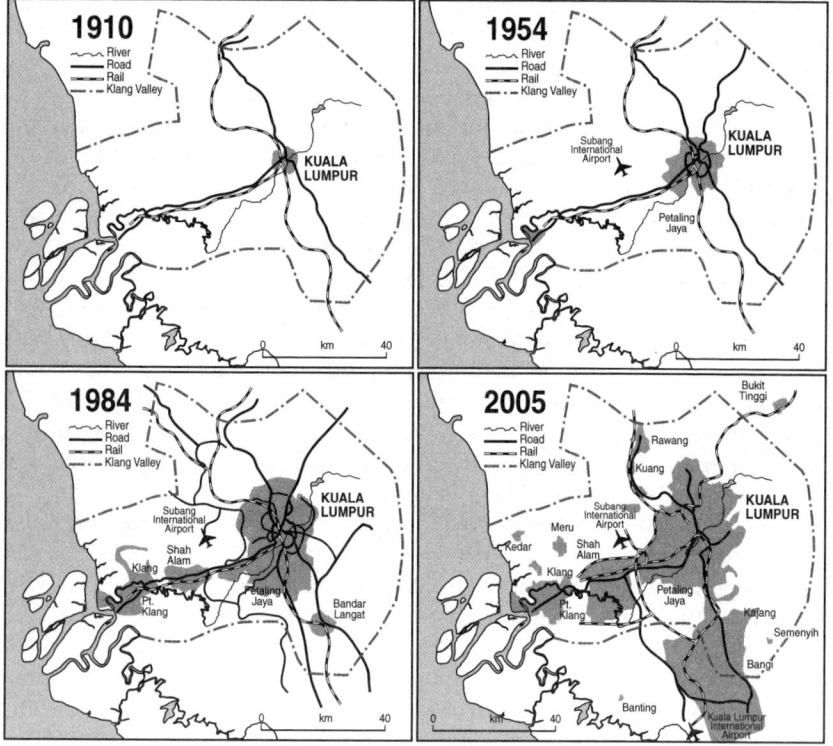

Figure 10.1 Kuala Lumpur city-region: development, 1910–2005 (Source: Various)

First World City: Kuala Lumpur 321

The small colonial town

Kuala Lumpur began as a mining town. In 1857 Chinese miners in Ampang discovered a rich source of tin cassiterite and a 6.4-kilometre overland route was laid to it from the confluence of the Gombak and Klang rivers (Chapter 6). At this convenient point developed the trading settlement of Kuala Lumpur, which in 1859 comprised just three huts. As immigrants arrived, foodstuffs and other mining essentials were imported 50 kilometres by river from Klang and tin was exported on the return leg (Lim 1978). Despite the weakness of indigenous Malay administration and conflicts between Chinese secret societies, the town's development proceeded at a rapid pace (Figure 10.2).

In 1874 further conflicts gave the British reason to intervene, ostensibly to safeguard the interest of its merchants in the Straits Settlements (Gullick 1988). The headquarters of the British administration remained in the royal capital of Klang but Kuala Lumpur's growth resumed once order had been restored. Under the guidance of the Chinese headman Yap Ah Loy, new corduroy-like timber roads were built from the settlement to surrounding

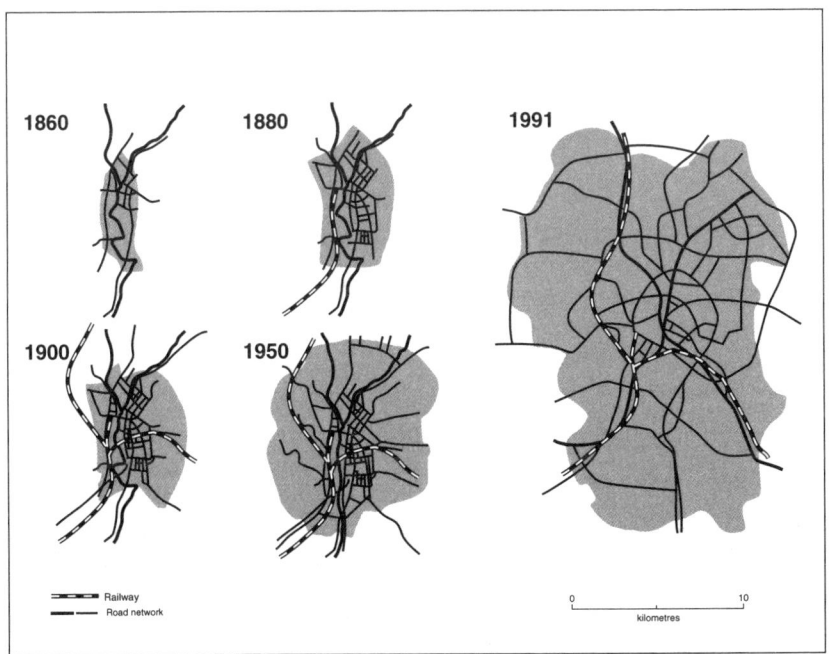

Figure 10.2 Kuala Lumpur: growth, 1860–1991 (*Source*: Based on Chin Kar Keong and Mahfix bin Omar 1994: 20)

mines. Continuing Chinese immigration raised the town's population to 2,600 in 1879. In the following year Kuala Lumpur replaced Klang as the capital of the state of Selangor. The move signalled British supremacy over local Malay rulers and a new balance between the colonial power and immigrant economic activities (Gullick 1988). Kuala Lumpur was still no more than a shanty settlement of adobe dwellings, served by narrow roads fit only for bullocks and horses.

Between 1880 and 1895 Kuala Lumpur was transformed from a raw Chinese mining centre into the 'neatest Chinese and Malay town within the colonial economy' (Gullick 1983). In 1883 a new wooden bridge over the Klang River allowed urban development on both banks. Then in 1886 the Kuala Lumpur–Klang railway was completed and dramatically cut the day-long journey by steam launch, pony and foot to just one hour. Extension of the railways with feeder roads to new mining districts made the new capital the hub for the surrounding tin mining areas and diverted trade along the Kuala Lumpur–Klang axis. Increasing trade led to more permanent buildings using bricks and tiles from local kilns. The street pattern was replanned with a rectilinear grid. In 1890 a lattice girder bridge replaced the wooden one, with two other bridges being constructed upstream and downstream. A Sanitary Board was made responsible for street cleaning and lighting, administration of markets, road construction, maintenance and signposting. By 1895 the town had a small centre with Chinese shophouses and other commercial premises serving a population of 25,000 (Figure 10.3).

After Kuala Lumpur had become the capital of the Federated Malay States in 1896, improved rail and road networks enabled it to extend its administrative and commercial influence over neighbouring states. Plantations including tea, pepper, gambier and tapioca diversified its tin-mining economy – experiments with coffee were unsuccessful. As feeders to the railway network, roads contributed to the commercial success of first rubber plantations and later oil palm estates in the vicinity of the capital. Accessibility was improved by relocation of the port from Klang, some 22 kilometres up river, to deeper water at Port Swettenham, south of the estuary (Lim 1978; Butcher 1979). Draught restrictions nevertheless continued to rule out development of an entrepot in competition with Singapore or Penang. Completion of the trunk rail line between Singapore, Kuala Lumpur and Butterworth/Georgetown facilitated shipment of Selangor's exports through these main deepsea ports.

Economic development intensified residential segregation between different ethnic groups (Lee and Shamsul 1984). The Chinese were located in shophouses within the central area's narrow street pattern. Pedestrians abounded, particularly in Petaling Street, with its food stalls and lines of waiting carriages, gharries and rickshaws – the first motor car was introduced in 1902 (Gullick 1983). The demand for cheap labour by the railways

First World City: Kuala Lumpur 323

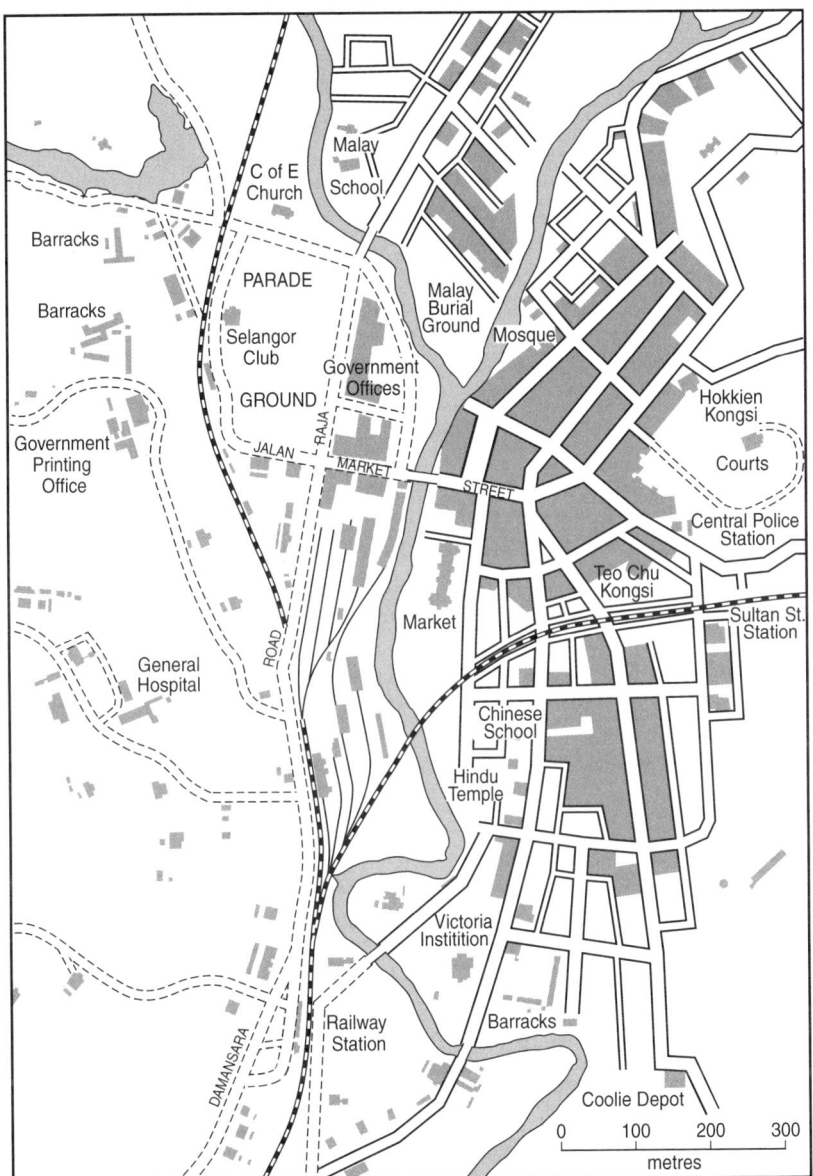

Figure 10.3 Kuala Lumpur, 1895 (*Source*: Based on Gullick 1988: 83 and other sources)

and the Public Works Department for road construction led to an influx of Indian labour that resided near their respective workshops. Malays located in the north at Kampong Bahru, an agricultural settlement established in 1899 on land reserved for their exclusive use by the colonial administration. European officials and businessmen built bungalows with their own gardens in outlying, cleared jungle areas in the undulating hills west of the city (Harrison 1923; Concannon 1955; Gullick 1983). Pony traps were used to travel along the laterite roads to their offices and clubs in town.

By the 1910s Kuala Lumpur was enjoying prosperity from rubber exports (Concannon 1955). Increasing numbers of brick shophouses were constructed to accommodate a population of 47,000 in 1911. However, unlike Singapore (303,000) and Penang (101,000), Kuala Lumpur was still too small to warrant an electric tramway system. Road transport was reliable and motorized vehicles proliferated after the First World War. Although a law was enacted in 1913 to prevent the transfer of Malay land to non-Malays, this had little initial effect on Kuala Lumpur as, with the exception of Kampong Bahru, it only affected rural areas.

Between 1921 and 1936 Kuala Lumpur's population increased from 80,000 to 120,000. Much of this growth occurred during the rubber boom of the 1920s, when roads and housing estates were completed and cars became widespread among the European community. The first government town planner was appointed in 1922 and schemes were introduced for zoning land use, widening streets and realigning the river to prevent flooding, despite which the town was inundated to the depth of one metre in 1926. A town plan was completed in 1933 but the depression halted urban development. Another plan to direct and control building development was gazetted in 1939 and used initially by the Sanitary Board – then the Town Board and later the Municipality – as the legal basis for accepting or refusing building or land-use applications (Concannon 1960). Public transport now relied upon 'mosquito buses' which stemmed from the practice of converting Ford car chassis into small seven-seater omnibuses (Rimmer 1986a: 115). In 1937 the government amalgamated bus owners into eight companies: the British General Transport Company served the inner city and seven Chinese companies linked the city to outlying rural areas.

In the 1930s Kuala Lumpur was still a classic colonial town. 'A small group of foreign administrators and commercial officials lived in large spacious houses in the hilly parts of the city, while a great mass of the population were assembled in ethnically and culturally segregated communities on the lowlands' (McGee 1963: 184). Kuala Lumpur was 'a city of European government buildings, European and Oriental banks and businesses, and Chinese and Indian traders and workers with occasional residual patches of the otherwise submerged Malay world' (Lim 1978: 99). It was also a low-rise town: the Suleiman Building (1930) and some retail stores were the tallest at just four storeys (Concannon 1960). Around the capital there was little agriculture because most land had been taken up for rubber plantations. Compared with the paddy-growing lowlands of Java, Luzon and the lower Central Plain of Thailand, rural densities were low (McGee 1990). Urban development in the lightly populated region of plantation agriculture was therefore highly concentrated on Kuala Lumpur.

Third World city, conurbation and corridor

During the Japanese occupation (1941–45), Kuala Lumpur began to acquire features of the Third World city as an influx of rural migrants led to widespread squatting. The returning British administration could not restore a sedate, well-ordered status quo. By 1947 Kuala Lumpur had a census population of 175,000, comprising Chinese (64 per cent), Indians (18 per cent), Malays (11 per cent) and Europeans, Eurasians and other nationalities (7 per cent). The outbreak of the Emergency in 1947 led to further rapid growth in population with the influx of 100,000 people into the city between 1947 and 1951. Hasty construction of 10,000 houses for 85,000 people sought to relieve the housing crisis. Meanwhile, in 1946 Kuala Lumpur had become capital of the Malayan Union and in 1948 of its successor, the Federation of Malaya (Singapore remained a Crown Colony). Now also a municipality, Kuala Lumpur experienced a building boom as office blocks and hotels between five and ten storeys high were built in the already overcrowded central area originally demarcated in 1895 (Lim 1978; Concannon 1960).

In 1953 the colonial government responded to the problems of squatter housing and poorly located industrial sites by establishing the British-style new town of Petaling Jaya 12 kilometres from Kuala Lumpur. The 1,214-hectare satellite was designed for a population of 70,000 (Concannon 1955; Johnstone 1983). However, the original intention of building houses for squatters and slum dwellers was soon reversed by the demand that the administrators of Petaling Jaya should break even financially. This resulted in the provision of better-quality suburban homes in other sections of Petaling Jaya for an emerging middle class, who commuted to work in Kuala Lumpur (McGee and McTaggart 1967). There was little impact on the rate of illegal squatting by rural migrants in and around the town area.

At Independence in 1957 Kuala Lumpur's population of 316,000 was less than one-third that of Singapore (1.4 million). The Malay population had doubled since 1947 but Kuala Lumpur, like Singapore, was still essentially a Chinese town with small, albeit important, minority groups. A striking preponderance of males reflected the concentration of immigrants (notably Chinese) in the inner areas of the city.

The installation of an indigenous Malay administration quickly changed the character of the city. The nationalist aim was to transform Kuala Lumpur from an outpost of imperialism into a centre of Malay culture (McGee 1963). This involved creating a 'city-within-a-city' on the lines of Vatican City or the Acropolis (Concannon 1960). The task was daunting. A start was made by replacing colonial street names with Malayan ones: High Street became Jalan Bandar, Victoria Avenue became Jalan Sultan Hishamuddin, Circular Road became Jalan Pekeliling, and so on. There followed construction of symbolic national buildings on sites overlooking the city, of which Parliament House, the National Mosque and the National Museum were the most notable.

In 1959 a United Nations physical planning expert was requested to produce a masterplan for the city, but the draft had little immediate impact. An appointed commissioner and advisory board replaced the fractious elected municipal council the following year (Hamzah 1965–66). By 1961 over 50 per cent of the population were living in squatter settlements scattered throughout the city. Those inhabited by Malays had a distinctive rural appearance with sizeable gardens of trees. At the same time, new buildings over ten storeys were being completed, including the Lee Wah Bank (1962) and the National Electricity Board Building (1966). These buildings highlighted the contrast between the modern commercial economy and the bazaar economy, a dualism also seen vividly in the contrast between the living conditions of the elite and those residing in slums and squatter settlements.

In 1963 Kuala Lumpur became capital of the enlarged nation of Malaysia. This stimulated the growth of government and commerce and the continued rural–urban migration of Malays, accelerating the development of squatter settlements on abandoned tin mines or flood-prone areas (McGee 1967). Kuala Lumpur's infrastructure was soon overloaded (Abd Rahim 1988). Public transport services were crammed and unreliable. Traffic congestion, water pollution and poverty made Kuala Lumpur difficult to manage. These characteristics were typical of the classic Third World city, but Kuala Lumpur differed because of the higher levels of car ownership and the fact that fewer people were below the poverty line.

Progressively, Kuala Lumpur's urban concentration gave way to sprawl made possible by an increase in the number of motor vehicles (Table 10.2). The result was a star-like urban form, reflecting the city's increasing dependence on the motor vehicle and long-distance commuting (Goh 1973; Lee Boon Thong 1995). Some tracts of Malay reservation land in Kuala Lumpur, most notably Kampung Bharu close to the city centre, became part of the urban fabric. Many Chinese mining and commercial settlements and 'new villages' built for the Chinese during the Emergency – such as Cheras – were progressively absorbed into Kuala Lumpur. A transport study recommended the development of inner and outer ring roads (TAMS 1963). The overall cost of the plan was too expensive, however, and the necessary foreign funds too difficult to acquire. In 1967 draft town plans and a masterplan were formulated but the May 1969 race riots in Kuala Lumpur overtook

Table 10.2 Greater Kuala Lumpur: Population by Conurbation, Klang Valley Corridor and agglomeration, 1947–2000 (thousands)

	1947	1957	1970	1980	2000
Kuala Lumpur	176	316	452	938	1,367
Petaling Jaya	2	17	101	218	460
Conurbation	*178*	*333*	*553*	*1,156*	*1,827*
Klang/Port Klang	47	76	114	196	503
Shah Alam	—	—	—	19	155
Klang Valley Corridor	—	—	*667*	*1,371*	*2,485*
Ampang	—	—	—	13	193
Batu Sembilan	—	—	—	—	68
Seri Kembangan	—	—	—	16	121
Selayang Baru	—	—	—	—	177
Agglomeration	—	—	—	*1,400*	*3,044*

Source: Lim (1978) and various other sources.

them before key issues involving transport, housing (and squatters) and ethnic residential segregation issues had been resolved (Loo 1973).

Meanwhile, by the early 1960s Kuala Lumpur faced a serious unemployment problem. This stemmed from the economy's shift from an agricultural to an industrial base. The government's import-substituting strategy concentrated urban development within the conurbation. By 1963 an industrial estate in Petaling Jaya had attracted over a hundred factories. These factories were able to take advantage of access by rail and road to Klang and Kuala Lumpur's potential labour force. However, because the formal sector could not accommodate the employment needs of migrants, there was much reliance on informal sector employment, particularly hawking and trishaw pedalling (McGee and Yeung 1977).

The National Economic Policy (NEP) of 1971, which was the government's response to the 1969 race riots, was the second watershed in urban policy. Malay migration to Kuala Lumpur was encouraged in order to dilute what had remained a predominantly Chinese city. In order to consolidate its control over the capital, in 1971 the Federal Government established the Urban Redevelopment Authority, which was not obliged to follow planning by the Town Hall. In 1972 control was transferred to an appointed administrator and advisory board and the following year Kuala Lumpur was designated a city. Although Kuala Lumpur and Petaling Jaya were combined into a conurbation, their land-use and transport developments were planned separately by City Hall and the state of Selangor. This was resolved in 1974 when Kuala Lumpur (including Petaling Jaya) became a Federal Territory of 243 square kilometres. Henceforth the Federal Government took a close interest in the city's development with increasing multi-agency involvement in its land use and transport.

Urban sprawl involved very uneven spatial development. While squatter settlements proliferated, new middle-class residential development centred on vacant areas within its inner and outer ring roads. Ethnic residential segregation persisted: Chinese in the centre, Malays in the north and Indians around the railway workshops and Public Works Department. Congestion was pronounced in the narrow streets of the old central area, aggravated by the construction of traffic-generating shopping and office complexes and the mixture of fast- and slow-moving vehicles.

In 1972 these problems called forth World Bank assistance. Most of the loan was expended on the six-lane Federal Expressway between Kuala Lumpur and Petaling Jaya. The balance was spent on a new transport study that recommended more expenditure on stage buses and on two ring roads – inner and outer – and improved intersections with twelve radial roads (WSA 1973). A second World Bank study in 1976 recommended a minibus system, priorities for high-occupancy vehicles carrying more than four occupants, increased parking charges for the central area and a Singapore-style area pricing scheme (Rimmer 1986a; Abd Rahim 1988). Apart from the minibuses, the public transport and the pricing proposals in the Second Kuala Lumpur Transport Project were never implemented. Separate bus lanes were introduced in 1977 but were abandoned due to strong pressure from private car users, among whom civil servants, planners and politicians were most influential. However, all the road proposals for Kuala Lumpur and Petaling Jaya were completed.

Klang Valley transport corridor

Because Petaling Jaya could no longer accommodate the population spillover from Kuala Lumpur's inner areas, in 1966 the Selangor Development Corporation began to develop a second new town, Shah Alam (Rimmer and Cho 1994; Lee Boon Thong 1995). Being located between Kuala Lumpur and the outport of Klang, Shah Alam expedited the development of a Klang Valley transport corridor. With a parallel railway and upgraded highway, the corridor quickly attracted industrial development. The growing need for industrial and service employees on a casual or full-time basis boosted the corridor's population through rural-to-urban migration on a circular or permanent basis. State and private housing was constructed in both Shah Alam and Klang, where land was much cheaper than in Kuala Lumpur. In 1978 Shah Alam became the new state capital of Selangor, as well as the site of the Institute of Technology, Southeast Asia's largest mosque and the Proton car plant (KNS 1983).

In 1972 the *Klang Valley Regional Planning and Development Study* recommended the development of a polycentric city region to comprise the existing Kuala Lumpur conurbation and the towns of Shah Alam, Klang, Rawang and Bangi/Kajang (SCP 1973/1974; Table 10.2). This proposal recognized the emergence of an urban corridor in the Klang Valley, which was concentrated along the 50-kilometre Kuala Lumpur–Port Klang transport spine. Hitherto much of Kuala Lumpur's expansion had been accommodated in medium-density housing estates built by the private sector on its urban fringe. In the mid-1970s the new towns of Sungei Way and Subang Jaya were laid out between Petaling Jaya and Shah Alam to infill the 'superlinear' city between Kuala Lumpur and Port Klang. As Klang was now within easier commuting distance of Kuala Lumpur and land was cheaper, it too attracted industrial development and a consequent housing boom. However, the rapid development of the Klang Valley also increased erosion and flooding, created health hazards from heavily polluted watercourses and increased congestion on major traffic arteries stemming from the growth in commuting (Aiken and Leigh 1975). Traffic was entirely road-based and three new expressways – Kuala Lumpur–Shah Alam, the New Klang Valley and the South Klang – accommodated its growth. The commuter railway between Kuala Lumpur and Klang underwent a series of closures and revivals during the 1970s and 1980s until electrification in 1994 guaranteed its continuous operation. A mass rapid transit system was mooted to integrate the individual cities of the Klang Valley but never implemented.

After the late 1970s, a new southern corridor was opened up towards the town of Seremban. Sites were made available for export-oriented industrial development based on foreign investment. A key component was establishment of the free-standing town of Bangi and the associated new campus of the National University of Malaysia.

During the 1980s a parallel southern corridor pushed development towards Sepang. This corridor attracted the Agricultural University of Malaysia, the Malayan Banking Training Institute, the National Electricity Board Training Institute and the Palm Oil Research Institute. Expressways also brought Port Dickson and Seremban closer to Kuala Lumpur. Less progress was made with a northern corridor though Rawang, 30 kilometres from Kuala Lumpur, which had been designated as a growth centre. Planned population targets in these low-rise new towns were not met until low-cost housing and inter-city public transport services were provided (JICA 1986).

The development of these urban corridors involved the absorption of rural communities, transforming lifestyles and economies, a process complicated by Malay reservation land not being transferable to non-Malays (Brookfield et al. 1991). Some of these areas have been brought into the urban fabric as land for immigrant worker housing. Others have remained agricultural or vacant, resulting in a sharp juxtaposition with commercial areas and high-rise condominiums on unreserved land. As polycentric development occurred mostly over plantations and waste tin-mining land rather than rice-farming areas, it was easier to attract large-scale economic activities.

Structure Plan and privatization

After a sharp recession in the mid-1980s, the Malaysian economy regained strength and by the 1990s was once again buoyant. The Kuala Lumpur Stock Exchange (KLSE) became one of the Asian-Pacific region's leading bourses in terms of market capitalization. Construction accelerated into a speculative property boom. Business and commercial sectors invested heavily in 'intelligent'/'smart', high-rise office buildings, luxury houses and hotels (Arthukorala 1998). The resultant traffic congestion led to road upgradings, bypasses and flyovers. These improvements brought temporary relief ahead of a strategy that was intended to relocate high traffic-generating activities in residential estates south of the city, notably Bukit Jalil, Bandar Tun Razak, Cheras and Sungai Besi (Rimmer and Cho 1989). Although City Hall desired to reposition Kuala Lumpur among Asia-Pacific cities, cumbersome and inconsistent physical planning policies borrowed from Britain were inadequate for the task (Azman et al. 1996).

The first plan covering the physical infrastructure of the Federal Territory of Kuala Lumpur was developed by City Hall and gazetted in 1984. It prescribed a polycentric city with four sub-centres at Damansara (west), Wangsa Maju (northeast), Bukit Jalil (southwest) and Bandar Tun Razak (south) (Dewan Bandaraya 1982; Wahab 1991). These were to be linked by a comprehensive system of highways and expressways, complemented by a scheme for dispersing traffic in the central area (Figure 10.4). Manufacturing plants were to be relocated in outlying areas, slums were to be eliminated, and squatter settlements were to be replaced with low-cost, multi-storey housing built by the private sector, often on disused mining land (Dillon 1994). The redistributive objectives of the National Economic Policy were to be met by upgrading urban amenities in Malay reservations.

The Structure Plan applied the government's new privatization policy to Kuala Lumpur's road transport and highway networks. The first toll road, Jalan Kuching, came into operation in 1987, followed by the second toll road at Jalan Cheras in 1991. Although the imposition of tolls sparked a minor riot at the Cheras booth, collection was extended to the three ring roads and twelve radial roads leading to neighbouring towns. Metramac Sdn Bhd, then a subsidiary of an United Engineers (Malaysia) Bhd's affiliate, was granted the franchise to levy tolls for 25–35 years in return for completing a schedule of works, including widening roads to six-lane urban expressways, building new roads and constructing interchanges (Cheong 1997). The aim was to accommodate the expected growth in traffic from the city's rapidly expanding fringe. Between 1983 and 1993 the number of motor vehicles registered in Kuala Lumpur increased from 236,000 to 714,000 (Chin and Mahfix 1994).

Privatization was also extended to public transport. Opportunities for private investment included building a light rail transit system, improving

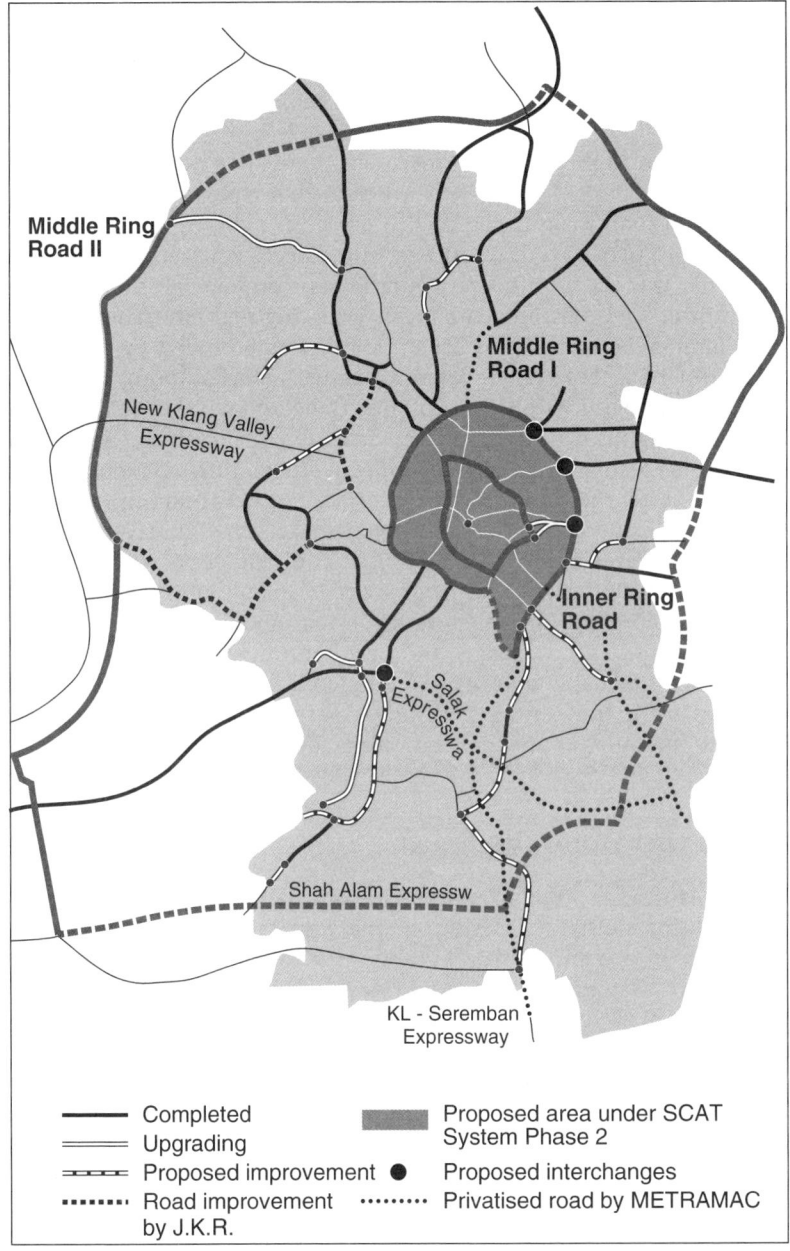

Figure 10.4 Kuala Lumpur: Structure Plan, 1982 (*Source*: Based on Dewan Bandaraya 1982; Rimmer 1986: 223)

bus services with larger vehicles and supplementary minibuses, and building parking stations. Key developments were the elevated, 24-kilometre STAR (Sistem Transit Aliran Ringan) light rail system and the automated 29-kilometre PUTRA (Projek Usahasama Transit Ringan Automatik) light rail system (Figure 10.5). However, given the massive investment in new road projects and the high proportion of single-occupancy vehicles on major arterial roads, the desired 40 per cent switch from private to public transport was little more than wishful thinking.

The structure plan gave most attention to the 18-square-kilometre central area that was the administrative, business, commercial, financial and tourist centre. Two distinct sub-areas were highlighted. The heritage central commercial area – two square kilometres bounded by Jalan Tun Perak, Jalan Cheng Lok, Lower Jalan Ampang. Jalan Kampung Attap and Jalan Sulaiman – with its links to colonial trade and commerce and shophouses was designated for redevelopment and gentrification. The Central Market became a shopping mall, the railway offices a first-class hotel, and the Majestic Hotel the National Art Gallery. Major new buildings included the Dayabumi Complex (offices and retail space) and Putra World Trade Centre. The newer 'Golden Triangle' business district – bounded by Jalan Ampang (intersection with Jalan Tun Razak), Jalan Sultan Ismail, Jalan Raja Chulan and Jalan Bukit Bintang – was expected to maintain its up-market, high-rise retail/office complexes and condominiums and to expand at the expense of affordable housing and public space. Some deconcentration of activities from the central planning area peripheral sub-centres was anticipated, though no mechanisms were identified for achieving this objective.

Towards a First World city-region

The old structure plan, though appropriate for a national capital and growth centre, did not satisfy the vision of the Prime Minister, Dr Mahathir Mohamad, of establishing Kuala Lumpur as a global city-region by 2020, when Malaysia was expected to attain developed-country status. In 1991 planning powers were removed from City Hall to a special panel chaired by the Prime Minister. Land-use regulations and development control standards could now be overridden if deemed necessary to reorganize the city to meet the demands of the global economy. The prime minister thus became the master urban planner for the whole city-region.

In providing the physical paraphernalia of a global city-region, the Federal Government relied heavily on public–private partnerships. These involved Malay (*bumiputra*) entrepreneurs with strong connections to the United Malays National Organisation (UMNO), such as Halim Saad of the strategic investment group Renong Bhd (Cheong 1997). Renong owned more than one-third of United Engineers (Malaysia) Bhd, which built much of Kuala

First World City: Kuala Lumpur 333

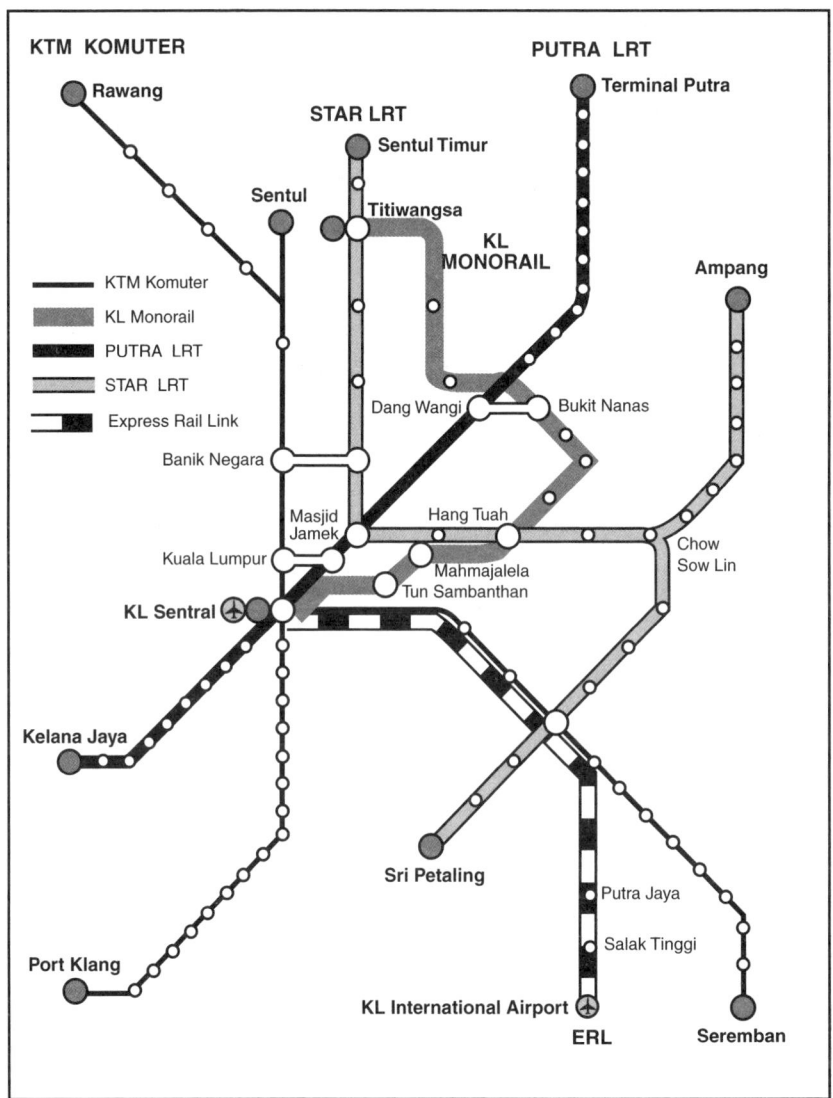

Figure 10.5 Kuala Lumpur: railways, 2000 (*Source*: Based on Kiat.net 2002)

Lumpur's new infrastructure, including toll roads, the National Sports Complex and the Games Village for the XVI Commonwealth Games, light rail transit and new airport buildings.

The public sector's role in these partnerships was to remove planning controls previously exercised by City Hall, so that 'high-tech' mega-projects could be 'fast-tracked' to improve the city-region's image (Morshidi and Ghazali 1999). These included the 421-metre Kuala Lumpur Tower and the projected 1.9-kilometre Kuala Lumpur Linear City and the 'cyber-township' of Bandar Sri Permaisuri (Corey 2000). Underwriting the high-tech mega-projects was the task of private property and real estate groups, such as the Renong Group, Berjaya, Lion and Seri Kuda. Until the Asian crisis, these were all eager to enhance their corporate image with strategic, national-international high-rise citadels (Arthukorala 1998; Gomez 1999, 2000). These groups also had strong political connections and access to finance and land banks within or close to the mega-projects. The mega-projects were also conducive to expanding the information technology sector to attract global control functions and producer services catering to an international clientele.

The southern corridor

The entrepreneurial Federal Government sought initially to reinforce the multi-corridor structure of the Kuala Lumpur city-region through a series of isolated infrastructural developments. These included the new West Port facilities at Port Klang, the location of the second Proton car plant at Serendah, the new town of Bandar Bukit Beruntung in the north and the new international airport at Sepang in the south. Subsequently, the Federal Government repackaged its proposals in a more integrated plan focusing on the Kuala Lumpur southern corridor towards Seremban and Port Dickson. The logic of the new plan was to reposition the Kuala Lumpur city-region into the emerging global web of cities by giving it a new image and identity. Its ambitious objective was to market the Kuala Lumpur city-region as a competitive alternative to Singapore for international service activities requiring 'smart' offices with state-of-the-art communications.

The key to the design of the southern corridor was three costly mega-projects: the Kuala Lumpur City Center (KLCC), the Kuala Lumpur International Airport (KLIA) and the Multimedia Super Corridor (MSC) (Figure 10.6). Collectively, the three mega-projects required a massive investment in new infrastructure by the government and private sector. Large projects included new expressways/tollways, street widening and underground and elevated roads and the 70-kilometre Express Rail Link between the KLIA (Sepang) and Kuala Lumpur Sentral Station (Brickfields). Funding relied heavily on an infusion of borrowed money through the private real estate development industry.

The Kuala Lumpur City Center project was largely a joint enterprise between the state oil company, Petronas, and Mai Holdings and its affiliates

First World City: Kuala Lumpur   335

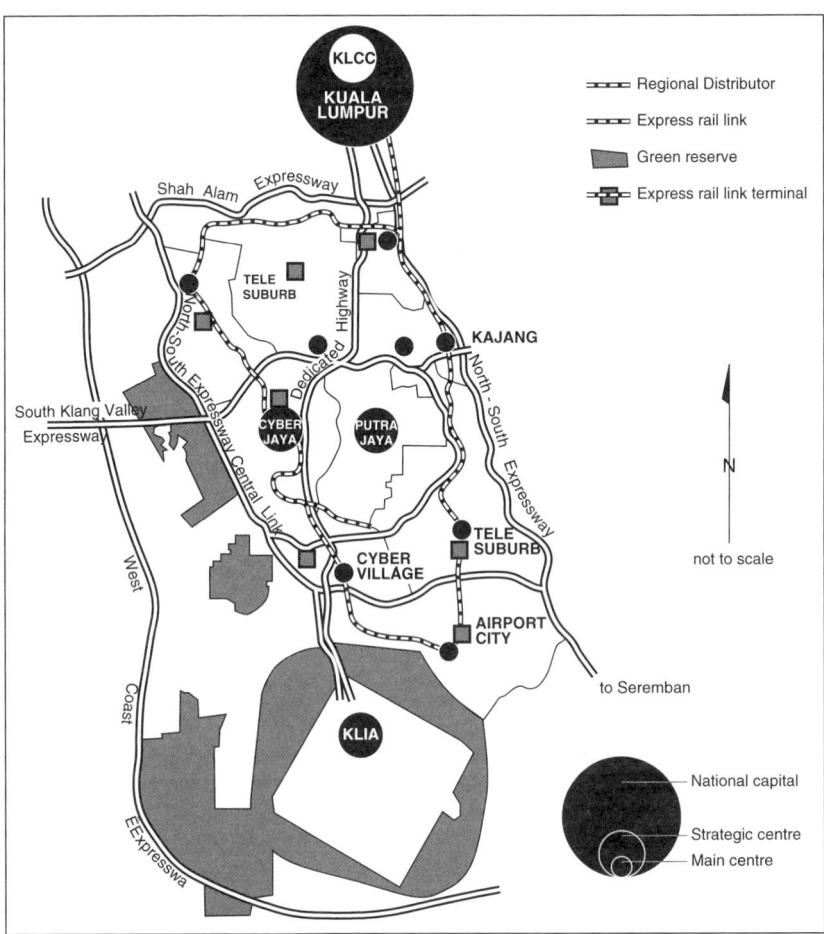

Figure 10.6 Kuala Lumpur City-region: mega-projects, late 1990s (*Source*: Based on FDTCP 1997 and other sources)

(Gomez and Jomo 1999; Morshidi and Ghazali 1999). Ananda Krishnan, a close associate of the prime minister, owned the latter. He was also controller of Biniriang, Malaysia's satellite television enterprise through his holding company, Seri Kuda. Originally, the 40-hectare site, located on the edge of the Golden Triangle, was a racecourse owned by the Selangor Turf Club. The three-phase project was designed by the Californian firm Klages, Carter, Vail & Partners. This city-within-a-city was to be completed within 15 to 20 years of the commencement date in 1993. The first phase designed for business and commercial activities included the 88-storey Petronas Twin Towers designed by the American-trained Argentinian, Cesar Pelli, which are linked by a sky bridge on the 41st and 42nd floors. Besides the 30-storey Esso Tower, the Mandarin Oriental Hotel and the Kuala Lumpur Hotel, the project also included a landscaped park intended to be evocative of London's Hyde Park and New York's Central Park. Here also comparisons were intended with London's Canary Wharf and New York's Battery Park City. The aim was to show the world that Kuala Lumpur was at last a First World city.

The Kuala Lumpur International Airport and 'Airport City' (incorporating Nilai and Salak Tinggi new towns) were designed to anchor the southern growth corridor from Kuala Lumpur to Port Dickson. In 1998 the US$2.5 billion airport was opened on a 100-square-kilometre site at Sepang with an annual capacity for 25 million passengers. It was designed to replace the existing Abdul Aziz International Airport at Subang and to reinforce Kuala Lumpur's claim for regional hub status (Hussein 1997). A dedicated 60-kilometre highway was built for air passengers between the airport and city centre with an employee bus service using the North–South Expressway linking the old and new airport sites (Sherwood et al. 1998). An express rail link has been completed and is expected to attract one-third of the passenger traffic from the highway. Substantial spin-offs in development projects adjacent to the new road and rail infrastructure have been anticipated with the mushrooming of new townships and lavish urban resorts. For example, the Mines Resort City has a grand canal running through the centre of its shopping complex that, in turn, incorporates Snow Town complete with an ice-skating rink, Malaysia's biggest indoor theme park.

The Multimedia Super Corridor is 15 kilometres wide and 50 kilometres long (750 square kilometres) (Figure 10.6). Larger than the island of Singapore, the area incorporates the existing road and rail route between the city centre and airport. Under the direction of the Multimedia Development Corporation (MDC), the high-tech information corridor is designed for tele-services, remote data services, electronic banking and state-of-the-art products such as wafer fabrication, high-tech packaging, component manufacturing, electronic publishing and cine-animation services. Seven planning precincts have been designated: Airport City, Cyberjaya, Cyber Village, Tele-Suburb, Research and Development Center, High-Tech Park and Putrajaya. Cyberjaya is projected as a leading multimedia centre, which is to provide a vision of a new commercial community on a 7,000-hectare site (FDTCP 1997). Putrajaya is the new federal administrative centre covering 4,000 hectares. It is designed as 'an intelligent city' of 250,000 people offering electronic government, high-quality housing and leading-edge multimedia estates. A feature of the Putrajaya's detailed garden city concept is the promotion of public transport using light rail transit and park-and-ride facilities to connect with commuter rail services to the city centre (Hussein 1997; WSA 1998).

These major land-use decisions are intended to reshape the city-region on the Silicon Valley model. The resultant new types of global economic and social activities have brought two changes to the Kuala Lumpur city-region's existing urban structure. Office development housing, producer services and the international service sector are being dispersed from Kuala Lumpur's old, congested core to prestigious high-rise buildings in the Kuala Lumpur City Center adjacent to the outer ring road and the Golden Triangle. A semi-urban corridor is emerging between Kuala Lumpur and the Kuala Lumpur International Airport to accommodate horizontal development within the Multimedia Super Corridor. Several large developers are holding land banks adjacent to this area. The Corridor's effect on land use is already spilling over into the neighbouring state of Negri Sembilan with Seremban seeking to become a megahub in its own right (Lim 1997).

The transformation of Kuala Lumpur into a First World city-region has not gone smoothly. After the property boom of the 1990s, the Asian crisis exposed the risks of relying on a pegged exchange rate and short-term capital inflows (Arthukorala 1998). Speculative capital was quickly withdrawn as the share market and the local currency plunged. The resultant downturn in the urban economy badly affected the real estate and property sector and resulted in a glut of available office space. In addition, construction of several mega-projects was rescheduled. These included the Berjaya Star City retail complex, the Kuala Lumpur Gateway, Mid Valley City and the Kuala Lumpur Sentral station complex, a transport hub bringing together commuter, intercity and airport express networks.

The big three projects underpinning the southern corridor survived the crisis. Release of new land parcels of the Kuala Lumpur City Center project was temporarily withheld. State funds were pumped into the new airport. The Multimedia Super Corridor nominally continued on schedule. When the STAR and PUTRA light rail systems experienced severe financial problems in 2001, they were taken over by the government and leased back to the companies. These actions have validated the arguments of critics of the Federal Government's 'development-at-any-cost' mentality. Many Malay reservations, densely settled squatter areas and old-style retailing complexes have been poorly integrated into the new infrastructure. Air and water pollution stemming from the new infrastructure and associated industrial development and traffic has intensified in Kuala Lumpur (Lee and Shamsul 1995). Critics have argued that Third World lifestyles and First World infrastructure do not mix. Rather than the government squandering money on sponsoring a handful of mega-projects, its funds would have been better spent on meeting the basic housing and transport needs of all of its citizens.

Conclusion

Kuala Lumpur is different from other capital cities in Southeast Asia. Its development as a city system and its dominance within the Malaysian system of cities diverged from the experience of Jakarta, Manila and

Bangkok (Chapters 8 and 9). Although Kuala Lumpur's planners had pretensions about transforming the city into a second Singapore, they were unable to replicate the island republic's highly regulated approach to transport and land-use activities (Chapter 7). As a result Kuala Lumpur has become a special case: a planned, road-based and low-density city resembling a mini-Los Angeles.

Kuala Lumpur's variance from other Southeast Asian cities stems from differences in physical size, timing of development, the nature of nation–hinterland relationships and the role of government. Before 1957 Kuala Lumpur was very much a city in search of a nation. It was not even the first city in its own hinterland because it ranked after Singapore and Georgetown (Penang). Kuala Lumpur's rise to prominence therefore has grown with Malaysia and over the past 40 years the city has come to reflect the nation. Apart from the special case of Singapore, West Malaysia has always been the most prosperous part of Southeast Asia. Underemployment and poverty, the bases of Third World urban misery, have always been relatively low. In fact, Malaysia as a whole has been a destination for migrants ever since independence. Even without public intervention, poverty would have been much less extreme in Kuala Lumpur than in Bangkok, Jakarta and Manila.

Government intervention diverted Kuala Lumpur from a Third World trajectory. The Malaysian government had the necessary capital and a strong motivation to invest in Kuala Lumpur's infrastructure. Unlike Bangkok, Jakarta and Manila, its infrastructure has kept pace with development. The city was small enough and the government sufficiently well coordinated to provide a good town-planning framework. Planners were able to establish an extensive set of new towns as land was cheap because former tin land was virtually useless and plantation agriculture under old trees produced low returns. Heavy spending on infrastructure accommodated the accompanying massive reliance on the motor vehicle.

After 1957 the predominance of the Chinese in Kuala Lumpur has diminished. Increasingly, Kuala Lumpur has become a city of the Malay middle class. Malays have moved from *kampung* into suburban housing and apartments. The new town model (Petaling Jaya and beyond) has met this demand.

Casting Kuala Lumpur as an Asian Los Angeles therefore makes sense in terms of income per capita, availability of land, public infrastructure and vehicle technology. It raises the interesting counterfactual position of what might have happened if a new Bangkok, Jakarta or Manila had been constructed post-1970s on a greenfield site.

Afterword

How does our mapping and analysis of the emergence of Southeast Asia's international network of cities, sub-regional hinterlands and capital cities help us to understand modern Southeast Asia? The simplest answer is that it offers a powerful way of viewing the real world from outside the intellectual straightjacket of the nation-state. The contested discourse of globalization points to the need, while the reciprocal duality of global–local suggests the way, but the solution has been evasive. In Europe the supra-national organization of the European Union has become a practical way to transcend the nation-state without destroying the framework of national government. As a region, Southeast Asia is more amorphous and the loose ASEAN10 plus East Timor grouping only weakly binds its member nations. Yet for centuries the region has had kingdoms and states, dynamic cities and religious, commercial and political networks. Only nations are new.

Here is the key. Nations were forcefully inseminated into Southeast Asia by the violent intrusion of colonialism and its often-violent rejection. Anderson (1983) has greatly influenced the literature by his idea of the nation as 'imagined community'. We have shown that the nations of Southeast Asia were also actual, constructed communities. Each took on a shape and structure moulded by the technologies and organization of modern transport and communications in the service of the state and associated private enterprise. Colonialism left behind a physical and human heritage, as well as effective modern states, engines that would be harnessed to the new nations.

The colonial heritage also involved specific patterns of interaction between urban enclaves and the wider global economy, and between those cities and their hinterlands. Since colonial infrastructure and its associated settlement pattern was tied to highly specific modes of primary production for export, it hardly could be expected to survive the transition of independent nations to an industrial mode of export production. As nation-states have embraced industrialization, the links between ports and extensive hinterlands have declined in relative importance. Road networks may better articulate hinterlands, but they have also become more 'local'. The global links are now highly concentrated upon the metropolitan regions around the main (and mostly capital) cities.

As a subsequent study will elaborate, the capital cities of Southeast Asia can be *re-imagined* as contiguous urban space, with Singapore at the centre. From there a businessperson can notionally commute within a day to and from any of the main cities of Southeast Asia (Figure A.1). Flight time varies from 45 minutes (Kuala Lumpur) to two hours (Bangkok and Surabaya). Allowing around one hour each way for transfer to and from the airport and

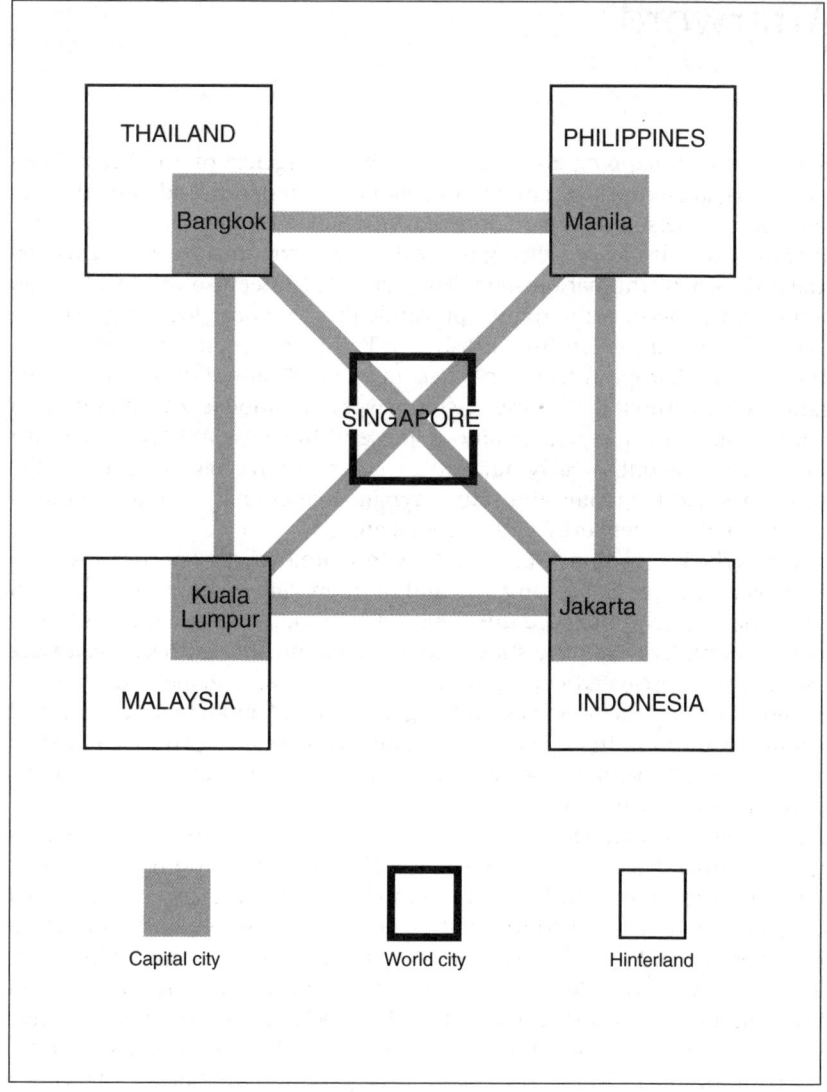

Figure A.1 Southeast Asia: contiguous urban space centred on Singapore (*Source*: Figure III.3)

up to an hour for check-in, door-to-door travel time is three to five hours in each direction. Transfer and waiting time are no longer wasted because mobile phones, faxes and the Internet allow business people to be productive, even if they are not involved in face-to-face meetings. The airport lounge becomes an extension of the office.

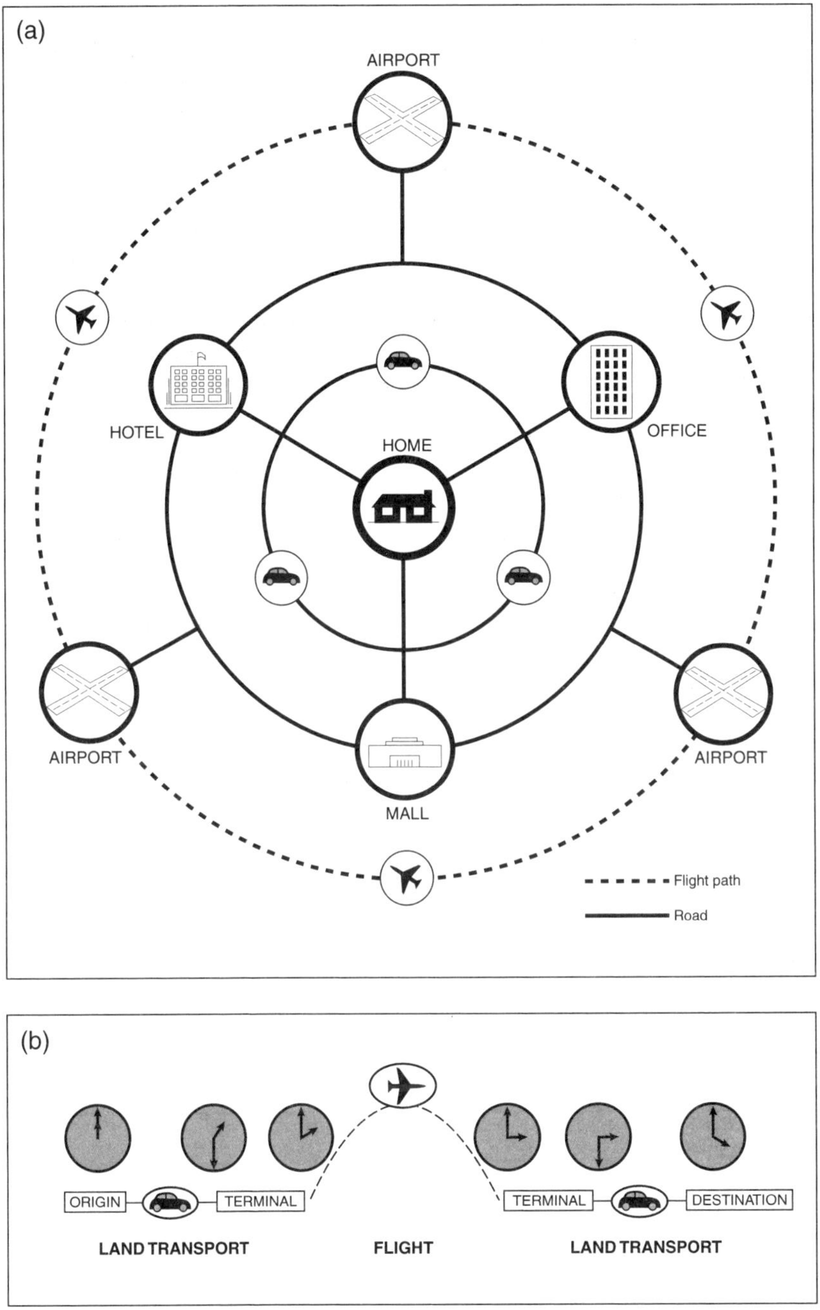

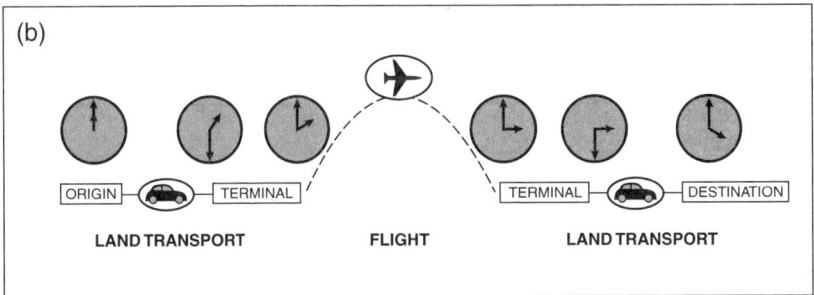

Figure A.2 (a) Physical links between home, hotel, mall, office and airport; (b) time lapse diagram showing time spent between origin and destination including travel to airport terminal and flight time

This situation for a 'typical' Southeast Asian capital is portrayed schematically in Figure A.2. Transfer time from home/office to the airport (notionally 60 minutes) may be taken as coinciding with the outer urban fringe. Between cities lies a time–space lapse of in-flight time plus waiting time at both ends. These two or three hours in transit are spent in the tube-like passageways of airport lounges and aircraft. The territory overflown is of no consequence to the traveller. The urban space of each city has common features that facilitate ease of movement. Except for the golf course, virtually all the spaces within which business people move will be air-conditioned spaces, whether home, car, office, hotel, shopping mall or the airport itself. This upper layer of the transnational Southeast Asian metropolis has become a single, seamless comfort zone (Dick and Rimmer 1999). For the rich, tropical Asia has become as controllable as winters in Europe or North America.

Members of the middle-class urban elite seldom travel far into the hinterland beyond the city. Assuming an average speed of 40 kph, in a few hours it is possible to travel about 100 kilometres into the hinterland by road or rail, somewhat more in directions served by toll roads. In reality, few need to travel so far. Notwithstanding the proliferation of four-wheeled vehicles in Southeast Asian cities, most owners are perpetual urban dwellers, who use access to high-speed air transport and telecommunications to extend their urban space, not into the hinterland but into adjacent cities and countries.

Telecommunications have brought the business cores of capital cities in Southeast Asia into instantaneous contact with world cities. The spatial outcome is that these essentially similar city-cores are stacked pancake-like on top of each other (Figure A.3). While travelling by air between them involves a time-lapse, this is not the case in telecommunications and, in effect, the cores of Singapore, Jakarta, Manila, Bangkok, Kuala Lumpur, Hong Kong, London, Tokyo and New York are virtually one.

Although the forces of globalization – trade, investment and telecommunications – impinge on everyone in Southeast Asia, only the elite have access to the full range of freedoms. Ordinary people can in theory gain access to intercity and international telecommunications, but the cost of even the cheapest international air travel puts it beyond their reach. The exceptions are those who enjoy a hard currency income, such as seafarers, air crew and international guest workers, but even for these people the freedom of mobility is tightly constrained. The poor mostly have to make do in an uncontrolled environment, now hotter, more crowded and more polluted than ever.

In countries that have made the transition to democracy, this spatial pattern of income and wealth gives rise to considerable tensions. 'One person, one vote' means that the poor have the same voting rights as the rich. It also means that the poor living in the hinterland have some veto power over the

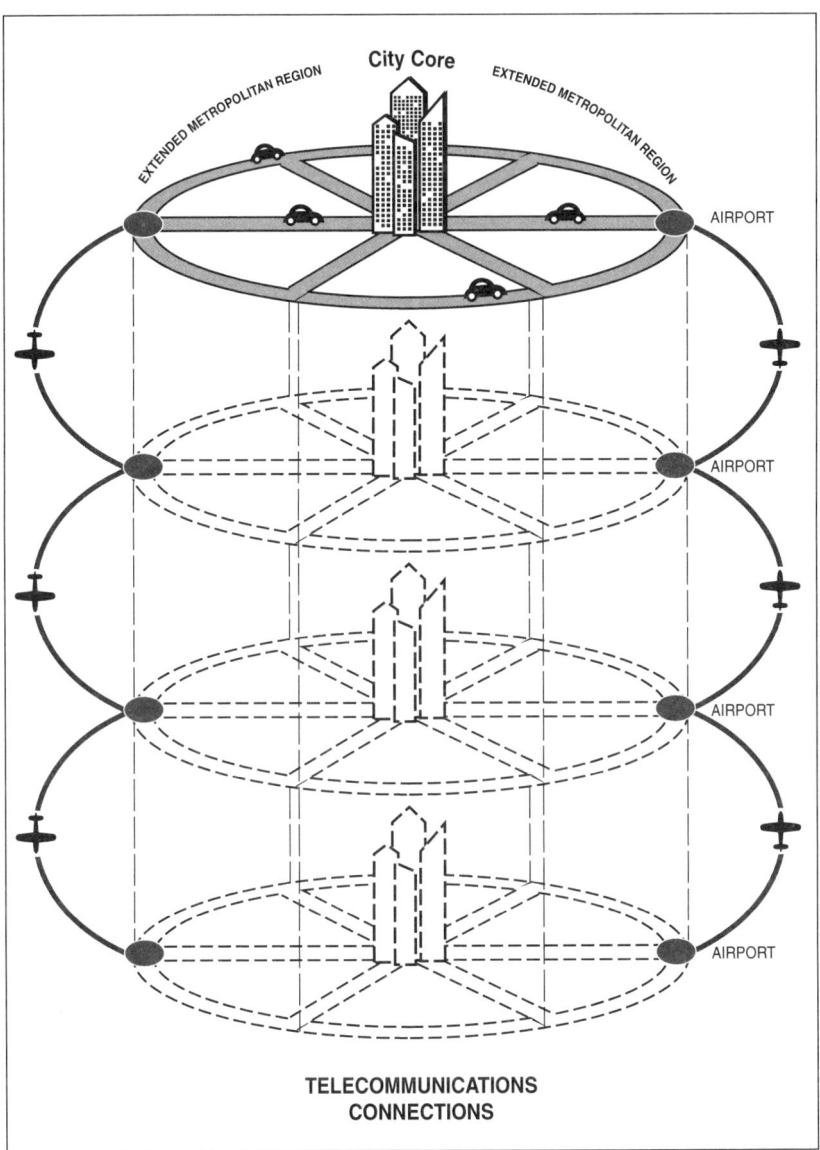

Figure A.3 Instantaneous telecommunications between city-cores has produced a 'pancake-like' urban structure

smaller total population living in the capital city. The 'national' elite has therefore to 'buy off' rural interests. Hegemony must be negotiated. Political hierarchies must be constructed, or the nation will disintegrate.

The uncomfortable truth is, therefore, that nation-states are ideological hegemonies controlled rather clumsily by those who most benefit from them. Urban middle-class elites, who congregate in capital cities, use their economic and political strength to sustain comfortable lifestyles of conspicuous consumption and, as ever, to defend them against the urban and rural poor. Without the leverage of the state and its claims upon the nation, these elites would be dissipated. Since the notion that the poor should know their place and support the rich has gone out of fashion, the ideology of nationalism is rather useful. Like any other ideology, it is most effective when opaque. It is a powerful but impermanent glue. What matters more in the long run is the articulation of the state and its associated economy, whether these reinforce or cut across the ideological hegemonies.

Bibliography

ABC (1995), *OAG World Airways Guide*, Bedfordshire: Reed Travel Group, ABC International Division.
Abd Rahim, M.N. (1988), 'Public transport planning in Malaysia', *Occasional Paper No. 14*, Keele: University of Keele, Department of Geography.
Abeyasekere, S. (1987), *Jakarta: a History*, Singapore: Oxford University Press.
Adam, Ahmat B. (1995), *The Vernacular Press and the Emergence of Modern Indonesian Consciousness, 1855–1913*, Ithaca: Cornell University, Southeast Asia Program.
Adas, M. (1974), *The Burma Delta: Economic Development and Social Change on an Asian Rice Frontier, 1852–1941*, Madison: University of Wisconsin Press.
Adas, M. (1989), *Machines as the Measures of Men: Science, Technology, and Ideologies of Western Dominance*, Ithaca: Cornell University Press.
ADB (1972), *Southeast Asian Regional Transport Survey*, prepared by Arthur D. Little Inc. and Associated Consultants, 4 vols, Singapore: Asian Development Bank.
Ahvenainen, J. (1981), *The Far Eastern Telegraphs*, Helsinki: Suomalainen Tiedeakatemia.
Ahvenainen, J. (1986), 'Telegraphs, trade and policy: the role of the international telegraphs in the years, 1870–1914', in W. Fischer, R.M. McInnis and J. Schneider (eds), *Papers of the IX International Congress of Economic History, Part II: 1850–1914*, Wiesbaden: Franz Steiner Verlag.
Aiken, S.R. and Leigh, C.H. (1975), 'Malaysia's emerging conurbation', *Annals of the Association of American Geographers*, 65 (4), 546–63.
Aldrich, R.J. (2000), *Intelligence and the War against Japan: Britain, America and the Politics of Secret Service*, Cambridge: Cambridge University Press.
Allen, C. and Mason, M. (eds) (1984), *Tales from the South China Seas: Images of the British in South-East Asia in the Twentieth Century*, London: Futura (first published 1983).
Allen, D.F. (1951), *Report on the Major Ports of Malaya submitted on 29th May 1950*, Kuala Lumpur, Singapore: Government Printing Office.
Allen, D.F. (1953), *Report on the Minor Ports of Malaya*, Singapore: Government Printer.
Allen, J. and Hamnett, C. (eds) (1995), *A Shrinking World?: Global Unevenness and Inequality*, Oxford: Oxford University Press.
d'Almeida, W.F. (1864), *Life in Java*, vol. 1, London: Hurst and Blackett.
Alpha Research (2001), *Thailand in Figures, 7th edn 2001–2002*, Bangkok: Alpha Research.
Aluit, A.J. (1995), *By Sword and Fire: the Destruction of Manila in World War II, 3 February –3 March 1945*, Manila: Bookmark.
Anderson, B.R.O'G. (1983), *Imagined Communities: Reflections on the Origins and Spread of Nationalism*, London: Verso.
Anderson, J.L. (1997), 'Piracy in the Eastern Seas, 1750–1850: some economic implications', in D.J. Starkey, E.S. van Eyck van Heslinga and J.A. de Moor (eds), *Pirates and Privateers: New Perspectives on the War on Trade in the Eighteenth and Nineteenth Centuries*, Exeter: University of Exeter Press.
Andrews, J.M. (1935), *Siam: 2nd Rural Economic Survey, 1934–1935*, Bangkok: W.H. Mundie, The Bangkok Times Press.
Anon (1987), *Where Monsoons Meet: a People's History of Malaya*, Kuala Lumpur: INSAN (Institute of Social Analysis).

Anon (1996a), 'Transportation infrastructure in the capital being improved as step to make Indonesia a 'lifestyle superpower'', *The Wheel Extended*, No. 96 (September), 19–21.

Anon (1996b), 'Asian 'chaos' has potential for the future; the key of effective use of public transport', *The Wheel Extended*, No. 96 (September), 26–8.

Arthukorala, Prem-chandra (1998), 'Malaysia', in R.H. McLeod and R. Garnaut (eds), *East Asia in Crisis: From Being a Miracle to Needing One?*, London: Routledge.

Askew, M. (1994), *Interpreting Bangkok: the Urban Question in Thai Studies*, Bangkok: Chulalongkorn Press.

Atkinson, F.W. (1905), *The Philippine Islands*, Boston: Ginn.

Atlas of the Philippine Islands (Atlas de Filipinas) (1900), Manila Observatory/United States Coast and Geodetic Survey, Washington, DC: Government Printing Office.

Azman Haji Awang, Haldane, J.F., Mahbob Salim and Abdullah Abdul Rahman (1996), 'Urban planning', in K.S. Jomo and Ng Suew Kiat (eds), *Malaysia's Economic Development: Policy and Reform*, Petaling Jaya: Pelanduk Publications.

Bach, J. (1976), *A Maritime History of Australia*, Sydney: Pan Books.

Bachrach Motor Company, Inc. (1930), *Consolidated Road Maps of the Philippines*, Manila: Percy Warner Tinan.

Bagwell, P.S. (1988), *The Transport Revolution*, London: Routledge.

Baker, W.J. (1970), *A History of the Marconi Company*, London: Methuen.

Bangkok Post, daily newspaper, Bangkok.

Bappenas (1991), Regional Aspects of Industrialization in Indonesia, Discussion Paper No. 23, Jakarta: Badan Perencanaan Pembangunan Nasional.

Batson, B.A. (1984), *The End of the Absolute Monarchy in Siam*, Singapore: Oxford University Press.

Baum, S. (1999), 'Social transformations in the global city: Singapore', *Urban Studies*, 36 (7), 1,095–1,117.

Beckett, J. (1982), 'The defiant and the compliant: the Datus of Magindanao under colonial rule', in A.W. McCoy and E.C. de Jesus, *Philippine Social History*, Sydney: Allen and Unwin.

Beddow, M. (2001), 'Top 30 ports', *Containerisation International*, March, 93.

Bernad, M.A. (1974), *The Western Community of Manila: a Profile*, Manila: National Historical Commission.

Bernstein, H.T. (1987), *Steamboats on the Ganges*, Calcutta: Orient Longman.

Berry, B.J.L. (1964). 'Cities as systems within systems of cities', *Papers and Proceedings of the Regional Science Association*, 13, 147–64.

BFTB (1996), *Vietnam: Overview of the Infrastructure (Covering: Roads, Rail, Inland-water Transport, Seaports, Air Transport, Energy)*, OBCE/BDBH – Study 96/14, Brussels: Belgian Foreign Trade Board.

Bird, I.L. (1883), *The Golden Chersonese and the Way Thither*, reprinted 1967 with introduction by Wang Gungwu, Kuala Lumpur: Oxford University Press.

Blainey, G. (1966), *The Tyranny of Distance*, Melbourne: Sun Books.

Blussé, L. (1986), *Strange Company: Chinese Settlers, Mestizo Women and the Dutch in VOC Batavia*, Dordrecht: Foris.

Boer, G.J. de (1997), *De Nederlandse Blauwpijpers*, Alkmaar: de Alk.

Boer, G.J. de, Mulder, A.J., Slettenaar, H.A. and Velterop, E.W.J. (1994), *Koninklijke Java–China–Paketvaart Lijnen (KJCPL)*, Alkmaar: de Alk.

Boer, M.G. de and J.C. Westermann (1941), *Een Halve Eeuw Paketvaart, 1891–1940*, Amsterdam: De Bussy.

Bogaers, E. and Ruijter, P. de (1986), 'Ir. Karsten and Indonesian town planning, 1915–1940', in P.J.M. Nas (ed.), *The Indonesian City*, Dordrecht: Foris.

Boomgaard, P. (1989), *Children of the Colonial State: Population Growth and Economic Development in Java, 1795–1880*, Amsterdam: Free University Press.
Bouinais, A. and Paulus, A. (1885), *L'Indo-Chine Française Contemporaine: Cochinchine, Camboge, Tonkin, Annam*, Paris: Challamel.
Bowie, K.A. (1992), 'Unraveling the myth of the subsistence economy: textile production in nineteenth-century northern Thailand', *The Journal of Asian Studies*, 51 (4), 797–823.
Bowring, John (Sir) (1859), *The Philippine Islands*, London: Smith, Elder.
BPS, Statistik Industri (1995), data file, Jakarta: Biro Pusat Statistik.
BPS (1992), *Statistical Pocketbook* [Statistik Indonesia] 1990, Jakarta: Biro Pusat Statistik.
BPS, Supas (1995a), *Penduduk Jakarta: Hasil Survei Penduduk Antar Sensus 1995*, Series S2.09, Jakarta: Biro Pusat Statistik.
BPS, Supas (1995b), *Penduduk Jawa Barat: Hasil Survei Penduduk Antar Sensus 1995*, Series S2.10, Jakarta: Biro Pusat Statistik.
BPS (2000), *Penduduk Indonesia: Hasil Sensus Penduduk 2000*, Seri RBL1.1, Jakarta: Biro Pusat Statistik.
BPS (2001), *Statistik Indonesia 2000*, Jakarta: Badan Pusan Statistik.
BRGO (1941), *Burma Railways Time Table*, Rangoon: Burma Railways General Offices.
Bristow, R. (1992), 'The origins of the Singapore land use planning system', *Occasional Paper No. 32*, Manchester: Department of Planning and Landscape, University of Manchester.
Broeze, F. (ed.) (1989), *Brides of the Sea: Port Cities from the 18th–20th centuries*, Sydney: New South Wales University Press.
Broeze, F. (ed.) (1997), *Gateways of Asia in the 13th–20th Centurie*s, London: Kegan Paul International.
Brookfield, H., Abdul Samad Hadi and Zaharah Mahmud (1991), *The City in the Village: the In-situ Urbanization of Villages, Villagers and their Land around Kuala Lumpur, Malaysia*, Singapore: Oxford University Press.
Brown, I. (1993), 'The end of the opium farm in Siam, 1905–7', in J. Butcher and H. Dick (eds), *The Rise and Fall of Revenue Farming: Business Elites and the Emergence of the Modern State in Southeast Asia*, New York: St Martin's Press.
Brown, I. (1997), *Economic Change in South-East Asia, c.1830–1980*, Kuala Lumpur: Oxford University Press.
Brown, I. (1998), *The Elite and the Economy in Siam, c.1890–1920*, Singapore: Oxford University Press.
BRS (1944), *Burma Pamphlets No. 4: Burma Rice*, Calcutta, Longmans Green for Burma Research Society.
BRS (1946), *Burma Pamphlet No. 9: Burma Facts and Figures*, London: Longmans Green.
Brugmans, I.J. (1950), *Tachtig Jaren Varen met de Nederland, 1870–1950*, Amsterdam: S.M. Nederland.
BT, *The Business Times*, daily newspaper, Singapore [http://business-times.asia1.com.sg]
Buchanan, I. (1972), *Singapore in Southeast Asia: an Economic and Political Appraisal*, London: G. Bell and Sons.
Buckley, C.B. (1867), *An Anecdotal History of Old Times in Singapore 1819–1967*, new edn 1984 with an introduction by C.M. Turnbull, Singapore: Oxford University Press.
Bureau of Navigation (1912), *List of Rivers of the Philippine Islands with Information as to Location, Navigability, Shelter Afforded and Adjacent Population*, Manila: Bureau of Printing.
Burger, D.H. (1939), *De Ontsluiting van Java's Binnenland voor het Wereldverkeer*, Wageningen: Veenman.

Butcher, J.G. (1979), *The British in Malaya, 1880–1941: the Social History of a European Community in Colonial South-East Asia*, Kuala Lumpur: Oxford University Press.

Caldwell, J.A. (1974), *American Economic Aid to Thailand*, Lexington: Lexington Books.

Calinao, B.P. (1990), 'A social–environmental approach to the development of informal settlements in Metropolitan Manila, Philippines', PhD thesis, State University of New York, Syracuse.

Campbell, A.L. (1993), *The Manila Club: a Social History of the British in Manila*, Manila: St Paul's Press.

Campbell, J.G.D. (1904), *Siam in the Twentieth Century: Being the Experiences and Impressions of a British Official*, London: Edward Arnold.

Campbell, R. (1986), *Teak-Wallah: the Adventures of a Young Englishman in Thailand in the 1920s*, reprinted, Singapore: Oxford University Press.

Campo, J.N.F.M. à (1992), *Koninklijke Paketvaart Maatschappij: Stoomvaart en staatsvorming in de Indonesische archipel, 1888–1914*, Hilversum: Verloren.

Cangi, E.C. (1997), *Faded Splendour, Golden Past: Urban Images of Burma*, Kuala Lumpur: Oxford University Press.

Cant, R.G. (1973), An Historical Geography of Pahang, Monograph No. 4, *Monographs of the Malaysian Branch, Royal Asiatic Society*, Singapore: Malaysian Branch Royal Asiatic Society of Great Britain and Ireland.

Carter, A.C. (ed.) (1904), *The Kingdom of Siam*, New York: G.P. Putnam's Sons.

Cavender, H.M. (1933), 'The development of Philippine interisland shipping', *The American Chamber of Commerce Journal*, (June) [USNA, BIA: 456: with 2893/38].

Cebu (c. 1996), *Cebu: Economic Fact Book 1996*, Cebu: Cebu Investment Promotion Center.

CEI (v. 12a). See Korthals Altes (1991).

Census 1930. *Volkstelling 1930*, Departement van Landbouw, Nijverheid en Handel (later Economische Zaken), Batavia, 1933–1936 [8 vols] [III: Native Population, East Java (1933), VI: Europeans (1935), VII: Chinese and Other Foreign Orientals (1935), VIII: Summary (1936)].

Central Bureau of Statistics (CBS) (1947), *Statistical Pocketbook of Indonesia 1941*, Batavia: Kolff.

Cervero, R. (1998), *The Transit Metropolis: a Global Inquiry*, Washington, DC: Island Press.

Chandran, J. (1971), *The Burma–Yunnan Railway: Anglo-French Rivalry in Mainland Southeast Asia and South China*, Athens: Ohio University Center for International Studies.

Chandran, J. (1977), *The Contest for Siam, 1889–1902: a Study in Diplomatic Rivalry*, Kuala Lumpur: Penerbit Universiti Kebangsaan Malaysia.

Chang, T.C. (2000), 'Singapore's Little India: a tourist attraction as a contested landscape', *Urban Studies*, 37 (2), 343–66.

Cheng, S.H. (1968), *The Rice Industry of Burma, 1852–1940*, Kuala Lumpur: University of Malaya Press.

Cheng, S.H. (1991), 'Economic change and industrialization', in E.C.T. Chew and E. Lee (eds), *A History of Singapore*, Singapore: Oxford University Press.

Cheong, S. (1997), *Bumiputera Entrepreneurs in the KLSE: Volume Two*, Kuala Lumpur: Corporate Research Services.

Chin Kar Keong and Mahfix bin Omar (1994), 'Planning of urban expressways in Kuala Lumpur', *The Wheel Extended: a Toyota Quarterly Review*, 89, 19–25.

Chin, Hoong Chor (1998), 'Urban transport planning in Singapore', in B.K.P. Yuen (ed.), *Planning Singapore: From Plan to Implementation*, Singapore: Singapore Institute of Planners.

Chua Beng-Huat (1985), *Designed for Living: Public Housing Architecture in Singapore*, Singapore: Housing and Development Board.
Chua Beng-Huat (1989), *The Golden Shoe: Building Singapore's Financial District*, Singapore: Urban Redevelopment Authority.
Chua Beng-Huat (1997), *Political Legitimacy and Housing: Stakeholding in Singapore*, London: Routledge.
Chua Beng-Huat (1998), 'World cities, globalisation and the spread of consumerism: a view from Singapore', *Urban Studies*, 35 (5–6), 981–1000.
Chua Beng-Huat and Edwards, N. (eds) (1992), *Public Space: Design, Use and Management*, Singapore: Singapore University Press.
Ciputra (c.1996), *The Developer Ir. Ciputra: Vision and Innovation*, Jakarta: Ir. Ciputra Foundation.
CIT (1971), 'Road–rail problems in the early thirties', *Chartered Institute of Transport: Malaysia Section Journal*, 1971 Issue, 1–18.
CIY (1987–2002), *Containerisation International Yearbook* (annual), London: Informa.
CMPS (1971), *The Urban Renewal and Development Project, Singapore: For the United National Development Programme (Special Fund)*, Singapore: Crooks Michell Peacock Stewart Graphics and Printing Division (five parts and summary volume).
Cobban, J.L. (1971), 'The city on Java: an essay in historical geography', PhD thesis, University of California (Berkeley), Ann Arbor: University Microfilms.
Cobban, J.L. (1993), 'Public housing in colonial Indonesia, 1900–1940', *Modern Asian Studies*, 27 (4), 871–96.
Cohen, S.B. (ed.) (1998), *The Columbia Gazetteer of the World*, New York: Columbia University Press.
Collis, M. (1966), *Raffles*, London: Faber and Faber.
Concannon, T.A.L. (1955), 'A new town in Malaya: Petaling Jaya, Kuala Lumpur', *The Malayan Journal of Tropical Geography*, 5, 39–43.
Concannon, T.A.L. (1960), *Building a City within a City*, Kuala Lumpur: Caxton Press. Reprinted from the *Straits Times*, 4 September 1957.
Corey, K. (2000), 'Electronic space – creating cyber communities in Southeast Asia', in M.I. Wilson and K.E. Corey (eds), *Information Tectonics: Space, Place and Technology in an Electronic Age*, Chichester: John Wiley and Sons.
Corpuz, A.G. (1989), 'Railroads and regional development in the Philippines: views from the colonial iron horse, 1875–1935', PhD thesis, Cornell University, Ithaca.
Cortes, R.M. (1990), *Pangasinan, 1901–1986: a Political, Socioeconomic and Cultural History*, Quezon City: New Day.
Cowan, C.D. (1961), *Nineteenth-Century Malaya: the Origins of British Political Control*, London: Oxford University Press.
Crawfurd, J. (1828), *Journal of an Embassy to the Courts of Siam and Cochin-China etc*, London: Henry Colburn.
Critchfield, R. (1970), *Hello Mister, Where are you Going?: the Story of Husen, a Javanese Becak Driver*, New York: Alicia Patterson Fund.
Cucherousset, H. (1927), 'Les chemins de fer de la péninsule Indochinoise', *Cahiers de la Société de Géographie de Hanoi*, No. 12: 29 pp (with maps).
Cushman, J.W. (1991), *Family and State: the Formation of a Sino-Thai Tin-mining Dynasty, 1797–1932*, edited by C.J. Reynolds, Singapore: Oxford University Press.
Cushman J.W. (1993), *Fields from the Sea: Chinese Junk Trade with Siam During the Late Eighteenth and Early Nineteenth Centuries*, Ithaca: Southeast Asia Program, Cornell University.
Daerah Khusus Ibu Kota Jakarta (DKI) (1972), *Some Data about Jakarta*, Jakarta: Pemerintah DKI.

Daerah Khusus Ibu Kota Jakarta (DKI) (1977), *Karya Jaya: Kenang-kenangan Lima Kepala Daerah Jakarta, 1945–1966*, Jakarta: Pemerintah DKI.

Daily Commercial News and Shipping List, daily newspaper, Melbourne.

Daniere, A.G. (1995), 'Transportation planning and implementation in cities of the Third World: the case of Bangkok', *Environment and Planning C: Government and Policy*, 13 (1), 25–45.

Davies, R.E.G. (1964), *A History of the World's Airlines*, London: Oxford University Press.

Davies, R.E.G. (1997), *Airlines of Asia since 1920*, London: Chrysalis Books.

De Corbigny, B. (1878), 'Huit jours d'Ambassade a Hué (Royaume d'Annam)', *Le Tour de Monde*, 35: 33–64 translated in W.E.J. Tips, *Cities of Nineteenth Century Colonial Vietnam: Hanoi, Saigon, Hue and the Champa Ruins*, Bangkok: White Lotus Press.

De Koloniale Roeping van Nederland (KRN) (1930), The Hague: Nederlandsch-Engelsche Uitgeversmaatschappij.

Del Testa, D.W. (2000), 'Some preliminary findings on the relationships of railroads to the economies of Tonkin and Annam Protectorates, French Indochina, 1919–1937', in J.-P. Bassino, J.D. Giacometti and K. Odaka (eds), *Quantitative Economic History of Vietnam: an International Workshop*, Tokyo: Institute of Economic Research, Hitotsubashi University.

Departement van Landbouw, Nijverheid en Handel (DLNH) (1926), *Landbouwatlas van Java en Madoera* (Agricultural Atlas of Java and Madura), Mededeelingen van het CKS no. 33, Batavia.

Desfeuilles, P. (1927), *Les Colonies Françaises: L'Indochine*, Paris: Éditions Pierre Roger.

Devins, H.B. (1905), *An Observer in the Philippines: or Life in Our New Possessions*, Boston: American Tract Society.

Dewan Bandaraya (1982), *Draf Pelan Struktur Kuala Lumpur* [Kuala Lumpur: Draft Structure Plan], Kuala Lumpur: Dewan Bandaraya.

Diaz, R. (1984), 'Public transport regulation: colorum vehicles', mimeo, Manila.

Dick, H.W. (1975), 'Prahu shipping in Eastern Indonesia', *Bulletin of Indonesian Economic Studies*, 11 (2), 69–107 and (3), 81–103.

Dick, H.W. (1981a), 'Urban public transport: part I', *Bulletin of Indonesian Economic Studies*, 17 (1), 66–82.

Dick, H.W. (1981b), 'Urban public transport: part II', *Bulletin of Indonesian Economic Studies*, 17 (2), 72–88.

Dick, H.W. (1987a), *The Indonesian Interisland Shipping Industry: an Analysis of Competition and Regulation*, Singapore: Institute of Southeast Asian Studies.

Dick, H.W. (1987b), 'Prahu shipping in Eastern Indonesia in the interwar period', *Bulletin of Indonesian Economic Studies*, 23 (1), 104–21.

Dick, H.W. (1993), 'The manufacturing sector', in Dick, Fox and Mackie (eds), *Balanced Development: East Java in the New Order*, Singapore: Oxford University Press.

Dick, H.W. (1996), 'The emergence of a national economy, 1808–1990s', in J.Th. Lindblad, *Historical Foundations of a National Economy in Indonesia, 1890s–1990s*, Amsterdam: North Holland.

Dick, H.W. (2000) 'Representations of development in 19th–20th century Indonesia: a transport history perspective', *Bulletin of Indonesian Economic Studies*, 36 (1), 185–207.

Dick, H.W. (2002), *Surabaya, City of Work: a Twentieth Century Socioeconomic History*, Athens: Ohio University Press.

Dick, H.W. and Forbes, D. (1992), 'Transport and communications: a quiet revolution', in A. Booth (ed.), *The Oil Boom and After: Indonesian Economic Policy and Performance in the Soeharto Era*, Kuala Lumpur: Oxford University Press.

Dick, H.W., Houben, J.H., Lindblad, J.T. and Thee Kian Wie (2002), *The Emergence of a National Economy: an Economic History of Indonesia, 1800–2000*, Sydney: Allen and Unwin.
Dick, H.W. and Rimmer, P.J. (1980), 'Beyond the informal/formal dichotomy: towards an integrated approach', *Pacific Viewpoint*, 21 (1), 26–41.
Dick, H.W. and Rimmer, P.J. (1998a), 'Beyond the Third World city: the new urban geography of Southeast Asia', *Urban Studies*, 35 (12): 2303–21.
Dick, H.W. and Rimmer, P.J. (1998b), 'Shared space, public transport and congestion: the new colonial city', *Malaysian Journal of Tropical Geography*, 29 (1), 1–10.
Dick, H.W. and Rimmer, P.J. (1999), 'Privatising climate: First World cities in Third World settings', in J. Brotchie, P.W. Newman, P. Hall and J. Dickey (eds), *East West Perspectives on 21st Century Urban Development*, Aldershot: Ashgate.
Dillon, R. (1994), 'Indian squatters in Kuala Lumpur: people in transition', in H. Brookfield (ed.), *Transformation with Industrialization in Peninsular Malaysia*, Kuala Lumpur: Oxford University Press.
Dilokwanich S. (1995), 'Industry, water and people in Greater Bangkok: a case study of Samut Prakan', unpublished PhD thesis, Department of Human Geography, The Australian National University, Canberra.
Dixon, C.J. (1977), 'Development, regional disparity and planning: the experience of Northeast Thailand', *Journal of Southeast Asian Studies*, 8 (2), 210–23.
Dixon, C.J. (1978), 'Settlement and environment in Northeast Thailand', *The Journal of Tropical Geography*, 46, 1–10.
Dixon, C.J. (1999), *The Thai Economy: Uneven Development and Internationalization*, London: Routledge.
Dobby, E.H.G. et al. (1955), 'Padi landscapes of Malaya', *The Malayan Journal of Tropical Geography*, 6, 1–94.
Dobby, E.H.G. et al. (1957), 'Padi landscapes of Malaya', *The Malayan Journal of Tropical Geography*, 10, 1–143 and plates.
Doeppers, D. (1984), *Manila, 1900–1941: Social Change in a Late Colonial Metropolis*, Monograph Series No. 27, New Haven: Southeast Asia Studies Program, Yale University.
DOTARS (2002), *Mekong Freight Logistics Study: a Study to Improve Freight Transport and Facilities in the Region Cambodia, Lao PDF & Vietnam [Draft]*, Canberra: Department of Transport and Regional Services [www.dotrs.gov.au/transinfra/mekong].
Drabble, J.H. (1973), *Rubber In Malaya, 1876–1922: the Genesis of the Industry*, Kuala Lumpur: Oxford University Press.
Drabble, J.H. (2000), *An Economic History of Malaysia, c.1800–1990: the Transition to Modern Economic Growth*, New York: St Martin's Press.
Duldulao, M.D. (1995), *Quezon City*, Manila: Japuzinni.
Duparc, H.J.A. (1972), *Trams en Tramlijnen: De Elektrische Stadstrams op Java*, Rotterdam: Wyt.
EBI (2002), *2002 Britannica Book of the Year*, Chicago: Encyclopaedia Britannica.
E.B.M. (1892), 'Boat life in Siam', *Directory for Bangkok and Siam 1892*, Bangkok: Bangkok Times.
ECAFE (1967), 'Economic and social progress seen resulting from a new highway built in Thailand', *Transport and Communications Bulletin*, No. 41, Bangkok: UN Commission for Asia and the Far East.
Edgerton, R.K. (1982), 'Frontier society on the Bukidnon Plateau, 1870–1941', in A.W. McCoy and E.C. de Jesus, *Philippine Social History*, Sydney: Allen and Unwin.

Edwards, N. (1992), 'The colonial suburb: public space and private space', in Chua Beng-Huat and N. Edwards (eds), *Public Space: Design, Use and Management*, Singapore: Singapore University Press.

Eggink, E.J. (1930), *Na 25 Jaar: Beknopt Gedenkschrift ter gelegenheid van het 25-jarig bestaan der Gemeente Batavia*, Batavia: Gemeente Batavia.

El Comercio (EC), daily newspaper, Manila.

Ellis, H.T. (1859), *Hong Kong to Manila and the Lakes of Luzon in the Philippine Isles in the Year 1856*, London: Smith, Elder.

Elson, R.E. (1984), *Javanese Peasants and the Colonial Sugar Industry: Impact and Change in an East Java Residency, 1830–1940*, Singapore: Oxford University Press.

Elson, R.E. (1994), *Village Java under the Cultivation System, 1830–1970*, Sydney: Allen and Unwin.

Elson, R.E. (1997), *The End of the Peasantry in Southeast Asia*, Macmillan, London.

Emerson, R., Mills, L.A., and Thompson, V. (1942), *Government and Nationalism in Southeast Asia*, New York: Institute of Pacific Relations.

Encyclopaedia Britannica (1993), 15th edn, Chicago.

ENI (1918–21), *Encyclopaedie van Nederlandsch-Indië*, 4 vols (various editors), The Hague: Martinus. Nijhoff.

ERTA (1991), *Annual Report 1991*, Bangkok: The Expressway and Rapid Transit Authority of Thailand.

Fairclough, G. (1995), 'Wrong turn: Bangkok car owners put their foot down', *Far Eastern Economic Review*, 158 (11), 63.

Fairclough, G. (1996), 'Motion sickness: little relief in sight for Bangkok's traffic ailments', *Far Eastern Economic Review*, 159 (7), 19.

Falkus, M. (1990), *The Blue Funnel Legend: a History of the Ocean Steam Ship Company 1865–1970*, London: Macmillan.

Falkus, M. (1993), 'Bangkok: from primate city to primate megalopolis', in T. Barker and A. Sutcliffe (eds), *Megalopolis: the Giant City in History*, London: George Allen and Unwin.

Falkus, M. (1996), 'Labour in Thai mining: some historical considerations', *Asian Studies Review*, 20 (2), 71–95.

FDTCP (1997), *Physical Planning: Guidelines for the Multimedia Super Corridor (MSC)*, Kuala Lumpur: Federal Department of Town and Country Planning, Malaysia.

Feeny, D. (1982), *The Political Economy of Productivity: Thai Agricultural Development, 1880–1975*, Vancouver: University of British Columbia Press.

FER (1912), 'The motor and good roads in the Philippines', *Far Eastern Review*, May, p. 440.

Fernando, M.R. (1996), 'Growth of non-agricultural activities in Java in the middle decades of the nineteenth century', *Modern Asian Studies*, 30 (1), 77–119.

Firman, T. (1997), 'Patterns and trends of urbanisation', in G.W. Jones and T.H. Hull (eds), *Indonesia Assessment: Population and Human Resources*, Canberra: Research School of Pacific and Asian Studies, Australian National University.

Fisher, C. (1950), 'Southeast Asia', in W.G. East and O.H.K. Spate (eds), *The Changing Map of Asia: a Political Geography*, London: Methuen.

Fletcher, M. (1958), 'The Suez Canal and world shipping, 1869–1914', *Journal of Economic History*, 18 (1), 556–73.

FMS (1924), *Federated Malay States Annual Report, 1924*, Kuala Lumpur: Government Press.

FMS (1930), *Federated Malay States Annual Report, 1930*, Kuala Lumpur: Government Press.

FMS (1939), *Malaya 1939 – Scale: Six Miles to an Inch*, Kuala Lumpur: Federated Malay States Survey Department.
FOM (1949), *Annual Report on the Federation of Malaya, 1949*, Kuala Lumpur: Government Press.
FOM (1950), *Annual Report on the Federation of Malaya, 1950*, Kuala Lumpur: Government Press.
Fossey, J., 2000, 'Packing a punch', *Containerisation International*, November, 54–9.
Fournerau, L. (1894), 'Bangkok', *Le Tour de Monde*, 68, 1–64 [Translation with introduction by W.J. Tips, *Bangkok in 1892*, Bangkok: White Lotus. 1998].
Friedman, M. (1977), *Friedman on Galbraith and Curing the British Disease*, Vancouver: Fraser Institute.
Fujita, M., Krugman, P. and Venables, A.J. (1999), *The Spatial Economy: Cities, Regions and International Trade*, Cambridge, MA: MIT Press.
Furnivall, J.S. (1931), *An Introduction to the Political Economy of Burma*, Rangoon: Burma Book Club.
Furnivall, J.S. (1957), 'Safety First: a study on the economic history of Burma', *Journal of the Burma Research Society*, 9 (1), 24–38.
Gardiner, P. (1997), 'Migration and urbanisation: a discussion', in G.W. Jones and T.H. Hull (eds), *Indonesia Assessment: Population and Human Resources*, Canberra: Research School of Pacific and Asian Studies, Australian National University.
GB (1850), *Abstract of Reports on the Trade of Various Countries and Places, for the Year 1854*, Board of Trade, London: Harrison and Sons.
GB (1856), *Abstracts of Reports on the Trade of Various Countries and Places for the Year 1854*, Papers presented to Parliament, London: Harrison and Sons.
GB (1892), *Siam: Report for the Year 1891 of the Trade of Bangkok*, Annual Series No. 1089, Diplomatic and Consular Reports on Trade and Finance, Foreign Office, London: HMSO.
GB (1895), *Siam: Report for the Year 1893 of the Trade of Bangkok*, Annual Series No. 1787, Diplomatic and Consular Reports on Trade and Finance, Foreign Office, London: HMSO.
GB (1896), *Siam: Report for the Year 1894 of the Trade of Bangkok*, Annual Series No. 2003, Diplomatic and Consular Reports on Trade and Finance, Foreign Office, London: HMSO.
GB (1901), *Trade and Shipping of South-East Asia*, Board of Trade, Parliamentary Papers, Cmd 324, London: HSMO.
GB (1923), *Report on Indochina, Economic Survey*, London: HMSO.
GB (1935), *Economic Conditions in Siam at the close of the fourth quarter, 1934*, Report by J. Bailey and H.R. Bird, Department of Overseas Trade, London: HMSO.
GB (1945), *Siam: Basic Handbook*, London: Foreign Office.
Gibbs, C.R.V. (1963), *British Passenger Liners of the Five Oceans*, London: Putnam.
Gibson-Hill, C.A. (1949), 'Cargo boats of the East Coast of Malaya', *Journal of the Malayan Branch of the Royal Asiatic Society*, 22 (3), 107–44.
Gibson-Hill, C.A. (1952), 'Tongkang and lighter matters', *Journal of the Malayan Branch of the Royal Asiatic Society*, 25 (1), 84–110.
Gibson-Hill, C.A. (1954), 'The steamers employed in Asian waters, 1819–39', *JMBRAS*, 27 (1), 120–62.
Gibson-Hill, C.A. (1958), 'Notes on the administration of the Singapore Post Office, 1819–67', *Journal Malayan Branch Royal Asiatic Society*, 31 (1), 145–62.
Ginsburg, N., Koppel, B. and McGee, T. (eds) (1991), *The Extended Metropolis: Settlement Transition in Asia*, Honolulu: University of Hawaii Press.

Ginsburg, N. and Roberts, C.F. (1958), *Malaya*, Seattle: University of Washington Press.
Girling, J. (1996), 'Interpreting development: capitalism, democracy, and the middle class in Thailand', *Studies of Southeast Asia No. 21*, Ithaca: Cornell University, Southeast Asia Program.
Gleeck, L.E. (1977), *The Manila Americans, 1901–1964*, Manila: Carmelo and Bauermann.
Goh Hock Guan (1973), 'Urbanisation–the Malaysian experience', in *KL Forum 73, July 1–6 1973*, Kuala Lumpur: Malaysian Institute of Planners and others, 67–73.
Gomez, E.T. (1999), *Chinese Business in Malaysia: Accumulation, Ascendance, Accommodation*, Richmond: Curzon.
Gomez, E.T. (2000), 'In search of patrons: Chinese business networking and Malay political patronage in Malaysia', in Chan Kwok Bun (ed.), *Chinese Business Networks: State, Economy and Culture*, Singapore: Prentice Hall.
Gomez, E.T. and Jomo, K.S. (1999), *Malaysia's Political Economy: Politics, Patronage and Profits*, 2nd edn, Cambridge: Cambridge University Press.
Gordon, R. (1891), 'The economic development of Siam', *Journal of the Society of Arts*, 39: 283–98.
Gorton, F. (1923), *Report on the Commercial Situation in Indo-China*, London: HMSO.
Goss, J.D. (1990), 'Production and reproduction among the urban poor of Metro Manila: relations of exploitation and conditions of existence', PhD thesis, University of Kentucky, Lexington.
Grava, S. (1972), 'The jeepneys of Manila', *Traffic Quarterly*, 24 (4), 465–83.
Guia Oficial de Filipinas (Guia) (1898), Manila.
Gullick, J.M. (1983), *Kuala Lumpur 1880–1895: a City in the Making*, Kuala Lumpur: Pelanduk Publications. Reprinted from the *Journal of the Malaysian Branch Royal Asiatic Society*, 26 (4), 1966.
Gullick, J.M. (1988), *The Story of Kuala Lumpur (1857–1939)*, Singapore: Eastern Universities Press.
Haan, F. de (1910), *Priangan: De Preanger-Regentschappen onder het Nederlandsch Bestuur tot 1810*, Batavia: Bataviaasch Gennotschap van Kunsten en Wetenschappen.
Hafner, J.A. (1969), 'Report no. 1: transport development and geographical change in the Central Plain of Thailand', *Research Project no. 30/3 – Case Study of Competition between Waterways, Road and Rail Transport*, Bangkok: Applied Scientific Research Corporation of Thailand.
Hafner, J.A. (1973), 'The spatial dynamics of rice milling and commodity flow in central Thailand', *The Journal of Tropical Geography*, 37, 30–8.
Halsema, E.J. (1916), 'Pampanga: past and present', *Bureau of Public Works Quarterly Bulletin*, 4 (1), 17–24.
Halsema, J.J. (1991), *E.J. Halsema, Colonial Engineer: a Biography*, Manila: New Day.
Hamzah S. (1965–66), 'The structure of Kuala Lumpur: Malaysia's capital city', *The Town Planning Review*, 36: 125–38.
Harrison, C.W. (1923), *An Illustrated Guide to the Federated Malay States*, 4th edn, London: Malay States Information Agency.
Harrison, C.W. (1985), *An Illustrated Guide to the Federated Malay States (1923)*, London: The Malay States Information Society, reprinted 1985 with introduction by P. Kratoska by Oxford University Press, Kuala Lumpur.
Harrison, F.B. (1916), 'Message of the Governor-General Francis Burton Harrison to the Third Philippine Legislature relative to the proposed purchase of outstanding stock of the Manila Railroad Company by the Philippine Government', Manila: Bureau of Printing.

Harrison, J.L. (1918), 'Traffic density does not justify building roads for heavy trucks in Philippines', *Engineering News-Record*, 16 May [USNA, BIA: (box 398) 2146A/2].
Hartendorp, A.V.H. (1958), *History of Industry and Trade of the Philippines*, Manila: American Chamber of Commerce.
Harvey, D. (1989), *The Condition of Postmodernity: an Enquiry into the Origins of Cultural Change*, Oxford: Blackwell.
Headrick, D.R. (1981), *The Tools of Empire: Technology and European Imperialism in the Nineteenth Century*, Oxford: Oxford University Press.
Headrick, D.R. (1988), *The Tentacles of Progress: Technology Transfer in the Age of Imperialism, 1850–1940*, Oxford: Oxford University Press.
Heuken, A. (1983), *Historical Sites of Jakarta*, 2nd edn, Jakarta: Cipta Loka Caraka.
HFA (1985), *The Bangkok Metropolitan Short Term Urban Transport Review*, Bangkok: Halcrow Fox and Associates.
HFA (1991), *Executive Report for the Seventh Plan Urban and Regional Transport (SPURT)*, Bangkok: Halcrow Fox and Associates with Pak Poy & Kneebone and Asian Engineering Consultants.
Higgins, B. and Savoie, D.J. (1995), *Regional Development Theories and Their Application*, New Brunswick, NJ: Transaction Publishers.
Hirsch, P. (1987), 'Deforestation and development in Thailand', *Singapore Journal of Tropical Geography*, 8 (2), 129–38.
Hirschman, A.O. (1958), *The Strategy of Economic Development*, New Haven: Yale University Press.
Ho Kim Hin, D. (1996), *The Seaport Economy: a Study of the Singapore Experience*, Singapore: Singapore University Press.
Ho Kong Chong and Lim Nyuk Eun, V. (1992), 'Backlanes as congested regions: construction and control of physical space', in Chua Beng-Huat and N. Edwards (eds), *Public Space: Design, Use and Management*, Singapore: Singapore University Press.
Holm, D.F. (1978), 'The role of the state railways in Thai history, 1892–1932', PhD thesis, Yale University (published by University Microfilms, Ann Arbor, 1980).
Hong, L. (1984), *Thailand in the Nineteenth Century: Evolution of the Economy and Society*, Singapore: Institute of Southeast Asian Studies.
Houben, V.J.H. (2002), 'Java in the 19th century: consolidation of a territorial state,' in Dick et al., *The Emergence of a National Economy: an Economic History of Indonesia, 1800–2000*, Sydney: Allen and Unwin.
Hudson, H. (1990), *Communication Satellites: their Development and Impact*, New York: The Free Press (Macmillan).
Huenemann, R.W. (1984), *The Dragon and the Iron Horse: the Economics of Railroads in China 1876–1937*, Cambridge: Harvard University Press.
Huff, W.G. (1994), *The Economic Growth of Singapore: Trade and Development in the Twentieth Century*, Cambridge: Cambridge University Press.
Hugo, G. (1996), 'Urbanization in Indonesia: city and countryside linked', in J. Gugler (ed.), *The Urban Transformation of the Developing World*, Oxford: Oxford University Press.
Hugo, G.J., Hull, T.H, Hull, V.J. and Jones, G.W. (1987), *The Demographic Dimension in Indonesian Development*, Singapore: Oxford University Press.
Hussein, J. (1997), '21st century airport: focus on the Kuala Lumpur international airport project', in Asian Strategy and Leadership Institute, *Malaysia Today: Towards the New Millennium*, London: Asean Academic Press.
Hyde, F.E. (1957), *Blue Funnel: a History of Alfred Holt and Company of Liverpool from 1865 to 1910*, Liverpool: Liverpool University Press.

IATA (1994–2001), *World Air Transport Statistics*, Geneva: International Air Transport Association.
ICAO (1986–2001), *Airport Traffic: Digest of Statistics*, Montreal: International Civil Aviation Organization.
Indische Verslag (IV) (annual 1931–41), Batavia: Landsdrukkerij.
Indonesian Statistical Yearbook (ISYB) (annual), Jakarta: Central Bureau of Statistics.
Ingram, J.C. (1955), *Economic Change in Thailand since 1850*, Stanford: Stanford University Press.
Ingram, J.C. (1971), *Economic Change in Thailand, 1850–1970*, Stanford: Stanford University Press.
Ireland, A. (1907), *The Province of Burma: a Report Prepared on Behalf of the University of Chicago*, Boston: Houghton Mifflin.
IRF (2001), *World Road Statistics 2001*, Geneva: International Road Federation.
Jackson, J.C. (1965), 'Chinese agricultural pioneering in Singapore and Johore, 1800–1917', *Journal of the Malaysian Branch of the Royal Asiatic Society*, 38 (1), 77–105.
Jakarta Post, daily newspaper, Jakarta [http://www.thejakartapost.com].
Janelle, D.G. (1969), 'Spatial reorganization: a model and a concept', *Annals of the Association of American Geographers*, 59, 348–64.
JDG (1970), *The Postwar Development of the Republic of Vietnam: Policies and Programs*, Joint Development Group, New York: Praeger.
Jellinek, L. (1991), *The Wheel of Fortune: the History of a Poor Community in Jakarta*, Sydney: Allen and Unwin.
Jenista, F.L. (1987), *The White Apos: American Governors in the Cordillera Central*, Manila: New Day.
Jennings, E. (1980), *Cargoes: a Centennial Story of the Far Eastern Freight Conference*, Singapore: Meridian.
Jerndal., R. and Rigg, J. (1998), 'Making space in Laos: constructing a national identity in a "forgotten" country', *Political Economy*, 17 (7), 809–31.
Jesus, E.C. de (1980), *The Tobacco Monopoly in the Philippines: Bureaucratic Enterprise and Social Change, 1766–1880*, Manila: Ateneo de Manila University Press.
JICA (1986), *Klang Valley Transportation Study: Draft Final Report. Summary of Findings and Recommendations – December 1986*, Kuala Lumpur: Japan International Cooperation Agency.
JICA (1990), *The Study on Medium- to Long-Term Improvement/Management Plan of Roads and Road Transport in Bangkok*, Bangkok: Japan International Cooperation Agency.
JICA (1991), *Road Development Study in the Southern Region: Draft Final Report, Volume 1 Main Text – July 1991*, Tokyo: Japan International Cooperation Agency.
JICA (1993a), *LISR: Master Plan Study on Luzon Island Strategic Road Network Development Project – Final Report – March 1993*, Tokyo: Katahira and Engineers International and Nippon Koei Co. Ltd for Japan International Cooperation Agency and Department of Public Works and Highways, Republic of the Philippines.
JICA (1993b), *HNDP: Highway Network Development Plan Study in Malaysia: Final Report – Main Volume – March 1993*, Tokyo: Japan International Cooperation Agency.
JIG (2002), *Jane's World Railways, Forty-third Edition 2001–2002*, Coulsdon: Jane's Information Group.
JLL (2000), *Property Research Paper Asia Pacific: Market Analysis Series; Asia Pacific Property Digest, The Market as at January 1998*, Hong Kong: Jones Lang LaSalle Research and Consultancy.
Johnstone, M. (1983), 'Housing policy and the urban poor in Peninsular Malaysia', *Third World Planning Review*, 5 (1), 249–71.

Joint Development Group (1970), *The Postwar Development of the Republic of Vietnam: Policies and Programs*, 3 vols, New York: Praeger.
Jomo, K.S. (1986), *A Question of Class: Capital, the State and Uneven Development in Malaya*, Singapore: Oxford University Press.
Jones, J.H. (1964), 'Economic benefits from development roads in Thailand', *Technical Note No. 15*, Bangkok: SEATO Graduate School of Engineering.
Journal of Philippine Statistics (JPS) (quarterly), Manila.
Juppenlatz, M. (1970), *Cities in Transformation: the Urban Squatter Problem of the Developing World*, Brisbane: University of Queensland Press.
Kahin, A.R. and Kahin, G.M. (1995), *Subversion as Foreign Policy: the Secret Eisenhower and Dulles Debacle in Indonesia*, New York: New Press.
Kamer van Koophandel en Nijverheid te Soerabaia (KKNS) (annual), *Jaarverslag*, Surabaya.
Kandiah, A. (1972), 'Transport scene in Malaysia', *Chartered Institute of Transport: Malaysia Section Journal, 1972 Issue*, pp. 37–48.
Kaothien, U. (1995), 'The Bangkok Metropolitan Region: policies and issues in the Seventh Plan', in T.G. McGee and I.M. Robinson, *The Mega-Urban Regions of Southeast Asia*, Vancouver: University of British Columbia Press.
Kaothien, U. and Webster, D. (1995), 'Thai city regions: the stage for Thailand's involvement in the new global economy', in *Cities and the New Global Economy: Conference Proceedings Volume 2*, Canberra: Australian Government Publishing Service.
Kaothien, U. and Webster, D. (2000), 'Globalization and urbanization: the case of Thailand', in S. Yusuf, W. Wu and S. Everett (eds), *Local Dynamics in an Era of Globalization: 21st Century Catalysts for Development*, Washington, DC: The World Bank.
Kasiraksa, W. (1963), *Economic Effects of the Friendship Highway*, Bangkok: SEATO Graduate School of Engineering.
Kaur, A. (1980), 'Road or rail? – Competition in colonial Malaya, 1909–1940', *Journal of the Malaysian Branch of the Asiatic Society*, 53 (2), 45–66.
Kaur, A. (1985a), *Bridge and Barrier: Transport and Communications in Colonial Malaya, 1870–1957*, Singapore: Oxford University Press.
Kaur, A. (1985b), 'Tracking the roots and routes of KTM', in Alias Mohamed (ed.), *Malayan Railway 100 Years 1885–1985*, Kuala Lumpur: AMW Communications Management.
Kaur, A. (1990), 'Working on the railway: Indian workers in Malaya, 1880–1957', in P.J. Rimmer and L.M. Allen (eds), *The Underside of Malaysian History: Pullers, Prostitutes, Plantation Workers...*, Singapore: Singapore University Press.
Kaye, B. (1960), *Upper Nankin Street Singapore: a Sociological Study of Chinese Households Living in a Densely Populated Area*, Singapore: University of Malaya Press.
Keyes, C.F. (1967), 'Isan: regionalism in northeastern Thailand', *Cornell Thailand Project Series Interim Reports Series No. 10, Data Paper No. 65*, Southeast Asia Program, Department of Asian Studies, Cornell University, Ithaca, New York.
Keyes, W.J. (1982), 'Metro Manila: a case study of policies towards urban slums', in MHS (1982), *Philippine Shelter System and Human Settlements*, Manila: Ministry of Human Settlements.
Khalid, A.A. (1993), 'Industrialization and urban development in Peninsular Malaysia, late 19th century to 1970's: continuity and change', in Lee Boon Hiok and K.S.S. Oorjitham (eds), *Malaysia and Singapore: Experiences in Industrialization and Urban Development*, Kuala Lumpur: University of Malaya.
Khoo Kay Kim (1985), 'Railway and the decline of traditional ports on the western coast of the peninsula', in Alias Mohamed (ed.), *Malayan Railway 100 Years 1885–1985*, Kuala Lumpur: AMW Communications Management.

Kiat.net (2002), Malaysia Kuala Lumpur Mass transit [Star, Putra, Monorail], http://www.kiat.net/malaysia/KL/transit.html.
Kidokoro T. (1992), 'Strategies for urban development and transport systems in Asian metropolises, focusing on Bangkok metropolitan area' (with comment by P.J. Rimmer), *Regional Development Dialogue*, 13 (3), 74–86.
Kinloch, C.W. (1987), *Rambles in Java and the Straits in 1852 by 'Bengal Civilian'*, Singapore: Oxford University Press.
KKRM (1984), 'Road development and policy', Kementerian Kerja Raya, Malaysia, in Proceedings of Conference entitled Bengkel Pengangkutan Nasional 5–6 March 1980, Kuala Lumpur Chartered Institute of Transport.
Knaap, G.J. (1989), *Transport, 1819–1940*, Changing Economy in Indonesia, vol. 9, Amsterdam: Royal Tropical Institute.
Knaap, G.J. (1996), *Shallow Waters, Rising Tide: Shipping and Trade in Java around 1770*, Leiden: KITLV.
KNS (1983), *Shah Alam Extension Plan*, Petaling Jaya: Kerajaan Negeri Selangor.
Kolb, A. (1978) 'Gross-Manila: Die Individualität einer Tropischen Millionenstadt', *Hamburger Geographische Studien*, vol. 34, Hamburg: Institut für Geographie, Universität Hamburg.
Koloniaal Verslag (KV) (1892), Appendix C: Overzichten betreffende den Oeconomischen Toestand van de Meeste Gewesten van Nederlandsch-Indië, s'Gravenhage: Landsdrukkerij.
Kompas, daily newspaper, Jakarta.
Kong, L. and Yeoh, B.S.A. (1994), 'Urban conservation in Singapore: a survey of state policies and popular attitudes', *Urban Studies*, 31 (2), 247–65.
Korff, R. (1986), *Bangkok: Urban System and Everyday Life*, Saarbrücken: Verlag Breitenbach.
Korthals Altes, W.L. (1991), *General Trade Statistics, 1823–1940*, Changing Economy in Indonesia, vol. 12a, Amsterdam: Royal Tropical Institute.
KOS (1936–37), *Report of the Financial Adviser on the Budget of the Kingdom of Siam for the Year B.E. 2479 (1936–1937)*, Bangkok: Kingdom of Siam.
KPM (1955), *Dienstregeling der N.V. Koninklijke Paketvaart Maatschappij voor het Jaar 1956*, Jakarta: Kolff.
Kraus, J. and Jongh, G.J. de (1910), *Verslag over de Verbetering van Haventoestanden van Soerabaia*, Batavia: Landsdrukkerij.
Krause, L.B., Koh Ai Tee and Lee Yuan (1987), *The Singapore Economy Reconsidered*, Singapore: Institute of Southeast Asian Studies.
Krausse, G.H. (1978), 'Intra-urban variation in kampung settlement of Jakarta: a structural analysis', *Journal of Tropical Geography*, 46, 1–26.
KRR (1975), *Bangkok Transportation Study: Final Report September 1975, Volume 1: Summary*, Dusseldorf: F.H. Kocks/Rhein Ruhr.
Krugman, P. (1996a), *Pop Internationalism*, Cambridge, MA: MIT Press.
Krugman, P. (1996b), 'Urban concentration: the role of increasing returns and transport costs', *International Regional Science Review*, 19 (1), 5–30.
Krugman, P. (1997), *Development, Economic Geography and Economic Theory*, Cambridge, MA: MIT Press.
Kubota, H. (1996), Bangkok: world's worst traffic jams and their causes', *The Wheel Extended: a Toyota Quarterly Review*, 96, 5–9.
Kuncoro, M. (2001), 'The economics of industrial agglomeration and clustering, 1976–1996: the Case of Indonesia (Java)', PhD thesis, University of Melbourne, Melbourne.

Kussendrager, B.J.L. (1841), *Natuur- en Ardrijkskundige Beschrijving van het Eiland Java*, Groningen: Oomkens.
Kuvanonda, K. (1969), *Effect of the Korat–Nong Khai Highway in Northeast Thailand on Rail Transportation*, Bangkok: Asian Institute of Technology.
Labita, A. (1995), 'Canberra aid helps finance upgrading of railway into Manila', *The Australian*, 23 May, p. 64.
Laird, D. (1961), *Paddy Henderson, 1834–1961: a History of the Scottish Shipping Firm*, Glasgow: George Outram.
Larkin, J.A. (1972), *The Pampangans: Colonial Society in a Philippine Province*, Berkeley: University of California Press.
Laxon, W.A. and Perry, F.W. (1994), *B.I.: the British India Steam Navigation Company Limited*, Kendal: World Ship Society.
Leckie, C.S. (1894), 'The commerce of Siam in relation to the trade of the British Empire', *Journal of the Society of Arts*, No. 2168, 42: 639–62.
Lee Boon Thong (1995), 'Challenges of superinduced development: the mega-urban region of Kuala Lumpur-Klang Valley', in T.G. McGee and I.M. Robinson (eds), *The Mega-Urban Regions of Southeast Asia*, Vancouver: University of British Columbia Press.
Lee Boon Thong and Shamsul Bahrin Tunku (1984), 'A review of urban renewal and planning in Kuala Lumpur, Malaysia', unpublished paper presented at the International Seminar on Urban Renewal, Hyderabad, 27–30 November.
Lee Boon Thong and Tengku Shamsul Bahrin (1995), 'Shifting potential urban surfaces: a regionopolis response to globalisation in Malaysia', in *Cities and the New Global Economy: Conference Proceedings Volume 2 – an International Conference Presented by the OECD and the Australian Government 20–23 November 1994, Melbourne, Australia*, Canberra: Australian Government Publishing Service.
Lee, En-han (1977), *China's Quest for Railway Autonomy 1904–1911: a Study of Chinese Railway-Rights Recovery Movement*, Singapore: Singapore University Press.
Lee, R. (1989), *France and the Exploitation of China: a Study in Economic Imperialism*, Hong Kong: Oxford University Press.
Lee, R. (1999), 'Tools of empire or means of national salvation? The railways in the imagination of western empire builders and their enemies in Asia', *Working Papers*, Institute of Railway Studies, York [http:www.york.ac.uk/inst/irs/irshome/papers/ robertl].
Lee Tsao Yuan (1995), 'The Johor–Singapore–Riau growth triangle: the effect of economic integration', in T.G. McGee and I.M. Robinson (eds), *The Mega-Urban Regions of Southeast Asia*, Vancouver: University of British Columbia Press.
Legge, J.D. (1973), *Sukarno: a Political Biography*, Harmondsworth: Pelican.
Leinbach, T.R. (1974), 'The spread of transportation and its impact upon the modernization of Malaya, 1887–1911', *Journal of Tropical Geography*, 39, 54–62.
Lim Heng Kow (1978), *The Evolution of the Urban System in Malaya*, Kuala Lumpur: Penerbit Universiti Malaya.
Lim, J. (1997), 'The birth of a megahub: Negeri Sembilan – southern gateway for the KLIA-Putrajaya growth corridor', *Real Estate Review*, 25, 4–11.
Lindblad, J.T. (1996), 'Between Singapore and Batavia: the Outer Islands in the Southeast Asian economy during the nineteenth century', in C.A. Davids, W. Fritschy and L.A. van der Valk (eds), *Kapitaal Ondernemerschap en beleid: Studien over economie en politiek in Nederland, Europa en Azie van 1500 tot heden*, Amsterdam: NEHA.
London, B. (1980), *Metropolis and Nation in Thailand: the Political Economy of Uneven Development*, Boulder, CO: Westview Press.

Loo, K. (1973), 'Problems of urban administration: the Malaysian experience', in *KL Forum 73, July 1–6 1973*, Kuala Lumpur: Malaysian Institute of Planners and others, 43–51.
LTA (1996), *A World Class Land Transport System: a White Paper Presented to Parliament, 2 January 1996*, Singapore: Land Transport Authority.
Mackie, J.A.C. (1974), *Konfrontasi: the Indonesia–Malaysia Dispute, 1963–1966*, Kuala Lumpur: Oxford University Press.
Mackinder, H.J. (1931), *Report on Port Swettenham, Federated Malay States, Presented by the President of the Board of Trade to Parliament by Command of His Majesty September 1931*, Imperial Shipping Committee, Cmd 3953, London: HMSO.
MacMicking, R. (1967), *Recollections of Manilla (sic) and the Philippines*, reprint. Manila: Filipiniana Book Guild.
Madrolle, M.C. (1930), *The Traveller's Handbook to French Indochina and Siam*, Paris and London: Librairie Hachette.
Makepeace, W. et al. (1921), *One Hundred Years of Singapore*, London: Murray.
Man, J. de (1984), *Recollections of a Voyage to the Philippines*, trans. E. Aguilar Cruz, Manila: National Historical Institute.
Manapat, R. (1991), *Some are Smarter than Others: the History of Marcos' Crony Capitalism*, New York: Aletheia.
Manarungsan, S. (1989), *Economic Development of Thailand, 1850–1950: Response to the Challenge of the World Economy*, IAS Monograph No. 42, Bangkok: Institute of Asian Studies, Chulalongkorn University.
Manila Railroad Company (MRC) (annual), *Report of the General Manager for the year ending December 31*, Manila: Bureau of Printing (1920–39).
MT, *The Manila Times*, daily newspaper, Manila.
Marina (1994), *Milestones and Visions*, Manila: Maritime Industry Authority.
Marina (1996), Philippine commercial fleet 1994, data print-out (June 1996), Manila: Maritime Industry Authority.
Marks, T. (1997), *The British Acquisition of Siamese Malaya, 1896–1909*, Bangkok: White Lotus.
Maung Kyaw (1954), 'The land utilization of Insein district, Burma', *Malayan Journal of Tropical Geography*, 2: 56–61.
McAndrew, J.P. (1990), 'From Friar Estates to Industrial Estates: the political economy of urbanization in two Philippine municipalities', PhD thesis, University of Amsterdam, Amsterdam.
McBeth, J. (1991), 'Back on track: Philippine railway to upgrade services', *Far Eastern Economic Review*, 151 (14), 50–1 (4 April).
McCoy, A.W. (1982), 'A queen slowly dies: the rise and decline of Iloilo City', in A.W. McCoy and E.C. de Jesus, *Philippine Social History*, Sydney: Allen and Unwin.
McCrae, A. and Prentice, A. (1978), *Irrawaddy Flotilla*, Paisley: James Paton.
McGee, T.G. (1963), 'The cultural role of cities: a case study of Kuala Lumpur', *The Journal of Tropical Geography*, 17: 178–96.
McGee, T.G. (1967), *The Southeast Asian City: a Social Geography of the Primate Cities of Southeast Asia*, London: Bell.
McGee, T.G. (1971), *The Urbanization Process in the Third World: Explorations in Search of a Theory*, London: Bell.
McGee, T.G. (1989), 'Urbanisasi or kotadesasi? Evolving urban patterns of urbanization in Asia', in F.J. Costa, A.K. Dutt, L.J.C. Ma and A.G. Noble (eds), *Urbanization in Asia*, Honolulu: University of Hawaii Press.
McGee, T.G. (1990), 'The state and urbanization in Asia: the emergence of desakota regions', unpublished paper presented at the Conference on States and

Development in the East Asian Pacific Rim, held at the University of California, Santa Barbara, 22–5 March.
McGee, T.G. (1991), 'The emergence of desakota regions in Asia: expanding a hypothesis', in N. Ginsburg, B. Koppel and T.G. McGee (eds), *The Extended Metropolis: Settlement Transition in Asia*, Honolulu: University of Hawaii Press.
McGee, T.G. (1995), 'The urban future of Vietnam', *Third World Planning Review*, 17 (3), 263–77.
McGee, T.G. (1998), 'Globalization and rural–urban relations in the developing world', in F.-C. Lo and Y.-M. Yeung (eds), *Globalization and the World of Large Cities*, Tokyo: United Nations University Press.
McGee, T.G. and McTaggart, W.D. (1967), *Petaling Jaya: a Socio-Economic Survey of a New Town in Selangor, Malaysia*, Pacific Viewpoint Monograph No. 2, Wellington: Department of Geography, Victoria University of Wellington.
McGee, T.G. and Yeung, Y.M. (1977), *Hawkers in Southeast Asia: Planning for the Bazaar Economy*, Ottawa: International Development Research Centre.
McLeod, R. and Garnaut, R. (1998), *East Asia in Crisis: From Being a Miracle to Needing One*, London: Routledge.
McGilvary, D. (1912), *A Half Century among the Siamese and the Lao: an Autobiography*, New York: Fleming H. Revell.
McLennan, M.S. (1980), *The Central Luzon Plain: Land and Society on the Inland Frontier*, Manila: Alemar-Phoenix.
MDS (1996), *Rancangan Struktur: Majlis Daerah Seremban and Pihakberkuasa Perancang Tempatan Daerah Seremban, 1995–2000 – Laporan Pemeriksaan*, Seremban: Majlis Daerah Seremban and Pihakberkuasa Perancang Tempatan Daerah Seremban.
Mears, L.A., Agabin, M.H., Anden, T.L. and Marquez, R.C. (1974), *Rice Economy of the Philippines*, Quezon City: University of Philippine Press.
Mehrtens, B. (1998), 'Bangkok makes tracks', *Asian Business*, 34 (3), 48–9.
MHB (1930), *The Port of Manila: a Year Book 1930*, Manila: Manila Harbor Board.
MHS (1982), *Philippine Shelter System and Human Settlements*, Manila: Ministry of Human Settlements.
Miller, E.W. (1947), 'Industrial resources of Indo-China', *Far Eastern Quarterly*, 9 (4), 396–407.
MIT (1995), *Strategic Planning for Metropolitan Bangkok: Draft Final Report*, Massachusetts Institute of Technology Consultant Team, Bangkok: Bangkok Metropolitan Administration, City Planning Division.
Mitchell, B.R. (1998), *International Historical Statistics: Africa, Asia & Oceania, 1750–1993*, 3rd edn, London: Macmillan.
MMUTIP (1981), *Metro Manila Urban Transport Improvement Project: Final Report*, Manila: Ministry of Transport and Communications.
MN (1988), *Marine News*, The World Ship Society, vol. 42, nos 2 and 4.
MOC (1975), *Year Book of Transport Statistics 1975*, Kuala Lumpur: Planning and Research Division, Ministry of Communications, and Highway Planning and Public Transport Unit, Ministry of Works and Public Utilities.
Morehead, F.T. (1944), *Burma Pamphlets No. 5: the Forests of Burma*, London: Longmans, Green for Burma Research Society.
Morshidi, S. and Ghazali, S. (1999), *Globalization of Economic Activity and Third World Cities: a Case Study of Kuala Lumpur*, Kuala Lumpur: Utusan.
MOT (1979–1990), *Buku Tahunan Statistik Pengangkutan Malaysia* [Yearbook of Transport Statistics Malaysia], Kuala Lumpur: Ministry of Transport.
MRT Review Team (1980), *Singapore's Transport and Urban Development Options: Final Report of the MRT Review Team*, Singapore: Republic of Singapore.

Mulder, A.J.J. (1991), *Koninklijke Paketvaart Maatschappij: Wel en Wee van een Indische Rederij*, Alkmaar: De Alk.
Murphey, R. (1953), *Shanghai: Key to Modern China*, Cambridge, MA: Harvard University Press.
Murray, M.J. (1980), *The Development of Capitalism in Colonial Indochina (1870–1940)*, Berkeley: University of California Press.
Muscat, R.J. (1990), *Thailand and the United States: Development, Security, and Foreign Aid*, New York: Columbia University Press.
Myint, H. (1971), *Economic Theory and Underdeveloped Countries*, New York: Oxford University Press.
Naerseen, T. van, Ligthart, M. and Zapanta, F.N. (1966), 'Managing Metropolitan Manila', in J. Rüland (ed.) (1996), *The Dynamics of Metropolitan Management in Southeast Asia*, Singapore: Institute of Southeast Asian Studies.
Nas, P.J.M. (1986), *The Indonesian City: Studies in Urban Development and Planning*, Dordrecht: Foris.
NCB (1992), *A Vision of an Intelligent Island*, The IT 2000 Report, March 1992, Singapore: National Computer Board.
NESDB (1991a), *Southern Seaboard Development Project: Conceptual Master Plan. Executive Summary*, Bangkok: National Economic and Social Development Board.
NESDB (1991b), *National Urban Development Policy Framework: Final Report*, two vols, Bangkok: National Economic and Social Development Board.
NESDB (1992), *The Southern Seaboard Development Project: Inception Report*, Bangkok: National Economic and Social Development Board.
NESDB (2000), *Pab Ruan karn Pattana Seik ta kit Lae Sangkok Thai* [Overview of Thai economic and social development], Bangkok: National Economic and Social Development Board.
Nguyen Dinh Dau (1998), *From Saigon to Ho Chi Minh City: 300 Year History*, Saigon: Land Service Science and Technics Publishing House.
Nippon Yusen Chosa Guru-pu (2001), *Sekai no contena senpaku oyobi shuko jokyo* [World Container Fleet Operations], Tokyo: Nihon Kaiun Shukaisho [Japan Shipping Exchange].
Nontawasee, P. (ed.) (1988), *Changes in Northern Thailand and the Shan States, 1886–1940*, Comparative Research Award, Report No. 1, Singapore: Southeast Asian Studies Program.
North, D. (1958), 'Ocean freight rates and economic development, 1750–1913', *Journal of Economic History*, 18 (1), 537–55.
NSO (2002), *Key Statistics of Thailand*, Bangkok: National Statistical Organisation, Office of the Prime Minister.
NTPP (1982), *Final Report: Part III (Roads)*, Manila: National Transportation Planning Project (August).
Nwe, T.T. (1998), 'Yangon: the emergence of a new spatial order in Myanmar's capital city', *Sojourn*, 13 (1), 86–113.
Nyoe, U.T. and Khin, D.M.M. (eds) (1957), *The Burma Year-Book and Directory, 1957–58*, Rangoon: The Student Press.
NYT (1931), 'Mile-high motor road reaches Igorot land', *New York Times*, New York, 26 April [USNA, BIA: 2146a/4a].
Ockey, J.S. (1992), 'Business leaders, gangsters, and the middle class: societal groups and civilian rule in Thailand', PhD thesis, Cornell University, Ann Arbor: University Microfilms International.
Officieele Reisgids [Official Timetable] (1939), 28th edition (2 No. 1939–1 Mei 1940), No place or publisher given.

Ohmae, K. (1995), *The End of the Nation State: the Rise of Regional Economies*, New York: The Free Press.
Olszewski. P. and Tay, R. (1996), 'Road infrastructure and development in Singapore and Malaysia', *Asia Pacific Journal of Transport*, 1 (1), 1–14.
Onderzoek naar de Mindere Welvaart der Inlandsche Bevolking op Java en Madoera (OMW) (1906), *Samentrekkingvan de Afdeelingsverslagen over de Uitkomsten der Onderzoekingen naar het Vervoerwezen in the Residentie Soerabaja*, Batavia: Van Dorp.
Onderzoek naar de Mindere Welvaart der Inlandsche Bevolking op Java en Madoera (OMW) (1907a), *Uitkomsten der Onderzoekingen naar het Vervoerwezen: Overzicht van het Vervoerswezen*, vol. IVa, Batavia: Kolff.
Onderzoek naar de Mindere Welvaart der Inlandsche Bevolking op Java en Madoera (OMW) (1907b), *Uitkomsten der Onderzoekingen naar het Vervoerwezen: De Tarieven der Staatsspoorwegen op Java door A.W.E. Weijerman*, vol. IVb, Part II: appendix 8, Batavia: Kolff.
Ooi Jin-Bee (1969), 'Singapore: the balance sheet', in Ooi Jin-Bee and Chiang Hai Ding (eds), *Modern Singapore*, Singapore: University of Singapore.
Osborne, M.E. (1975), *River Road to China: the Mekong River Expedition 1866–1873*, New York: Liveright.
Osborne, M.E. (2000), *The Mekong: Turbulent Past, Uncertain Future*, New York: Atlantic Monthly Press.
Ouyyanont, P. (1997), 'Bangkok's population and the Ministry of the Capital in early 20th century Thai history', *Southeast Asian Studies*, 35 (2), 240–60.
Ouyyanont, P. (1999), 'Physical and economic change in Bangkok, 1851–1925', *Southeast Asian Studies*, 36 (4), 437–74.
Owen, N.G. (1971), 'The rice industry of mainland Southeast Asia 1850–1914', *Journal of the Siam Society*, 59 (2), 78–143.
Owen, N.G. (1984), *Prosperity without Progress: Manila Hemp and Colonial Life in the Colonial Philippines*, Berkeley: University of California Press.
Paauw, D.S. (1963), 'From colonial to guided economy', in R. McVey (ed.), *Indonesia*, New Haven: Yale University, Southeast Asian Studies Program.
PADECO (1994), *Regional Technical Assistance on Promoting Subregional Cooperation among Cambodia. The People's Republic of China, Lao PDR, Myanmar, Thailand, Vietnam–Subregional Transport Sector Study: Final Report – October 1994, PADECO Co., Ltd*, Manila: Asian Development Bank.
Pallegoix, J.B. (1854), *Description du Royaume Thai ou Siam*, Paris: Mission de Siam.
Panikkar, K.M. (1943), *The Future of South-East Asia*, London: Allen and Unwin.
Papendrecht, A. Hoynck van (1933), *De Zeilvloot van Willem Ruys Jan Danielszoon en de Rotterdamsche Lloyd*, Rotterdam.
Parkinson, C.N. (1964), *British Intervention in Malaya, 1867–1877*, Kuala Lumpur: University of Malaya Press.
Patanapanich, T. (1964), *Economic Effects of the East–West Highway*, Bangkok: SEATO Graduate School of Engineering.
Patterson, A. (1999), 'Burgeoning Bonifacio: the Global City', *Real Investments*, December 1999/January 2000, 20–4.
Pearn, B.R. (1939), *A History of Rangoon*, Rangoon: American Baptist Mission Press.
Pelzer, K. (1945), *Pioneer Settlement in the Asiatic Tropics*, New York: American Geographical Society.
Pendleton, R.L. (1943), 'Land use in Northeastern Thailand', *Geographical Review*, 33: 15–41.
Pendleton, R.L., with the assistance of Kingsbury, R.C. and others (1962), *Thailand: Aspects of Landscape and Life*, New York: Duell, Sloan and Pearce.

Perry, M., Kong, L. and Yeoh, B. (1997), *Singapore: a Developmental City State*, Chichester: John Wiley and Sons.
Phang Sock Yong (1992), *Housing Markets and Urban Transportation: Economic Theory, Econometrics and Policy Analysis for Singapore*, Singapore: McGraw-Hill Book Co.
Philippine Almanac (PA) (annual), Manila.
Philippine Commission (Philcom) (annual), *Report to the President of the United States*, Washington, DC: Government Printing Office, 1900–1915.
Philippine Commission (Philcom), Act 1112 of 11/4/04.
Philippine Islands, Bureau of Commerce and Industry (PIBCI) (*c*.1931), *Statistical Bulletin of the Philippine Islands, 1929*, Manila: Bureau of Commerce and Industry.
Philippine Islands, Bureau of Commerce and Industry (PIBCI) (*c*.1932), *Statistical Bulletin of the Philippine Islands, 1930*, Manila: Bureau of Commerce and Industry.
Philippine Islands, Governor-General of the (PIGG) (annual), *Report of the Governor-General of the Philippine Islands to the President of the United States*, Washington, DC, 1916–1935.
Philippine Islands, Governor-General of the (PIGG) (1927), *Report of the Governor-General of the Philippine Islands*, Appendix G (Report of the Advisory Committee on Interisland Shipping), Washington, DC: Government Printing Office.
Philippine Islands, High Commissioner to the (PIHC) (1939), *Second Annual Report of the United States High Commissioner to the Philippine Islands Covering the Year 1937*, Washington, DC: Government Printing Office.
Philippine Islands, High Commissioner to the (PIHC) (1943), *Fourth Annual Report of the United States High Commissioner to the Philippine Islands covering the Fiscal Year 1938/39*, Washington, DC: Government Printing Office.
Philippine Islands, Insular Collector of Customs (PIICC) (1931), *Annual Report, 1930*, Manila: Bureau of Printing.
Philippines, Bureau of Public Works (PBPW) (1950), 'Philippine highways: a study of the Highway Transport System requirements and recommendations', Highway Planning Survey, Manila: BPW/U.S/ Bureau of Public Roads (June).
Philippines, Census (1992), *1990 Census of Population and Housing*, Manila: National Census and Statistics Office.
Philippines, High Commissioner to the (PHC) (1947), *Seventh and Final Report of the High Commissioner to the Philippines covering the period from September 14, 1945 to July 4, 1946*, Washington, DC: Government Printing Office.
Philippines Herald Year Book (PHYB) (1934), Manila: Philippines Herald.
Philippines Ports Authority (1995), *Annual Statistical Report, 1994*, Manila, data print-out.
Philippines Statistical Yearbook (PSYB) (annual), Manila: National Statistical Coordination Board.
Phongpaichit, P. (1990), *The New Wave of Japanese Investment in ASEAN*, Singapore: ASEAN Economic Research Unit, Institute of Southeast Asian Studies.
Phongpaichit, P. and Baker, C. (1995), *Thailand: Economy and Politics*, Kuala Lumpur, Oxford University Press.
Phongpaichit, P. and Baker, C. (1996), *Thailand's Boom!*, Sydney: Allen and Unwin.
Phongpaichit, P. and Baker, C. (2000), *Thailand's Crisis*, Singapore: Institute of Southeast Asian Studies.
Phongpaichit, P., Pirayarangsan, S. and Treerat, N. (1998), *Guns, Girls, Gambling, Ganja: Thailand's Illegal Economy and Public Policy*, Chiang Mai: Silkworm Books.
Pinches, M.D. (1984), 'Anak-Pawes, children of sweat: class and community in a Manila shanty town', PhD thesis, Monash University, Clayton (Vic).
Pinder, D.A. (1997), 'Deregulation policy and revitalization of Singapore's bunker supply industry: an appraisal', *Maritime Policy and Management*, 24 (3), 219–31.

Poboon C., Kenworthy, J., Newman, P., Barter, P. (1995), 'Bangkok: anatomy of a traffic disaster', *Working Paper No. 47*, Perth: Asia Research Centre on Social, Political and Economic Change, Murdoch University, Western Australia.

Potter, D.M. (1996), *Japan's Foreign Aid to Thailand and the Philippines*, New York: St Martin's Press.

Prinsep, G.A. (1830), *An Account of Steam Vessels and of Proceedings Connected with Steam Navigation in British India*, Calcutta: Government Gazette Press.

PSA (2000), *Annual Report 2000*, Singapore: PSA Corporation.

Punpuing S. (1996), 'Commuting behaviour patterns in Bangkok', unpublished PhD thesis, Centre for Resource and Environmental Studies, The Australian National University, Canberra.

QBBPW (1915), *Quarterly Bulletin of the Bureau of Public Works*, Manila.

Raffles, Sir S. (1819) Letter to John Taylor, 9 June 1819, MSS, National Library of Singapore.

Raguraman, K. (1997), 'International air cargo hubbing: the case of Singapore', *Asia Pacific Viewpoint*, 38 (1), 55–74.

Ramaer, R. (1994), *The Railways of Thailand*, Bangkok: White Lotus Press.

Ramsay, J.A. (1976), 'Modernization and centralization in Northern Thailand', *Journal of Southeast Asian Studies*, 7 (1), 16–32.

Randolph, R.B. (1986), *The United States and Thailand: Alliance Dynamics, 1950–1985*, Berkeley: University of California, Institute of East Asian Studies.

Reed, R.R. (1967), 'The colonial origins of Manila and Batavia: desultory notes on nascent metropolitan primacy and urban systems in Southeast Asia', *Asian Studies*, 5, 543–62.

Reed, R.R. (1977), *Colonial Manila: the Context of Hispanic Urbanism and Process of Morphogenesis*, University of California Publications in Geography vol. 22, Berkeley.

Reid, A.J.S. (1969), *The Contest for North Sumatra: Atjeh, the Netherlands and Britain, 1858–1898*, Kuala Lumpur: Oxford University Press.

Reid, A.J.S. (1988), *Southeast Asia in the Age of Commerce, 1450–1680: the Land Below the Winds*, New Haven: Yale University Press.

Reid, A.J.S. (1992), 'Economic and social change, ca. 1400–1800', in N. Tarling (ed.), *The Cambridge History of Southeast Asia*, Cambridge: Cambridge University Press.

Reid, A.J.S. (1993), *Southeast Asia in the Age of Commerce, 1450–1680: Expansion and Crisis*, New Haven: Yale University Press.

Reitsma, S.A. (ed.) (1925), *Gedenkboek der Staatsspoor- en Tramwegen in Nederlandsch-Indië, 1875–1920*, Weltevreden: Topographische Inrichting.

Republik Indonesia (1952), *Kotapradja Djakarta Raya*, Jakarta: Kementerian Penerangan.

Rigg, J. (1997), 'Land-locked Laos: Dilemmas of development at the edge of the world', *Geopolitics and International Boundaries*, 2 (1), 153–74.

Rimmer, P.J. (1971a), *Transport in Thailand: the Railway Decision*, Department of Human Geography Publication HG/6 (1971), Canberra: Research School of Pacific Studies, The Australian National University.

Rimmer, P.J. (1971b), 'Government influence on transport decision-making in Thailand', in G.J.R. Linge and P.J. Rimmer (eds), *Government Influence and the Location of Economic Activity*, Department of Human Geography Publication HG/5 (1971), Canberra: Research School of Pacific Studies, The Australian National University.

Rimmer, P.J (1973), 'Freight transport in Thailand', in Robert Ho and E.C. Chapman (eds), *Studies of Contemporary Thailand*, Department of Human Geography Publication HG/8 (1973), Canberra: Research School of Pacific Studies, Australian National University.

Rimmer, P.J. (1982a), 'Theories and techniques in Third World settings: trishaw pedallers and towkays in Georgetown, Malaysia', *Australian Geographer*, 15 (3), 147–59.

Rimmer, P.J. (1982b), 'Urban public transport in smaller Malaysian towns: Threat to the trishaw industry', *Malaysian Journal of Tropical Geography*, 5 (June), 54–66.

Rimmer, P.J. (1986a), *Rikisha to Rapid Transit: Urban Public Transport Systems and Policy in Southeast Asia*, Sydney: Pergamon.

Rimmer, P.J. (1986b), 'Changes in transport organisations within Southeast Asian cities: petty producers to statutory corporations', *Environment and Planning A*, 18 (12), 1559–80.

Rimmer, P.J. (1989), 'Structure, conduct and performance of the rickshaw industry in East and Southeast Asian cities, 1869–1939', in *Transport Policy. Management & Technology Towards 2001: Selected Proceedings of the Fifth World Conference on Transport Research, Yokohama 1988, Volume III Challenges Facing Transport in Urban and Regional Development and Transport in Developing Countries*, Ventura: Western Periodicals, C597–612.

Rimmer, P.J. (1990), 'Hackney carriage syces and rikisha pullers in Singapore: a colonial registrar's perspective on public transport, 1892–1923', in P.J. Rimmer and L.M. Allen (eds), *The Underside of Malaysian History: Pullers, Prostitutes, Plantation Workers...*, Singapore: Singapore University Press.

Rimmer, P.J. (1991), 'A tale of four cities: competition and bus ownership in Bangkok, Jakarta, Manila and Singapore', *Transportation Planning and Technology*, 15 (2/4), 231–52.

Rimmer, P.J. (1995), 'Urbanization problems in Thailand's rapidly industrializing economy', in Medhi Krongkaew (ed.), *Thailand's Industrialization and its Consequences*, New York: St Martin's Press.

Rimmer, P.J. (1998), 'Ocean liner shipping services: corporate restructuring and port selection/competition', *Asia Pacific Viewpoint*, 39 (2), 193–208.

Rimmer, P.J. (1999), 'The Asia-Pacific Rim's transport and telecommunications systems: spatial structure and corporate control since the mid-1980s', *GeoJournal*, 48 (1), 43–65.

Rimmer, P.J. (2002), 'Asian-Pacific transport and communications in a global context: implications for economic and technical cooperation', unpublished paper presented at the ECOTECH/NGO/CEO Academic Seminar of APEC Ocean-related Ministerial Meeting Towards the Sustainability of Marine and Coastal Resources, 22–3 April, Seoul.

Rimmer, P.J. and Cho, G.C.H. (1981), 'Urbanization of the Malay States since Independence: evidence from West Malaysia, 1957 and 1970', *Journal of Southeast Asian Studies*, 9 (2), 349–63.

Rimmer, P.J. and Cho, G.C.H. (1989), 'Locational stress on Kuala Lumpur's urban fringe: Bandar Tun Razak case studies', in World Conference on Transport Research, *Transport Policy, Management and Technology Towards 2001: Selected Proceedings of the Fifth World Conference on Transport Research, Yokohama 1988, Volume I, The Role of the Public Sector in Transport*, Ventura: Western Periodicals.

Rimmer, P. and Cho, G. (1994), 'Locational stress in Shah Alam', in H. Brookfield (ed.), *Transformation with Industrialization in Peninsular Malaysia*, Kuala Lumpur: Oxford University Press.

Rimmer, P., Dick, H., Parapak, J. and Siregar, M. (1994), 'Facilitating trade: problems on the periphery', in *Expanding Horizons: Australia and Indonesia into the 21st Century*, Canberra: AGPS for Dept of Foreign Affairs & Trade.

Rimmer, P.J. and Dick, H.W. (2000), 'To plan or not to plan: Southeast Asian cities tackle transport, communications and land use', in S. Yusuf, W. Wu and S. Everett (eds), *Local Dynamics in an Era of Globalization*, Washington, DC: World Bank.

Rimmer, P.J. with Davenport, S.M. (1996), 'Flying from Empire to Commonwealth, 1919–1994', in Y.M. Yeung (ed.), *Global Change and the Commonwealth*, Hong Kong: Hong Kong Institute of Asia-Pacific Studies, Chinese University of Hong Kong.

Roachanakanan T. (1999), 'Bangkok and the Second Bangkok International Airport: politics of planning and development', unpublished PhD thesis, Department of Geography, The Faculties, The Australian National University, Canberra.

Robequain, C. (1944), *The Economic Development of French Indo-china*, trans. by I.A. Ward with supplement by J.R. Andrus and K.R.C. Greene, London: Oxford University Press.

Robinson, H. (1964), *Carrying British Mails Overseas*, New York: New York University Press.

Robinson, T. (1953), *Piekerans van een Straatslijper*, Bandung: Masa Baru.

Rodrigue, J.-P. (1994), 'Transportation and territorial development in the Singapore Extended Metropolitan Region', *Journal of Tropical Geography*, 15 (1), 56–74.

Roschlau, M.W. (1986), 'Public transport in the provinces: a study of innovation, diffusion and conflict in the Philippines', unpublished PhD, Department of Human Geography, Australian National University, Canberra.

Rosendale, P. (1978), 'The Indonesian Balance of Payments, 1950–1966: Some New Estimates', PhD thesis, Australian National University, Canberra.

RSRS (1947), *Royal State Railways of Siam: Fiftieth Anniversary*, Bangkok: Nai Udon Panpipatana.

Rüland, J. and Ladavalya, M.L.B. (1996), 'Managing Metropolitan Bangkok: Power contest or public service?', in J. Rüland (ed.), *The Dynamics of Metropolitan Management in Southeast Asia*, Singapore: Institute of Southeast Asian Studies.

Sadka, E. (1968), *The Protected Malay States: 1874–1895*, Kuala Lumpur: University of Malaya Press.

Sahni, J.N. (1953), *Indian Railways: One Hundred Years, 1853 to 1950*, New Delhi: Ministry of Railways.

Sandhu, K.S. (1969), *Indians in Malaya: Some Aspects of their Immigration and Settlement (1786–1957)*, Cambridge: Cambridge University Press.

Sato, S. (1994), *War, Nationalism and Peasants: Java under the Japanese Occupation, 1942–1945*, Sydney: Allen & Unwin.

Savage, V.R. (1992), 'Street culture in colonial Singapore', in Chua Beng-Huat and N. Edwards (eds), *Public Space: Design, Use and Management*, Singapore: Singapore University Press.

Savage, V.R. and Yeoh, B.S.A. (1993), 'Urban development and industrialization in Singapore: an historical overview of problems and policy responses', in Lee Boon Hiok and K.S.S. Oorjitham (eds), *Malaysia and Singapore: Experiences in Industrialization and Urban Development*, Kuala Lumpur: University of Malaya.

Schrieke, B. (1957), *Ruler and Realm in Early Java*, The Hague: Van Hoeve.

Scott, J.G. Sir (1921), *Burma: a Handbook of Practical Information*, 3rd edn rev., London: Daniel O'Connor.

Scott, James (1998), *Seeing like a State: How Certain Schemes to Improve the Human Condition have Failed*, New Haven: Yale University Press.

SCP (1973/1974), *Klang Valley Regional Planning and Development Study for the International Bank for Reconstruction and Development and the Government of Malaysia and the state of Selangor: Final Report*, Parts I, II, London: Shankland Cox Partnership.

Seah Chee Meow D. (1975), 'Some key issues in Singapore's domestic transportation: Who gets where, when and how?', *Department of Political Science Occasional Paper Series No. 18*, Singapore: University of Singapore.

Segers, W.A.I.M. (1987), *Manufacturing Industry, 1870–1942*, Changing Economy in Indonesia, vol. 8, Amsterdam: Royal Tropical Institute.

Seidenfaden. E. (1928), *Guide to Bangkok with Notes on Siam*, 2nd Edition, Bangkok: Royal State Railways of Siam [Reprinted Singapore, Oxford in Asia Paperbacks, 1994].

Sharp, L., Hauck, H.M., Chanlehka K., and Textor, R.B. (1953), *Siamese Rice Village: a Preliminary Study of Bang Chan, 1948–49*, Bangkok: Cornell Research Center.

Sherwood, C.P.K., Liew Choon Kong, L. and Nasron Adil Ibrahim (1998), 'Kuala Lumpur International Airport public transport system', in *The International Conference on Transportation into the Next Millennium: Proceedings 9–11 September 1998*, Singapore: Centre for Transportation Studies, Nanyang Technological University.

Short, D.E. and Jackson, J.C. (1971), 'The origins of an irrigation policy in Malaya: a review of developments prior to the establishment of the Drainage and Irrigation Department', *Journal of the Malaysian Branch of the Royal Asiatic Society*, 44 (1), 78–103.

Sidhu, J.S. (1965), 'Railways in Selangor 1882–1886', *Journal of the Malaysian Branch of the Royal Asiatic Society*, 38 (1), 6–22.

Siew Kuan Wai (1987), 'Malaysia: road transport', in P.J. Rimmer (ed.), *The ASEAN–Australian Transport Interchange: Proceedings of the 1985 Australian Workshop for Senior ASEAN Transport Officials*, Canberra: National Centre for Development Studies, The Australian National University.

Silcock, T.H. (1967), 'Outline of economic development 1945–65', in T.H. Silcock (ed.), *Thailand: Social and Economic Studies in Development*, Canberra: ANU Press.

Silverstein, J. (1964), 'Politics and railroads in Burma and India', *Journal of Southeast Asian History*, 5 (1), 17–20.

Singapore (1956), *Commission of Inquiry into the Public Passenger Transport System*, [The Hawkins Report], Singapore: Government Printing Office.

Singapore (1966), *Committee to Review the Licensing of Taxis and Taxi Drivers* [The Tan Report], Misc. (Singapore Parliament), 3 of 1966, Singapore: Government Printer.

Singapore (1974), *Management and Operations of Singapore Bus Service Ltd*, Report of Government Team of Officials, Singapore: Singapore National Printers.

Singapore Municipality (1892–1939), *Administration Report (Annual)*, Hackney Carriage and Jinrikisha Department (later Vehicle Registrations Office and Vehicle Registration Department), Singapore: Fraser and Neave/Straits Times/Government Printer.

Singapore, DOS (1960–1998), *Yearbook of Statistics Singapore*, Singapore: Department of Statistics.

Singapore, Legislative Assembly (1956), *Report of a Court of Inquiry into a Dispute between the Singapore Traction Company Limited and Its Employees*, Sessional Paper No. S.L. 10 of 1956, Singapore: Government Printing Office.

Singapore, Registry of Vehicles (1970), *Reorganisation of the Motor Transport Service of Singapore in Singapore, Paper presented to Parliament…21st* July 1970, Cmd 21 of 1970, Singapore: Government Printing Office.

Skinner, G.W. (1957), *Chinese Society in Thailand: an Analytical History*, Ithaca: Cornell University Press.

Smithies, M. (1986), *Old Bangkok*, Singapore: Oxford University Press.

Smyth, H.W. (1898), *Five Years in Siam: From 1891–1896*, 2 vols, London: John Murray.

South China Morning Post, daily newspaper, Hong Kong.

Spate, O.H.K. (1943), *Burma Pamphlets No. 2: Burma Setting*, London: Longmans Green for Burma Research Society.
Spate, O.P. (sic) (1958), 'Beginning of industrialisation in Burma', *Economic Research Project, Reprint Series No. 1*, Rangoon: Department of Economics, Statistics and Commerce, University of Rangoon. (Reproduced from *Economic Geography*, 17 (1), 1941).
Spate, O.H.K. (1991), *On the Margins of History: From the Punjab to Fiji*, Canberra, National Centre for Development Studies, Australian National University.
Spate, O.H.K. and Trueblood, L.W. (1942), 'Rangoon: a study in urban geography', *Geographical Review*, 32 (1), 56–73.
SS and FMS (1939), *Malayan Year Book, 1939*, Statistics Department, Straits Settlements and Federated Malay States, Singapore: Government Printer.
SSS (1938), *Handbook of Sailings of Straits Steamship Company, Ltd and of Companies Operating in Association*, Singapore: Mansfield & Co. (March).
Staatsspoorwegen in Nederlandsch-Indië (SSNI) (1933), *Verslag over het Jaar 1932*, Batavia: Landsdrukkerij.
Stadsgemeente Batavia (1937), *Batavia als Handels-, Industrie- en Woonstad* [Batavia as a Commercial, Industrial and Residential Center], Batavia: Kolff.
Staple, G. (ed.) (1998–2001), *TeleGeography*, annual, Washington, DC: TeleGeography Inc.
Sternstein, L. (1964), 'Settlement in Thailand: patterns of development', unpublished PhD thesis, Department of Geography, Research School of Pacific Studies, Australian National University, Canberra.
Sternstein, L. (1976), *Thailand: the Environment of Modernisation*, Sydney: McGraw-Hill.
Sternstein, L. (1977), 'Internal migration and regional development: the Khon Kaen development centre of Northern Thailand', *Journal of Southeast Asian Studies*, 8 (1), 106–16.
Sternstein, L. (1982), *Portrait of Bangkok*, Bangkok: Bangkok Metropolitan Administration.
Sternstein, L. (1995), 'Bangkok Metropolitan Region: population change from the censuses of 1960, 1970, 1980, 1990', *Working Paper*, Bangkok: Kiat Consultants Ltd Partnership.
Stifel, L.D. (1973), 'The growth of the rubber economy of southern Thailand', *Journal of Southeast Asian Studies*, 4 (1), 107–32.
Stocks, S.P. (1998), 'Alignment design for the Bangkok elevated road and train system', *Proceedings of the Institute of Civil Engineers: Transport*, 129 (2), 95–101.
Stone, R.L. (1973), *Philippine Urbanization: the Politics of Public and Private Property in Greater Manila*, DeKalb: Northern Illinois University, Center for Southeast Asian Studies.
Straits Settlements (1938), *Committee Appointed by His Excellency the Governor of the Straits Settlement to Enquire into and Report on the Present Traffic Condition in the Town of Singapore* [The Trimmer Report], Straits Settlements Legislative Council, No. 76 of 1938, Singapore: Government Printing Office.
Straits Settlements, Tramways Commission (1921), *Proceedings and Report of the Commission to Inquire whether the Public are or are not being afforded the full benefit of the Tramways* [The Barrett–Lennard Report], Singapore: Government Printing Office.
The Straits Times, daily newspaper, Singapore.
Suehiro A. (1989), *Capital Accumulation in Thailand, 1855–1985*, Tokyo: Centre for East Asian Cultural Studies.
Suwat W. (1994), 'Bangkok transport: a way forward', *The Wheel Extended: a Toyota Quarterly Review*, 87, 19–24.

Takaya Yoshikazu (1987), *Agricultural Development of a Tropical Delta: a Study of the Chao Phraya Delta*, trans. P. Hawkes, Honolulu: University of Hawaii Press.

TAMS (1963), *Kuala Lumpur Transportation Study*, Kuala Lumpur: Tippets-Abbett-McCarty-Stratton.

Tan, S.K. (1967), 'Sulu under American Military Rule, 1899–1913', *Philippine Social Science and Humanities Review*, 32, 1–187.

Tarvisin, N. and Suwarnarat, K. (1996), 'City study of Bangkok', in *Megacity Management in the Asian and Pacific Region: Policy Issues and Innovative Approaches, Volume II City and Country Case Studies*, Manila: Asian Development Bank and United Nations/World Bank Urban Management Programme for Asia and the Pacific.

Tasker, R. (1990), 'Smell of excess', *Far Eastern Economic Review*, 150 (48), 52–3.

Tate, D.J.M. (1979), *The Making of Modern South-East Asia*, 2nd edn, Kuala Lumpur: Oxford University Press.

Tay, S.S.C, Estanislao, J.P. and Soesastro, H. (2001), *Reinventing ASEAN*, Singapore: Institute of Southeast Asian Studies.

Taylor, J.G. (1983), *The Social World of Batavia: European and Eurasian in Dutch Asia*, Madison: University of Wisconsin Press.

TCI (1959), *A Comprehensive Evaluation of Thailand's Transportation System Requirements: a Report to H.E. the Minister of Communications, Government of Thailand*, Washington, DC: Transportation Consultants Inc.

Textor, R.B. (1961), *From Peasant to Pedicab Driver: a Social Study of Northeastern Thai Farmers Who Periodically Migrated to Bangkok and Became Pedicab Drivers*, Cultural Report Series No. 9, New Haven: Southeast Asia Studies Program, Yale University.

TFHS (1975), *Housing Policy Guidelines for the MBMR (Manila Bay Metropolitan Region): Technical Report*, Manila: Task Force on Human Settlements (March).

Than, M. and Rajah, A. (1996), 'Urban management in Myanmar, Yangon', in J. Rüland (ed.), *The Dynamics of Metropolitan Management in Southeast Asia*, Singapore: Institute of Southeast Asian Studies.

Théry, R. (1931), *L'Indochine Française*, Paris: Les Éditions Pittoresques.

Thet, A.T. (1989), *Burmese Entrepreneurship: Creative Response in the Colonial Economy*, Stuttgart: Steiner Verlag Wiesbaden.

Thompson, V. (1967), *Thailand, The New Siam* [1st pub. 1941], 2nd edn, New York: Paragon Book Reprint Corporation.

Tiglao, R. (1994a), 'Do I hear 50 billion?: Philippine developers plot bids for prime urban tract', *Far Eastern Economic Review*, 1 September.

Tiglao, R. (1994b), 'A Foot in the Door', *Far Eastern Economic Review*, 22 September.

Tijdschrift voor Nederlandsch-Indië (TNI), 12/74, Varia: Voetreizigers van Samarang naar Batavia, pp. 483–4.

Tinker, H. (1967), *The Union of Burma: a Study of the First Years of Independence*, 4th edn, London: Oxford University Press.

Torre, V.R. de la (1981), *Landmarks of Manila, 1571–1930*, Manila: Filipinas Foundation.

Tregonning, K.G. (1965), 'The origin of the Straits Steamship Company in 1890', *Journal of the Malaysian Branch of the Royal Asiatic Society*, 32 (1), 274–89.

Tregonning, K.G. (1966), 'Singapore and Kuala Lumpur: a politico-geographical contrast', *Pacific Viewpoint*, 7 (2), 238–41.

Tregonning, K.G. (1967), *Home Port Singapore: a History of Straits Steamship Company Limited, 1890–1960*, Singapore: Oxford University Press.

Trocki, C.A. (1979), *Prince of Pirates: the Temenggongs and the Development of Johor and Singapore, 1784–1885*, Singapore: Singapore University Press.

Trocki, C.A. (1990), *Opium and Empire: Chinese Society in Colonial Singapore, 1800–1910*, Ithaca, Cornell University Press.

TSD (1953), *The Siam Directory 1953–54*, Bangkok: Thai Inc.
Tsou, Pao-chun (1967), *Urban Landscape of Kuala Lumpur: a Geographical Survey*, Singapore: Nanyang University.
Tuck, P. (1995), *The French Wolf and the Siamese Lamb: the French Threat to Siamese Independence, 1858–1907*, Bangkok: White Lotus.
Twang, P.Y. (1998), *The Chinese Business Elite in Indonesia and the Transition to Independence, 1940–1950*, Kuala Lumpur: Oxford University Press.
Umbgrove Commissie (1862), 'Stukken betreffend de Suikercultuur op Java', in *Handelingen de Tweede Kamer van de Staten Generaal*, Zittingen 1862/63, Bijlage I, No. 11.
UN (1995), *World Urbanization Prospects: the 1994 Revision – Estimates and Projections of Urban and Rural Populations and of Urban Agglomerations*, New York: United Nations.
UN (2001), *World Urbanization Prospects: the 1999 Revision*, New York: United Nations.
UNDP (1997), *Draft Report: Review of HCM City Master Problems – National Project VIE/95/051 'Strengthening the Capacity for Urban Management and Planning in Ho Chi Minh City'*, Ho Chi Minh City: United Nations Development Program.
Unger, D. (1998), *Building Social Capital in Thailand: Fibres, Finance and Infrastructure*, Cambridge: Cambridge University Press.
United Nations Centre for Human Settlements (2001), *Cities in a Globalizing World: Global Report on Human Settlements in 2001*, London: Earthscan Publications.
URA (1991), *Living the Next Lap: Towards a Tropical City of Excellence*, Singapore: Urban Redevelopment Authority.
Usher, D. (1967), 'The Thai rice trade', in T.H. Silcock (ed.), *Thailand: Social and Economic Studies in Development*, Canberra: ANU Press.
Van Beek, S. (1995), *The Chao Phya: River in Transition*, Kuala Lumpur: Oxford University Press.
Van Beek, S. (1999), *Bangkok Then and Now*, Bangkok: AB Publications.
Van der Heide, J.H. (1903), 'The economic development of Siam during the last half century', *The Journal of the Siam Society*, 3, 74–101.
Van der Heide, J.H. (1906), *General Report on Irrigation and Drainage in the Lower Menam Valley*, Bangkok: Ministry of Agriculture.
Van Grunsven, L. (2000), 'Singapore: the changing residential landscape in a winner city', in P. Marcuse and R. van Kempen (eds), *Globalizing Cities: a New Spatial Order?*, Oxford: Blackwell.
Vatikiotis, M. (1995), 'Going under: Bangkok decides on a subway – again', *Far Eastern Economic Review*, 158 (40), 84–5.
Versteeg, W.J. (1860), *Nieuwe levens Etappe-Kaart van Java en Madoera* [place: publisher not given].
Viraphol, S. (1977), *Trade and Profit: Sino-Siamese Trade, 1652–1853*, Cambridge, MA: Harvard University Press.
Vries, E. de (1931), *Landbouw en Welvaart in het Regentschap Pasoeroean: Bijdragen tot de Kennis van de Sociale Economie van Java*, Wageningen: Veenman.
Waal, E. de (1879/1880), *Onze Indische Financiën*, vols III (1879), IV (1880): *Zeemacht en Aanverwante Zaken*, s'Gravenhage: Martinus Nijhoff.
Wahab, I. (1991), *Pengangkutan Bandar di Malaysia*, Kuala Lumpur: Dewan Bahasa dan Pustaka, Kementerian Pendidikan Malaysia.
Walker. A. (1997), 'The legend of the Golden Boat: regulation, transport and traders in north-western Laos', PhD thesis, Department of Anthropology, RSPAS, Australian National University, Canberra.
Walker, A. (1999), *The Legend of the Golden Boat: Regulation, Transport and Traders in the Borderlands of Laos, Thailand, China and Burma*, Richmond: Curzon Press.

Ward, M.W. (1960), 'The trade of the ports of Malaya', unpublished PhD thesis, University of Minnesota.
Warr, P.G. (1993), 'The Thai economy', in P.G. Warr (ed.), *The Thai Economy in Transition*, Cambridge: Cambridge University Press.
Warren, J.F. (1981), *The Sulu Zone, 1768–1980*, Singapore: Singapore University Press.
Warren, J.F. (1986), *Rickshaw Coolie: a People's History of Singapore, 1880–1940*, Singapore: Oxford University Press.
Warren, J.F. (1995), 'A strong stomach and flawed material: Towards the making of a trilogy', Singapore, 1870–1940', *Southeast Asian Studies*, 33 (2), 245–64.
Watanabe, H. (1996), 'Government response to urban transport problems in Asia – Thailand: a Case Study', *The Wheel Extended: a Toyota Quarterly Review*, 97, 5–13.
Wayte, M.E. (1959), 'Port Weld', *Journal of the Malayan Branch of the Royal Asiatic Society*, 33 (1), 154–67.
Webster, D. (1999), Infrastructure development policies and economic development: a perspective from Thailand, unpublished paper presented to workshop on ASEAN infrastructure planning and management, Bangkok, 30 September.
Weitzel, A.W.P. (1860), *Batavia in 1858: Or Schetsen en beelden uit de hoofdstad van Nederlandsch-Indië*, Gorinchem: Noorduijn.
Wernstedt, F.L. (1957), *The Role and Importance of Philippine Interisland Shipping and Trade*, Southeast Asia Program Data Paper No. 20, Ithaca: Cornell University.
Wernstedt, F.L. and Spencer, J.E. (1967), *The Philippine Island World: a Physical, Cultural and Regional Geography*, Berkeley: University of California Press.
Wickenden, P.F. (1994), 'Development of a plan for the rehabilitation of the railway in Cambodia', *Transport and Communications Bulletin for Asia and the Pacific, No. 64*, Bangkok: United Nations Economic and Social Commission for Asia and the Pacific.
Wijeyewardene, G. (1973), 'Hydraulic society in contemporary Thailand', in Robert Ho and E.C. Chapman (eds), *Studies of Contemporary Thailand*, Department of Human Geography Publication HG/8 (1973), Canberra: Research School of Pacific Studies, Australian National University.
Willis, K.D. and Yeoh, B. (1998), 'The social sustainability of Singapore's regionalisation drive', *Third World Planning Review*, 20 (2), 203–21.
Willmott, W.E. (1967), *The Chinese in Cambodia*, Vancouver: Publications Centre, University of British Columbia.
Wilson, C.M. (1993), 'Revenue farming, economic development and government policy during the Early Bangkok Period, 1830–92', in J. Butcher and H. Dick (eds), *The Rise and Fall of Revenue Farming: Business Elites and the Emergence of the Modern State in Southeast Asia*, New York: St Martin's Press.
Wilson, G.W., Bergmann, B.R., Hirsch, L.V. and Klein, M.S. (1966), *The Impact of Highway Investment on Development*, Washington, DC: World Bank.
Winichakul, T. (1994), *Siam Mapped: a History of the Geo-body of a Nation*, Honolulu: University of Hawaii Press.
Wolters, O.W. (1967), *Early Indonesian Commerce: a Study of the Origins of Srivijaya*, Ithaca: Cornell University Press.
Wong Lin Ken (1960), 'The Trade of Singapore, 1819–1869', *Journal of the Malaysian Branch of the Royal Asiatic Society*, 33 (4) (passim).
Wong Lin Ken (1965), *The Malayan Tin Industry to 1914: With Special Reference to the States of Perak, Selangor, Negri Sembilan and Pahang*, Tucson: University of Arizona Press.

Wong Lin Ken (1978), 'Singapore: its growth as an entrepot port, 1819–1941', *Journal of Southeast Asian Studies*, 9 (1), 50–84.
Wong, A.K. and Yeh, S.H.K. (eds) (1985), *Housing a Nation: 25 Years of Public Housing in Singapore*, Singapore: Maruzen Asia.
Wongsuphasawat, L. (1997), 'The extended Bangkok Metropolitan Region and uneven industrial development in Thailand', in C. Dixon and D. Smith (eds), *Uneven Development in South East Asia*, Aldershot: Ashgate.
Worcester, D.C. (1898), *The Philippine Islands and their People*, New York: Macmillan.
World Bank (1996), 'Republic of Indonesia: Strategic Urban Roads Infrastructure', *Staff Appraisal Report No. 15295-IND*, Washington, DC, June.
World Bank (1997a), *Socialist Republic of Vietnam: Inland Waterways and Port Rehabilitation Project: Staff Appraisal Report – September 25, 1997*, Washington, DC: East Asia Transport Sector Unit, East Asia and Pacific Regional Office.
World Bank (1997b), *World Development Report 1997: the State in a Changing World*, New York: Oxford University Press.
World Bank (1999), *Vietnam Moving Forward: Achievements and Challenges in the Transport Sector*, Washington, DC: East Asia Transport Sector Unit, East Asia and Pacific Regional Office.
World Bank (2000), *Entering the 21st Century: World Development Report 1999/2000*, Washington, DC: World Bank.
World Bank (2002), *World Development Report 2002: Building Institutions for the Market*, Washington, DC: World Bank.
WRI (2002), Watershed Research Institute [http:www.wri.org/watersheds/indexhtml].
Wright, A. (1910), *Twentieth Century Impressions of Burma: Its History, People, Commerce, Industries, and Resources*, London: Lloyd's Greater Britain.
Wright, A. and Breakspear, O.T. (eds) (1903/1908), *Twentieth Century Impressions of Siam: Its History, People, Commerce, Industry and Resources*, London: Lloyd's Greater Britain Publishing Company (Reprinted Bangkok: White Lotus, 1994).
Wright, A., and Cartwright, H.A. (1908), *Twentieth Century Impressions of British Malaya: Its History, People, Commerce, Industries and Resources*, abridged edn reprinted 1989, Singapore: Graham Brash.
Wright, G. (1991), *The Politics of Design in French Colonial Urbanism*, Chicago: University of Chicago Press.
Wright, L.R. (1970), *The Origins of British Borneo*, Hong Kong: Hong Kong University Press.
WSA (1973), *Urban Transport Policy and Planning Study for Metropolitan Kuala Lumpur*, New Haven: Wilbur Smith and Associates.
WSA (1998), *Putrajaya Holdings Sdn Bhd, Putrajaya Transport Planning Study*, Kuala Lumpur: WS Atkins (Malaysia) Sdn Bhd.
Yearbook of Philippine Statistics (YPS) (1940), Manila: Bureau of Printing.
Yearbook of the Philippine Islands (YBPI) (1920), Manila: Chamber of Commerce of the Philippine Islands.
Yeh, S.H.K. and Laquian, A.A. (1979), *Housing Asia's Millions: Problems, Policies and Prospects for Low-Cost Housing in Southeast Asia*, Ottawa: International Development Research Centre.
Yeoh, B.S.A. and Huang, S. (1998), 'Negotiating public space: strategies and styles of migrant female domestic workers in Singapore', *Urban Studies*, 35 (3), 583–602.
Yeung, Wai-Chung H. and Savage, V.R. (1996), 'Urban imagery and the main street of the nation: the legibility of Orchard Road in the eyes of Singaporeans', *Urban Studies*, 33 (3), 473–94.

Yordphol T. (1997), 'Bangkok's traffic crisis: can "demand management" cool it?, *The Wheel Extended: A Toyota Quarterly Review*, 98, 2–3.

York, F.W. and Phillips, A.R. (1996), *Singapore: a History of its Trams, Trolleybuses and Buses: Volume One, 1880s to 1960s*, Croydon: DTS Publishing.

Yuen, B. (1998), 'Singapore: the planned city state', in B. Yuen (ed.), *Planning Singapore: From Plan to Implementation*, Singapore: Singapore Institute of Planners.

Yun, H.A. (1999), 'The Singapore state takes charge: strategizing for the new hi-tech service society', *European Planning Studies*, 7 (2), 189–205.

Zaharah binti Hj. Mahmud (1970), 'The period and the nature of "traditional" settlement in the Malay peninsula', *Journal of the Malaysian Branch of the Royal Asiatic Society*, 43 (2), 81–113.

Web sites

http://www.bartleby.com/151/a105.html
http://www.cebu-airport.de.tt
http://www.kiat.net/malaysia/KL/transit.html
http://www.tvhistory/tvindex.html.
http://www.wri.org/watersheds/indexhtml.

Archival sources

Algemene Rijksarchief (ARA), Tweede Afdeeling, The Hague:

Koninklijke Paketvaart Maatschappij (KPM): Jaarverslag van de Directie in Indië (annual, 1891–1966) and Doos 908 (Heap Eng Moh).

Oost Java Stoomtram Maatschappij (OJS), Doos 24: Vereenigde Nederlandsch-Indische Stoomtram Maatschappijen (VNIST), Nota, 23/12/26.

Oost Java Stoomtram Maatschappij (OJS) Doos 24: Motorverkeer, Governor-General (GG), 9/7/25.

National Archives of Thailand:

NAT (1919), Thai integrated land-transport planning: Some documents in the National Archives of Thailand, Tha Wasukri, Bangkok, Archival file Rama 6. SB (suam bukkhom) 001/9.

NAT (1925), Thai integrated land-transport planning: Some documents in the National Archives of Thailand, Tha Wasukri, Bangkok, Archival file Rama 6. SB (suam bukkhom) 2.4/5.

United States National Archive (USNA), Washington, DC: Bureau of Interinsular Affairs (BIA) (Record Group 350) (cited by file no./item):

674: 12957/2: Proposal for Transportation of Philippine Government Mails and Passengers and Freight over Interisland Transportation Route No. 1 (Advertisement), Manila, 10/6/05.

674: 2893/40: Governor-General to Secretary of War, cable 11/11/27.

863/49: Report of the Committee on Coastwise Laws, 1903–1904 (16/8/04).

2146/17: Chief Engineer Officer, Report on road work in the Philippine Islands by the military authorities since 1899, Manila, 4 May 1903.

2146/38: Roads in the Philippine Islands: Present Condition and Proposed Betterment, Report, 15 March 1909.

2399/5: Governor-General to Secretary of War, 20/9/10.
6300/5 (Box 627): Press release (copy of), 22/12/02.
18343/22 (box 809): Secretary of State to Mr Isaac S. Dement re. automobiles in use in the Philippine Islands, 20/7/11.
13931/156 (with): Board of Public Utility Commissioners, Case 220, 31/7/15.
13931A/288A: Board of Public Utility Commissioners to the Governor General, 21/7/14.
13931A/289: Governor General to Secretary of War, 1/8/14.
13931A/341A: Manila Railway Company, Plan of Reconstruction, November 1906.
13931A/994: Brief History of the Manila Railroad Company, 10/9/37.
13931D/19, 'Traffic Manager Outlines Railroad's Role in P.I. Economic Development', clipping, 24/1/40.
26640/43: Public Service Commission Order No. 1, 13/12/26.

Index

Abaca 24, 75, 100–1, 106, 116, 136, 146
Abercrombie, Sir Patrick 240
Aboitiz Lines 116
accessibility 145, 272
Aceh War 84
aeroplanes: *see* aviation
agglomeration economies 6, 225; *see also* cities, urbanization
Agno River 136
agribusiness 302–3, 306
Agricultural University of Malaysia 329
agriculture 75, 77, 119, 126, 151, 153, 166, 171, 177, 179–80, 183, 186, 189, 190, 195, 197, 199, 203, 209, 219, 222, 234, 300, 308–9, 318, 324, 327, 329, 338; *see also* rice
aid, foreign 97, 113, 132, 147, 166, 178, 184, 281
Air Express Co. 58
Air France 55
Air India 58
Air New Zealand 253
Air Orient 55
Air Thai 58
air transport: *see* aviation
airconditioning 71, 134, 227, 286, 287, 305
airlines: *see* by company
Akyab (Sittwe) 170
Alexandria 50–1
Alfred Holt & Co. 54, 86–7, 90, 92–3, 95, 97
Allied South East Asia Command 7
Alor Star 216
Ambon 87, 274
America: *see* United States
American President Lines 253
Ampang 321, 332
Amphoe Pakkret 313
Amsterdam 55, 86–7, 275
Andaman Sea 159
Anderson, B.R.O. 339
Ang Mo Kio 244, 246
Anglo-Burmese War 48

Anglo-Dutch Treaty 83
Anjer 120, 123
Annam 156, 186–8, 191–3, 198–200, 204; *see also* Central Vietnam 186–8, 214–16
Aparri 113, 148
APEC 36
Arabs 13, 231
Arakan Flotilla Company 172
Aranyaprathet 168, 184
Arayat 138, 141
Area License Scheme 247, 251
area traffic control 305
ASEAN 3, 8, 26, 35, 36, 223, 316, 339
Asian crisis 36, 286, 288, 290, 309, 315–16, 318, 334, 337
Asian Development Bank (ADB) 184, 271
Asian Dollar Market 248
Asia(n)-Pacific 5, 21, 35, 250, 330
Asiatic Petroleum Company 235; *see also* Shell
Asoke Road 306
Association of Southeast Asian Nations: *see* ASEAN
Australia 3, 8, 9, 10, 11, 15, 18, 21, 26, 36, 41, 44, 47–8, 51, 55, 58, 59, 61, 70, 84, 90, 92–3, 97, 186, 192, 201, 252–3, 257
Automobile Club 129, 143, 266
automobiles: 40, 71, 83, 129, 144, 198, 226, 252, 265, 310, 324, 338; assembly of 129, 334; mobility 151; numbers (registration) 174, 209, 272, 280, 287, 294, 306, 312, 326, 328; first vehicles 237, 300, 322; parking of 244, 269, 305–6, 313, 328, 332; pooling 247; private 66, 68, 143, 202, 239, 247, 257, 269, 272; *see also* roads
Ava (Mandalay) 62, 70, 155, 161, 172, 174
aviation 40, 54–9, 173, 214, 239, 250–1, 295; aircraft 55, 71, 104; airlines, 21, 55–9, 175, 229, 236;

airports, 246, 281, 286–7, 334, 337, 339–40, 342; mails 17, 55, 58, 175; open skies policy 252; passengers, 29, 175, 177, 252, 336; *see also* by airline
Ayala Corporation 268–9, 271, 285
Ayutthaya 161, 164, 166, 170, 289, 290, 315

Baguio 58, 141, 144–6, 266
baht 158, 185, 309
baht economic zone 185
Baiyoke Tower xv, 309
bajaj 286
Bali 26, 83, 88, 135, 231
Bandung 8, 123, 129, 132, 222, 257, 281
Bandung Institute of Technology (ITB) 281
Bangi 328–9
Bangkapi 298, 308, 313
Bangkok xv, xviii, 27, 32, 34, 45, 54–5, 58, 70, 184, 187, 195–6, 225, 227, 289–318 *passim*, 338–42; and railways 64, 163–6, 168–71, 180, 208, 214; and roads 66, 68, 174, 178, 180; as regional hub 21, 185, 252; economy of 222–4, 286; government quarter 306; Municipal Act (1933) 295; population of 24, 29, 219–22; trade of 14, 19, 86, 88, 90, 158–61, 177, 180, 196; traffic congestion in 181, 303, 305, 309–10, 312
Bangkok and Vicinity 309
Bangkok Bank Group 302
Bangkok Elevated Road and Train System (BERTS) 311
Bangkok Mass Transit Authority 305, 310
Bangkok Metropolitan Administration (BMA) 303, 306, 308, 315
Bangkok Metropolitan Region (BMR) 309, 313, 315, 318
Bangkok Metropolitan Transport Authority 311
Bangkok Municipality 295, 303
Bangkok Traffic Management Project 305
Bangkok Transportation Study 305
Bangna 298, 310

banking and finance 36, 44, 233, 240, 248–50, 282, 302, 306, 309, 313, 337
Banque de l'Indo-Chine 300
Banyuwangi 120, 123
Bassein 62, 170
Bataan 106, 146
Batam 222, 254
Batamindo Industrial Park 254
Batangas 106, 138, 148, 153, 287
Batavia Freight Conference 88
Batavia: *see* Jakarta
Bataviasche Electrische Tram-Mij 278
Bataviasche Verkeers Mij (BVM) 279
Battambang 67, 162, 172, 174
Batu Pahat 197, 204
Bay of Bengal 9, 10, 41, 90, 155
becak 134, 280, 282–3, 288
Behn Meyer & Company 86, 88
Bekasi 135, 287
Belawan 89, 92, 98
bemo 282
Bencoolen: *see* Bengkulu
Bengal 3, 7; *see also* India
Bengkulu 67, 83
Benguet Road 144, 146; *see also* Baguio
Bentleys 45
Berjaya 334, 337
Berjaya Star City 337
Berry, B.J.L. 5
Bhamo 158, 161
Bicol 100–1, 112–13, 136, 138, 143, 146, 148, 151
bicycles 70–1, 134, 173, 198–9, 205, 238, 280, 286, 293
Bien Hoa 215, 316
bill of lading: *see* transhipment
Bintan 254
Bird, Isabella 190
Blue Funnel Line: *see* Alfred Holt & Co.
Boat Quay 235–6
Bogaardt, T.C. 86, 88, 90; *see also* Straits Steamship
Bogor 44, 61, 119–20, 122–3, 135, 278, 287
Bombay (Mumbai) 16, 32, 33, 41, 50, 54, 59, 161, 240
borders 6, 29, 155, 162–3, 185–7; of Malaya 62, 196, 211, 214; of Indochina 62, 67, 162, 168; of Thailand 62, 67, 162, 168, 179, 199,

borders – *continued*
  203, 205, 211; cross-border links 28–9, 39, 185, 254; *see also* Golden Triangle, nation, periphery
Borneo Company 160, 171
Borneo 45, 50, 107, 110, 160, 171, 23; *see also* Kalimantan, Sabah, Sarawak
Boulton & Watt 48
Bowring Treaty 3, 51, 158, 289, 292
Brantas River 119, 125
bridges 138, 177, 194, 294; construction of 160, 184–5, 265, 316; in Bangkok 292–5, 303; in Singapore 238; road 142, 178, 205, 214, 292, 303; wooden 322
Brisbane 45
Britain, British 8, 39, 47–8, 54, 64, 67, 101, 115, 141, 196, 204, 215, 237, 253, 281, 293, 321; Empire 3, 7, 9–10, 32, 41, 44–5, 50–1, 55–7, 59, 61, 70, 81, 83–4, 86, 90, 155, 160–6, 186–7, 192, 195–7, 300, 322; Post Office 48; Royal Navy 50, 231, 243; shipping 48, 50–4, 87–8, 90, 93–4, 97, 110; town planning 330; trade 7, 9, 11, 13, 15–17, 50, 81–4, 93–4, 97, 101, 107, 158, 190, 234–5; and Siam 158, 166, 170, 187, 195–6; in Malaya 3, 8, 62, 186–7, 189–90, 192, 194–7, 203–5, 215, 321–2, 325; in Singapore 9, 44, 82–3, 95, 186, 229–30, 235–6, 243
British India Steam Navigation Company (BI) 84, 90
British Indian Submarine Telegraph Company 41
British Military Administration (BMA) 204
British Petroleum (BP) 242
British rule 61, 82, 197
Brunei 3, 5
Brunel, I. 41
Buddhism 6
buffalo 14–15, 119, 174, 189–91, 202, 236, 238, 322; *see also* carts
Bugis 231–2, 235, 245
Buitenzorg: *see* Bogor
Bulacan 263, 287
bull(ock) cart: *see* cart
bungalows 260, 262, 280

bunkering: *see* coaling
Burachat, Prince 168–9, 173–4
bureaucracy 107, 120, 173, 274–5, 281, 293, 295, 309
Burma, Burmese 3, 6, 10, 13–14, 17, 24, 26, 28, 48, 68, 75, 90, 93, 179–80, 183–5, 253; colony of 7, 13–14, 61–2, 155–6, 160–1, 163–4, 169–74, 191, 195–6, 298; Union of (1948) 300: Inland Water Transport Board (Irrawaddy Section) 175, Inland Water Transport Nationalization Act 175; military rule (1962) 308, 316, 318; Socialist Republic of (1974–1988) 308; *see also* Myanmar
Burma Road 68, 174
Burma Railways 175
Burnham, Daniel H. 263
Bus de Gisignies, L.P.J. du 122
buses 67, 129, 130, 144–5, 173–4, 183, 238, 251, 266, 287, 294, 303, 332; air-conditioned 305; express 209; inter-city 134, 209; trolley 239–40; companies 237–8, 247, 304–5, 324; private 246, 304–5; public 134, 151, 173, 237, 282–3, 300, 304, 310–11; -lanes 305; passengers 132, 175, 177, 204–5, 270; services 146, 209, 239, 247
Butterworth 222, 322; *see also* Penang

Cabanatuan 137, 141
cables 38, 41–4; optic fibre 251; submarine 41, 47, 83; *see also* telegraphs, telecommunications
cable tram: *see* tramways
cabotage 102
Cagayan River 136–7
Calcutta 9–10, 16, 24, 32–5, 44, 48, 50–1, 56, 58–9, 70, 195, 219; *see also* Bengal
California 3, 7, 335
Caltex Oil 239, 242
Cam Ranh Bay 214
Cambodia 3, 62, 64, 155–6, 161–6, 168–9, 172–5, 183–5, 214, 253, 308
Cameron Forbes 142
Cameron Highlands 201, 203
Camino Real 138
Campbell, George Murray 164

Canada, Canadian 64, 241
canals 336; in Bangkok 290, 292, 294, 297–8, 300, 306, 308; in Thai Central Plain 158, 160–2, 171; *see also* Kra Isthmus
canoes 71, 136, 260
capital inflow 258, 285; *see also* foreign investment
capitalism 153, 181, 186, 197, 211, 215
cars: *see* automobiles
caravans 9, 122, 155, 159, 161, 169, 173
carriages 59, 70, 119, 121, 138, 260, 266, 278, 292, 322
carromata 265–6
carts 66, 121–2, 125, 127, 129, 134, 138, 141–3, 159, 164, 168, 171, 178, 189–91, 195, 202, 236, 280, 282, 293
cassava: *see* tapioca
Cathay Pacific Airways 58
Catholic orders 136, 260
cattle 14–15, 106, 110, 122, 126, 139, 145, 159, 198
Cavite 138, 148, 263, 264, 287
Cebu 58, 61, 79, 100, 104, 107, 110, 112, 115–6, 147, 222, 257; *see also* Visayas
Celebes: *see* Sulawesi
Centerpoint 309
Central Plain (Luzon) 61, 136, 139, 141, 143, 145–8, 151, 153
Central Plain (Thailand) 24, 26, 79, 161, 170, 177, 186, 264, 269, 287, 298, 325
Central Provident Fund (Singapore) 241, 249
Central Vietnam 79, 156, 186–8, 214–16; *see also* Annam, Vietnam
Ceylon: *see* Sri Lanka
Chachoengsao 164, 166, 168, 313, 315
Changi 236, 246, 252; International Airport 251–2
Chao Phraya Multipolis 316, 318
Chao Phraya River 196, 289, 294–5, 297–8, 303, 305
Charoen Phokphand 302
Cheras 326, 330
Chettiars 232
Chiang Mai 62, 79, 158–61, 164, 166, 168–9, 171, 174, 183

Chiang Mai Treaty 160
Chiang Saen 157, 159, 164
China Navigation Company 90, 93
China 3, 7–13, 15–16, 19, 21, 26, 28–9, 32–6, 40, 50–1, 62, 64, 66–8, 70, 75, 84, 86, 90, 93, 95, 98, 158–9, 161, 164, 168, 174, 184–5, 195, 202, 229, 232, 253, 325
Chinese (Overseas) 6, 66, 70, 173, 237, 260, 275, 279, 285, 300, 325; bus owners 239–40; business networks 4, 13–15, 82–3, 92, 94, 156, 190, 248, 257, 324; clans 237; immigrants 10, 13, 163, 172, 186, 188, 232–3, 289–90, 293, 321–2; labourers 166, 168, 172; miners 187, 189–90, 194–5, 321–2; secret societies 186, 189, 321; shipowners 54, 86, 88, 90, 92–6, 102–3; town quarters 232, 235–6, 240, 245, 249, 259, 274, 278, 294–5, 298, 306, 322, 327–8, 338; traders 6, 9–10, 12–14, 16, 32, 34, 36, 45, 54, 82–3, 86, 90, 92, 94, 107, 156, 159, 197, 199, 201, 231, 244, 257, 324; truckowners 202–3
Cholon: *see* Saigon
Chonburi 315
Christianity 141; *see also* Catholic
Chulalongkorn, King 164, 168, 293
cigarettes 295; cigars 259–60; *see also* tobacco
cinema 71, 280
Ciputra, Ir 285
cities 5, 71, 219, 222–7, 229, 232, 239–40, 253, 255, 272, 275, 278, 289–90, 302, 308–9, 312–13, 318, 336–7, 339, 344; capital 219, 222–3, 225–6, 258, 275, 339, 344; city-state 5, *see also* Singapore; colonial 259–67, 274–80, 309, 321–5; commercial 289–90; dual 289; extended metropolitan region 222; First World 36, 225, 227, 256, 289, 309, 312, 319, 332, 336–7; Fourth World 289; garden 244, 258, 336; global 219, 224–6, 240, 250, 332; industrial 239; mega- 222, 225, 257, 272; metropolitan 75, 225, 257–8; new enclave pattern 222; newly-industrializing 225, 309;

cities – *continued*
  post-colonial 289; provincial 222; royal 227, 289; systems of 219–22, 229, 342–3; Third World 224–5, 227, 229, 258, 289, 302, 308, 319–20, 325–6, 337–8; world 29–32, 36, 254; *see also* towns
class system 64, 266
cloves: *see* spices
clubs 264, 269, 274, 324
coaches: *see* carriages
coal 15, 40, 51, 54, 89, 92, 103, 198, 214, 235; coaling station 50
Coast Guard 101, 105
coastal shipping: *see* shipping
Cochinchina 3, 9, 155–6, 161, 163, 172, 174, 187, 198; *see also* Vietnam
coffee 11, 24, 75, 84, 119, 121, 127, 192, 194, 280, 322
Cojuangco, Eduardo 271
Cold War 7–8, 40
Colombo 35, 54, 241
Colonial Highway One 199, 204, 214; *see also* Vietnam
colonialism 6, 7, 38, 47, 54, 61, 64, 70, 71, 81–2, 88, 101, 107, 113, 117, 120, 141, 152, 225, 245, 257–9, 266, 278, 289, 309, 325
commercialization 166, 214
communications: *see* telecommunications, telegraph, telephone
Communist Party 8, 178, 204, 270, 318
Compagnie des Messageries Fluviales 172, 175
Compagnie des Messageries Maritimes 51, 86
Compania de los Tranvias de Filipinas 262
Compania Maritima 101
competition 77, 129, 130
Computer Integrated Terminal Operations System (CITOS) 253
Confrontation 95
congestion 68, 135, 148, 209, 236, 239, 247, 266, 270, 272, 286, 287, 289, 293, 300, 303, 305, 315, 318, 326, 328, 329; *see also* roads
Conrad, Joseph 246

containerization 98, 101, 106, 112, 114; *see also* ports, shipping
copra 24, 75, 89, 114, 116, 146, 234
core regions 27, 79, 114, 116–17; *see also* cities
Cores de Vries, W. 84
corridor 215, 315, 325, 329, 336; traffic 21, 313; multi- 334; technology 250, 336–7; transport 183–4, 211, 216, 246, 319–20, 328–9
corruption 97, 305
corvée 120–1; *see also* Cultivation System
cotton 16, 156, 162; *see also* textiles
coup d'état 289, 295
Cultivation System 84, 119–23, 126–7, 158
Cyber Village 336
Cyberjaya 336

Da Nang (Tourane) 184, 198, 215
Daendels, H.W., Governor-General 40, 66, 120, 138, 275
Dagupan 61, 137, 139, 147, 152
Daihatsu 282
Dalat 199
Damansara 191, 330
Damortis 144, 146
dams 183–4
Darwin 41
Davao 58, 103–4, 110, 112–13, 147, 257
Davao Chamber of Commerce 103
Dayabumi Complex 332
De Jongh, G.J. 89
decentralization 132, 183, 297
Decentralization Law 278
decolonization 17, 94, 222
deforestation 141, 184, 190–1
Deli 41, 86–9, 92; *see also* Belawan
Democracy Movement 305; *see also* Thailand
Democratic Republic of Vietnam: *see* Vietnam
demography: *see* population
depression 70, 94, 114, 130, 134, 139, 280
deregulation: *see* regulation
desa-kota model 75, 224, 260, 283
Devawongse, Prince 164

Development Bank of Singapore, DBS   243
Don Muang Airport   295, 297
Dona Paz   105
Dong Ha   204, 214
Dover   41, 50
drainage   166, 199, 264, 279
drug trade   316; *see also* Golden Triangle
dualism   71, 225, 286–9, 294, 318, 326
Dummler, Hr   278
Dusit   295, 298
Dutch rule: *see* Netherlands; *see also* colonization
Dutch: *see* Netherlands

East Asia   7, 17, 19, 21, 33, 36, 81, 86, 92–3, 153, 171, 234–6, 239, 316
East Asiatic Company, Danish   54, 90, 204
East Coast Malay(si)a: *see* Malay Peninsula
East India Company (EIC)   7, 9, 50, 83, 230
East Indies Ocean Steamship Company (EIOSS)   87–8, 90
East Malaysia: *see* Sarawak, Sabah
East Sumatra: *see* Sumatra
East Timor   3, 81, 95, 339
Eastern Indonesia   27–8, 83, 88, 114–16; *see also* periphery
Eastern Seaboard   100, 183, 313
Eastern Shipping Company   90
education   198, 211, 248, 263, 274–5, 279, 283, 287
Egypt   41, 50
electric tram: *see* tramways
electricity   39–40, 71; *see also* railways, telecommunications, tramways
electronics   183, 249, 313
Electronic Road Pricing (ERP)   251
elephants   66, 159, 171, 174, 189, 190, 195, 202, 293
elite   144, 258–9, 266–7, 269, 272, 281, 283, 286, 288, 312, 342, 344; *see also* middle class
Emergency: *see* Malayan Emergency
employment: *see* corvée, labour
enclave   154, 259, 269, 287, 289; *see also* cities

England   41, 55, 159; *see also* Britain
entrepot   86, 89, 187, 198, 234, 250, 319, 322; *see also* free port, free trade, Singapore
entrepreneurs   13, 64, 101, 173, 185, 236, 244, 265, 294, 298, 332
erosion   125; *see also* deforestation, siltation
Esso   242, 336
estates, landed   186; *see also* plantations
estero   259–60, 264; *see also* drainage, garbage
Eurasians   325
Europe(ans)   3, 8, 21, 24, 38, 40, 127, 158, 168, 293, 295, 306, 339; aviation 55, 58; business 170, 201, 235–6, 249, 298; colonialism 13, 24, 47, 94, 201, 227, 231–2, 257–9, 274, 280, 324; railways 59, 64, 68, 78, 171–2; shipping 9, 13–14, 16–17, 19, 33, 50–1, 54, 86, 93–5, 97–8, 158, 235, 253; telecommunications 41, 44–5, 47, 249, 275; trade 10–11, 13, 82–4, 86, 156, 234, 236; tramways 266; urban residents 45, 47, 55, 66, 70, 122, 132, 134, 144, 190, 202, 231–2, 234, 258–60, 263, 265, 274–5, 278–80, 324–5
European Union   8, 21, 339
expatriates   258, 269, 272, 285; *see also* Americans, Europeans
export-oriented policy   243, 329, 249
export-oriented manufacturing: *see* manufacturing
exports, non-oil   11, 82, 84, 86–9, 112, 114, 170, 177, 192, 197, 207, 222, 239, 250, 300, 322; *see also by* commodity, trade
Express Rail Link   334, 336
Express Transport Organization (ETO)   178
Expressway and Rapid Transit Authority   305, 310
expressways: *see* roads
Extended Bangkok Metropolitan Region (EMBR)   315; *see also* Bangkok
extended metropolitan region: *see* cities

Factories: *see* manufacturing
family planning   283; *see also* population

fares, passenger; airline 58; bus 305; interisland shipping 104–6, 113; railway 64, 125–6, 128–30, 132, 166, 201, 251, 311; tramway 237, 265, 280, 282; *see also* by mode
Federal Express 250
Federated Malay States (FMS): *see* Malaya
Federated Malay States Railway 197; *see also* Malayan Railway
feeder traffic 192, 311; *see also* railways, roads, shipping
ferries 48, 195, 201, 207, 254; interisland 106, 113; river 142, 174, 177–8, 266, 289, 293; vehicular 67
Filinvest Corporate City 271
firewood 127, 129, 141, 207, 235
First Pacific group 271
First World city: *see* cities
First World War 45, 54, 66–7, 88, 89, 103, 114, 129, 141, 168, 237, 324
fish(ing) 126, 136, 156, 186, 187, 198, 205, 207; dried 14, 83, 197–8, 204
floods 166, 283; *see also* drainage, estero
Floridablanca 141, 146
foodstuffs 192, 195, 202, 233, 321; *see also* agriculture, rice
Ford Motors 66, 174, 324
foreign investment 222, 271, 290, 316, 329; *see also* multinational corporations
France, French 8, 17, 45, 47, 51, 54–5, 59, 93, 158, 162; in Indo-China 3, 7, 16, 44, 47, 59, 62, 70, 161, 163–6, 169, 174–5, 187, 191, 198–9, 204, 215, 227, 298, 300
Fraser's Hill 201
free ports 13, 36, 54, 87, 89, 110, 207, 230, 234; *see also* ports, Sabang, Singapore
free trade 13, 36, 81–2; free trade agreements 36
freeways: *see* expressways, tollroads
freight 180, 207, 311; air 29, 31, 112, 252; rail 64, 66, 123, 126, 128, 130, 132, 134–5, 139, 151, 169–71, 183, 194, 199, 201, 211, 214; river 158, 160–1, 169, 172, 177, 183; road 66, 135, 159, 175, 237–8, 293; sea 50, 84, 92, 102, 104, 179; rates, by rail 128, 170, 173, 197, 204, by river 125, 191, by sea 54, 98, 105, 128, by truck 66, 180, 202; *see also* by mode
freight forwarder, multinational 250, 252
French: *see* France
Friendship Bridge 184
Friendship Highway 68, 177, 179
fruit 24, 136, 141, 145, 198, 260

Galle 51
gambier 83, 190, 194, 234, 322
garbage collection 264, 266, 269, 275, 278
Garden City movement 232; *see also* cities
garden suburbs 265, 280; *see also* suburbs
Garuda Indonesian Airlines 58, 95, 135
gated communities 259, 269, 313; *see also* cities
gateway policy 35, 98; *see also* shipping, Indonesia
Gemas 195–6
gems 155, 298
General Motors 129
General Transport Company 324
Geneva Accords 300
Georgetown (Penang) 319, 322, 338
German(y) 16, 40, 54, 55, 86, 88, 93, 110, 164, 166, 305
Geylang Serai/Eunos Area 244
gharry 237–8, 293, 322
globalization xv, 4, 5, 13, 19, 29, 36–8, 116, 151, 225, 257, 274, 339–44; *see also* cities, world
glutinous rice: *see* rice
Golden Shoe District 248, 250
Golden Triangle 6, 28, 332, 335, 337
Gombak River 321
goods movement 4–5, 36, 120, 185, 226, 237, 260, 300; *see also* freight
government: *see* by country
Grand Palace 290, 293
Great Depression 239, 295
Great Indian Peninsula Railway Company 59
Great Post Road 40
Great Strike 240
Greater Bangkok Plan (2533) 297

Greater Jakarta: *see* Jabotabek
Greater London Plan   240
Greater Mekong Sub-region (GMS)   184
gross regional domestic product (GRDP)   26, 223
growth triangle   27; *see also* SIJORI
guerillas   66, 145, 281; *see also* insurgency
guest workers   342
Gulf of Thailand   15, 178
gunboats   100
gutta percha   41, 83

Hackney carriage: *see* gharry
Haiphong   198, 215
handcarts   238, 300; *see also* hawkers
Hanoi   34, 62, 64, 67, 174, 198, 215, 219, 298
Hapag-Lloyd   95
harbours, natural   187, 192, 207, 278; *see also* ports
Hat Yai   196, 208, 211
Hatta, Moh.   281
hawkers   283, 292; *see also* streetsellers, informal sector
headquarters, business (BHQ)   249
Headrick, D.R.   40, 41, 47, 59
health, public   264, 283, 287; *see also* drainage, garbage, malaria
Heap Eng Moh S.S. Co.   93
Hébrard, Ernest   300
Henderson Line   54
highlands   67, 127, 132, 144–6, 179
High-Tech Park   336
Highway No. 1: *see* Colonial Highway One
highways: *see* roads
hill station   58, 144; *see also* highlands, by place
hinterlands   5, 24, 36–9, 64, 71, 73–216 *passim*, 219, 222, 224, 226, 256, 283, 287, 309, 338
Hitachi Zosen   242
Ho Chi Minh City: *see* Saigon
Ho Chi Minh Trail   214
Ho Hong S.S. Co.   93
Holts: *see* Alfred Holt & Co.
Hong Kong   3, 10, 13, 14–19, 21, 27–9, 31–6, 41, 44, 45, 51, 55, 58, 70, 81, 82, 89, 93, 96–8, 100–1, 107, 110
Hong Kong & Shanghai Bank   236

horses   59, 142, 159, 322; packhorses   145, 159, 192; posthorses   40, 120–2; -back   292; -cabs   265; *see also* carriages, carromata, carts, mules, ponies, sado, tramways
horsetram: *see* tramways
Hotel Indonesia   281–2
housing, estates, public   240, 249, 267, 269, 285, 287, 308, 313, 324
Howard, Ebenezer   244
Hualampong   294–5
hubs   33; airline   21; container   19; distribution   252; global   250, 253; logistics   252; regional   336
Hue   67, 79, 191, 198
Hukbalahap   269

IATA   17, 21, 58
ice   280
identity, national   2, 58, 274–5; regional   6–8
Ilocos   75, 136–9, 143, 146–7, 151
Iloilo   79, 100–1, 104, 110, 112, 147; *see also* Visayas
immigrants: *see* migration, Chinese, Indians
Imperial Airways   55, 236
Imperial Shipping Committee   202
imperialism   38, 40, 70, 71, 163–6, 172; *see also* colonialism
import substitution   208, 243, 302, 306, 327; *see also* manufacturing
imports   82, 114, 129, 179, 197, 202, 207, 234, 295
independence   94, 107, 147, 267, 281
India   21, 26, 40–1, 50–1, 55, 58, 61, 64, 90, 144, 187, 235, 253, 293, 298; and trade networks   7–15, 33–4, 82–3; Indians   13; in Bangkok   295; in Burma   300; in Malay(si)a   197–8, 325, 328; Singapore   232–3, 245, 249; *see also* Klings, South India, Tamils
Indian Ocean   8, 13, 45
Indochina   14–17, 26–8, 47, 54, 62, 67, 163, 169–72, 174–5, 183, 198, 215, 235, 298; *see also* Vietnam, Cambodia, Laos
Indo-European Telegraph Company   41
Indonesia, Republic of (1945)   xv, 6–7, 9–11, 27, 36, 47–8, 58, 222–4, 258, 281–2; Directorate-General of Sea

Indonesia, Republic of (1945) – *continued*
  Communications 96–8, 105; land transport 132, 134–6; Ministry of Shipping 96; shipping of 95–9;
  see also Jakarta, Java, Netherlands Indies
industrial estates 243, 250, 253, 254, 271, 286, 287, 313, 316, 327
Industrial Revolution 38–9, 48, 70, 152
industrialization 9, 24, 33, 38, 75, 111, 114, 116, 117, 153, 181, 216, 219, 222, 225, 246, 256, 259, 283, 287, 327, 339; *see also* manufacturing
informal sector 258, 271, 282, 289, 302, 313, 315; *see also* small-scale sector
Information Age 4
information flows 4–5, 36, 38–40, 47, 51, 71, 100, 185, 219, 223, 226, 231; *see also* telecommunications
information technology (IT) 4, 38, 44, 250–4, 334, 336; *see also* telecommunications
infrastructure 64, 70, 141, 147, 153, 191–2, 197, 215, 227, 339; *see also* ports, railways, roads, sewerage; also by city
innovation: *see* technology
insurgency 8, 113, 115–16, 179, 269; *see also* guerillas
Intergovernmental Group on Indonesia (IGGI) 97
interisland shipping: *see* shipping
interisland trade: *see* trade
Internet 29, 31, 37, 48, 51, 71, 153, 251–2, 340
intra-Asian trade: *see* trade
investment: *see* foreign investment, infrastructure, ports
Ipoh 192, 195, 207, 211, 319
Irrawaddy Flotilla Company 161, 172, 175
Irrawaddy River 62, 75, 155–61, 174–5, 191
irrigation 125, 141, 166, 172, 197, 199; *see also* agriculture, canals, rice
Islam 6, 111, 326
IT2000 252

Jabotabek 222–3, 257, 287; *see also* Jakarta
Jakarta 8–11, 17, 21, 29, 34, 44–5, 47–8, 84, 87–8, 119, 122, 227, 257–9, 274–88 *passim*; airport 17, 21, 34, 55, 132, 135; economy of 26–9, 31, 114, 153, 222–3, 289; infrastructure 120, 122, 129, 135, 279, 338; population of 24, 29, 132, 222, 257–8, 283, 286, 318; public transport 61, 68, 70, 123, 278; seaport 9, 31, 61, 88, 95, 98, 100, 278; trade of 11–15; traffic 135, 257, 280, 282, 287
Jakarta Transport (PPD) 282
Japan(ese) 3, 7–8, 11–12, 15–16; occupation by 24, 94, 114, 134, 147, 175, 205, 222, 227, 258, 267, 280, 282, 325
Java 3, 6, 9, 21, 58–9, 117–36 *passim*, 152–4; economy of 11–13, 15, 24, 26–8, 75, 78–9, 82–4; international networks 21, 41, 44–5, 50–1, 55, 58; public transport 70; and railways 59, 61–2, 64, 78; and roads 66–7
Java–China–Japan Line (JCJL) 92–5
Jawatan Kereta Api 132; *see also* Java, railways
jeepney 134, 151, 270, 272, 282
Jervois, Sir William 190
jitney 70, 280; *see also* bemo, jeepney, opelet
Johor Baru 195, 208–9, 211, 216, 222, 254
Johore 9, 36, 62, 100, 187, 190, 195–7, 199, 201, 203, 207, 238, 254–5
Johore Causeway 238
Jumbo jet 19, 58–9; *see also* aviation
jungle 83, 119; *see also* deforestation, highlands, teak, timber
junks 14, 54, 207, 298; *see also* shipping, sail
Jurong, port of 243, 246; *see also* Singapore
Jurong Industrial Estate 241
Jurong Town Corporation (JTC) 241, 243

Kalimantan 3, 9, 27–8, 67, 83, 84, 90, 92–3, 95, 114, 116
Kallang 231–2, 235–6, 239, 244
kampung 194, 274, 278, 283, 286, 338
Kaohsiung 19, 31, 112
Karachi 41, 55
Kedah 186–7, 195–9, 205, 207, 216

Kediri 119, 121, 123, 127
Kelantan 64, 90, 187, 195–6, 201, 205
Keppel group 95
Keppel Harbour (New Harbour) 235–7
kerosene 235; *see also* oil
Khaw family 195
Khon Kaen 62, 174, 179
Khong Falls (Khone Falls) 162, 164, 172, 174–5, 184
Kinta Valley 189
Kipling, Rudyard 246
Klages, Carter, Vail & Partners 335
Klang 189, 190–2, 194, 321–2, 327–9
Klang River 189, 191, 194, 321–2
Klang Valley 208; Corridor 319, 322, 328–9
Kling coolies 237
Klong Toey 297–8, 308
KNILM 58
Kobe 19, 24, 31, 33, 61
Kompong Cham 184
Kompong Som: *see* Sihanoukville
Koninklijke Java–China Paketvaart Lijnen: *see* Java–China–Japan Line
Koninklijke Luchtvaart Mij (KLM) 55
Koninklijke Paketvaart Mij (KPM) 87, 236
Korat: *see* Nakhon Ratchasima
Korean War 95
Kota Bharu 199, 201, 204, 207, 208
Kra Isthmus 75, 186, 195–6, 211
Kratie 162, 174, 184
Krishnan, Ananda 335
Krung Kasem Canal 292–3
Kuala Lumpur xv, 21, 29, 32, 34, 58, 62, 112, 187, 189, 191–2, 201, 202–4, 207–9, 211, 214, 219, 222, 225, 227, 318, 319–38 *passim*, 339–40; Airport City 336; Central Market 332; City Center (KLCC) 334, 337; City Hall (Dewan Bandaraya) 327, 330, 332, 334; city-region 334; conurbation 319–20, 325, 327–8; Federal Territory 327, 330; Gateway 337; International Airport (KLIA) 334, 336; Linear City 334; municipal government 324; Public Works Department 324, 328; Sanitary Board 322, 324; Sentral Station (Brickfields) 334, 337

Kuala Lumpur Stock Exchange (KLSE) 330
Kuala Lumpur Tower 334
Kuala Lumpur–Klang Railway 194, 322
Kuantan 199, 201, 203–4, 207–9
Kussendrager, B. 119, 121

La Union 136, 146
labour 13, 66, 70, 75, 77, 117, 119, 139, 189, 197–8, 201, 259, 260, 269, 309, 329, 342
Labuan 3, 83
Laem Chabang 100, 183, 313
Laguna 138, 148, 153, 272, 287
Laguna de Bay 136–7, 153, 263
Lampang 161, 166, 168, 171, 174, 183
land frontier 75; *see also* agriculture
land transport: *see* automobiles, buses, carts, horses, railways, roads, tramways
land use 241, 248, 306, 308, 313, 318; rural 141, 241, 298, 308; urban 227, 230, 235, 240, 241, 248, 251, 270, 297, 306, 308, 313, 315, 318, 324, 327, 332, 336, 338
land use and transport 240, 256, 289, 308, 315, 318, 327
Lang Son 174
Laoag 146–7
Laos 3, 6, 155–6, 159, 161–2, 166, 172, 174–5, 178–9, 183–5, 204
Larut Valley 189
Lashio 62, 68
Latin America 39, 55
law and order 189–90, 264
Leckie, John Stuart 160–1
Legaspi 146, 152
Leyte 110, 113, 116
licensing: *see* regulation
light rail, Kuala Lumpur 330, 332; *see also* STAR, tramways
Light Rail Transit (LRT), Manila 271–2
lighterage 107, 207, 243, 260
Lion group 206
Lippo group 285
Litchfield Whiting Browne & Associates 297
Little India 232, 245
livestock: *see* cattle, poultry, swine 82, 83
logistics 100, 250, 252–3, 315

London   36, 50–1, 55, 68, 160, 229, 240, 248, 250, 336, 342
lorries: *see* trucks
Los Angeles   16, 29, 151, 226, 320, 338
Luang Prabang   159, 162, 164, 172, 174, 184
Lufthansa   55
Luzon   3, 21, 24, 27–8, 40–1, 50, 61–2, 67, 71, 75, 78, 79, 82, 100–1, 104, 107, 110, 112–13, 116–17, 121, 136–54 *passim*, 192, 259, 269, 272, 287, 325
Luzon Bus Lines   146

Mactan   112; *see also* Cebu
Madras (Chennai)   41, 59, 70
Madura   75, 119, 129, 132
Mahathir, Dr Mohamad, Prime Minister   209, 319, 332
mails   13, 17, 32, 44, 48, 50, 55, 71, 83, 87, 101, 104, 120, 122, 172, 175; contracts for   54, 87; *see also* shipping
Main Street   32–4, 36
maize   110, 172, 180, 198
Makassar   87, 89, 98, 115, 222
Makati   268–73
Malabon   263–4
Malacañang   260
Malacca   3, 7, 9, 10, 13, 82–3, 88, 94, 186–7, 189–90, 194, 199, 202, 204–5, 209, 215–16, 232, 257
Malacca Strait   9, 13, 67, 88, 94, 186, 232
Malang   123, 127
Malang Stoomtram Mij   127
malaria   264, 275
Malay Archipelago   6, 9, 81–116 *passim*
Malay Peninsula   3, 9, 83, 95, 156, 186–216 *passim*, 254, 319; East Coast   64, 67, 90, 187, 191, 196, 198, 207, 215; *see also* Malaysia
Malay States, Federated (FMS)   187, 194, 197, 319, 322; Unfederated (UMS)   187
Malaya, Federation of (1948)   204–5, 325–6; *see also* Malaysia
Malayan Airways Limited   58, 95, 207
Malayan Banking Training Institute   329
Malayan Communist Party   204
Malayan Emergency   204–5, 208, 326

Malayan Railway (*Keretapi Tanah Melayu*)   194, 201, 235; corporatization   214
Malayan Stevedoring and Transportation Company   207
Malayan Union   204, 325
Malaysia, Federation of (1963), Malaysia   3, 6, 8, 10, 17, 19, 21, 23, 26–8, 36, 58, 75, 81, 95, 98–100, 114, 116, 205–16 *passim*, 222, 234, 251, 254–5, 257, 319, 329–30, 332, 335–8; Federal Government of   327, 332, 334, 337; Highway Authority   209; Malaysia Plan, Second   209; National Electricity Board   326, 329; Urban Redevelopment Authority, Malaysia   327
Malaysian Airlines System (MAS)   58
Malaysian International Shipping Corporation   98
Malaysia–Singapore Second Expressway
malls, shopping   271, 283, 285, 287; *see also* retailing
Mandalay: *see* Ava
Mandarin Road   67, 191
Manila   12, 40, 48, 227, 257–74 *passim*, 278–88 *passim*, 342; airport   17, 34, 55, 58, 112; economy   222–3; hinterland of   61, 110, 136, 139, 141, 143, 145–6, 151–2, 265; infrastructure   264, 268–9, 278; Municipal Board (1916)   267; population   24, 29, 219, 222, 257–8, 260, 269, 283, 318; public transport   68, 70, 270, 280; seaport   16, 31, 51, 100–1, 103, 106–7, 110, 112, 136–7, 146–7, 153, 259; traffic   138, 142–3, 147–8, 151, 257, 265–6, 272
Manila Automobile Club   143, 266
Manila Electric Railway   265
Manila Golf Club   296
Manila Hotel   147, 264
Manila Port Terminal   147, 218
Manila Railroad Company (MRC)   141, 145–6
Manila Railway Company   61, 152
Manila Suburban Railways Company   265
Mansfield & Co.   86
manufacturing   154, 183, 208, 232, 239, 242–4, 249, 252–3, 257, 260, 295, 298, 306, 309, 313, 330; *see also* industrialization, textiles

Marconi, Guglielmo 45
Marcos, F., President 105, 113, 270–1
Marcos, Imelda 267, 270, 283
marine engineering: *see* shipbuilding
marine motors 103; *see also* shipping
marine safety 104–5; *see also* shipping
Maritime Industry Authority (Marina) 105
markets 119, 125–6, 134–5, 232; local 119, 125, 126–8, 132, 134–6, 141, 170–1, 178, 180, 265, 274–5, 282, 294, 306; *see also* banking and finance
market economy 13–14, 19, 24, 36, 72, 77, 116, 126–7, 136, 153, 156, 160, 171, 179, 227, 233, 283, 309; *see also* capitalism, commercialization
Marseilles 50, 55, 86, 120
Mass Rapid Transit (MRT), Singapore 70, 247, 250–1, 312, 329
mass transit: *see* railways, tramways
Maugham, Somerset 246
McLeod, Neil 101, 286
McMicking, Joseph 268
megacity: *see* city
mega-projects 305, 319, 334, 337
Mekong Commission 184
Mekong Project 155, 183
Mekong River 67, 158, 161, 184–5; sub-region 29
Melbourne 45, 50, 59, 68, 70, 98
Meralco 265–6, 270
mercantilism 81, 100
Mersing 201, 203–4, 207
Messageries Maritimes: *see* Compagnie des
Metro Manila 113, 148, 151, 222, 270, 271; *see also* Manila
Metro Manila Commission (MMC) 270
Metro Manila Transit Corporation 271
Metro Pacific Corporation 271
Metrotren 272
Mid Valley City 337
middle class 66, 72, 223–5, 257–9, 266–7, 269, 272, 283, 286–8, 308, 312–13, 315, 325, 328, 342, 344; *see also* elite
Middle East 41, 117, 198
migration 3, 24, 136, 139, 166, 232, 240, 244, 264, 315; circular 126, 135, 148; labour 156, 197, 249, 259, 282–3, 327; rural–urban 24, 67, 75, 208–9, 219, 222, 224, 227, 257–8, 269, 281–3, 297, 308, 325–6, 328, 338; transport stimulus to 110, 190, 197, 199; *see also* Chinese
Minburi 306, 313
Mindanao 3, 24, 27, 58, 61, 67, 81, 100, 103, 106–13, 116, 147, 215
Mindanao Motor Line 147
minerals: *see* mining
Mines Resort City 336
minibus 305, 328, 332
mining 145, 162, 171, 186, 190–2, 194, 197–8, 203–4, 215, 234, 321, 322, 330; *see also* oil, tin
Mobil Oil 242
mobility 68, 117, 125–6, 135, 151–2, 265, 280, 287; *see also* migration
Monetary Authority of Singapore 249
money politics 272
monopoly 77–8, 107, 270; *see also* regulation
Mons 295
monsoons 6, 13, 83, 88, 100, 138, 265; *see also* seasonality
Morse, Samuel 40–1, 44
Morse code 41, 44
mosquito bus: *see* buses
mosquitoes 264, 275; *see also* malaria
motor bus: *see* buses
motor car: *see* automobiles
motor transport: *see* automobiles, buses, vehicle, registration
motorcycles 132, 134, 143, 151, 178, 238, 287, 306, 310, 316
Moulmein 62, 159–61, 164, 170, 184
Mountain Province 145
Muar 197, 201
mules 144, 159, 173–4; *see also* horses
Multimedia Development Corporation (MDC) 336
Multimedia Super Corridor (MSC) 334–7
multinational corporations 4, 31, 36, 44, 97, 243, 248, 302, 313, 315
municipal administration 227, 237–9, 265, 278–9, 295, 303, 324–5; *see also* by city
muslim: *see* Islam
My Thuan Bridge 184
Myanmar, Union of (1988) 3, 184–5, 289–90, 316, 318; State Law and

Myanmar, Union of (1988) – *continued*
 Order Council (SLORC) renamed State Peace and Development Council (SPDC) (1997)   316; *see also* Burma
Myint, Hla   77, 156
Myitkyina   62, 171
Mytho   62, 163

Naga   146, 148, 152
Nagasaki   12, 41
Nai Lert   294
Nakhon Pathom   309, 315
Nakhon Phanom   159
Nakhon Ratchasima (Korat)   58, 62, 68, 159, 164, 166, 168–70, 177, 179, 183, 222
Nakhon Sawan (Paknampo)   155, 158, 160–1, 170–1, 180
Nakhon Si Thammarat   196
Nan   155, 159, 173
Nanyang Technological University   251
nation(-state)   xv–xvi, 3–5, 219, 289, 344, 339
national carriers: *see* aviation, shipping
National Economic Policy (NEP)   327, 330
national economies   4–5, 25–9, 75, 81, 114, 116, 178, 257
National Highway, Indonesia   112–13, Philippines   148, 181; *see also* roads
national income accounting   4; *see also* GDP
National Iron and Steel Mills   241
National Liberation Front   308
national planning   208, 216, 227, 316
National University of Malaysia   329
nationalism   2, 58, 66, 81, 264; *see also* identity, nation
nationalization   5, 95
Nederlandsch Stoomvaart Maatschappij 'Oceaan' (NSMO)   87, 88
Nederlandsche Scheepvaart Unie (NSU)   116
Nederlandsch-Indische Spoorweg Maatschappij (NISM)   123, 125–6, 128
Nederlandsch-Indische Stoomtram Maatschappij (NISTM)   123, 278
Nedlloyd Lines   95, 116
Negri Sembilan   189, 197, 199, 337
Neptune Orient Lines   242, 253
Netherlands (Dutch)   44, 59, 70, 97, 134, 166, 280–1; airlines   15, 17, 55; enterprise   153, 215, 239; imperialism   81, 87–8, 152, 230, 232; shipping   15–16, 86–9, 91–5, 97, 115–16, 236; town planning   227, 244, 274–5, 281; trade of   11–12; *see also* Netherlands Indies
Netherlands Indies, colony of   3, 7, 9, 11, 66–7, 107, 113, 120, 123, 278; colonial rule of   7, 13, 47, 84, 88–9, 122, 130, 274–5, 288n; shipping of   50–1, 84–6, 90–2, 101; telecommunications   44–5, 47; trade of   13, 83–4, 114; Public Works Department   129, 279; Road Traffic Act (1933)   130; *see also* Indonesia, Republic of; Java
Netherlands Dredging & Engineering Company (NEDECO)   97
Netherlands Indies Steam Navigation Company (NISN)   84–7, 92, 278
Netherlands Trading Company (NHM)   84
networks   5, 19; airline   15, 35, 38–9, 54–5, 58, 153, 337; commercial   6, 36, 89, 115; railway   39, 59, 61, 66, 71, 116, 123, 128, 130, 141, 146, 153, 175, 197, 201, 203, 247; regional   6, 12, 14, 81, 83; roads   39, 66–7, 112–13, 120–1, 135, 138, 148, 153, 155, 174, 177, 180–1, 197, 201, 208, 211, 214–16, 251, 294, 297, 303, 310–11, 316, 322, 330, 339; shipping   15, 35, 38–9, 51, 54, 86, 97, 99, 112, 1125, 153; telecommunications   35, 38–40, 44–5, 122, 153, 158, 226, 251–2; urban public transport   39, 68, 132, 145, 265–6, 272, 278, 282
New Harbour (Keppel Harbour)   235
New Order   97, 134, 283, 287
New Road (Charoen Krung)   293, 306
new towns   244, 250–1, 325, 328–9, 334, 336, 338
new villages   326
New York   36, 48, 51, 68, 229, 336, 240, 248, 250, 309, 319, 336, 342
newly industrializing city-region: *see* cities
Newly Industrializing Economies (NIE)   183, 215, 243, 313

newspapers   47, 122, 274, 278
Ngo Dinh Diem   300
Nha Trang   198, 215
Nippon Yusen Kaisha   54
Nong Khai   58, 159, 164, 169, 174, 178–9, 183–4
non-oil exports   98; *see also* exports
Norddeutscher Lloyd (NDL)   54, 86–90, 95
North America   38, 40, 47, 68, 70, 306, 342; railways in   64, 78; shipping routes to   19, 93, 253; trade of   10, 13, 16, 26, 82–4, 98, 249; *see also* United States
North Vietnam   156; *see also* Tonkin, Vietnam
Northern Focal Economic Area   215
Nueva Ecija   137, 146

Ocean S.S. Co.: *see* Alfred Holt & Co.
Oei Tiong Ham   93
Ohmae, K.   4
oil   40, 89, 92, 103, 106, 130, 144, 183, 191, 202, 205, 208, 211, 214, 234–5, 239, 242, 295, 297–8, 303, 334; refineries   242
oil palm   197, 208, 209, 322
Olano, Larrinaga & Co.   54
omnibus: *see* buses, horses
opelet   280, 282
opium   16, 29, 32, 83–4, 159, 174, 189, 190, 192, 194, 235
Opium War   3
Orchard Road   247–8
Orient Express   214, 260
Orient Line   54, 242, 253
Overseas Containers Ltd (OCL)   97
Overseas Chinese: *see* Chinese
oxen   66, 121–2, 128, 159, 164, 173, 177, 180, 195, 293; *see also* carts

P&O   13, 32, 50–1, 55, 84, 86, 90, 97, 101, 235
Pacific Mail   51
Pacific War   7, 58, 67, 173, 186, 203, 204, 214, 215, 222, 225, 239, 282, 289, 297, 300; *see also* Japanese occupation, Second World War
packhorses: *see* horses
Padang Besar   64, 196, 201
paddy: *see* rice

Pahang   64, 67, 187, 196, 199, 201, 203, 205, 209
Paknampo: *see* Nakhon Sawan
Pakse   159, 171, 174; bridge at   184
Pakubuwana, Sultan   129
palanquin: *see* sedan chair
Palapa   48
Palembang   9, 58, 67, 84, 222
Palm Oil Research Institute   329
Pampanga   136, 143, 146, 263, 287
Pampanga River   137
Pan-American Airways (Pan-Am)   55
Pan-Asianism   8
Pangasinan   143, 146
Pan-Philippine (Maharlika) Highway   113; *see also* roads
Papua New Guinea   3
para-transit   306
Paris   45, 50, 120
park-and-ride   305
parking: *see* automobiles
Pascual, Enrile, Governor-General   138
Pasir Gudang   254
Pasir Panjang   236, 243, 253
Pasuruan   122–3
paths   67, 121, 138, 142, 145, 189; *see also* horses, mules, ponies, pedestrians, porters
Pathet Lao   184
Pathum Thani   309
patrol officers   145
Payar Lebar   239, 246
Payne–Aldrich Act   13, 107, 146
Pearl of the Orient   110; *see also* Manila
pedestrians   121–2, 125, 141–2, 189, 238, 240, 248, 250, 266, 272, 278, 280, 286, 294, 310, 312, 322
pedicabs   297; *see also becak*, trishaw
Pegu   160
Pelli, Cesar   335
Pelni   96–7, 99
Penang   3, 7, 9, 10, 32, 34, 41, 51, 54, 58–9, 62, 68, 70, 83, 86–92, 94, 96, 186–9, 192, 194–7, 199, 202, 204, 207–13, 215–16, 222, 319, 322, 324, 338
Peninsular and Oriental Steam Navigation: *see* P&O
Peninsular Malaysia: *see* Malay Peninsula
People's Action Party (PAP)   229, 318

People's Homesite Corporation, Manila 267
pepper 83, 190, 322
Perak 187, 189, 191–2, 194–5, 197, 199, 216, 332
periphery 27–9, 79, 114, 116, 126
Perlis 187, 195–6
Petaling Jaya 319, 325, 327–9, 338; *see also* Kuala Lumpur
petrochemicals 183, 242, 313
petrol rationing 205
petroleum: *see* oil
Petronas 319, 334–5
Petronas Twin Towers xv, 335
Phayathai 295, 313
Phibun Songkhram, Field Marshall 297
Philippine Aerial Taxi Company (PATCO) 58, 146
Philippine Airlines (PAL) 58
Philippine Coast Guard 105
Philippine Commission 142, 263, 265
Philippines Islands, colony of 3, 13–14, 41, 45, 54, 59, 61, 64, 66, 70, 90, 100–5, 130, 141–3, 146, 192; Assembly 102–3, 266; Collector of Customs 102; Insular Government 102, 141–2, 263; Insular Purchasing Agent 101; Public Service Commission 104
Philippine National Railways (PNR) 151–2
Philippine Railway Company 61
Philippine Tramway Company 262
Philippines, Republic of (1946) 5–8, 17, 19, 21, 26–7, 35–6, 58, 71, 75, 79, 81, 105–13 *passim*, 115–16, 134, 136, 147, 151, 153, 222, 249, 258; Bureau of Public Works 138, 142; Capital City Planning Commission 267; Congress of 268; Ministry of Human Settlements 270; National Housing Authority 271; National Urban Planning Commission 267
Philips Corporation 47
Phnom Penh 8, 62, 67, 161, 163, 172, 174, 184
Phrae 159, 161, 171
Phrakhanong 298
Phuket 195, 211
Phya Tak 290
pigs: *see* swine

pilferage 125, 134
pineapples 194, 234
piracy; *see also* shipping 3, 50, 86, 100, 107
planning 229, 240–1, 251, 269, 287, 315, 318, 324, 326, 330; colonial 227; national 216; town 227, in Singapore 229–30, 253, in Manila 259, 267–9, in Jakarta 274, 281–3, in Kuala Lumpur 319–20, 327, 332, 336, 338; *see also* cities, national planning
plantations 123, 127, 130, 153, 190, 197–8, 322
Pontianak 87, 90, 92
ponies 141, 322; *see also* horses
pony trap 128, 324; *see also* cart
population 21, 29, 75, 111, 117, 139, 152–3, 219, 222–3, 227, 232, 240, 242, 244, 246, 257, 259, 269, 272, 280–1, 284, 286, 289, 294, 297–8, 300, 308, 315–16, 319, 324–5, 328, 344; *see also* migration
Port Dickson 195, 201, 216, 329, 334, 336
Port Klang 29, 34, 100, 191, 214, 319, 329, 334
Port of Singapore Authority (PSA) 19, 98, 242–3, 252–3
Port Swettenham 100, 191, 192, 194–5, 197, 201–2, 207, 216, 319, 322; *see also* Port Klang
Port Weld 190, 192, 195
port-city: *see* city
porter 119, 121–2, 129, 136, 142, 145, 159; *see also* paths, pedestrians
ports 9, 32–4, 187, 207, 214, 254, 295; container 19–20, 31, 100, 112, 189; industrial 106, 241; military 177–9; railway 192–5, 197, 202; investment in 88–9, 97, 100, 107, 190, 196, 235–6; 239, 242, 264, 297–8, 322; transhipment in 98, 202, 229; *see also* entrepot, free ports, by place
Post Office Savings Bank 241
post roads 120–3, 138; *see also* roads
postal service 293
poverty 72, 75, 269, 271, 280
prahu 13, 54, 83, 86, 94, 119, 125, 127, 128, 235; *see also* shipping, sail
Prai 62, 192, 194–6, 207

Prajadhipok, King (Rama VI) 295
Priangan 119, 120–1
principalities 117, 119–23, 163
privatization 211, 214, 330
producer services 248–9, 254, 309, 334, 337
Proton car 328, 334
public health 279; *see also* drainage, malaria, sewerage
public space 249, 266
public transport 238, 240–1, 247, 251, 265, 269, 281–2, 294, 304, 316, 324, 326, 328–30, 332, 336; *see also* buses, railways, tramways
public works 138, 142, 143, 146, 160, 197, 264, 279; *see also* infrastructure
Pulau Brani 235, 243
Pulau Bukom 235, 242
Punggol 250–1
purchasing power 71
purchasing power parity (PPP) 26–7
Pusan 19, 29
PUTRA (Projek Usahasama Transit Ringan Automatik) 332, 337
Putra World Trade Centre 332
Putrajaya 336

Qantas Airways 55, 236, 252
Queenstown 244
Quezon, Manuel 112, 147, 267

radio 45, 47, 58
radio telegraphy 45, 47
Radio-Coloniale 47
Raffles, Sir Stamford 9, 13, 229–30, 232, 235–7, 246, 248
Raffles City 246
Raffles Hotel 246
Raffles Square (Commercial Square) 231–2
railways 15, 37–8, 51, 59–66, 67–8, 78, 89, 121, 123, 132, 135, 160, 163, 177–8, 192, 195–7, 201, 205, 215, 293–4, 322, 328; branch 146, 171, 196; elevated 311–12; express 179, 196, 336; feeder traffic 66, 130, 139, 145, 178; freight 145–6, 166, 170; gauge 164, 191–2, 196; high-speed 154, 214; light 171, 330, 332, 334, 336–7; mass transit 311, 318; narrow gauge 62, 163, 168–9; private 123, 128, 132, 191, 195; state 78, 126–8, 130, 132, 141, 153, 166, 168, 173, 177, 183, 203, 279; strategic 162, 166; bridges 177, 194, 295; impact of 59, 64, 123, 125, 127, 153, 170–1; connections to 196, 278; construction of 61–2, 168, 170, 198; dieselization 40, 151, 211, 214; hill surcharge 173; junctions 195–6; monopolies 66, 77, 152; networks 39, 59, 61, 66, 71, 116, 123, 128, 130, 141, 146, 153, 175, 197, 201, 203, 247; offices of 332; passengers 139, 166, 171, 201, 294, 205, 211, 294; ports 192, 194–5, 202, 207; rates 170, 173, 204, 207; stations 66, 129, 134, 177, 199; workers 66, 172, 197, 322; *see also* tramways
Rama I, King 290
Rangoon 24, 44, 54, 62, 66, 68, 70, 93, 160–1, 163, 170, 172, 175, 219, 225, 289–91, 298–300, 308–10; renamed Yangon 316, 318; Municipal Commissioner 300
Rangsit 161, 170, 306
Rawang 135, 328, 329
Rayong 313, 315
real estate 272, 283, 309
reclamation 235, 260, 264
Red River 24, 62, 67, 75, 156, 187, 191, 198, 219
refrigeration 280
regions: *see* cities, core region, hinterlands, periphery
registration: *see* vehicles
regulation 66, 78, 105; of interisland shipping, in Indonesia 96–8, 105; in Philippines 101–6; Board of Public Utility Commissioners 102; Board of Rate Regulation 102; Board of Transportation 105; certificate of public convenience 102; Public Utilities Commission 105; Presidential Task Force on Interisland Shipping 106; *see also* aviation, railways, roads, tramways
remote areas: *see* periphery
Renong Bhd 332
renovation (*doi moi*) 316; *see also* Vietnam
Republik Indonesia: *see* Indonesia, Republic of

retailing 271; *see also* malls
revenue farming 190, 275; *see also* taxes
Riau 27, 36, 50, 87, 100, 254
ribbon development 233, 315, 318
rice 13, 75, 83, 86, 88, 114, 136, 141, 146, 155–6, 160–2, 166, 169–70, 172, 180, 186–7, 190–1, 194–5, 198–9, 201–2, 207, 216, 224, 234, 260, 289, 293, 295, 298, 306, 309, 324, 329; *see also* agriculture
ricemills 141, 170, 180, 298
rickshaw 70, 237, 239, 265, 293–4, 300, 322
ringroads: *see* roads
rivers 37, 61–2, 66–8, 75, 77, 81, 83, 88, 90, 92, 103, 116–17, 119, 123, 125, 155, 159–63, 166, 169–70, 172, 174, 177, 180, 183–7, 190–2, 194, 198, 242, 289–90, 292–8, 305, 316, 324; buses 305; navigability 116, 125, 160–2; traffic, in Thai Central Plain 158–9, 172, 180, 289, 294, in Malay Peninsula 189, 195, 321–2; *see also* canals
road pricing 251, 305
road–rail competition 67, 70, 125, 135, 169; *see also* freight, passengers
roads (all types) 29, 39, 59, 66–7, 77, 155, 183–6, 190; all weather 143, 145, 147, 177, 179; accident rates 209; competition 174, 201; condition of 174, 192, 279; construction of 148, 150, 178, 202, 204–5, 209, 211, 294, 310, 322, 324; expressways (access-controlled roads) 209, 211, 214, 246, 255, 271, 305, 310, 315, 320, 328–30, 334; East North–South Expressway 209, 211; South Klang Expressway 328; West Coast Expressway, 211; Coast Expressway 211; Federal Expressway 328; feeder 77, 123, 130, 143, 155, 169, 197, 270, 203, 322; highways (principal roads) 37, 67–8, 77, 113, 120–2, 135, 143, 148, 152, 168, 174, 179, 204–5, 209, 211, 214, 279, 283, 315, 328, 336; in Malay(si)a 203–11; in Central Vietnam 214–16; in Java 119, 121–2, 125, 135–6, 154–5; in Philippines 61, 112–13, 117, 138, 141–2, 147–8, 150–2, 154, in Thailand 158–60, 168–71, 173–4, 177–9; maintenance of 67, 121, 138, 154; networks of 67–8, 77, 113, 199, 201, 208, 216; post roads 120–3; surfaces 138, 147, 275; tollroads 71, 287, 334; trunk 186, 203; urban 233, 236, 246, 251, 254, 266, 270, 279, 281, 286, 287, 292, 293–4, 297–8, 300, 303, 305, 310–11, 313, 315–16, 321–2, 324, 326–30, 334, 336–9; *see also* congestion
Roi Et 159, 169
Rotterdam 19, 86
Rotterdamsche Lloyd (RL) 86
Royal Automobile Association of Siam 174
Royal Interocean Lines (RIL) 95
Royal Navy: *see* Britain
rubber 13, 16, 24, 75, 83, 90, 93, 95, 127, 156, 186, 196–8, 201–2, 207, 209, 215, 234, 236, 239, 242, 322, 324

Sabah 27, 45, 88, 95, 205
Sabang 45, 88–9
sabotage 139, 205
Sadikin, Ali 283
sado 280
sago 83, 234
Saigon (-Cholon) 14, 17, 19, 24, 34, 41, 44, 47, 51, 55, 62, 67, 70, 90, 92, 160, 163, 169, 170–1, 174–5, 184, 198, 214–5, 219, 222, 225, 227, 289–90, 298, 300–1; renamed Ho Chi Minh City 308, 316–18
sail craft: *see* prahu; shipping, sail
Salim Group 254
salt 14, 83, 92, 155, 173, 191
Salween River 160
Samar 110, 113, 116
Sampeng 293–5
Samsen 295, 298
Samut Prakan 294, 306, 309, 313
Samut Sakhon 309, 315
San Francisco 16, 51, 55, 68
San Miguel group 271
Sandakan 45, 88, 89
Saraburi 68, 164, 177, 315, 316
Sarawak 27, 47, 83, 95, 205
Sarinah department store 282
Sarit Thanarat, Field Marshall, PM 177, 297

satellite town   242, 244, 259, 271, 285–7, 300, 316, 325; *see also* cities, new towns
Sattahip   177–9
Savannakhet   162, 172, 174, 184, 204
sawmills   295, 298; *see also* timber
schools: *see* education
seafarers   115, 342
seaplanes   232, 236; *see also* aviation
seasonality   139; *see also* monsoons
seaworthiness: *see* marine safety
Second Indochina War   183
Second World War   47, 55, 68, 83, 132, 175, 234
sedan chair   138, 191, 237
Selangor   187, 189, 191, 194, 197, 202–3, 322, 327–8, 335
Selangor Development Corporation   328
Selangor Turf Club   335
Selatar   236, 250
semaphore   40
Semarang   45, 61, 88, 93, 119–23, 135, 222
Sembawang   236, 243
Sepang   329, 334, 336
Seremban   194, 201, 209, 216, 329, 334, 337
Seri Kuda   334–5
sewerage   264, 269, 275
Shah Alam   328–9
Shanghai   3, 16, 19, 24, 31–4, 36, 41, 45, 51, 101, 219, 236, 237, 238, 240, 253
Shanghai Electric Construction Co. Ltd   238
Shell, Royal Dutch   83, 97, 144, 235, 239, 242
shipping   12–17, 81–113 *passim*, 198, 214, 294; bulk   106, 207; coastal   62, Malay Peninsula   187–91, 196, 198, 201, 204–5, 207, and railways   192–5, at Singapore   235, 236, 239; container   19, 29, 33, 98, 101, 106, 112, 187, 242, 252–3, and feeders to 86–7, 97–8, 236; interisland (Indonesia and Singapore)   82–100; interisland (Philippines)   100–13; passengers   84, 87, 95, 147; sail   83, 100   103–4, 119, 232; steamship networks   32, 48–54; *see also* by port, shipping company

shophouses   232–3, 236, 240, 248, 293, 306, 313, 315, 322, 324, 332; shophouse
shops; *see also* retailing
shortwave   47; *see also* radio
Siam(ese), kingdom of, 3, 6, 9, 13, 51, 75, 90, 92, 100, 155–73 *passim*, 174, 177, 195–6, 289, 290, 302, 306; Crown Property Bureau   293; Royal Irrigation Department   166; Royal Ordnance Factory   298; Royal Palace   292; Royal State Railways   164, 168–9, 173; *see also* Thailand
Siam Canals, Land and Irrigation Company   161
Siam Motors   302
Siam Square   306
Siam Steam Navigation Company   90
Siamese Airways   177
Siem Reap   162
Siemens   41
Sijori (Singapore–Johore–Riau)   254
silk   159, 161–2
siltation   125, 141, 275; *see also* deforestation, rivers
Singapore: 3–36 *passim*, 41, 44–5, 47, 48, 50–1, 54–5, 58–9, 62, 66, 68, 70, 75, 81–100 *passim*, 112, 114–16, 174, 186–7, 189, 190–1, 194–7, 199, 201–5, 207–8, 214–15, 219, 222–3, 225, 227, 229–256 *passim*, 275, 279, 315–16, 318–19, 322, 324–5, 328, 334, 336, 338–9; colony of: Municipality 237–8; Singapore Harbour Board 236; Singapore Improvement Trust 236, 244; Greater Singapore   27, 36, 222, 254–6; Republic of (1965): Civil Aviation Authority   252; Economic Development Board (EDB)   243; Housing and Development Board 240, 244; Land Transport Authority 251; Ministry of Communications 240; Ministry of Finance   249; Ministry of Trade and Industry   249; National Computer Board   251–2; National Grain Board   241; National Information Technology Plan   251; Urban Redevelopment Authority (URA)   244, 250; Greater Singapore 27, 36, 222, 254–6
Singapore Bus Services (SBS)   247

Singapore Inc. 252
Singapore International Airlines (SIA) 59, 244, 252–3
Singapore River 231, 233, 235–7, 243, 245, 248
Singapore Science Park 249
Singapore Sugar Industries 241
Singapore Technologies 254
Singapore Telecom (SingTel) 244, 253
Singapore Traction Company 240
Singapore Unlimited 253
Sittwe (Akyab) 170
sleds 71, 168
SLORC: see Myanmar
slums 240, 244, 250, 269, 288, 308, 325–6, 330
smallholders 186, 196–7, 201, 209; see also rice, rubber
small-scale sector 71, 172, 270, 283; see also informal sector
smuggling 95, 204; see also blockade, shipping, trade
social mobility 171
Socialist Republic of Vietnam: see Vietnam
Soeharto, President 8, 136
Solo: see Surakarta
Solo River 119, 123, 125
Songkhla 195–6, 208, 211
Soon Bee S.S. Co. 93
Sorsogon 146, 148, 151
South Africa 92–3, 192
South Asia 6, 8, 21, 24, 249, 253
South China Sea 9, 27, 136, 155, 191, 215
Southeast Asia, concept of 4, 6–7, 35, 36n; demography 21–4; economic geography 8–21, 24–9
South East Asian Treaty Organization (SEATO) 7–8
South India 197, 231, 237; see also India, Tamils
South Korea 8, 36, 98, 243, 250
South Vietnam 8, 300, 308; see also Cochinchina, Saigon, Vietnam
Southern Seaboard Development Program 211
Spain, Spanish 3, 12, 40, 50, 59, 61, 71, 81–2, 100–2, 104, 107, 138–9, 141–2, 144–5, 151–3, 227, 259–60, 264, 275

Spanish Chamber of Commerce 104
spices 83, 114
squatters 240, 244, 258–9, 269, 281, 283, 289, 302, 308, 313, 316, 325–8, 330; see also slums
Sri Lanka 13, 32, 51, 83, 249
Standard Vacuum Oil 235
STAR (Sistem Transit Aliran Ringan) 332, 337
Star Alliance 253
state, formation of 163–6, 195; see also nation-state
State and City Planning Project 241, 246–8
steam engines, origins 48, 59; compound 54
steam launches 136, 166
steamships 13–15, 32, 39, 48–54; see also shipping
steamtram: see tramways
Stephenson, George 59
Stoomvaart Maatschappij 'Nederland' (SMN) 54, 86–8, 116
Straits of Malacca 9, 13, 88, 94, 186, 232
Straits produce 83, 234, 236
Straits Settlements 7, 13, 50, 186–7, 194, 203–4, 238–9, 321; see also Malacca, Penang, Singapore
Straits S.S. Co. 58, 90, 93–5, 194, 199, 204, 236
Straits Trading Co. 194, 235
streetsellers 281; see also hawkers
Subang 329, 336
subsidies 54, 84, 87, 101, 104; see also mail contracts, railways, regulation, shipping
suburbs 70, 132, 232–3, 237, 258, 260, 262, 264–5, 269, 279–81, 294, 298, 300, 308–9, 313, 325, 338; see also garden suburbs, cities
Suez Canal 13, 16, 33, 41, 51, 54, 82, 86, 97, 225, 235, 275
sugar 66, 152–3; agriculture 24, 114, 126–7, 130, 139, 146, 194, 207, 306; trade in 11, 14–15, 75, 83–4, 92, 106–7, 119, 127–8, 136, 145–6, 153, 156, 180, 191–2, 207
sugar mills 10, 66, 100, 110, 119–20, 122–3, 125, 127, 141, 152, 161, 263
Sukarno, President 8, 135, 281–3

Sukumvit Road 308, 313
Sulawesi 27, 67, 79, 83, 86, 115, 231
Suleiman Building 324
Sulu 50, 81, 83, 86, 100, 107, 116
Sumatra 24, 27, 45, 66–7, 83–4, 86–90, 93, 95, 114, 116, 135, 153, 186, 197, 215
Sunda Strait 13, 67, 88, 135, 232
Sungai Kolok 196, 201, 205
Sungei Ujong 187
Sungei Way 329
Surabaya 34, 45, 50, 61, 68, 88–9, 92, 98, 119, 120–5, 127, 131–2, 135, 153, 222–3, 257, 278, 339
Surakarta 61, 117, 119, 121, 123, 125, 129
Swettenham, Sir Frank 189, 191, 194
swine 14, 106, 166, 179
Sydney 32, 35, 50, 54–5, 59, 68, 98, 219

Tabacalera 101
Taiping 45, 189–90, 192, 195
Taiwan 7, 19, 21, 28, 35–6, 98, 243, 250–1, 253, 309, 313
Tak (Raheng) 159, 161, 164, 171, 290
Taksin, King 290
Tamils 192, 198
Tampin 195
Tampines, New Town 244, 251
Tanglin 232
Tanjong Pagar 232, 235, 242–3
Tanjong Pagar Dock Company 235
Tanjung Pelepas 100, 254
tanneries 190, 295, 298
tapioca 180, 194, 234, 322
tariffs: *see* fares, freight rates
taxes 142, 144, 147, 154, 163, 278, 283; *see also* revenue farming
taxis 70, 209, 240, 246, 266, 280, 294, 306
teak 66, 75, 119, 126, 155–6, 160–2, 169–70, 298; *see also* sawmills
technological change 13, 37–72 *passim*, 153, 225; *see also* by city, mode, region
technology parks 250–1, 334–7
telecommunications 19, 21, 37, 38, 40–8 *passim*; Bangkok 292–3; Jakarta 274; Java and Luzon 154; Singapore 239, 248, 250–2; Thailand 168, 177, 195; urban 226–7, 339, 342; *see also* by mode

telegraphs 40, 44, 47, 51, 71, 122, 153, 158, 162, 266
telephone 45, 58, 71, 153, 173, 278, 298
television 47–8, 219, 335
Telok Anson (Telok Intan) 192, 194–5
Telok Ayer 231, 235–6, 243
temples 198, 290
textiles 13, 83–4, 155, 159, 161, 183, 234, 243, 295, 298, 306, 309
Thai Tobacco Monopoly 298
Thailand, kingdom of xvii–xviii, 3, 5–8, 10, 13, 21, 24, 26–8, 36, 48, 54–5, 58, 61–2, 64, 67–8, 75, 79, 100, 155–85 *passim*, 186–7, 195–6, 201, 205, 207–8, 211–16, 222, 235, 253, 289–318 *passim*, 325; democratization 173–4, 180–1, 289, 295, 302–3; Highways Department (Thailand) 174; Ministry of the Capital 293, 295; Ministry of the Interior (MOI) 295, 304; National Economic and Social Development Board 310, 315; National Housing Authority 313; Town and Country Planning Act 308; Central Plain 14, 26, 79, 163, 166–8, 170–1, 298; North 28, 62, 79, 156, 159–60, 162, 169–73, 177, 180, 294; Northeast 28, 58, 62, 155–6, 159, 161–2, 164–6, 169–71, 174, 177–81, 183–5, 294; South (Peninsular) 158, 166, 186–7, 195–6, 201, 205, 207–8, 211–13, 295
Thakhek 199, 204
Thanlyin-Kyautan Industrial Zone 316
Third World 24, 224–5, 227, 229, 258, 280, 287, 289, 300, 302, 308, 319–20, 325–6, 337–8; *see also* cities, technology
Thon Buri 168, 290, 303, 306; Municipality 295
Three Pagodas Pass 175, 195
timber 66, 110, 127, 207, 235, 242; *see also* sawmills, teak
time–space convergence 37
timetables 64, 130, 191
Timor 3, 45, 55, 81, 84, 95, 339
tin 11, 85, 156, 186–7, 189–92, 195–6, 201–2, 204, 209, 215, 234–5, 239, 242, 321–2, 326, 329, 338; smelter 194–5; *see also* mining

Titanic 45, 105
Toa Payoh 244
tobacco 24, 75, 84, 86–8, 136, 162
Tokyo 19, 21, 24, 29, 31, 33–5, 45, 61, 68, 70, 229, 240, 248, 250, 281, 342
tollroads: *see* roads
tongkang 207, 216; *see also* lighterage
Tonkin 3, 15, 156, 187, 198; *see also* Vietnam
Tonle Sap 162
Toungoo 163, 172
Tourane: *see* Da Nang
town planning: *see* planning, towns
towns 62, 251; colonial 298, 309, 320, 324; *see also* cities, new towns, satellite towns
tracks: *see* paths
trade unions 66
trade 89, 174; interisland 110, 113–14, intra-Asian 11–14, 16, 32; *see also* exports, free trade, imports 107, 159
traffic: Bangkok 289, 309–10, 312, 315; Jakarta 286; Kuala Lumpur 326, 328, 330, 337; Luzon 145, 147; Malaya 202, census 209, passenger 201; Manila 265; Singapore census 238; Thailand census 168; *see also* automobiles, congestion, vehicles
traffic signals 300, 303, 305
trails: *see* paths
trains: *see* railways
tramways 300; cable 68; horse-drawn 66, 68, 262–3, 265, 278–9; electric 68, 70–1, 265, 278–81, 294, 324; rural in Java 66, 123–30; steam 39, 237, 278–9; urban, in Singapore 237, in Manila 265–6, 270, in Jakarta 278–83, in Bangkok 294, 303–4; *see also* light rail
transhipment 88–9, 92, 95, 98, 106–7, 202, 252
Transindochinois: *see* railways, Annam 62, 64
Trans-Island Bus Services (TIBS) 247
transnational corporations: *see* multinationals
transport: corridor 183, 211, 216, 246, 319, 328; costs 37, 70, 77, 103, 123, 129, 151, 152, 159, 179; infrastructure 146–7, 208–9, 253, 309, 318; networks 146, 205, 319, 330; planning 300, 316; public 151, 272, 274; urban 39, 68, 70, 236, 262, 310, 326; *see also* aviation, cities, railways, roads, shipping, tramways
Transport Revolution 39, 134
Trengganu 90, 187, 195–6, 201, 204, 207
tricyle 238, 270
trishaw, 239, 327; *see also* becak
trolley bus: *see* buses
trucks 67; Bangkok 304–5, 311; Java 129–30, 287; Ho Chi Minh City 316; Luzon 61, 145–7; Malaya 201–2, 205; Manila, 266–7; Singapore 238; Thailand 173–5, 177, 179–80, 183; Vietnam 214; *see also* freight
trunk roads 186, 203; *see also* roads
Tumpat 201, 207
typhoons 116, 151, 269
tyres 142, 197, 202

Ubon Ratchathani 164, 166, 169–71, 179–80
Udon Thani 178–9
Unfederated Malay States (UMS): *see* Malaya
United East India Company (VOC) 7, 83
United Engineers (Malaysia) Bhd (UEM) 211, 330, 332
United Kingdom: *see* Britain
United Malays' National Organisation (UMNO) 204, 318, 332
United Nations (UN) 4, 241, 244, 326
United States (American) 7, 9, 16, 17, 19, 36, 44, 51, 55, 58, 70, 95, 97–8, 110, 134, 249, 253, 270, 272; and aviation 17, 55, 58; consultants from 285, 297, 335; in Indochina 8, 68, 178–9, 214, 300, 302–3, 308; in Philippines 13, 61–2, 67, 71, 81, 101–7, 138–9, 141–9, 151–3, 227, 263–7, 300; United States Operations Mission (USOM) 178
University of the Philippines 267
uplands: *see* highlands
urban enclaves 255, 339
urban hierarchy 5, 211
urban planning: *see* cities, planning, population

urban renewal 244, 249; *see also* cities, planning, slums, squatters
urban sprawl 267, 272, 274, 318; *see also* cities, population
urbanization 5–6, 24, 29, 135, 222, 224, 287
Uttaradit 158–9, 164, 166

Van der Heide, J. Homan 166, 170
vegetables 136, 141, 145, 198, 260
vehicles, three-wheel 162, 266, 282; four-wheel 342; movements of 135, 312; numbers of 68, 143, 286, 319–20; registration of 143, 147, 208; *see also* by type
Venice 289, 297
vent for surplus 77, 126, 156
Verne, Jules 51
Victoria Point Agreement 90
Vientiane 162, 178, 184
Vietnam War 68, 179, 214, 302
Vietnam 3, 7, 21, 26, 61, 289, 308; *see also* Annam, Cochinchina, Hanoi, Saigon, Tonkin
villages 128, 259; *see also* agriculture
Vinh 198, 204
Virgin Airlines 253
Visayas 3, 27, 50, 61, 100, 103, 110, 112–13, 115, 269
Vung Tau 215, 316

walking: *see* pedestrians
Wangsa Maju 330
warehousing 252–3, 292; *see also* logistics
Washington DC 41, 319
water transport: *see* canals, rivers, shipping

waterways, inland 156, 162, 169–72, 175, 177, 180; *see also* canals, rivers
Watt, James 48
Wearne Brothers 58
Wearne's Air Services Ltd 204
Wernstedt, F.L. 104, 107, 110
West Malaysia: *see* Malay Peninsula
West Port 334: *see* Port Klang
Westerners 71, 266, 292; *see also* by country
wet season: *see* monsoons
Winsemius, Dr Albert 241
wireless: *see* radio
women 126, 202, 243, 249, 259–60, 267, 275, 278, 309
wood products 239, 309; *see also* sawmills, timber
Woodlands 246, 250
working class 172, 233
World Bank 105, 135, 177, 181, 223, 242, 270–1, 305, 310, 328
world city: *see* cities
World Trade Center (Bangkok) 309
Wright, Orville and Wilbur 54

Yangco S.S. Co. 263
Yangon: *see* Rangoon
Yangtze River Delta 253
Yap Ah Loy 321
Yishun 244, 246
Yogyakarta 61, 117, 121, 123, 129
Yokohama 19, 24, 29, 33, 34, 51, 61
Yom River 155, 160–1
Yunnan 28, 62, 64, 155, 159, 173–4, 184–5

Zamboanga 100, 107, 110–11, 113, 147